Dear Brigid,

Tha

AF683230

friendship, support, collaborations and for being such an inspiration here at Vanderbilt.

I wish you all the best at Duke and hope that this book will be reminding you that zebrafish is a useful system to study development

Your energizing presence will be greatly missed at Vandy

Lila Solnica-Krezel

Nashville 10-4-2002

Results and Problems in Cell Differentiation

Series Editor:
W. Hennig

40

Springer
Berlin
Heidelberg
New York
Barcelona
Hong Kong
London
Milan
Paris
Tokyo

Lilianna Solnica-Krezel (Ed.)

Pattern Formation in Zebrafish

With 95 Figures, 17 in Color

Springer

Dr. Lilianna Solnica-Krezel
Dept. of Biological Sciences
Vanderbilt University
VU Station B 351634
Nashville, TN 37235-1634
USA

ISSN 0080-1844
ISBN 3-540-43576-X Springer-Verlag Berlin Heidelberg New York

Library of Congress Cataloging-in-Publication Data

Pattern formation in zebrafish / Lilianna Solnica-Krezel (ed.).
p. cm. – (Results and problems in cell differentiation ; 40)
Includes bibliographical references (p.).
ISBN 354043576X (alk. paper)
1. Zebra danio – Development. 2. Pattern formation (Biology) I. Solnica-Krezel, Lilianna, 1961– II. Series.

Springer-Verlag Berlin Heidelberg New York
a member of BertelsmannSpringer Science+Business Media GmbH

http://www.springer.de

Printed in Germany

The reproduction costs for colour figures have in part been defrayed by GENETOOLS, LLC (Philomath, Oregon, USA). The company's support is gratefully acknowledged.

Production: PRO EDIT GmbH, Heidelberg, Germany
Cover concept: Meta Design, Berlin, Germany
Cover Production: design & production, Heidelberg, Germany
Typesetting: SNP Best-set Typesetter Ltd., Hong Kong
Printed on acid-free paper SPIN 10830724 39/3130/So – 5 4 3 2 1 0

Preface

What makes an ideal model system to study embryonic development? The answer to this question has varied in the course of scientific quest, depending on the available experimental approaches. Yet, at the beginning of the new millennium, most of the mainstream model organisms have one feature in common, they are amenable to genetic analysis. Indeed, the genetic approach has proven remarkably effective in the dissection of complex biological phenomena, starting with the regulation of the lactose operon in *Escherichia coli*, the decision between the lysogenic and lytic cycles of the lambda phage and the yeast cell cycle control. The genetic screens for mutations affecting embryonic development of a fruit fly, *Drosophila melanogaster*, carried out by Christiane Nüsslein-Volhard, Eric Wieschaus and Edward Lewis, marked the turning point in experimental approaches to embryonic development of more complex organisms. The resulting mutants facilitated identification of the key regulatory molecules, while their phenotypic classes hinted at the underlying regulatory mechanisms. This work not only provided a paradigm for genetic dissection of other developmental processes, but also inspired many budding embryologists throughout the world.

Soon thereafter, George Streisinger noted that, among vertebrates, the zebrafish, *Danio rerio*, offered many features facilitating the dissection of nervous system development and function. Among these, transparency of the embryo, ease of husbandry, and the highly prolific nature make the zebrafish a good genetic model system. The pioneering work of George Streisinger was continued by his colleagues Chuck Kimmel, Judith Eisen, John Postlethwait, Monte Westerfield, and Jim Weston in Eugene, Oregon (USA). These studies laid the methodological foundation for using the zebrafish, and their experimental and intellectual elegance sparked an interest in this model system in the scientific community.

The last decade witnessed an explosion of zebrafish research, culminating in the zebrafish joining the ranks of "mainstream models for human development and disease". This was the result of a concerted, multi-pronged, collaborative effort between many laboratories throughout the world. Two large-scale genetic screens for zygotic embryonic lethal mutants in the Nüsslein-Volhard laboratory in Tübingen (Germany) and the Driever laboratory in Boston (USA) yielded thousands of mutations in hundreds of genes. The results of these screens, which were published in a special volume of *Development* in 1996, demonstrated the feasibility of the large-scale, forward genetic approach in zebrafish and in vertebrates, and promised that all aspects of embryonic development are susceptible to mutations. Concurrent work in the Postlethwait laboratory in Eugene (USA) and in the Jacob, Fishman, and Driever groups in Boston (USA) assembled the first zebrafish genetic maps, thus paving the road towards molecular characterization of identified mutant loci. Simultaneously, the efforts of tens and then hundreds of labs focused on isolating the zebrafish homologues of known genes from other vertebrate and invertebrate organisms, providing crucial descriptions of normal development, generating genomic resources, and methods for sophisticated cellular and embryological analyses, as well as devising new mutagenesis procedures. These methods have been described in a two-volume set of *Methods in Cell Biology*, published in 1999 by Academic Press, providing a much needed resource for an increasing number of scientists using zebrafish.

In this book, we present the current view of the mechanisms underlying the early patterning and morphogenetic processes in zebrafish. The book was written 5 years after the conclusion of the first, large-scale screening effort, at a time when phenotypic and molecular characterizations of many mutant loci had been accomplished. We deliberately avoided describing methods and tools, our main objective being to provide a synthesis of what we have learned so far, and to identify the key questions for the next decade of research.

The book is divided in four parts with the first two parts devoted to the early inductive and morphogenetic events, respectively. The third part focuses on patterning and differentiation of neural tissues, whereas the fourth part discusses aspects of organogenesis. We start with the earliest inductive and patterning events in the zebrafish blastula and gastrula. The chapter by Takuya Sakaguchi, Toshiro Mizuno, and Hiro Takeda is concerned with the formation and developmental role of the yolk syncytial layer. This fascinating, extra-embryonic structure serves not only as a source of energy and supplies for the developing embryo, but also has important roles in patterning the overlying blastoderm. The next two chapters describe our current understanding of the formation and patterning of the germ layers, with David Kimelman and Alex Schier focusing on mesoderm, whereas Rachel Warga and Didier Stainier address endoderm development. They review insights from zebrafish into Nodal signaling that in many ways contributed to the rapid progress in this field. Formation of the Spemann-Mangold gastrula organizer and its roles in embryonic polarity specification are the topics of the chapter by Masahiko Hibi, Toshio Hirano, and Igor Dawid. Matthias Hammerschmidt and Mary Mullins further elaborate the problem of anteroposterior and dorsoventral patterning with special emphasis on the role of bone morphogenetic proteins in this process. This part culminates with Chris Wright and Marnie Halpern considering the origin of left-right embryonic asymmetry. There are several designed overlaps between these chapters, highlighting the interconnectedness of the early inductive events. Indeed, zebrafish studies clearly show that Nieuwkoop center activity encompasses not only the dorsal yolk syncytial layer, but also dorsal marginal blastomeres. Furthermore, Spemann-Mangold organizer activity is not confined to dorsal mesoderm but is also present in ectoderm and possibly endoderm.

The second part of the book presents our current understanding of gastrulation, the morphogenetic process transforming the two-dimensional blastula fate map into the three-dimensional vertebrate body plan. From the chapter by Don Kane and Richard Adams, we learn about epibolic movements that spread blastoderm around the yolk cell, and about the internalization of mesodermal and endodermal precursors. Lila Solnica-Krezel and Mark Cooper address the convergence and extension movements that narrow germ layers from back to belly, while elongating them from head to tail. It becomes clear that fish gastrulation entails a complex repertoire of cell behaviors, some not encountered in the well-characterized frog gastrula. In the last chapter of this part, Erez Raz and Nancy Hopkins describe specification of primordial germ cells in the blastoderm margin and their fascinating migration towards the future gonad. These chapters highlight the notion that combining mutational analyses with tracking cell movement behaviors in living embryos is a powerful approach for the dissection of morphogenetic processes.

The third part of the book, focusing on neural development, begins with an overview of early neural patterning by Steve Wilson, Michael Brand, and Judith Eisen. Work on zebrafish revealed the existence of new signaling centers that pattern neuroectoderm and provided novel insights into how spatiotemporal regulation of canonical Wnt signaling regionalizes neuroectoderm along the anteroposterior axis. In the next chapters, Robert Kelsh and Dave Raible discuss specification of the neural crest, whereas Bruce Appel and Ajay Chitnis address the mechanisms determining neuronal cell identity. The chapter by Christine Beattie, Michael Granato, and John Kuwada concludes this part with a discussion of axon guidance mechanisms.

Amenability to investigate early embryonic patterning was the initial attraction of using the zebrafish. However, in the last part of this book, we hope to illustrate that this system also has much to offer for studying organogenesis. Caroline Brennan, Sharon Amacher, and Peter Currie discuss formation of somites and their subsequent patterning and differentiation. Genetics of cardiovascular system development and function are described by Brant Weinstein, Debbie Yelon, and Mark Fishman. Although the zebrafish and mammalian kidney differ significantly, Iain Drummond makes a case that analyses of zebrafish pronephros will provide important insight into the early stages of mammalian kidney development. The book closes with a discussion of mechanisms involved in eye development by Stephen Easter and Jarema Malicki.

The current knowledge of the morphological, cellular, and genetic events of early zebrafish development summarized in this volume establishes a solid foundation for the next phase of scientific inquiry. New generations of genetic screens for mutations altering fluorescently marked organs and tissues, new methods of cell tracking, implementation of morpholino antisense oligonucleotide technology to provide a greatly needed reverse genetic tool, and finally the emergence of the zebrafish genome sequence will allow us to pose more sophisticated questions in the next decade. We hope that this book will serve as a useful resource in this endeavor.

The making of this book in many ways reflects the collegial atmosphere and synergistic interactions among the zebrafish scientific community, being the fruit of discussions and collaborative efforts between researchers from many countries. It is our hope that this volume will help maintain the tradition of collaboration and scientific discourse in the zebrafish community.

Lila Solnica-Krezel
March 2002

Acknowledgements. I would like to express my gratitude to all colleagues who contributed to the book, first for embracing the idea of collaborative chapters, next for sharing their knowledge, valuable time and effort during writing, and finally for patiently bearing with a somewhat over-eager editor. Springer-Verlag editors Sebastian Granderath and Ursula Gramm were very understanding and helpful with coordinating and editing this volume. Sponsorship of publishing color illustrations in the book by Gene Tools, LLC, is greatly appreciated. Special thanks to my friends Bruce Appel and Mark Cooper, and to my family, for encouraging me to take on this ambitious task and for their continued support. I am also grateful to the members of my research group for their infinite enthusiasm for zebrafish development, and for graciously tolerating my chronic lack of time during the last year. Work in my laboratory and efforts on this volume have been supported by grants from National Institutes of Health, March of Dimes Birth Defects Foundation and Pew Scholars Program in Biomedical Sciences.

Contents

I. Germ Layer Formation and Early Patterning

Formation and Patterning Roles of the Yolk Syncytial Layer
Takuya Sakaguchi, Toshiro Mizuno, Hiroyuki Takeda

Mesoderm Induction and Patterning
David Kimelman, Alexander F. Schier

The Guts of Endoderm Formation
Rachel M. Warga, Didier Y. R. Stainier

III. Neural Development

Patterning the Zebrafish Central Nervous System
Steve W. Wilson, Michael Brand, Judith S. Eisen

Specification of the Zebrafish Neural Crest
Robert N. Kelsh, David W. Raible

Neurogenesis and Specification of Neuronal Identity
Bruce Appel, Ajay Chitnis

Cellular, Genetic and Molecular Mechanisms of Axon Guidance in the Zebrafish
Christine E. Beattie, Michael Granato, John Y. Kuwada

IV. Aspects of Organogenesis

Somitogenesis
Caroline Brennan, Sharon L. Amacher, Peter D. Currie

Cardiovascular System
Deborah Yelon, Brant M. Weinstein, Mark C. Fishman

I. Germ Layer Formation and Early Patterning

Formation and Patterning Roles of the Yolk Syncytial Layer

Takuya Sakaguchi[1], Toshiro Mizuno[2], and Hiroyuki Takeda[1,3]

1 Introduction

Zebrafish develop in a typical manner of teleosts (for review, see Driever 1995). The animal/vegetal axis is specified during oogenesis and sperm can enter the egg through a single entry site (micropyle) located at the animal pole. When eggs are laid, yolk and cytoplasm are intermixed. After fertilization, cytoplasm streams to the animal pole as it segregates from the yolk. About 30 min after fertilization, the cytoplasm forms the blastodisc at the animal pole and surrounds the vegetal yolk mass as a thin cytoplasmic layer (Fig. 1A,B). The cleavages take place atop a large yolk mass. During mid-blastula transition, activation of zygotic transcription (10th cleavage in zebrafish) coincides with the generation of the first three separate lineages of the embryo (Fig. 1C,D). Two of these lineages are extra-embryonic, the enveloping layer (EVL) forming the outer surface of the blastoderm, and the yolk syncytial layer (YSL) that covers the animal part of the yolk cell, respectively. The third lineage, termed the deep cell layer, will form the embryo proper.

One unique and fascinating feature of early teleost development is the formation of the YSL at the surface of the yolk, peripheral to and underneath the blastoderm. The syncytium, which was originally known as the "periblast", was renamed "yolk syncytial layer" by Trinkaus (1992). The YSL has great importance for teleost development. Since it separates the yolk from the embryo (deep cell layer), all nutrients from the yolk must pass through it to reach the embryo. Indeed, analyses with the zebrafish anemia mutant, *weissherbst*, revealed that Ferroportin1 expressed in the YSL plays a crucial role in transport of iron from the yolk to the embryo (Donovan et al. 2000). In addition to the role in nutrient transport, the YSL has been implicated in epiboly movement (Trinkaus 1993). More recently, the YSL has been found to play an important role in induction and patterning of the mesoderm and endoderm (Mizuno et al. 1996 1999b; Ober and Schulte-Merker 1999; Rodaway et al. 1999). Thus,

[1] Division of Early Embryogenesis, National Institute of Genetics, Yata, Mishima 411-8540, Japan
[2] Graduate School of Science, Osaka City University, Sumiyoshi, Osaka, 558-8585, Japan
[3] Department of Biological Sciences, Graduate School of Science, University of Tokyo, Bunkyo, Tokyo 113-0033, Japan

Results and Problems in Cell Differentiation, Vol. 40
L. Solnica-Krezel (Ed.): Pattern Formation in Zebrafish

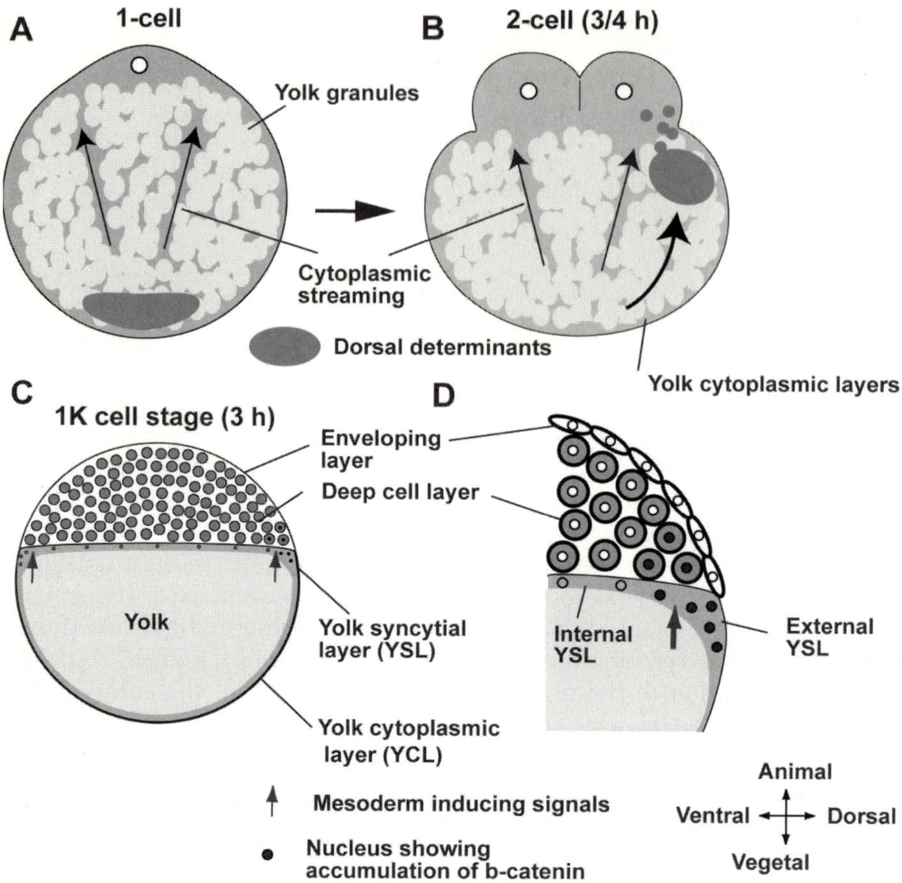

Fig. 1A–D. Early development of zebrafish. The development of zebrafish embryos from 1-cell to mid-blastula stages is represented by schematic drawings of mid-sagittal sections of the embryos. The embryos are oriented with future dorsal to the right and developmental times are in hours postfertilization at 28 °C. **A, B** In the zygote, yolk (*white*) and cytoplasm (*gray*) are mixed, but separate during the first 2 h of development by cytoplasmic streaming to the animal pole (*arrows*). Cytoplasmic determinants that are required for dorsal development (*dorsal determinants*) may be present in the vegetal region at the time of fertilization. They are then translocated to the future dorsal sides in a microtubule-dependent manner during early cleavage stages. **C, D** The 1000(1k)-cell embryo represents the mid-blastula. The different embryonic and extra-embryonic lineages can be clearly distinguished at this stage: deep cell layer (*embryo proper*), enveloping layer (*EVL*), yolk syncytial layer (*YSL*) and yolk cytoplasmic layer (*YCL*). The YSL is further divided into the internal and external YSLs. The *arrows* in **C** represent mesoderm- and endoderm-inducing signals from the YSL. High magnification of the dorsal region is shown in **D**. Nuclear localization of β-catenin (*solid circle*) is detected in dorsal blastomeres as well as the external YSL the YSL has attracted the attention of more fish biologists than ever before and has become the target of various studies. In this chapter, we will discuss the formation and the activities of the zebrafish YSL, focusing on its roles in germ layer formation and its patterning.

2 Formation of the YSL

In zebrafish, the initial nine cleavages occur synchronously at 15-min intervals (Kimmel et al. 1995). The first five cleavage planes are usually vertical in orientation and alternate at right angles to one another. The sixth cleavage plane is horizontal, thus producing two tiers of blastomeres. These early cleavages are meroblastic and the marginal vegetal blastomeres maintain large cytoplasmic bridges with the yolk cell (Fig. 2A,B). Formation of the YSL begins by fusion of the marginal blastomeres with the yolk cell at mid-blastula stage.

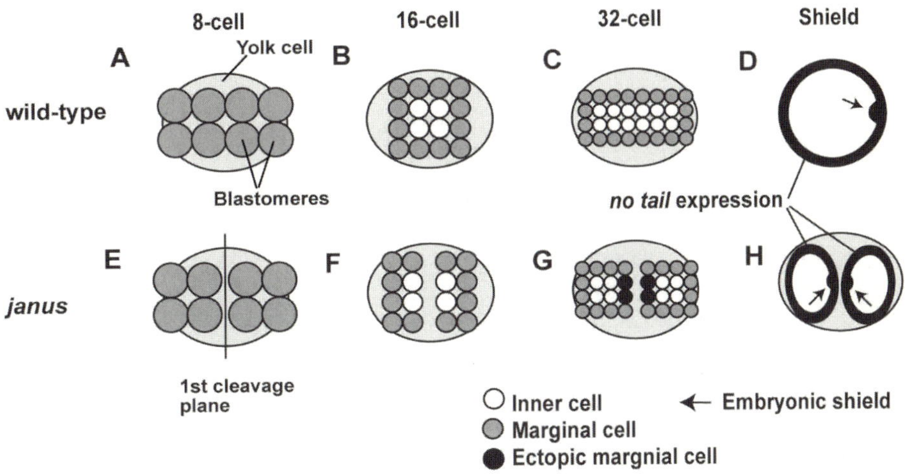

Fig. 2. Cleavage pattern and possible marginal zone formation in wild-type (**A–D**) and *janus*-mutant embryos (**E–H**). Schematic drawings viewed from the animal pole are shown. **A–D** The cleavages are meroblastic so that initially all blastomeres (2–8-cell stages) and later the marginal blastomeres (16–512-cell stages) maintain cytoplasmic connections with yolk cell (marked as *gray circles*). From the 16-cell stage on, the central blastomeres (*open circles*) lose cytoplasmic connections and are completely separated from the yolk (**B, C**). At the shield stage, *no tail*, a pan-mesodermal marker, is expressed in the blastoderm margin (indicated by the *solid black zone*; **D**). Most of the marginal zone is mainly constituted by cells that stem from the lineages of the early marginal cells. **E–H** In *janus*-mutant embryos, a separation of the blastoderm into two halves frequently occurs within the first four cleavages and each half separately develops until fusion occurs at the end of gastrulation (Abdelila and Driever 1997). The separation results in ectopic formation of marginal cells that would normally contribute mainly to the animal-pole region (*solid circle* in **G**). At the shield stage, *no tail* expression is detected in the entire margin of each half including the ectopic region. (After Solnica-Krezel et al. 1995; Abdelila and Driever 1997)

Interestingly, YSL formation appears to be correlated with the number of cell divisions and a loss of cytoplasmic bridges.

Using both dye injections into blastomeres and direct examination of blastomere cleavages in live embryos, Kimmel and Law (1985b) described in detail the process of YSL formation in zebrafish. Between the 9th and 10th cleavages (512- to 1024-cell stages), the lower cell borders adjacent to the yolk cell fade and disappear as the marginal cells collapse. When the cells complete 10th mitosis, the nuclei are detected within the YSL and are lined up in a single ring around the cellular blastoderm (Fig. 3A,B). Once incorporated, YSL nuclei

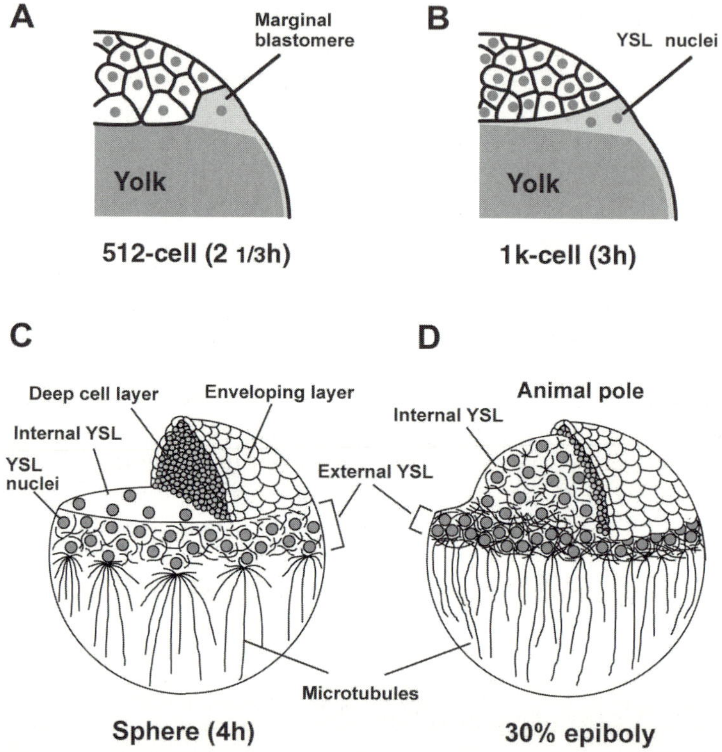

Fig. 3. Schematic drawings of YSL formation (**A, B**) and changes in organization of the blastoderm and the yolk cell at the onset of epiboly (**C,D**). At the 512-cell stage (complete of 9th cleavage) in zebrafish, dividing blastomeres at the margin of the blastoderm are still in cytoplasmic confluence with the yolk cell (**A**). As the 10th cell cycle begins, the cytoplasm and nuclei of marginal blastomeres collapse into the yolk cell, thus forming the YSL (**B**). Just before the onset of epiboly (**C**, *sphere*), the yolk cell is furnished with two types of microtubular arrays: a network in the microtubules present in the YSL, and animal-vegetal-oriented microtubules in the YCL originating in the YSL. Epiboly starts with the yolk cell bulging towards the animal pole, a process called "doming" (**D**). The external YSL has contracted and exhibits densely packed YSL nuclei. As epiboly proceeds, the YSL with the nuclei and intercrossed microtubule network will spread vegetally, while the YCL with the animal-vegetal microtubule system will decrease in size (see text). **C** and **D** were adapted from Solnica-Krezel and Driever (1994)

divide more rapidly than their deep cell cousins and undergo a series of metachronous divisions eventually becoming postmitotic at the 12th cycle (about 4h of development; Kane et al. 1992). After 10th cleavage, usually no further contribution of blastomeres to the YSL is observed.

YSL formation is accompanied by a loss of cytoplasmic bridges between the yolk cell and blastoderm. Until the formation of the YSL, marginal blastomeres remain joined to the yolk cell through the cytoplasmic bridges. However, the bridges are disconnected as soon as the YSL is formed: injection of the fluorescent tracer fluorescein-dextran (FD) into the yolk cell after the YSL has been formed usually labels the yolk cell specifically. FD is expected to diffuse through cytoplasm but to be too large to pass through gap junctions.

With respect to the lineage that contributes to the YSL, whether a blastomere will participate in YSL formation appears to depend on its marginal position within the embryo rather than on its lineage. At the 64-cell stage, usually 20 marginal blastomeres remain connected to the yolk cell (Fig. 2A,B; Kimmel and Law 1985a). It could be that all of the progeny of the marginal cells participate in YSL formation. However, this possibility is unlikely because injection of FD into marginal cells at 64- to 256-cell stages produces both a labeled YSL and many labeled cells within the blastoderm (Kimmel and Law 1985b), suggesting that not all, but only a subset of the progeny of early marginal cells contribute to the YSL. Direct microscope observation supports the result of the lineage-tracing experiment (Kimmel and Law 1985b).

When YSL formation completes at mid-blastula stage, three compartments of continuous cortical cytoplasm can be distinguished in the yolk cell (Fig. 1C,D). A thin, anuclear yolk cytoplasmic layer (YCL) surrounds the bulk of the yolk mass with the vegetal pole. The YSL can be further divided into two parts. One is the external YSL, located between the YCL and the blastoderm rim, a relatively thick belt of cytoplasm populated by the YSL nuclei. The other is internal YSL that is located beneath the blastoderm, comprising a thinner cytoplasm. The internal YSL is finally populated by nuclei derived from the external YSL.

3 Epibolic Movement and the YSL

During epiboly, the deep cells, the enveloping layer (EVL) and the YSL lineages vegetally expand eventually to cover the entire yolk cell. The beginning of epiboly is marked by a major morphological change in the yolk cell, which bulges toward the animal pole in a process called "doming" (Fig. 3C, D; Warga and Kimmel 1990). Subsequently, the YSL spreads toward the vegetal pole, with the YCL being diminished.

Studies of another teleost, *Fundulus heteroclitus*, suggested a crucial role of the YSL in epiboly (for details, see Kane and Adams, this Vol.). The blastoderm isolated from a *Fundulus* embryo does not undergo epiboly unless it is attached to the YSL. The YSL, in contrast, undergoes epiboly in the absence of the blastoderm (Trinkaus 1951). Epiboly of the YSL is at least partially driven by microtubule-dependent forces. In zebrafish, the yolk cell exhibits two distinct types of microtubule arrays that change their organization during epiboly (Fig. 3C; Solnica-Krezel and Driever 1994). In the external YSL, microtubules radiate from centers associated with the YSL nuclei, forming an extensive network of intercrossing microtubules. In the YCL, however, an array of microtubules extends along the animal-vegetal axis to the vegetal pole. This array in the YCL appears to emanate from the marginal blastomeres at the early-blastula stage and, later, from the microtubule network of the YSL. During epiboly, the YSL expands vegetally with its intercrossing microtubule network, while the set of animal-vegetal microtubules becomes shorter as the YCL diminishes (Fig. 3C,D; Solnica-Krezel and Driever 1994). Analyses of epiboly in embryos treated with nocodazole (depolymerizing reagent of microtubules) indicate that microtubules are involved in YCL regression and epibolic movement of the YSL nuclei, but only partially required for the remaining aspects of epiboly such as epiboly of the blastoderm (Strähle and Jesuthasan 1993; Solnica-Krezel and Driever 1994).

4 Dorsal Determinants in Teleost Yolk Cell

One of the major questions in early teleost development is how the dorsoventral axis becomes specified. Specification of the dorsoventral axis is best understood in amphibians. In *Xenopus*, the sperm entry point indicates the side of the egg destined for ventral structures, with the opposite side fated for the dorsal structures. In contrast, it is not possible to distinguish morphologically between the regions corresponding to dorsal and ventral sides until at least blastula stages (Schmitz and Campos-Ortega 1994). Do the mechanisms similar to those discovered in amphibians exist in fish? It has been clearly shown in amphibian embryos that the cytoplasmic determinants that specify the dorsal fate are present in the eggs just after fertilization (reviewed by Gerhart et al. 1989; Heasman 1997). These determinants, so-called "dorsal determinants", are located around the vegetal-pole region at the time of fertilization. They are then translocated to the prospective dorsal side before the first cleavage (Fujisue et al. 1993) and, thus, incorporated into future dorsal blastomeres. In teleosts, however, several studies have determined that the dorsoventral axis in zebrafish is not associated with the orientation of early cleavage planes or the position of the blastomeres (Kimmel and Warga 1987a; Abdelilah et al. 1994; Helde et al. 1994). Thus, formation of dorsoventral asymmetry is unlikely to be caused by the unequal partitioning of cytoplasmic determinants during initial divisions of the blastomeres. Consistent with this, *Fundulus* embryos develop normally when one blastomere is removed at the 2- and 4-cell stages (Hoadley 1928). Thus, a current hypothesis in fish is that dorsal determinants, if any in a fish embryo, may be transported to the future dorsal side at later stages.

Experiments with goldfish provided the first evidence for the presence of dorsal determinants in fish. In goldfish embryos, *Carassius auratus*, a deletion of the vegetal half of the yolk cell equatorially at the cleavage stage produced a series of specific malformations with dorsal deficiencies (Tung et al. 1945; Mizuno et al. 1997; Yamaha et al. 1998). Similarly, in zebrafish, removal of the vegetal half of the yolk cell by squeezing or ligaturing off just after fertilization caused hyper-ventralized embryos (Fig. 4; Mizuno et al. 1999a,b; Ober and Schulte-Merker 1999). Mizuno et al. (1999b) also demonstrated that the sensitivity to removal of vegetal hemisphere changes during the early cleavage stages. The frequency of the severe dorsal defects was highest when the removal was completed prior to 30 min postfertilization (the 1st cleavage completes within 45 min in zebrafish), and it gradually decreased in operations performed at later stages. Eventually, most of the operated embryos developed normally when the operation was done after 4- and 8-cell stages (Fig. 4). The simplest interpretation of these results is that, as observed for amphibian

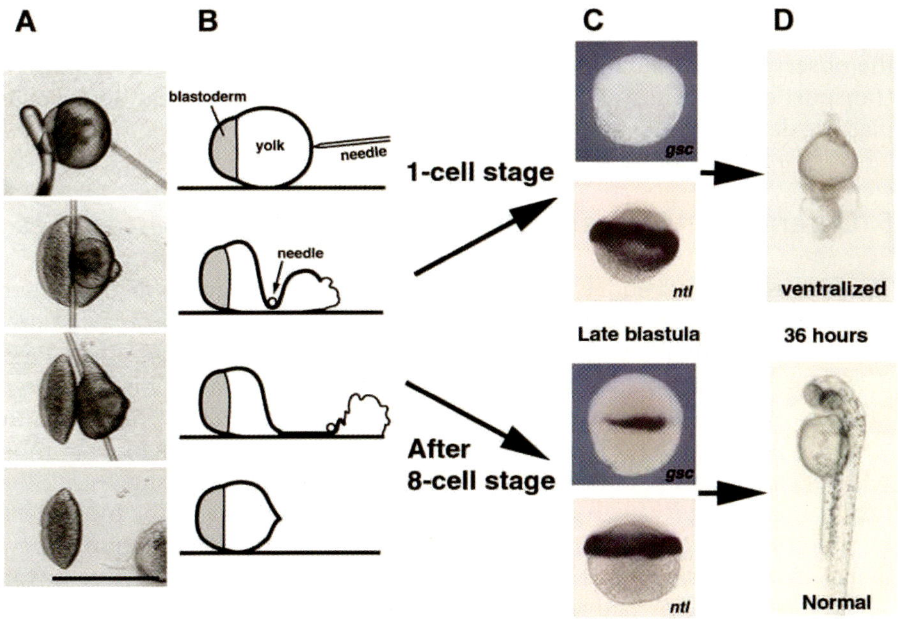

Fig. 4A–D. Removal of the vegetal yolk hemisphere. **A** Process of the operation. The vegetal yolk mass is squeezed out though a small hole made in the vegetal yolk membrane. The operation is finished in a few seconds. **B** Schematic representations of the operation shown in **A**. When the operation is done before 1-cell stage, the operated embryo frequently develops no dorsal tissues. In this embryo, *no goosecoid* (*gsc*) expression is detected at late-blastula stage, while *no tail* (*ntl*), a pan-mesodermal marker, is normally expressed (*upper two pictures* in **C**). The operated embryo finally exhibits a typical ventralized phenotype (bilateral symmetry) at a later stage (*upper picture* in **D**). In contrast, when the operation is done after the 8-cell stage, most of the operated embryos normally develop in terms of gene expression (*lower two pictures* in **C**) and external morphology (*lower picture* in **D**)

embryos, dorsal determinants, originally localized around the vegetal pole, are translocated away toward the animal pole during early cleavage stages (Fig. 1A,B). Consistent with this idea, Jesuthasan and Strähle (1997) observed the transport of cytoplasmic substances from the vegetal pole to the blastoderm at cleavage stages in zebrafish. They proposed that the cytoplasmic movement, which is mediated by a transient array of microtubules in the yolk cortex, is required for the axis specification. This hypothesis was supported by the observation that the disruption of microtubules soon after fertilization produced dorsal-deficient embryos at a high frequency (Jesuthasan and Strähle 1997).

In amphibian embryos, the microtubule system has also been implicated in translocation of dorsal determinants, and the disruption causes severe defects in dorsal development (reviewed by Gerhart et al. 1989), suggesting a conserved mechanism between frog and fish. However, translocation of the determinants in amphibians takes place during the 1st cleavage through cortical rotation in which the egg cortex rotates about 30° with respect to the underlying subcortical cytoplasm. In contrast, asymmetric rearrangement of the cytoplasm has not been observed in teleosts, probably due to the presence of a large yolk mass that may prevent a rotation of the cytoplasmic layer. In spite of this, the observed microtubule-mediated movement in fish eggs could function as transport of the determinants and result in asymmetric distribution of cytoplasmic determinants.

5 Determinants Function in the YSL and Dorsal Blastomeres

The results reported above, however, did not answer the question as to where dorsal determinants function in teleosts, in the YSL or dorsal blastomeres, or both? Experiments with embryo manipulations favor the idea that dorsal determinants exert their effects through the YSL. Transplantation of younger rainbow trout blastoderms onto gastrula-stage yolk cells (YCs) from which the blastoderm had been removed led to development of dorsal structures by those cells located atop the dorsal side of the host YC (Long 1983), suggesting that the localized activity of the YSL determines the dorsal fate in the blastoderm in a non-cell-autonomous manner. Recent yolk-cell transplantation experiments in zebrafish further strengthened this idea; zebrafish yolk cell, from which the blastoderm had been mechanically removed, induced dorsal-specific genes such as *goosecoid* (*gsc*) in the surrounding cells when transplanted into the animal-pole region of the host (Fig. 5A,B). It was also shown that the dorsal-inducing ability of the yolk cell is dependent on dorsal determinants that are present in the vegetal hemisphere at early cleavage stages (Fig. 6; Mizuno et al. 1999b). Furthermore, the *bozozok/dharma/nieuwkoid* gene, which encodes a transcription factor required for the formation of the organizer, is expressed in the dorsal YSL at the early-gastrula stage and was shown to act in a non-cell-autonomous manner (Koos and Ho 1998; Yamanaka et al. 1998). Indeed, it was shown that specific expression of the

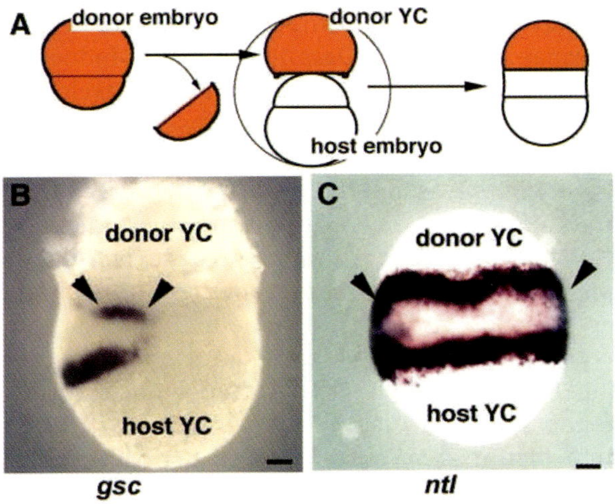

Fig. 5A–C. Transplantation of the yolk cell (*YC*). **A** Schematic representation of the operations. The labeled YC, from which the blastoderm has been removed, is transplanted atop the animal pole region of an unlabeled embryo. **B** Induction of *gsc* expression by the transplanted YC. **C** Induction of *ntl* expression by the transplanted YC. *Scale bar* 100 μm

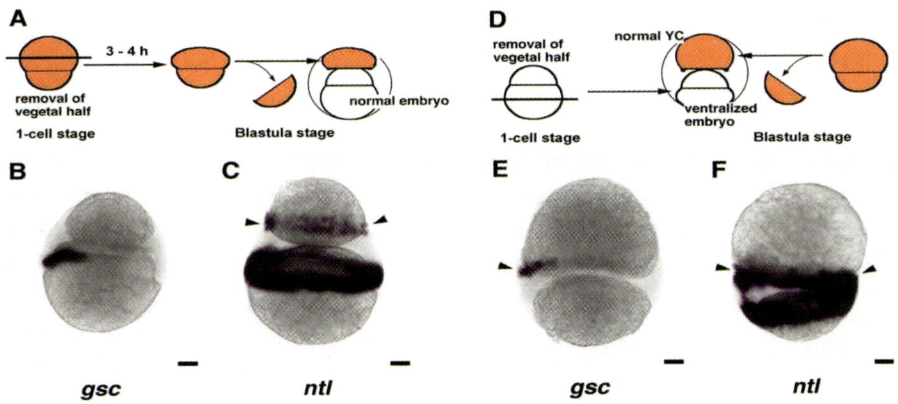

Fig. 6A–F. Induction of *gsc* and *ntl* expression by the transplanted yolk cell (*YC*). **A–C** Labeled YC from vegetal-removed embryo (operated on 5–20 min postfertilization) was transplanted into the normal host blastula. **A** Schematic of the experiment. At the early-gastrula stage, no ectopic expression of *gsc* is observed (**B**), whereas ectopic *ntl* expression (*arrowheads*) is induced (**C**). **D–F** Labeled normal YC was transplanted into vegetal-removed host blastula (operated at 5–20 min postfertilization). **D** Schematic of the experiment. At the early-gastrula stage, ectopic expression of *gsc* (**E**, *arrowhead*) was induced in vegetal-removed host cells (unlabeled) near the donor YSL. Note that endogenous expression of *gsc* is not detected (**E**). Ectopic expression of *ntl* (**F**, *arrowheads*) is also detected (**F**) in the vegetal-removed blastoderm. *Scale bar* 100 μm

bozozok/dharma/nieuwkoid gene in the YSL of *boz* mutants rescued the mutant phenotype, indicating that function of the *bozozok/dharma/nieuwkoid* gene in the YSL is sufficient for normal dorsal development (Fekany et al. 1999).

Transplantation experiments support the idea that the dorsal YSL in teleosts is equivalent to the amphibian Neiuwkoop center that is present in the dorsovegetal region of the endoderm and is responsible for induction of the Spemann organizer. Despite the dorsal-inducing ability of the YSL, whether or not the YSL is required for dorsal development was not addressed until recently. However, the following observations suggest that the blastoderm also incorporates dorsal determinants and autonomously develops dorsal fate. First, Jesuthasan and Strähle (1997) observed that polystyrene beads that were injected into the vegetal hemisphere of the yolk were able to enter into the marginal blastomeres at cleavage stage. Throughout early cleavage stages, cytoplasm continues to stream from the yolk cell into the attached blastomeres via cytoplasmic bridges, and some marginal blastomeres remain in cytoplasmic confluence with the yolk cell until after the mid-blastula transition (Cooper and Kimmel 1998). Second, although nuclear translocation of β-catenin, which is thought to mediate the Wnt-related signaling cascade required for activation of dorsal-specific genes, is first detected in the dorsal YSL, it is also observed in dorsal blastomeres adjacent to the YSL (Schneider et al. 1996). Finally, and more decisively, RNase injection into the YSL, which is expected to abolish nearly all zygotic activities of the YSL by degradation of endogenous mRNAs in the YSL, does not disrupt organizer formation (Fig. 7; Chen and Kimelman 2000). Probably, dorsal determinants are delivered to a region that will give rise to both the dorsal YSL and dorsal blastomeres and the YSL plays a cooperative role with the blastoderm to develop a dorsal fate. This is further supported by observations that, in the yolk cell transplantations, ectopic expression of *gsc* was usually much weaker compared with endogenous expression in the transplanted hosts (Mizuno et al. 1996, 1999b). Furthermore, the RNase injection experiment described above demonstrated that proper

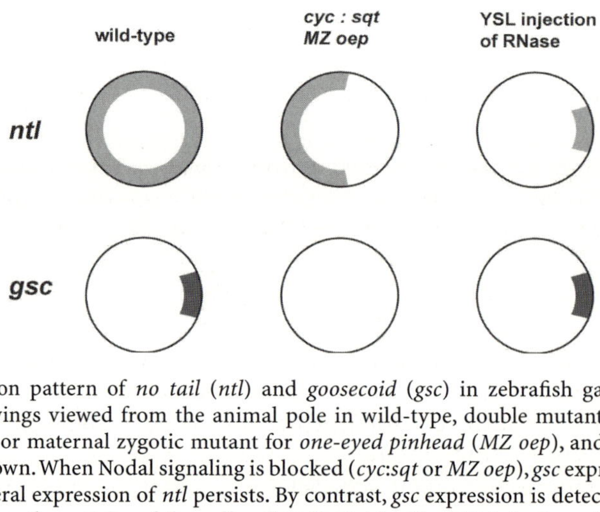

Fig. 7. Expression pattern of *no tail* (*ntl*) and *goosecoid* (*gsc*) in zebrafish gastrula embryos. Schematic drawings viewed from the animal pole in wild-type, double mutant for *cyclops* and *squint (cyc:sqt)* or maternal zygotic mutant for *one-eyed pinhead* (*MZ oep*), and RNase-injected embryos are shown. When Nodal signaling is blocked (*cyc:sqt* or *MZ oep*), *gsc* expression is absent, while ventrolateral expression of *ntl* persists. By contrast, *gsc* expression is detected in the dorsal blastomeres when the activity of the yolk cell is diminished by YSL injection of RNase

regulation of dorsal mesodermal genes requires YSL activity; the expression of *bozozok/dharma/nieuwkoid* was upregulated in the dorsal blastoderm after RNase injection into the YSL (Fig. 7; Chen and Kimelman 2000).

6 The Role of the YSL in Mesoderm and Endoderm Formation

Like induction of a dorsal fate, the zebrafish YSL is able to induce mesoderm and endoderm in a non-cell-autonomous manner. This was first suggested by experiments in which the entire yolk cell was removed from the blastoderm of many teleosts, *Fundulus*, loach and medaka. The operation resulted in severe malformations including mesodermal defects (Oppenheimer 1936; Kostomarova 1969; Bozhkova et al. 1994; Hyodo et al. 1996). However, because a depletion of the yolk cell caused non-specific effects on embryonic development, due to mechanical damages to the deep cell layer and/or deprivation of yolk-cell derived nutrition, the results seemed inconclusive. Again, the yolk-cell transplantation experiments described above provided more direct evidence. The isolated yolk cell induced the ectopic expression of mesodermal (*ntl*) as well as dorsal markers (*gsc*) when transplanted to the host blastula (Fig. 5B; Mizuno et al. 1996, 1999b; Ober and Schulte-Merker 1999). Early endodermal marker, *gata5*, is also induced by the transplanted yolk cell (Rodaway et al. 1999). These ectopic expression domains were always observed in the host blastomeres in close vicinity to the transplanted YSL, suggesting that the YSL directly induces mesoderm and endoderm in the overlying blastoderm. As in the case of dorsal induction, the requirement of the YSL was examined by YSL injection of RNase. When injected at the 1k-cell stage, the embryo did not develop any mesoderm and endoderm derivatives except for the most dorsal ones, indicating a requirement for YSL activity in mesoderm and endoderm formation (Fig. 7; Chen and Kimelman 2000). Furthermore, the induction of ventrolateral mesoderm and endoderm was found to require YSL RNAs between 1k- (3 h) and sphere stage (4 h; Chen and Kimelman 2000). Taken together, it is reasonable to say the YSL is necessary and sufficient to induce at least ventrolateral mesoderm and endoderm in zebrafish (Fig. 1C).

Recently, genetic analyses in mouse and zebrafish have shown that TGF-β signaling by Nodal family ligands is required for mesoderm and endoderm formation (Schier and Shen 2000). Mouse *Nodal* mutants lack a primitive streak and most mesoderm (Conlon et al. 1994), and, similarly, zebrafish double mutants for *Nodal*-related genes, *cyclops* and *squint*, also lack axial mesoderm and a wide range of dorsolateral mesoderm (Fig. 7; Feldman et al. 1998). Consistently, the marginal blastomeres, mesoderm and endoderm precursors, express both *squint* and *cyclops*, while *squint* expression is also detected in the YSL (Erter et al. 1998; Feldman et al. 1998; Rebagliati et al. 1998b; Sampath et al. 1998). At the moment, the role of Squint in the YSL is not clear because a *squint* mutation does not affect early marginal mesoderm development (Feldman et al. 1998). Interestingly, however, the expressions of *cyclops* and *squint* in marginal blastomeres were also eliminated by RNase injection into the YSL, indicating that the expressions of *Nodal* genes in the blastoderm are also regulated by YSL activity through unknown mechanisms (Chen and Kimelman 2000). Furthermore, the ventrolateral mesoderm persists in *cyclops/squint* double mutants (Fig. 7; Feldman et al. 1998), suggesting that unknown factors act in parallel with Nodal signaling to specify this tissue (Shimizu et al. 2000).

7 Localized Inducing Activities Within the YSL

As described above, the YSL consists of the two continuous regions, internal and external YSLs, demarcated by the site at which the tight junctions between EVL and the YSL form (Fig. 1D; Cooper and Kimmel 1998). Nuclei in the two YSL regions seem to have distinct properties because they exhibit different gene expression profiles (see below). Furthermore, mesoderm and endoderm are induced only in the marginal region juxtaposed to the external YSL as a circumferential ring. Since all blastomeres are competent to express mesodermal markers in response to mesoderm inducers (Ober and Schulte-Merker 1999), endogenous signals must be locally restricted to the external YSL. So far, there are few reports that address these mechanisms. Localization of the activity could be achieved by interaction between the marginal blastomeres and the YSL. The analysis of zebrafish maternal-effect mutation *janus* may provide some insights (Fig. 2E-H). In the mutant embryos, the cleavage-stage blastoderm is frequently segregated into two halves that undergo separate development until fusion occurs at the end of gastrulation (Abdelilah et al. 1994). Intriguingly, when the two separated blastoderms detach sufficiently enough, ectopic mesodermal marker, *ntl*, is induced near the gap region, where the ectopic marginal zone is formed (Fig. 2H; Abdelilah and Driever 1997). Therefore, mesoderm induction might depend on the interaction between the morphological unique marginal positions and the underlying YSL. At the moment, we do not know whether the YSL close to the ectopic marginal region exhibits the features of the external YSL in terms of the morphology (e.g., thick syncytial layer) and gene expression (e.g., *squint*). Further analysis of zebrafish mutations will make clear the mechanism with which to localize inducing activities within the YSL.

8 Is Visceral Endoderm in Mammal Equivalent to Teleost YSL?

Recent studies in mice suggest that extra-embryonic endoderm, especially anterior visceral endoderm (AVE), has an important role in early embryonic patterning, especially in the anterior head induction (Beddington and Robertson 1999). Many key regulators obtained from *Xenopus* embryos were shown to be expressed in the AVE, e.g., *hex*, *hesx1*, *lim1*, *otx2* and *cer-l*. Intriguingly, the zebrafish *hex* (recently renamed *hhex*) gene is expressed in the dorsal YSL at the gastrula stage (Ho et al. 1999; Liao et al. 2000). Overexpression experiments have demonstrated that Hex in the YSL antagonizes ventralizing factors, Bmp2b and Wnt8, in the overlying blastoderm (Ho et al. 1999). Similarly, Hex in mouse embryos may serve to inhibit the expression of *Bmp* genes in the AVE and, in a non-autonomous manner, in the anterior part of the embryo (Beddington and Robertson 1999). Together with the fact that the YSL, the extra-embryonic lineage, has various inducing activities on the overlying blastoderm (e.g., mesoderm, endoderm and axial structures), the YSL could be the equivalent tissue of mouse visceral endoderm (Thomas et al. 1998). It is also worth noting that critical regulators in zebrafish endoderm development, such as *faust/gata5* and *casanova/226D7*, are expressed both in endodermal precursor cells and the YSL, suggesting that teleost YSL has some characteristics of endoderm (Dickmeis et al. 2001b; Kikuchi et al. 2001; Reiter et al. 2001a; Sakaguchi et al. 2001). Cloning

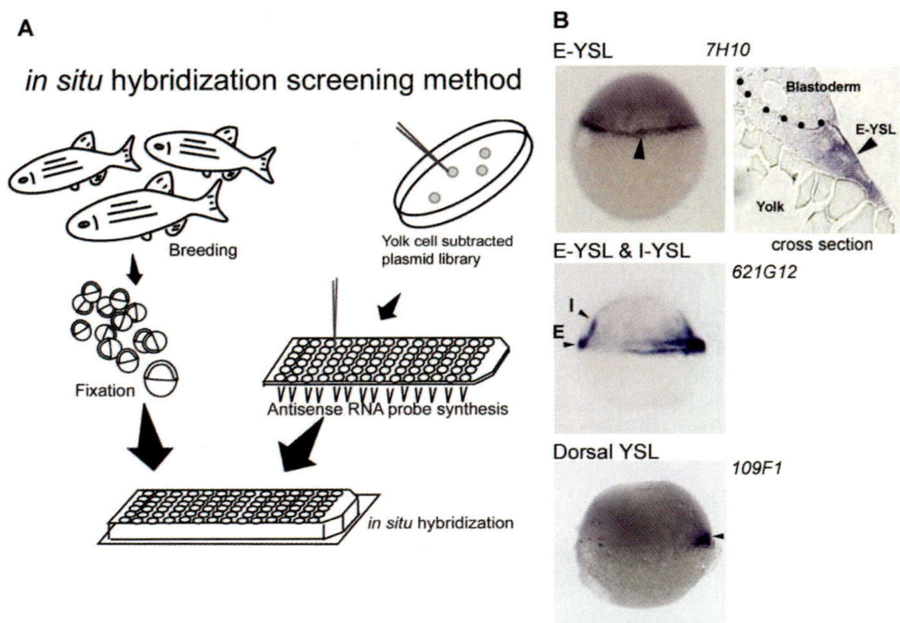

Fig. 8A,B. In situ hybridization screening of genes expressed in the YSL. **A** Schematic representation of the experiment. Colonies from a subtracted cDNA library are randomly picked up, grown in culture medium and directly subjected to PCR amplification of insert cDNA fragments. The amplified fragments are used as a template for probe synthesis. All procedures including hybridization and probe synthesis are carried out in a multi-titer plate. **B** Typical expression patterns of isolated genes. *Top* (*7H10* at 30%-epiboly) the entire external YSL (*E-YSL, arrow*), *middle* (*621G12* at 50%-epiboly); internal-YSL (*I-YSL, arrow*) and E-YSL. *Below* (*109F1* at 50%-epiboly) dorsal E-YSL (*arrow*)

and expression analyses of more genes expressed both in mouse AVE and zebrafish YSL will be required to further test these notions.

9 Searching for Genes Specifically Expressed in the YSL

As discussed above, there must be novel key factors expressed in the YSL, e.g., endogenous inducers for ventrolateral mesoderm and endoderm and transcription factors that regulate *Nodal* genes. In addition, factors that regulate cell movements during epiboly might be expressed in the YSL, because RNase injection into the YSL severely disrupts epiboly movements and yolk cell lifts off the blastoderm (Chen and Kimelman 2000). To further understand the molecular nature of the YSL, we have developed a screening method to isolate genes expressed in the YSL (T. Sakaguchi and H. Takeda, in prep.). The method involves in situ hybridization in a multi-titer plate and a subtracted cDNA library in which yolk-cell-expressed cDNAs are enriched (Fig. 8A). So far, we have screened about 600 clones, and obtained more than 30 independent clones showing YSL expression. Although obtained clones are partial fragments that frequently include 3′-non-coding regions, we have identified previously described YSL genes such as *bozozok/dharma/nieuwkoid* (Yamanaka et al. 1998) and *mtx1* (Hirata et al. 2000) and many novel clones that have no similarity to any known genes. At present, we estimate that we have obtained more than a dozen novel genes expressed in the YSL. Among them, some genes exhibit restricted pattern of expression in the entire or dorsal external YSL, while others are widely expressed in both the external and internal YSL (Fig. 8B), revealing a regional specificity within the YSL. The functional analyses of these genes using injection of synthetic RNAs or morpholino-modified antisense oligonucleotides are now underway.

Mesoderm Induction and Patterning

David Kimelman[1] and Alexander F. Schier[2]

1 The Origin of Mesoderm

The mesoderm arises prior to gastrulation as a ring at the equator of the embryo (Fig. 1A). This ring of cells sits on top of an extra embryonic structure called the yolk syncytial layer (YSL). The YSL forms when the most marginal blastomeres collapse into the yolk cell at the 10th cleavage (Kimmel and Law 1985b). The cells closest to the YSL develop into either mesoderm or endoderm, whereas cells further away from the YSL develop only as mesoderm (Kimmel et al. 1990; Warga and Nusslein-Volhard 1999).

At approximately 6h after fertilization, gastrulation begins. The major role of gastrulation is to move the mesoderm (and the endoderm) from the outside of the embryo into the inside. Three types of cell movements are critical for the rearrangement of mesodermal progenitors and the formation of the embryo (for more information on this subject, see Kane and Adams; Solnica-Krezel and Cooper, this Vol.). First, the mesoderm around the entire equator moves from the outer layer (epiblast) into an inner layer (the hypoblast), thus placing the mesoderm inside the future ectoderm of the embryo (Fig. 1B; Warga and Kimmel 1990). While much of the mesoderm moves inside during the gastrula stages, the process of internalization continues throughout somitogenesis, until all of the cells that will become the tail muscle have moved inside (Kanki and Ho 1997). Mesodermal cells not only move inside, but also undergo a second specific set of movements called convergent extension (for more information, see Solnica-Krezel and Cooper, this Vol.). These movements bring the mesodermal cells to the dorsal side of the embryo, where the embryo proper will form. In addition, this process drives the elongation of the embryo along the anterior-posterior axis. A third movement, called epiboly, includes the vegetal movement of all the cells on the outside of the embryo and results in the complete envelopment of the yolk by the developing embryo. Together, involution, convergent extension and epiboly coordinate the movement of the

[1] Department of Biochemistry, Box 357350, University of Washington, Seattle, Washington 98195-7350, USA
[2] Developmental Genetics Program, Skirball Institute of Biomolecular Medicine, Department of Cell Biology, New York University School of Medicine, New York, New York 10016, USA

Results and Problems in Cell Differentiation, Vol. 40
L. Solnica-Krezel (Ed.): Pattern Formation in Zebrafish

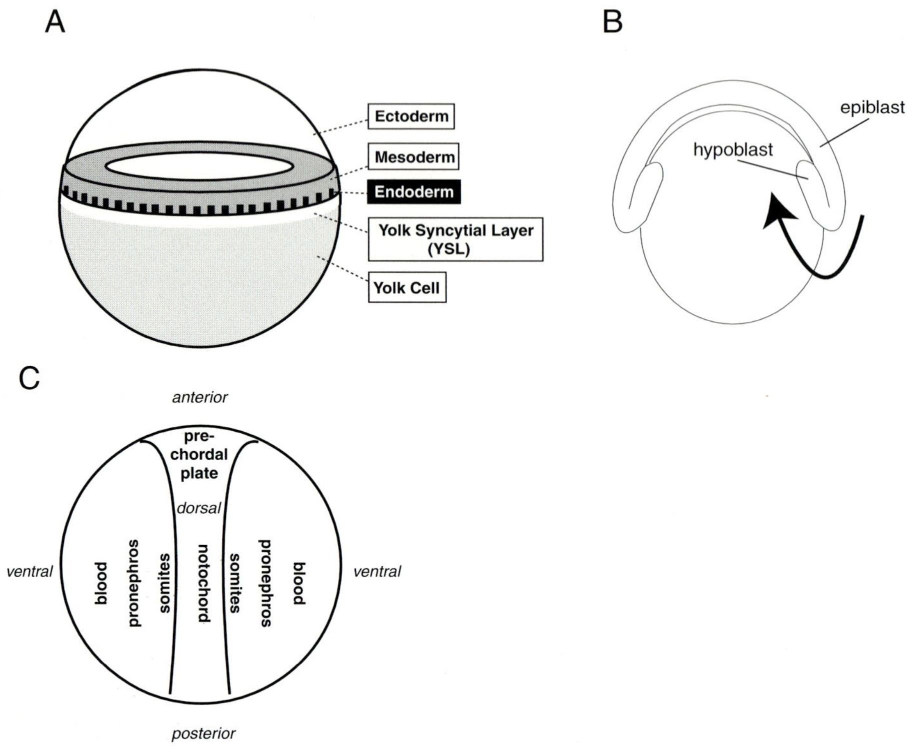

Fig. 1A–C. Fate map and gastrulation of the early zebrafish embryo. **A** Fate map of a pre-gastrula zebrafish embryo, with the germ layers indicated. The ectoderm has been rendered transparent to show that the mesendoderm forms in a ring around the equator, with the center of the ring developing as ectoderm. The cells closest to the YSL will develop as either mesoderm or endoderm. **B** Mesodermal (and endodermal) cells move from the outer layer (the epiblast) into the inner layer (the hypoblast). This view shows a section through the lateral regions of a gastrula-stage fish embryo. The movement of cells from the epiblast to the hypoblast continues throughout somitogenesis. **C** Fate map of a post-gastrula embryo showing the major mesodermal derivatives. The view is on to the dorsal side

mesodermal cells so that mesodermal derivatives develop in the correct position. By the end of gastrulation, the precursors for these different mesodermal cell types are arranged along the dorsal-ventral and anterior-posterior axes, with axial mesoderm (developing into notochord and the future head mesoderm, which is called the prechordal plate) being the most dorsal tissue type and paraxial mesoderm (future somites), intermediate mesoderm (pronephros), and lateral plate mesoderm (blood and endothelial cells of blood vessels) being more dorsolateral, lateral and ventral mesodermal derivatives, respectively (Fig. 1C). Correct movement of the mesodermal cells is also required for their differentiation into a wide variety of cell types, including muscle, heart, blood, dermis, cartilage, and bone.

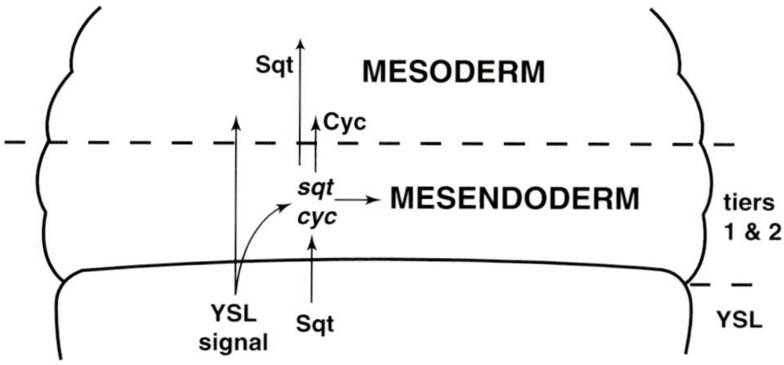

Fig. 2. Mesoderm-inducing signals. Sqt and a putative unidentified inducing signal are expressed in the YSL. The main target of these signals is the activation of *sqt* and *cyc* in the tier 1 and 2 blastomeres, which induce the mesoderm and endoderm in this region. Sqt and Cyc also induce mesoderm in the overlying cells, although Sqt is able to travel a greater distance than Cyc. In the absence of Sqt and Cyc, the YSL signal is only able to induce tail mesoderm

2 Induction of Mesoderm by Intercellular Signals

The mesoderm is initially induced at the equator of the embryo by signals arising from the YSL, followed by the activation of inducing signals expressed in the mesendodermal blastomeres closest to the YSL (Fig. 2). Two lines of evidence support a critical role for the YSL. First, Mizuno and colleagues removed the blastomeres from an embryo, leaving the YSL and yolk. When they placed this YSL/yolk cell on top of the animal pole of another embryo, they observed that a ring of ectopic mesoderm was induced in the prospective ectodermal cells of the host embryo, demonstrating that the YSL/yolk cell has a mesoderm-inducing capacity (Mizuno et al. 1996). A later experiment showed that the YSL/yolk cell can also induce endoderm (Rodaway et al. 1999). In a second approach, transcripts were eliminated from the YSL/yolk cell by the injection of RNase into the YSL just after it formed. This not only prevented the formation of endoderm and all but the most dorsal mesoderm, but also eliminated the expression of the mesendoderm-inducing signals described below (Chen and Kimelman 2000). These lines of evidence suggest that the YSL is both necessary and sufficient for the formation of the mesendoderm.

The reason that the mesoderm forms in a ring and not as a sheet that covers the YSL is because the YSL appears to release inducing signals in a ring at its outer edge (Mizuno et al. 1996; Ober and Schulte-Merker 1999; Rodaway et al. 1999). Interestingly, the YSL nuclei are located at the outer edge of the YSL at the time when mesoderm is first induced (Kimmel and Law 1985b), and the experiments with RNase suggest that the YSL signal is activated soon after the embryonic genome becomes transcriptionally active (Chen and Kimelman 2000). Thus the YSL signal may be transcribed by the YSL nuclei or locally activated by another factor expressed by the YSL nuclei while they are present in a ring at the outer edge of the YSL, and immediately sent to the overlying blastomeres. The nature of the YSL signal is not yet known. In frogs, which have many parallels to zebrafish, a maternal transcription factor named VegT is localized to the vegetal region of the embryo during oogenesis, and it activates the transcription of several *nodal* family members in the prospective mesendodermal region (reviewed in Kimelman and Griffin 2000; Schier and Shen 2000; Schier 2001). A zebrafish ortholog of *vegT* has been identified, but it is only transcribed zygotically and is not present maternally (Griffin et al. 1998). In addition, mutations in this gene (*spadetail*) do not prevent the initial formation of the mesendoderm (Griffin et al. 1998), indicating that a different mechanism is used to establish the future mesendodermal region in fish than in frogs (all mutants described in this chapter are listed in Table 1).

Table 1. Mutations affecting zebrafish axis formation and patterning

Mutation	Gene product	Molecular role	Phenotype
Signaling pathways			
cyclops	Nodal	Nodal signal	Cyclopia
squint	Nodal	Nodal signal	Cyclopia, dorsal mesoderm defects
cyclops;squint			No endoderm and head/ trunk mesoderm
one-eyed pinhead	EGF-CFC	Nodal co-receptor	No endoderm and head/ trunk mesoderm
schmalspur	FAST1/FoxH1	Transcription factor	Dorsal mesoderm defects
swirl	BMP2b	BMP signal	Severely dorsalized
chordino	Chordin	BMP inhibitor	Weakly ventralized
ace	FGF8	FGF signal	No cerebellum
wnt8	Wnt 8	Wnt signal	No ventral and posterior structures
Transcription factors			
bozozok	Homeodomain	Transcriptional repressor	Variable loss of dorsal mesoderm and forebrain
vox/vent	Homeodomain	Transcriptional repressor	Severely dorsalized (double mutants)
no tail	T-box	Transcriptional activator	Lack of tail mesoderm and notochord
spadetail	T-box	Transcriptional activator	Lack of trunk mesoderm
Transcription factor/signaling pathways double mutants			
bozozok;squint			Severe head reduction and loss of notochord
bozozok;chordino			Range from complete loss of head and trunk to severe reduction of head and loss of notochord
Unknown			
ichabod	?	β-Catenin localization?	Variably ventralized

3 Nodal Signaling

A major breakthrough in understanding mesoderm induction came with the identification of two zygotically expressed mesoderm-inducing agents as members of the Nodal subfamily of the TGF-β superfamily of signaling factors (reviewed in Schier and Shen 2000; Schier 2001; Schier and Talbot 2001). These two factors, Squint (Sqt) and Cyclops (Cyc), are functionally redundant and expressed in partially overlapping patterns (Fig. 2). *Sqt* is expressed earlier than *cyc*, and is found in both the YSL and in the future mesendoderm (Erter et al. 1998; Feldman et al. 1998; Rebagliati et al. 1998a). In contrast, *cyc* is expressed only in the mesendoderm (Feldman et al. 1998; Rebagliati et al. 1998a; Sampath et al. 1998). While elimination of either *sqt* or *cyc* produces relatively minor defects in the initial induction of the mesoderm (Heisenberg and Nusslein-Volhard 1997; Feldman et al. 1998; Hatta et al. 1991b; Chen and Schier 2001), the double mutant lacks all endoderm and all head and trunk mesoderm, including blood, heart, pronephros, notochord and prechordal plate (Feldman et al. 1998). Interestingly, these double mutants still produce tail mesoderm, which may be induced directly by the YSL signal. *cyc;sqt* mutants affect the expression of marginal genes already during blastula stages, demonstrating that these signals act before gastrulation in the specification of mesodermal and endodermal progenitors. In the absence of *cyc* and *sqt*, marginal cells do not involute. Dorsal marginal cells acquire neural fates instead of dorsal mesodermal fates, while ventral and lateral marginal cells contribute exclusively to the tail (Feldman et al. 2000). These results indicate that Cyc and Sqt are essential zygotic signals required for mesoderm and endoderm induction.

Cyc and Sqt appear to be able to activate the same set of target genes, indicating that both signals activate the same downstream signaling pathway. Moreover, both factors activate target genes in a concentration-dependent manner. High levels of Sqt and Cyc are required to activate genes such as *goosecoid*, whereas lower levels are sufficient to activate genes such as *no tail* or *bhikhari* (Chen and Schier 2001; Gritsman et al. 2000). Interestingly, however, Sqt can activate target genes directly in distant cells while Cyc acts only locally (Chen and Schier 2001). Sqt thus has the properties of a secreted morphogen.

It has been hypothesized that Nodal signals act similarly to the TGF-β signal Activin and directly activate the Activin receptor serine/threonine kinase. Surprisingly, however, extracellular co-receptors of the EGF-CFC family are required for receptor activation by Nodal signals, but not by Activin (Fig. 3; Gritsman et al. 1999; Schier and Shen 2000; Shen and Schier 2000; Schier 2001; Schier and Talbot 2001; Yeo and Whitman 2001). This was first revealed in genetic studies that showed that the membrane-attached protein One-eyed pinhead (Oep), a member of the EGF-CFC family, is required for cells to respond to Nodal signals (Zhang et al. 1998b; Gritsman et al. 1999). Maternal-zygotic *oep* mutants display a very similar phenotype to *cyc;sqt* mutants and

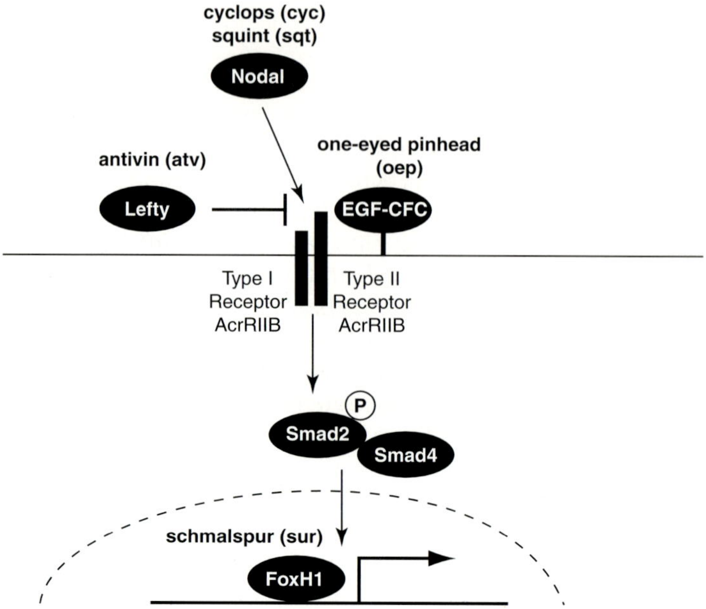

Fig. 3. Model for the Nodal-signaling pathway in zebrafish. The Nodal signals Cyclops and Squint activate, in the presence of the putative EGF-CFC co-receptor One-eyed pinhead, Activin receptors such as ActRIB and ActRIIB. This leads to the phosphorylation of Smad2 and its association with Smad4, the FAST/Fox protein Schmalspur and other transcription factors. Members of the Lefty family act as inhibitors of the pathway

are unresponsive to the overexpression of Cyc or Sqt. In contrast, activated forms of the type I Activin receptor ActRIB and the transcription factor Smad2 can rescue various aspects of the *oep* phenotype (Gritsman et al. 1999). In addition, the extracellular domain of the type II Activin receptor ActRIIB can block the effects of ectopic Sqt (Meno et al. 1999). Cross-linking studies in *Xenopus* animal caps have shown that the mouse EGF-CFC protein Cripto can bind to ActRIB, and that mouse Nodal can bind to a complex containing ActRIB, Cripto and ActRIIB (Yeo and Whitman 2001). The genetic and biochemical results lead to the model that Oep acts as an essential extracellular co-receptor for Nodal signals to activate Activin receptors such as ActRIB and ActRIIB, which then leads to the phosphorylation and activation of Smad2 (Fig. 3; Gritsman et al. 1999; Schier and Shen 2000; Schier 2001). The use of Oep as a co-receptor could spatially and temporally constrain Nodal signaling: cells expressing Oep are responsive to Cyc and Sqt signals, whereas cells lacking the co-receptor are protected from the effects of these potent inducers. It is unclear, however, if this strategy is used during mesoderm induction since *oep* is widely expressed during the time when mesoderm is induced.

Overexpression of TGF-β factors of the Lefty family (also called Antivin) in wild-type embryos induces a phenotype similar to *cyc;sqt* or maternal-zygotic *oep* mutants (Bisgrove et al. 1999; Meno et al. 1999; Thisse and Thisse 1999; Schier and Shen 2000; Thisse et al. 2000; Schier and Talbot 2001). Furthermore, the effects of Lefty are counteracted by overexpression of Sqt or Cyc or the extracellular domain of the ActRIIB activin receptor (Meno et al. 1999). Since Nodal signals induce *lefty* expression, these results suggest that Lefty proteins act as competitive feedback inhibitors by binding to Nodal receptors such as ActRIIB (Fig. 3; Meno et al. 1999). As Nodal signals can autoregulate, Lefty antagonists ensure that Nodal signaling is not hyperactivated, restricting the amount of mesoderm.

The activation of the Nodal signaling pathway leads to the phosphorylation of Smad2 and its association with other transcription factors such as FoxH1/FAST1 (Whitman 1998) to regulate downstream genes (Fig. 3). FoxH1 is disrupted by mutations in the *schmalspur* locus (Brand et al. 1996a; Schier et al. 1996; Solnica-Krezel et al. 1996; Pogoda et al. 2000; Sirotkin et al. 2000b). Maternal-zygotic *schmalspur* mutants have a reduction of dorsal mesoderm, but are less affected than *cyc;sqt* or maternal-zygotic *oep* mutants. These results suggest that Schmalspur is required for only a subset of the effects of Nodal signaling.

4 Dorsal-Ventral Patterning of the Mesoderm

The future dorsal-ventral axis of the mesoderm is first established at fertilization. Determinants located at the vegetal pole of the egg prior to fertilization are transported to the future dorsal side of the embryo, most likely on microtubules (Jesuthasan and Strähle 1997). Experiments in which the vegetal pole is physically removed from the rest of the zebrafish embryo demonstrate that these determinants gradually move toward the future dorsal side of the embryo between the 1- and 8-cell stages (Mizuno et al. 1999b). In zebrafish, the determinants move at some point 20 min postfertilization (Ober and Schulte-Merker 1999), which is halfway through the first cell cycle, but it is not yet clear how long it takes for the determinants to move in this system.

One big mystery in zebrafish is how the future dorsal side of the embryo is specified. In *Xenopus*, the sperm cues the future dorsal side, with the dorsal determinants moving more or less to the opposite side of the embryo from which the sperm enters (DeRobertis et al. 2000; Moon and Kimelman 1998). The sperm pronucleus, with its associated microtubule-organizing center (MTOC), orients cortical microtubules so that the determinants move toward the dorsal side (Larabell et al. 1996; Rowning et al. 1997; Miller et al. 1999a). However, in zebrafish, the sperm enters at the very top of the animal pole through a specialized structure called the micropyle. It is not clear whether the sperm pronucleus/MTOC moves to one side of the embryo after it enters the egg, thus cueing the future dorsal side, or whether the dorsal determinants move randomly to one side of the egg after fertilization. While cortical microtubule arrays have been observed in the zebrafish embryo (Jesuthasan and

Strähle 1997), there is currently no evidence that they are oriented toward the dorsal side. In the final analysis, it is not at all important which side of the embryo becomes the future dorsal side; the important thing is that the determinants move to one side, and only one side, of the embryo.

The nature of the dorsal determinants are not known, but results from both fish and frogs (Nasevicius et al. 1998; Pelegri and Maischein 1998; Sumoy et al. 1999) have shown that they are activators of the Wnt intracellular signaling pathway (reviewed in Moon and Kimelman 1998; DeRobertis et al. 2000), which results in the localized accumulation of β-catenin in the nuclei on the future dorsal side of the embryo (Schneider et al. 1996), in the region where the organizer (shield) will form (for more information on the function and formation of the organizer, see Hibi et al., this Vol.; Schier 2001; Schier and Talbot 2001). Intriguingly, a maternal zebrafish mutant named *ichabod* has recently been described that shows the normal dorsal stabilization of β-catenin, but β-catenin does not enter the nuclei (Kelly et al. 2000), indicating that nuclear transport of β-catenin is under regulatory control. In both fish and frogs, it is still not clear whether the Wnt pathway is activated by a Wnt ligand or by a novel intracellular mechanism. Inhibition of the Wnt receptor causes defects in the formation of dorsal structures (Nasevicius et al. 1998; Sumanas et al. 2000), yet no dorsally localized Wnt ligand has been found in either system that is present specifically on the early dorsal side of the embryo.

The direct target genes of β-catenin are mostly not known, but include the transcription factor *bozozok* (*boz*; the gene was originally named *dharma* and *nieuwkoid*; Schier et al. 1996; Solnica-Krezel et al. 1996; Yamanaka et al. 1998; Fekany et al. 1999; Koos and Ho 1999; Solnica-Krezel and Driever 2001). The *boz* promoter has binding sites for Tcf/Lef factors (Ryu et al. 2001), which are DNA-binding transcription factors that interact with β-catenin (Bienz 1998). *Sqt*, which has a β-catenin-dependent dorsal domain of expression in addition to its pan-mesodermal expression discussed above, is also a potential direct target of β-catenin (Fig. 4; Erter et al. 1998; Feldman et al. 1998; Rebagliati et al. 1998a). *Boz* and *sqt* single mutants have a variable loss of dorsal mesoderm (Schier et al. 1996; Solnica-Krezel et al. 1996; Feldman et al. 1998), ranging from a loss of axial structures to near normal development. In contrast, the *boz;sqt* double mutants have a severe loss of dorsal structures, showing that these two factors have overlapping roles in patterning the dorsal mesoderm (Sirotkin et al. 2000a; Shimizu et al. 2000; Solnica-Krezel and Driever 2001). Among the genes regulated by Boz and Sqt are the Bmp antagonists *chordin* and *noggin* (Fig. 4; Miller-Bertoglio et al. 1997; Schulte-Merker et al. 1997; Fürthauer et al. 1999; Sirotkin et al. 2000a; Shimizu et al. 2000). In line with this, double mutants of *bozozok* and *chordin* cause a synergistic loss of head and trunk (Gonzalez et al. 2000).

On the ventral and lateral sides of the embryo, the intracellular signaling factors *bmp2b*, *bmp7*, and *bmp4* and the transcription factors *vox* and *vent* (also known as *vega1* and *vega2*) are active (Fig. 4; Hammerschmidt et al. 1996b; Kishimoto et al. 1997; Dick et al. 2000; reviewed in Nguyen et al. 1998b;

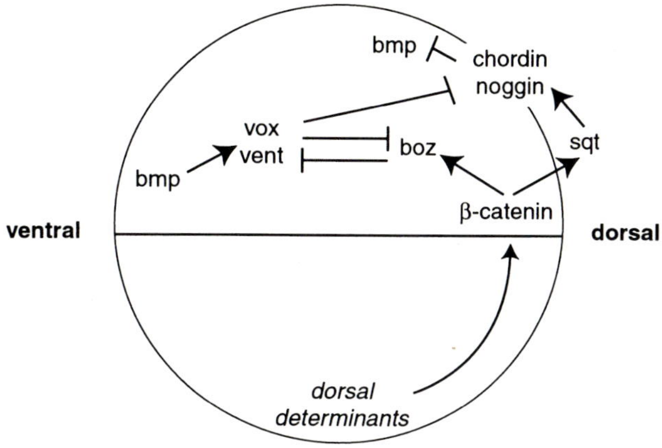

Fig. 4. Model for dorsal-ventral patterning of the mesoderm. Dorsal determinants are moved from the vegetal pole after fertilization to one side of the embryo, which will become the future dorsal side. These determinants cause the stabilization of β-catenin, which then activates target genes including *boz* and *sqt*. Boz is a repressor that prevents transcription of the Bmp target genes *vox* and *vent*, whereas Sqt is proposed to be an activator of the Bmp inhibitors, *chordin* and *noggin*. On the ventral side of the embryo, where β-catenin is not stabilized, Bmp activates the expression of *vox* and *vent*, which inhibit the transcription of *boz*, *chordin* and *noggin*

Kodjabachian et al. 1999; Kawahara et al. 2000a,b; Melby et al. 2000; Schmid et al. 2000). Vox and Vent are transcriptional repressors that function redundantly to block the transcription of organizer genes, including *boz* (Onichtchouk et al. 1998; Melby et al. 1999; Kawahara et al. 2000a,b; Imai et al. 2001). Boz is also a putative transcriptional repressor, which blocks the expression of *vox* and *vent* (Koos and Ho 1999; Kawahara et al. 2000a,b; Melby et al. 2000). Thus the dorsal-ventral patterning of the mesoderm can be viewed as the generation of an initial asymmetry in the early embryo by the Wnt/β-catenin pathway, which sets up during the gastrula stage two opposing forces: on the dorsal side, the transcriptional repressor Boz and the Bmp antagonists Chordin and Noggin, and on the ventral side, the transcriptional repressors Vox and Vent (Fig. 4). This antagonism is thought to set up a gradient of Bmp signaling across the dorsal-ventral axis, that patterns the mesoderm as well as the neurectoderm (Neave et al. 1997; Nguyen et al. 1998b; Barth et al. 1999; Connors et al. 1999; Hild et al. 1999; for more information on Bmp gradients and neural patterning, see Hammerschmidt and Mullins, this Vol.). Modulation of this genetic pathway through a variety of means changes the fate of the mesoderm across the dorsal-ventral axis. For example, *boz* and *chordin* mutants have reduced dorsal gene expression and expanded expression of ventral genes (Hammerschmidt et al. 1996b; Solnica-Krezel et al. 1996; Solnica-Krezel and Driever 2001), whereas reduction of *bmp* or *vox* and *vent* function results in embryos with the opposite changes in gene expression (Mullins et al. 1996; Imai et al. 2001). A very similar set of interactions also patterns the dorsal-ventral axis in frogs, although so far no *Xenopus boz* gene has yet been discovered. Instead, the function of Boz may be replaced by Goosecoid and/or Xiro1, which are also transcriptional repressors expressed on the dorsal side of the embryo (Cho et al. 1991; Gomez-Skarmeta et al. 2001).

5 Anterior-Posterior Patterning of the Mesoderm

Genetic and embryological data have shown that the zebrafish embryo can be divided into three principle domains: the head, trunk and tail. The border between the head and trunk within the mesoderm is the division between prechordal mesoderm and the somitic mesoderm and is marked by the expression of genes such as *goosecoid* (gsc; Stachel et al. 1993) within the prechordal plate and *no tail* (Schulte-Merker et al. 1992) within the trunk and tail mesoderm. The border between the trunk and tail domains is at the 17th somite, which is marked by the most posterior extension of the gut tube.

Anterior-posterior patterning of the mesoderm is best understood for dorsal (axial) mesoderm. The axial mesoderm develops anteriorly into the prechordal plate and posteriorly into the notochord (Melby et al. 1996). Fate mapping studies have shown that this arrangement is already apparent before gastrulation, when prechordal plate progenitors are located right at the dorsal margin, whereas notochord precursors reside further away from the margin (Gritsman et al. 2000). The differential expression of genes such as *gsc* in the prechordal plate fate domain and *floating head* (Talbot et al. 1995) in the notochord domain indicates that these precursor populations are at least partially specified. During involution, prechordal plate progenitors precede notochord progenitors and end up more anteriorly. As described above, an absence of Nodal signaling on the dorsal side changes the fate of cells from dorsal mesoderm to neural. Interestingly, Nodal signaling not only acts as a switch between neural and mesodermal fates, but differential Nodal signaling can also specify prechordal plate vs notochord fates (Gritsman et al. 2000; Thisse et al. 2000). High levels of Nodal signaling are required for prechordal plate specification, whereas lower levels are essential for notochord specification (Gritsman et al. 2000). This corresponds to the expression of *cyc* and *sqt* at the margin, overlapping with prechordal plate precursors, while notochord precursors are found a few cell diameters away (Fig. 5).

The posterior mesoderm, which includes both the trunk and tail, is under the control of signals from the FGF intercellular signaling pathway; embryos that are blocked for FGF signaling by the ectopic expression of a dominant-negative FGF receptor develop only a head, and completely lack a trunk and tail (Griffin et al. 1995). The key targets for the FGF signaling pathway are two members of the T-box transcription factor family, *no tail* (*ntl*) and *spadetail* (*spt*). *No tail* mutants develop a normal head and trunk, but completely lack the tail domain (Schulte-Merker et al. 1992, 1994; Halpern et al. 1993). In con-

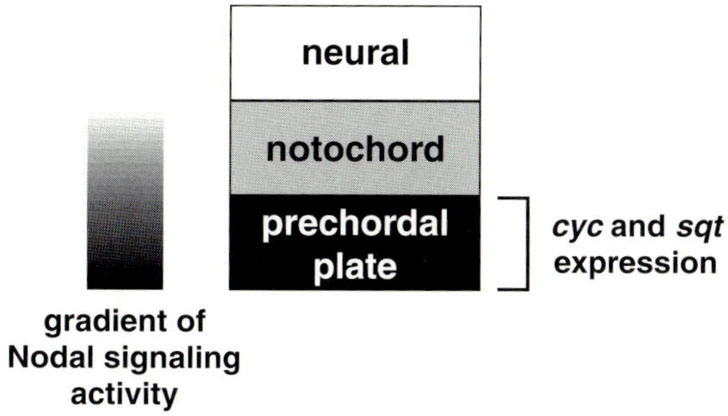

Fig. 5. Nodal signaling patterns the dorsal mesoderm. A gradient of Nodal signaling is proposed to subdivide the dorsal mesoderm into prechordal plate (high Nodal signaling), notochord (low Nodal signaling) and neural tissue (no Nodal signaling). The Nodal signals *cyc* and *sqt* are expressed in the domain corresponding to the prechordal plate

trast, *spadetail* (*spt*) embryos have severely defective trunks but relatively normal tails (Kimmel et al. 1989; Ho and Kane 1990; Griffin et al. 1998). Importantly, embryos lacking both *ntl* and *spt* function phenocopy the effect of blocking FGF signaling (Amacher et al. 2002). Thus *ntl* and *spt* have been proposed to be the key regulators of mesoderm induction in the trunk and tail, which carry out the function of FGF in the mesoderm (Griffin et al. 1998).

It is still not certain how many FGFs are important for the formation of the mesoderm. A mutation in *fgf8*, which is expressed in the posterior mesoderm as the trunk and tail develop (Fürthauer et al. 1997; Reifers et al. 1998), has a strong cerebellar defect, and thus it is named *acerebellar* (*ace*), but it has only very slight effects on the formation of the mesoderm (Reifers et al. 1998). However, treatment of wild-type and *ace/fgf8* embryos with a drug that inhibits FGF signaling has shown that the *ace/fgf8* fish are hypersensitive to the drug, indicating that *ace/fgf8* normally functions together with another *fgf* within the posterior mesoderm (K. Griffin and D. Kimelman, unpublished results). A good candidate for this is *gfgf*, which is also expressed throughout the posterior mesoderm (B. Draper and C. Kimmel, pers. comm.)

Why the posterior mesoderm is subdivided into a trunk and tail region is not yet clear since the somites that make up the trunk and tail are morphologically identical. However, the mesoderm of the trunk will produce ribs whereas the mesoderm of the tail will not. Thus the genetic boundary between these two regions might be used to subdivide the posterior mesoderm into a rib-bearing versus non-ribbed boundary (Kimelman and Griffin 2000).

Recent evidence suggests that the trunk/tail boundary may be established by Bmp signaling. Elimination of the early dorsal determinants by microdissection results in an embryo that lacks a head and trunk, but which forms a clear tail (Ober and Schulte-Merker 1999). Interestingly, these embryos begin forming somites at the same stage as control embryos begin to form their first tail somite (somite 18). Thus, in the absence of dorsal determinants, embryos can only form a tail. However, when the dorsal determinants were removed

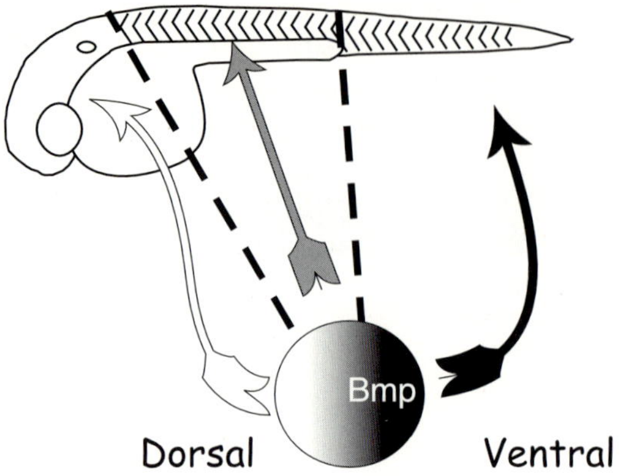

Fig. 6. Model for patterning of the trunk and tail by Bmp signaling. Bmps are proposed to act in a dorsal-ventral gradient and to impose patterning along the anterior-posterior axis of the embryo

from *swirl* embryos, which lack Bmp signaling (Kishimoto et al. 1997; Nguyen et al. 1998b), these embryos started forming somites at the same time as control embryos began to form their first trunk somite (somite 1; Ober and Schulte-Merker 1999). These results indicate that Bmp signaling is required for tail formation and that it suppresses trunk formation. This view is consistent with the analysis of mutants that affect the Bmp-signaling pathway, which show defects in the formation of the tail (Mullins et al. 1996).

Since Bmp activity has been proposed to form a dorsal-ventral gradient across the embryo (see preceding section and Hammerschmidt and Mullins, this Vol.), it is tempting to speculate that this gradient establishes the trunk/tail boundary (Fig. 6B). While the regions of the embryo that give rise to the trunk and tail have not been rigorously mapped, existing data suggests that the tail comes from the more ventral side of the embryo (Warga and Nüsslein-Volhard 1999; Feldman et al. 2000). Similarly, in the frog embryo, which is patterned by Bmp signaling much like the fish, the tail has been fate mapped to the ventral side of the embryo (Lane and Smith 1999). How Bmps might regulate the anterior-posterior axis remains an unsolved problem in zebrafish development.

6 Wnt Signals in Mesoderm Patterning

In addition to its role in establishing the early dorsal side of the embryo (see above), the Wnt/β-catenin pathway has another important role in patterning the mesoderm. The bicistronic *wnt8* locus is expressed throughout the ventrolateral mesoderm at the shield stage, and continues to be expressed in the tailbud (Kelly et al. 1995b; Lekven et al. 2001). A deletion that removes both *wnt8* genes leads to a dramatic decrease in ventrolateral mesoderm and a corresponding increase in dorsal axial mesoderm, indicating that the bicistronic *wnt8* locus plays a critical role in delimiting the boundary between the axial and paraxial mesoderm, as well as in establishing ventrolateral fates (Lekven et al. 2001). Similar effects are seen when the gene products of both *wnt8* genes are inhibited with antisense morpholino oligonucleotides, but there is no effect if only one of the pair of genes is inhibited (Lekven et al. 2001). Thus, the two *wnt8* genes are functionally redundant, which is likely to explain why a *wnt8* mutant was not isolated in the large-scale screens. There are similarities between the effects caused by mutants in the Bmp pathway and a loss of Wnt8 signaling, yet also significant differences. In both mutants ventrolateral and posterior mesodermal fates are lost, but the *wnt8* mutants lose paraxial fates whereas the *bmp* mutants show an expansion of paraxial mesoderm. It will be interesting to determine how the Bmp and Wnt pathways intersect to regulate the patterning of the trunk and tail mesoderm.

7 Future Directions

Having identified key players in mesoderm patterning, one of the future challenges is to determine how cells in different regions of the embryo integrate the many inputs they receive from different signaling pathways and transcription factors. For instance, how does a cell respond to graded levels of both Nodal and BMP signaling? This is further complicated by the complex movements of mesodermal cells which are constantly changing the positions of both the signaling and the responding cells. It will be crucial to study the dynamics of mesoderm formation in finer detail to determine how a cell integrates information over time and responds to novel cues. Moreover, we do not understand how a group of individual cells coordinates its decisions. Does this occur by exposure to the same global signals and/or by local cell-cell signaling and adhesive interactions? Finally, it remains a mystery how patterning information is translated into morphogenesis and differentiation. For instance, how does Nodal signaling lead to the involution of cells; how are dorsal-ventral and animal-vegetal patterning coupled to convergent extension; and how does epiboly influence fate decisions during gastrulation.

Answering these questions will require the isolation of novel genes affecting gastrulation (Bentley et al. 2000; Lekven et al. 2000; McCallum et al. 2000; Nasevicius and Ekker 2000; Peterson et al. 2000; Talbot and Hopkins 2000; Ma et al. 2001), the use of sophisticated imaging and labeling techniques (Cooper et al. 1999; Herbomel et al. 1999; Köster and Fraser 2001a) and the application of genomic resources (Talbot and Hopkins 2000).

Acknowledgements. The authors wish to thank K. Griffin and S-R. Chen for supplying figures. D.K.'s work on zebrafish is supported by NSF grant 0078303. A.F.S. is supported by the NIH and is a Scholar of the McKnight Endowment Fund for Neuroscience and the Irma T. Hirschl Trust.

The Guts of Endoderm Formation

Rachel M. Warga[1] and Didier Y.R. Stainier[2]

In this chapter, we will review the formation of the definitive endoderm, the population of cells that give rise to the lining of the digestive tract, its associated organs and the pharyngeal pouches. At the cellular level, we will describe the location and movement of endodermal cells from the onset of epiboly until the end of gastrulation. At the molecular level, we will discuss the genes associated with endoderm formation beginning with Nodal signaling. For convenience, we use the term involution, sometimes referred to as internalization; we also separate endoderm formation into the pre-involution (blastula) and post-involution (gastrula) periods although of course endoderm formation involves a continuous series of events. In addition, we refer to the cells that contribute to the endoderm as progenitors prior to their involution and precursors after their involution.

1 Endoderm Formation During the Blastula Period

1.1 Location of Endodermal Progenitors

Following cleavage and mid-blastula transition at the 512-cell stage, the blastula consists of three cellular domains, the enveloping layer of the blastoderm, the deep cells of the blastoderm and the yolk cell with its syncytial nuclei (Kane et al. 1992). The deep cells form the embryonic anlagen (Kimmel et al. 1990) and, for the remainder of this chapter, unless otherwise stated, we refer only to these cells. Fate mapping studies at the 1- to 2K-cell stage reveal two fate domains within the blastoderm, an animal domain whose cells will remain superficial and become the future ectoderm (Kimmel et al. 1990; Warga and Kimmel 1990), and a marginal domain whose cells will involute during gastrulation to become the future endoderm and mesoderm (Kimmel et al. 1990; Warga and Kimmel 1990; Stainier et al. 1993; Lee et al. 1994). Endoderm originates from only the most marginal cells of this latter domain (Fig. 1A; Warga and Nüsslein-Volhard 1999). However, early-blastula marginal cells may also give rise to mesoderm (Warga and Nüsslein-Volhard 1999), thus indeterminacy in cell fate seems to exist early between endoderm and mesoderm.

By the late-blastula stage, one or two cell divisions later, indeterminate cell fate has given way to restricted cell fate as late-blastula cells generally give rise to progeny of a single germ layer (Kimmel et al. 1990). Fate mapping analyses show that endodermal progenitors at the late-blastula stage are found no more than four cell diameters away from the blastoderm margin with most located within two cell diameters (Fig. 1C). While it appears that a majority of the endoderm derives predominantly from the future dorsal side (Fig. 1B; Warga and Nüsslein-Volhard 1999), this could be partly a consequence of the compaction movement occurring in the late blastula (see below). In addition to the

[1] Department of Biology, University of Rochester, Rochester, New York 14627, USA

[2] Department of Biochemistry and Biophysics, Programs in Developmental Biology, Genetics and Human Genetics, University of California, San Francisco, California 94143-0448, USA

Results and Problems in Cell Differentiation, Vol. 40
L. Solnica-Krezel (Ed.): Pattern Formation in Zebrafish

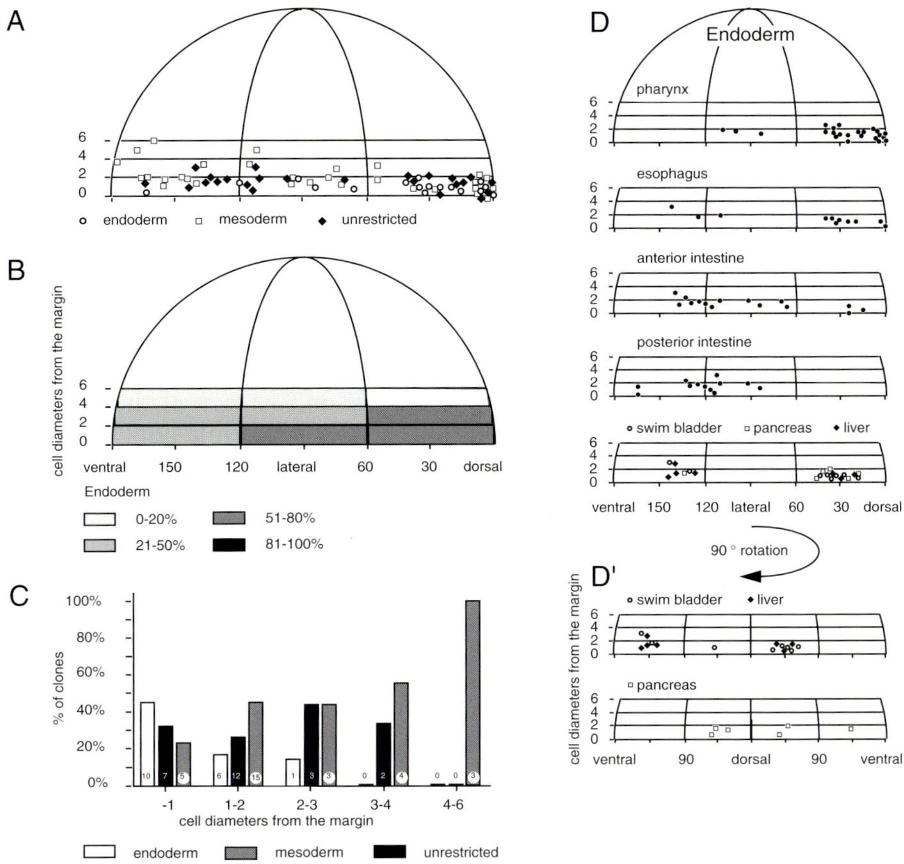

Fig. 1A–D. Cells at the blastoderm margin generate endoderm. **A** Position of endodermal, mesodermal and unrestricted clones at 40% epiboly from injections of single cell at the 1- to 2K-cell stage. **B–C** Summary data from the fate map clones illustrating location biases. **B** Percentage of clones giving rise to endodermal derivatives in relation to dorsal (0–60°), lateral (61–120°) or ventral (121–180°). **C** Percentage of endodermal, mesodermal and unrestricted clones in relation to distance from the blastoderm margin. **D–D′** Fate map positions of endodermal derivatives at 40% epiboly. **D** Endodermal derivatives presented as a compilation of both left and right sides of the blastulae. In the case of the liver, swim bladder and pancreas, we also present a rotated dorsal view (**D′**) to illustrate the asymmetries between the left and right side

endoderm, some marginal cells also give rise to a portion of the mesoderm (Warga and Nüsslein-Volhard 1999). Thus, endoderm and mesoderm progenitors appear to be intermingled to some degree along the margin of the blastoderm (Fig. 1A), suggesting bi-potential cell fate in the late-blastula stage.

1.2 Early Topographic Map of Endodermal Organs

Within the endodermal territory of the late-blastula-stage embryo there is a topographic map of the gut and its associated organs (Warga and Nüsslein-

Volhard 1999). In general, anterior gut progenitors are located more dorsally and posterior gut progenitors more laterally. Thus, pharynx: the endoderm of the pharyngeal cavity and pouches, comes predominantly from dorsal cells, esophagus: the endoderm directly posterior of the pharynx, from lateral cells, and intestine: endoderm of the trunk (sometimes referred to as the stomach) and hindgut region, from lateral and ventral cells (Fig. 1D). Oddly enough, endoderm progenitors that give rise to the gut-associated organs are located in two disparate locations on either side of the blastula. The liver and swim bladder originate from cells located dorsolaterally on the right, as well as from cells located ventrolaterally on the left, whereas the pancreas usually originates from cells in the inverse locations (Fig. 1D′; Warga and Nüsslein-Volhard 1999). Note that the location of these organs is also asymmetric in the adult. It remains to be determined whether populations on one side of the fate map represent the endodermal components of these organs while populations on the other side represent the mesenchymal components.

1.3 Organs of Ambiguity: the Hypochord and Forerunner Cells

Two additional cell types that are frequently stated to be endodermal are the hypochord and forerunner cells. Based on morphological studies, the hypochord is thought to derive from endoderm adjacent to the notochord (Eriksson and Lofberg 2000). However, fate mapping studies do not support this idea. Cells in the endodermal progenitor territory do not appear to contribute to the hypochord (Warga and Nüsslein-Volhard 1999), rather hypochord originates from cells that are located more distantly from the margin and that give rise exclusively to dorsal mesodermal fates such as notochord (Melby et al. 1996). In agreement with this observation, the midline of the zebrafish *deltaA* mutant has many more notochord cells with correspondingly fewer hypochord cells (Appel et al. 1999). Thus, hypochord seems to be a mesodermal rather than an endodermal derivative.

Forerunner cells are of more ambiguous origin. Cell tracing studies indicate that forerunner cells derive from dorsal marginal enveloping layer cells which separate from the blastoderm and move out onto the yolk cell (K. DelKanic and R.M. Warga, unpubl.). In addition, some forerunners may also derive from dorsal marginal deep cells (D’Amico and Cooper 1997). However, in keeping with the observation that forerunner and enveloping layer cells are related, in *half-baked* mutant embryos where epiboly of the deep cells is halted, epiboly of both the enveloping layer and forerunner cells occurs normally (Kane et al. 1996; Solnica-Krezel et al. 1996).

Once forerunners, this group of approximately 30 cells form an aggregate which precedes the vegetally advancing blastoderm margin (Cooper and D’Amico 1996; Melby et al. 1996). Later, these cells form a transient structure known as Kupfer’s vesicle and then become mesodermal tissues in the tail such as notochord and muscle (Melby et al. 1996; Warga and Nüsslein-Volhard 1999). Forerunners might therefore be considered mesodermal, except for their partially "extraembryonic" origin. Regardless, forerunner cells do share many similarities with the endoderm. Forerunner and endodermal cells often share lineages (Melby et al. 1996), often express the same genes (see below), and share a similar location and environment with respect to the yolk cell during gastrulation.

1.4 Cell Behavior of Endodermal Progenitors

To understand how location may determine the fate of the endodermal progenitors it helps to consider the dynamic cell rearrangements that occur during the blastula period (see also Kane and Adams, this Vol.). Between the late-blastula stage and the onset of gastrulation, extensive cell mixing occurs (Warga and Kimmel 1990). During this period, cells intercalate radially outwards forcing neighboring cells to scatter widely, leading to the thinning and spreading of the blastoderm. Because of this movement, a cell's position may change dramatically from the late-blastula to the early-gastrula stages. However, as radial intercalation is less extensive at the margin (Wilson et al. 1995), marginal cells are far less likely than other cells to be displaced. Thus endodermal progenitors are more likely than other cells to be continuously exposed to local signaling factors that are expressed in the yolk cell and have been shown to be important in endoderm formation (Rodaway et al. 1999; Chen and Kimelman 2000) or by the marginal cells themselves (Feldman et al. 1998). Moreover, because of an additional cell movement which causes cells to coalesce towards the dorsal organizer region before the onset of gastrulation, dorsal marginal cells have even less of a tendency to scatter (Warga and Nüsslein-Volhard 1998). Compaction movements, which may also explain the greater tendency for endoderm dorsally (Fig. 1B), would further prevent cells from moving away from local influences within the organizer region which may itself promote endodermal fate (Chen and Kimelman 2000; S. Dougan et al., submitted).

1.5 Genes Involved in Endoderm Formation

Nodal signaling is essential for both endoderm and mesoderm formation in zebrafish. As the Nodal pathway is discussed extensively elsewhere (see Kimelman and Schier, this Vol.), we will focus specifically on the genes involved in patterning the endoderm, first by discussing relevant early components of the Nodal pathway, and then by discussing endodermal effectors of Nodal signaling.

1.6 The Nodal Factors and Cofactors: Cyclops, Squint and One-Eyed Pinhead

Mutational analyses and ectopic expression studies demonstrate the critical importance of Nodal signaling in zebrafish endoderm formation. In embryos lacking function of both Nodal-related genes, *cyclops* (*cyc*) and *squint* (*sqt*), as well as those lacking maternal and zygotic function of the Nodal facilitator *one-eyed pinhead* (*oep*), endoderm is absent, as well as most mesoderm (Feldman et al. 1998; Gritsman et al. 1999). Conversely, a constitutively active form of the type-I TGFβ receptor, Taram-A, diverts cells to an endodermal fate (Peyrieras et al. 1998).

Given that Nodals are required for both endoderm and mesoderm formation, how does a cell decide which fate to choose? One hypothesis is that cell fate is determined by the amount of Nodal signaling a cell receives. In support of this hypothesis, embryos lacking only *cyc*, *sqt* or zygotic *oep* gene function show greater loss of endoderm than mesoderm (Schier et al. 1997; Warga and Nüsslein-Volhard 1999; S. Dougan et al., submitted). A similar selective perturbation of endoderm is observed in embryos misexpressing low levels of Antivin, a competitive inhibitor of the Activin/Nodal signaling pathway related to mammalian Lefty (Thisse and Thisse 1999; Thisse et al. 2000). Hence, cells receiving higher levels of Nodal signals may differentiate as endoderm whereas cells receiving lower levels may differentiate as mesoderm. This concept of concentration-dependent Nodal signaling determining cell fate is supported by the observation that Sqt can act as a morphogen for mesodermal fates (Chen and Schier 2001).

It should be noted, however, that this model might be too simplistic. Characterization of the homozygous *sqt* mutation in a heterozygous *cyc* background by fate map analyses and gene expression patterns reveals minimal defects in ventral mesendodermal fates, but severe defects in dorsal mesendodermal fates (S. Dougan et al., submitted). In addition, cell movement defects are only observed on the dorsal side of mutant embryos. Although these data are not inconsistent with the model described above that more Nodal signaling specifies endodermal fate, they do suggest a modified model whereby Nodal signaling may be primarily involved in determining dorsal fates which includes a large proportion of the endoderm as indicated by the fate map.

1.7 Nodal Versus Bmp Activity

Studies in *Xenopus* and mouse suggest that a balance of Nodal and Bmp signaling may regulate the formation of the definitive endoderm (Sasai et al. 1996). This may be by direct antagonism as Nodal and Bmp7 can form inactive heterodimers (Yeo and Whitman 2001). Zebrafish embryos lacking Bmp activity such as *swirl* (*bmp2b*) and *snailhouse* (*bmp7*) have no ventrally derived fates (Mullins et al. 1996; Kishimoto et al. 1997; Dick et al. 2000) and exhibit a greater number and expanded distribution of endodermal cells by the end of gastrulation (J. Alexander and D.Y.R. Stainier, unpubl. data). Embryos lacking the Bmp antagonist *dino* (*chordin*) have decreased amounts of dorsally derived fates (Hammerschmidt et al. 1996b; Schulte-Merker et al. 1997) and fewer endodermal cells (J. Alexander and D.Y.R. Stainier, unpubl. data). Together these results indicate that Bmps may also participate in endoderm formation and patterning in zebrafish embryos, although clearly more work is needed to understand whether this is via inhibition of Nodal signaling or through some other mechanism.

1.8 Nodal-Independent, Oep-Dependent Cell Motility

Nodal signaling in zebrafish depends on *oep* function (Gritsman et al. 1999). *oep* encodes a membrane-bound protein closely related to mouse Cripto and both proteins contain a characteristic EGF-like domain and a conserved cysteine-rich region known as the CFC domain (Zhang et al. 1998b; for more about Oep, see Kimelman and Schier, this Vol.). Biochemical analyses of Cripto indicate that its conserved CFC motif is necessary for interaction with the TGFβ receptor Alk4 and its EGF-like motif is important for Nodal binding to Alk4 (Yeo and Whitman 2001). As Activin, another TGFβ-related factor with similar inducing abilities to Nodal, does not require Cripto to activate downstream genes (Yeo and Whitman 2001), it is possible that Cripto helps the TGFβ receptor to differentiate between these two ligands.

In zebrafish, Oep may have an additional cryptic role in endoderm formation apart from regulating TGFβ activity. Cells lacking some or all *oep* function show specific defects in cell motility that do not resemble the defects observed in cells lacking both *sqt* and *cyc* function (R.M. Warga and D.A. Kane, in prep.). Notably, *oep* mutant cells form abnormal cell contacts cohering together, and appear unable to undergo an epithelial to mesenchymal transition upon involution. Also, movement of *oep* mutant cells is slower. In contrast, *sqt;cyc* mutant cells show none of these defects, but rather appear unable to involute normally. The abnormal coherence and cell movement of *oep* mutant cells is autonomous, as tested by cell transplantation. This is perhaps not surprising, as previous work has demonstrated an autonomous role for *oep* in endoderm formation (Gritsman et al. 1999). However, none of these cellular defects can be rescued by activating the TGFβ receptor through misexpression of Activin (R.M. Warga and D.A. Kane, in prep.). This is altogether a rather surprising result, as Activin does rescue the tissue defects of the *oep* mutant (Gritsman et al. 1999). Thus these results argue that *oep* function could have Nodal-independent roles and offer an interesting additional pathway for its involvement in endoderm formation perhaps through control of cell migration, an important component of cell fate determination.

1.8.1 Oep as a Component of EGF Signaling

Further support for a Nodal-independent role of *oep* in cell migration comes from studies in mammalian cell culture. Mouse mammary epithelial cells expressing elevated levels of Cripto, as do some forms of human breast cancer, display a more mesenchymal morphology and show increased cell motility

(Salomon et al. 1999). Conversely, mouse embryonic stem cells lacking Cripto function show defective migration and changes in adhesiveness (Xu et al. 1999). These aspects of Cripto function appear to be transduced by the EGF-related family of receptors (Bianco et al. 1999). In *Drosophila*, EGF receptor signaling guides the dorsal cell migration of specialized somatic follicle cells during oogenesis (Duchek and Rørth 2001).

Undoubtedly, there may be other pathways at play that are involved in determining endodermal fate in addition to the aforementioned Nodal- and EGF-signaling pathways. The identification of such potential pathways is an important goal for the future.

1.9 Effectors of Nodal Signaling: Casanova, Bonnie and Clyde, and Faust

Thus far we have largely focused on early components of Nodal signaling in endoderm formation, except for a small digression on EGF signaling. However, there are three mutations in zebrafish with specific endodermal defects that appear to affect downstream components of the Nodal pathway. These mutations are *casanova* (*cas*), *bonnie and clyde* (*bon*) and *faust* (*fau*), and the corresponding genes all encode transcription factors. Namely, *cas* encodes a novel Sox-related protein (Dickmeis et al. 2001b; Kikuchi et al. 2001); *bon* encodes a Mix-type homeodomain protein related to *Xenopus* Mixer (Henry and Melton 1998; Kikuchi et al. 2000); and *fau* encodes the zinc finger protein Gata5 (Reiter et al. 1999). As these three endodermal mutations offer insights that cannot be learned solely from misexpression studies, we focus the remainder of this section on the role of these loci in endodermal formation.

The expression of *cas*, *bon* and *fau* each begins during blastula stages and analyses of *oep*, *cyc* and *sqt* mutant embryos, as well as of embryos misexpressing Antivin or a constitutively active form of Taram-A, demonstrate that their expression is regulated by Nodal signaling (Alexander and Stainier 1999; Rodaway et al. 1999; Dickmeis et al. 2001b; Kikuchi et al. 2001; Reiter et al. 2001a; Sakaguchi et al. 2001). Cyc and Sqt, however, do not appear to play equivalent roles, as *bon* expression is altered in *sqt* mutants but is normal in *cyc* mutants (Alexander and Stainier 1999), in keeping with the idea that Cyc is required later in development (Erter et al. 1998; Sampath et al. 1998). It is not yet clear whether Nodal signaling directly regulates the expression of *cas*, *bon* and *fau*, although there is some indication of a close interaction with Bon, as studies in *Xenopus* show that some Mix family members interact directly with Smad2, a downstream component of the Nodal pathway (Germain et al. 2000), and Bon contains the same Smad interaction domain.

1.9.1 The Endodermal Phenotypes of *cas*, *bon* and *fau* Mutants and the Expression of These Genes in the Blastula

Embryos mutant for *cas*, *bon* and *fau* were originally identified by their abnormal heart morphology (Chen et al. 1996; Stainier et al. 1996). However, closer inspection revealed that each of these mutations also causes specific endodermal defects. While *cas* mutants, like *MZoep* and *cyc;sqt* doubles, lack all endodermal precursors with the exception of pharyngeal pouch endoderm as assessed by *nkx2.3* expression (J. Alexander and D.Y.R. Stainier, unpubl.), they do not show most of the other patterning defects exhibited by *MZoep*, and *cyc;sqt* mutants, indicating that *cas* function may be principally necessary for endoderm formation. *bon* and *fau* mutants, on the other hand, lack only some of the endodermal precursors (Reiter et al. 1999; Kikuchi et al. 2000). *bon* mutants show a more severe defect having only 10% of the normal amount of endodermal cells, whereas *fau* mutants show a less severe defect having ~60% of the normal amount. It is important to note in this regard that, while the *bon* mutations appear to represent null alleles, neither of the *fau* mutations analyzed to date appears to be a null (Reiter et al. 1999). Therefore, a complete understanding of the role of *fau* in endoderm formation will require the identification and analysis of complete loss-of-function alleles.

The expression of *cas*, *bon* and *fau* all initiate at about the same time and, by dome stage, all three genes are clearly expressed circumferentially in deep cells near the margin. *cas* and *bon* expression first comes on in dorsal marginal blastomeres, but their expression progressively expands around the margin (Alexander et al. 1999; Dickmeis et al. 2001b; Sakaguchi et al. 2001). This dorsoventral progression of expression may be due to the similar progression of *sqt* expression at about the same time (Erter et al. 1998; Feldman et al. 1998; Rebagliati et al. 1998a). No such pattern has been reported for *fau* expression, which instead appears to arise simultaneously everywhere.

While *cas*, *bon* and *fau* are all expressed around the circumference of the margin, they Fig. 2A–C differ in the extent of their expression along the animal-vegetal axis (Fig. 2A); *bon* is the most broadly expressed. At the onset of gastrulation, *bon* transcripts are detected in approximately 8–10 rows of cells, resembling the expression of *no tail* (Schulte-Merker et al. 1992), *spadetail* (Griffin et al. 1998), and *snail1* (Hammerschmidt and Nüsslein-Volhard 1993; Thisse et al. 1993), as well as that of other genes expressed in mesendodermal progenitors. By contrast, at the same stage, *fau* is expressed in approximately 3–4 rows of cells (Reiter et al. 1999; Rodaway et al. 1999) and, based on the fate map analyses discussed earlier (see Sect. 1.1), these cells include all endodermal and some mesodermal progenitors. *cas* expression is seen in only a subset of the *fau*-expressing cells. The presumption is that only endodermal progenitors express *cas*. This hypothesis, which remains to be tested, is based on the observation that during gastrulation *cas* expression appears to be in only endodermal and not in mesodermal precursors (Dickmeis et al. 2001b; Kikuchi et al. 2001; Sakaguchi et al. 2001).

cas, *bon* and *fau* are also expressed in the yolk syncytial domain of the blastula, underlying the blastoderm margin during the high and sphere stage. As in the deep cells, *bon* and *cas* show a dorsal onset of expression (Alexander et al. 1999; Dickmeis et al. 2001b; Kikuchi et al. 2001; Sakaguchi et al. 2001). Yolk cell expression of *bon* remains dorsal whereas yolk cell expression of *cas* progresses around the margin, to eventually resemble the yolk cell expression

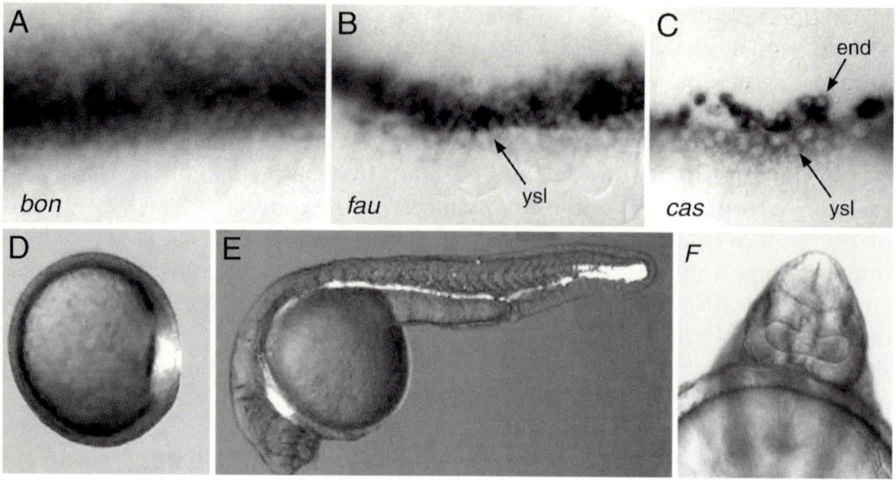

Fig. 2A–F. Comparison of *bon*, *fau* and *cas* expression at late-blastula stages. **A** *bon* appears to be expressed in all mesendodermal progenitors (8–10 rows of cells). **B** *fau* appears to be expressed in all endodermal and the most marginal mesodermal progenitors (3–4 rows of cells). In addition, it is expressed in the yolk syncytial layer (*ysl*). **C** *cas* appears to be expressed exclusively in the endodermal progenitors (*end*), as well as in the yolk syncytial layer (*ysl*). Cas can transfate dorsal mesodermal progenitors into endoderm. A marginal blastomere of a 16-cell stage embryo was injected with *cas* and *gfp* mRNA and examined at shield stage (**D**) and at 24-h postfertilization (hpf) (**E, F**). **E** GFP-expressing cells accumulate in the entire undifferentiated endoderm, from the level of the pharyngeal arches anteriorly to the anal opening in the posterior trunk region. More caudally, GFP-expressing cells are observed underlying the notochord in the anterior part of the tail, and in the notochord and tail bud more posteriorly. These fates are similar to those derived from the forerunner cells (Melby et al. 1996; Warga and Nüsslein-Volhard 1999). **F** High magnification view of the head of the embryo shown in **E**, revealing a cyclopic phenotype presumably caused by the transfating of dorsoanterior mesoderm cells into endoderm

of *fau*. *cas* is additionally expressed in a subset of dorsal marginal cells, which also express *sqt* (Erter et al. 1998; Feldman et al. 1998; Rebagliati et al. 1998a), and which later become the forerunners. It remains unclear whether dorsal marginal cells expressing *cas* are within the enveloping layer, as lineage data discussed earlier would suggest (see Sect. 1.3), but it is observed that, when *cas* is misexpressed in single animal pole cells, *sqt* expression is induced in only a few cells of the enveloping layer (Kikuchi et al. 2001).

1.9.2 Cas Is Sufficient To Convert Mesoderm into Endoderm

Ectopic expression of *cas* in presumptive mesodermal cells causes them to transfate into endoderm (Fig. 2D–F). Dorsal mesodermal cells mostly give rise to hatching gland and notochord (Kimmel et al. 1990; Melby et al. 1996; Warga and Nüsslein-Volhard 1999). When dorsal cells misexpress *cas*, they fail to express prechordal plate markers and instead express endodermal markers.

These cells later incorporate into digestive tract endoderm. The conversion of dorsal mesoderm into endoderm is further illustrated by cyclopia and U-shaped somites in such embryos (Kikuchi et al. 2001). Lateral mesodermal cells mostly give rise to muscle (Kimmel et al. 1990; Warga and Nüsslein-Volhard 1999) and ventral mesodermal cells normally give rise to tissues such as blood (Kimmel et al. 1990; Lee et al. 1994; Stainier et al. 1996; Warga and Nüsslein-Volhard 1999). When lateral and ventral cells misexpress *cas*, they lose the ability to form muscle or blood and express endodermal markers instead (Kikuchi et al. 2001).

Altogether, these data indicate that *cas* can initiate an endoderm-specific program in cells not normally fated to form endoderm, and, if these cells are dorsal, an apparently complete conversion to differentiated endodermal derivatives. This ability, however, appears to be restricted to only the mesoderm. When *cas* is misexpressed in animal pole blastomeres at the 256-cell stage, these cells do not seem to transfate from ectoderm into endoderm, but rather they populate various ectodermal locations such as the nervous system or epidermis of the yolk sac (Kikuchi et al. 2001). Thus it appears that only when *cas* is misexpressed near the margin can it redirect cells to an endodermal fate. An important task will be to identify the factors responsible for this spatial restriction. In this context, it is interesting to note that *cas* is able to induce in *MZoep* mutants expression of some early, but not later, markers of endoderm differentiation (Dickmeis et al. 2001b; Kikuchi et al. 2001; Sakaguchi et al. 2001). This indicates that an intact Nodal or EGF pathway is essential for the complete process of endodermal differentiation.

Less is known about the function of *bon* and *fau* in the embryo. However, misexpression of *fau* in wild-type embryos can lead to expanded endodermal gene expression (Reiter et al. 2001a) as well as to the formation of ectopic myocardial cells (Reiter et al. 1999).

1.10 A Molecular Pathway Leading to Endoderm Formation

The relationships between Cas, Bon and Fau have been addressed in several ways, and the resulting data, outlined below, suggest that Cas acts either in parallel to, or downstream of, Bon and Fau. The data are as follows: First, *bon;fau* double mutants contain far fewer endodermal cells than either mutant alone, suggesting that *bon* and *fau* act in parallel in endoderm formation (Reiter et al. 2001a). In agreement with this model, *bon* expression appears normal in *fau* mutant blastulae, and *fau* expression appears normal in *bon* mutant blastulae (Reiter et al. 2001). Second, misexpression of *bon* or *fau* in *cas* mutants fails to restore later markers of endodermal differentiation (Alexander and Stainier 1999; Reiter et al. 2001a), showing that *cas* is essential for *bon* and *fau* function. Third, *cas* misexpression can expand later markers of endodermal differentiation in *bon* and *fau* mutants (Kikuchi et al. 2001), showing that *cas* is able to function in the absence of *bon* or *fau* function.

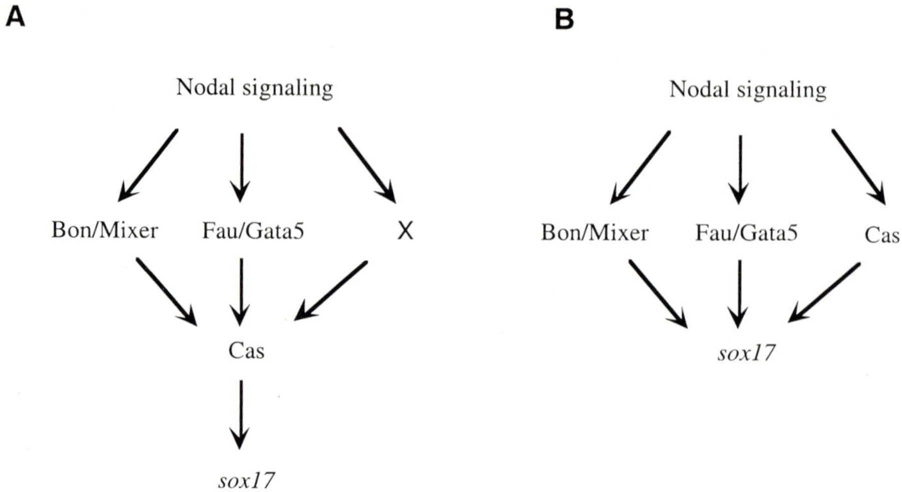

Fig. 3A,B. Genetic pathway of endoderm formation. Current data indicate that Cas acts either **A** downstream of or **B** in parallel to Bon and Fau to regulate *sox17* expression (see text, Sect. 1.9)

At least two different models could explain these results (Fig. 3): *bon*, *fau* and *cas* are regulated independently by Nodal signaling and function in parallel, or *bon*, *fau* and a third factor X act upstream of *cas*. (The third factor, X, is evoked to explain the fact that *cas* is expressed in only a subset of *bon*- and *fau*-expressing cells.) To distinguish between these two models, *cas* expression was examined in *bon* and *fau* mutants and it appeared to be altered. However, because it is difficult to visualize the very early expression of *cas*, it cannot be ruled out that its initial expression in these mutants is normal and only as cells fail to maintain endodermal identity does *cas* expression disappear. Thus, there may be a very narrow window of time when *cas* expression is normal in *bon* and *fau* mutants (Kikuchi et al. 2001). Another line of experiments was to determine whether *bon* and *fau* misexpression could induce *cas* expression in wild-type embryos. Again the results were inconclusive in the sense that *bon* and *fau* misexpression could induce *cas* expression but only in the marginal zone. Thus, it could be that Cas functions downstream of Bon and Fau and that the third factor X is restricted to the margin. Alternatively, the increase in *cas* expression observed in these experiments may merely reflect an increase in endoderm formation induced by *bon* and *fau* misexpression and not support a model where *cas* functions downstream of *bon* and *fau*. In summary, it is not yet clear whether Cas functions in parallel to, or downstream of, Bon and Fau, and molecular dissection of the *cas* promoter may help to further investigate this question.

Functional dissection of the *cas* promoter might also help elucidate perhaps the most intriguing question: what regulates early *cas* expression to emerge in only a scattered subset of marginal cells? One possibility is that, in concert with

Nodal regulation, endodermal progenitors are selected by a process of lateral specification similar to that proposed for the spacing of neuroblasts in the proneural clusters of *Drosophila* (reviewed by Greenwald and Rubin 1992). This Notch-dependent process has previously been demonstrated to regulate zebrafish primary neuron development (Haddon et al. 1998) as well as the decision between hypochord and notochord cell fate (Appel et al. 1999).

2 Endoderm Formation During the Gastrula Period

2.1 Formation of the Endodermal Layer

Germ layer and body axis formation occur during gastrulation, a period of stereotypical cell movements (see Kane and Adams; Solnica-Krezel and Cooper, this Vol.). In zebrafish, the mesendodermal layer (otherwise known as the hypoblast) forms as cells involute around the edge of the blastoderm (Warga and Kimmel 1990). Endodermal cells are among some of the earliest to involute (Warga and Kimmel 1990; Warga and Nüsslein-Volhard 1999). Because of its position and the relatively small number of progenitor cells, the majority of endoderm has already involuted by shield formation and few cells remain near the margin, but rather lie scattered within the hypoblast beneath the presumptive ectoderm (Warga and Nüsslein-Volhard 1999). During this early period of gastrulation, endodermal and mesodermal layers are not discernable morphologically (Warga and Kimmel 1990, see Fig. 6; Warga and Nüsslein-Volhard 1999), neither is there a reliable way to identify endoderm precursors except with putative marker genes.

During mid-gastrulation the endoderm becomes a morphological entity. Within the hypoblast, cells lying next to the yolk cell gradually alter their appearance, transforming from small spherical to large flat somewhat bipolar cells with numerous fine filopodia (Fig. 4A, B). Eventually, by the end of gastrulation, they form a sparse but uniform monolayer (Warga and Nüsslein-Volhard 1999). Cell-labeling studies demonstrate that only these characteristic cells give rise to definitive endodermal derivatives (Warga and Nüsslein-Volhard 1999) and, although the commitment of these cells has not been tested, these changes mark the first time the definitive endoderm can be identified morphologically. In contrast, the remainder of the cells in the hypoblast show less marked morphological changes during gastrulation and appear more closely spaced, have limited contact with the yolk cell and eventually give rise to only mesodermal derivatives (Warga and Nüsslein-Volhard 1999).

2.2 Spatial Allocation of Endodermal Precursors

The spatial location of endodermal precursors reflects both the positions from which these cells originate and the stereotypical cell movements occurring

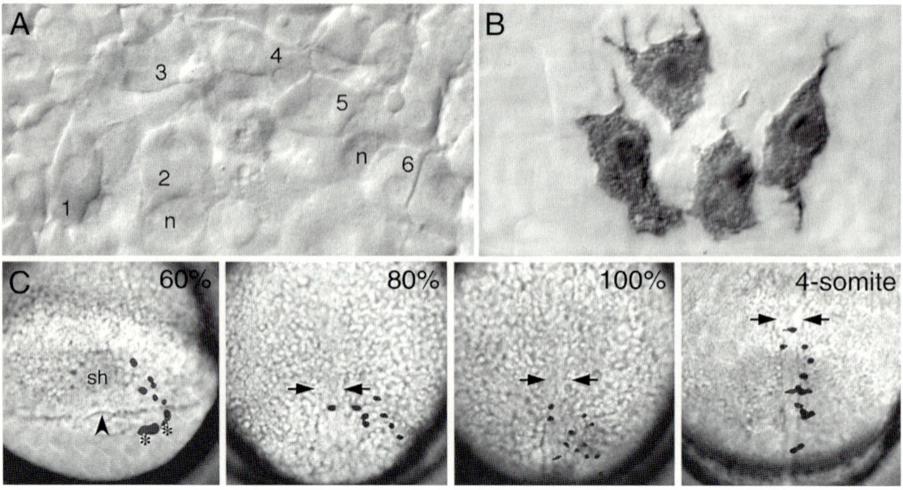

Fig. 4A–C. Cellular morphology and movement of gastrula endoderm (precursors). **A, B** Endodermal precursors at 100% epiboly. **A** Unlabeled cells in the live embryo located 45° away from the dorsal midline near the margin. In focus are six endodermal precursor cells (indicated with *numerals*) spaced roughly one-cell diameter apart from one another. These cells lie directly above the yolk cell, whose nuclei (*n*) are also in focus. Note that the long filopodia on cells 1, 4 and 5 are also in focus. **B** Labeled cells that were injected with fixable lineage tracer at the 2K-cell stage and later fixed and stained. These cells are located 30° away from the dorsal midline and in the posterior half of the embryo. The four cells of this clone have the distinctive flat appearance, somewhat bipolar shape and extensive filopodia of endodermal precursors. **C** Illustration of the convergence of endodermal precursors towards the dorsal midline from selected composite frames of a time-lapse video. At shield stage the clone consists of six deep cells and four enveloping layer cells (*asterisks*), just to the right of the embryonic shield (*sh*). The labeled deep cells have already involuted deep into the hypoblast layer. The *arrowhead* indicates the location of the blastoderm margin. By 80% epiboly, the cells in the clone have begun to extend apart from one another along the anterior-posterior axis of the embryo (enveloping layer cells are no longer in focus). In addition, the labeled cells have begun to converge towards the embryonic axis whose boundaries are indicated with *arrows* for this and subsequent panels. By 100% epiboly, more divisions have occurred and the labeled cells have spread and converged more. Because the view is tilted somewhat vegetal, the clone appears to be spreading down and not apart. By the 4-somite stage, the labeled cells now extend extensively along the anterior-posterior axis and have either reached the midline or lie adjacent to it. The more anterior cells were located in the prechordal plate (the dorsoanterior region of the hypoblast) and gave rise to pharynx, a commonly observed prechordal plate derivative, the more posterior cells gave rise to liver, swim bladder and stomach

during gastrulation. Endodermal cells originating on the dorsal side undergo very little convergence to the midline, but rather spread rapidly as a loose aggregate towards the animal pole in the prechordal plate region of the hypoblast (Melby et al. 1996, see Fig. 4B; Warga and Nüsslein-Volhard 1999). Endoderm derived more laterally upon involution tends to extend as a wing alongside the axis from the advancing hypoblast edge to the margin (Fig. 4C; Warga and Nüsslein-Volhard 1999), as very little convergence appears to occur during the first hour of gastrulation (Kane and Warga 1994; Schmitz and

Campos-Ortega 1994). When these lateral endodermal precursors begin to converge, they enter the midline in relative order of their distance from dorsal so that precursors located very close to dorsal may eventually contribute to endoderm in the prechordal plate as well as found endoderm located along the trunk. Precursors located further from dorsal tend to distribute more posteriorly along the trunk (Warga and Nüsslein-Volhard 1999). Ventrally derived endoderm initially appears to undergo very little anterior-posterior spreading; rather after it involutes it tends to spread laterally along the margin before converging (R.M. Warga, unpubl.). Many of these ventral precursors do not reach the midline until well after gastrulation and give rise to posterior endoderm. Thus the endodermal anlage forms as successive waves of precursors meet at the midline. This process is distinct from the formation of all other midline structures such as the floorplate, notochord or hypochord, as endoderm derives from cells located around the entire circumference of the blastoderm.

2.3 Genes Involved in Endoderm Formation: *cas, bon* and *fau* Revisited

As the endodermal progenitors begin to involute, a number of endodermal genes become active in addition to those discussed earlier: *cas, bon* and *fau*. In the gastrula, *cas* and *fau* expression are both maintained, although they differ in their extent of expression. While *cas* is expressed only in the endodermal precursors (Kikuchi et al. 2001), *fau* is also expressed in a subset of mesodermal cells fated to the anterior lateral plate mesoderm (Reiter et al. 1999; Rodaway et al. 1999). By contrast, *bon* expression is downregulated shortly after the onset of gastrulation, and becomes undetectable by 60% epiboly (Alexander et al. 1999).

As soon as endodermal progenitors involute, they begin to express *sox17* (Alexander and Stainier 1999). Like *cas*, *sox17* encodes a high mobility group (HMG) domain transcription factor and *Xenopus* Sox17 has been proposed to be an intrinsic regulator of endoderm formation (Hudson et al. 1997). In zebrafish, less is known about its biological function, but its expression closely parallels that of *cas* except that it is never expressed in the yolk syncytial layer. It is interesting to note that, based on the early gastrula expression of *cas* and *sox17*, endodermal precursors appear to originate from around the circumference of the margin, but with no greater dorsal tendency as would be expected from fate map data presented earlier. This discrepancy between the fate map and gene expression could be due to the cell movement of dorsal compaction which draws cells dorsally as dorsal cells compact together during the late-blastula stage (Warga and Nüsslein-Volhard 1998). As position relative to dorsal cannot be determined until shield stage, dorsal compaction could skew fates to appear more dorsal. Alternatively, *cas* and *sox17* could be expressed in only part of the endodermal cells on the dorsal side or in non-endodermal cells on the ventral side.

2.3.1 Cas May Regulate *sox17* Directly

Several lines of evidence suggest that the regulation of *sox17* by Cas may be direct. First, *cas* misexpression can induce *sox17* expression even in *MZoep* mutants, indicating that *cas* can induce *sox17* expression in a Nodal-independent manner (Dickmeis et al. 2001b; Kikuchi et al. 2001). Second, the regulation of endodermal *sox17* expression by Cas is cell-autonomous, as shown by cell transplantation experiments (Alexander et al. 1999), as well as by the finding that *cas* misexpression throughout the embryo, but not in the yolk cell, upregulates *sox17* expression (Sakaguchi et al. 2001). Finally, HMG domain binding sites are present upstream of *sox17* (Y. Kikuchi and D.Y.R. Stainier, unpubl.). Whether *bon* and/or *fau* also regulate *sox17* directly remains to be tested, although misexpression of *bon* and/or *fau* cannot restore significant levels of *sox17* expression in *MZoep* mutants (Y. Kikuchi and D.Y.R. Stainier, unpubl.). Perhaps *bon* and *fau* act in concert with *cas* to help regulate the temporal expression of *sox17*, which is remarkably coupled to endoderm entering the hypoblast and being exposed to a new environment.

2.3.2 The Fox/Forkhead Transcription Factors

Fox factors, formerly known as the Forkhead or HNF3 factors, belong to another family of proteins expressed by the endodermal precursors during gastrulation. They constitute a family of winged helix transcription factors which were first identified in *Drosophila* as being necessary for endoderm formation and have since been shown to be necessary in vertebrate endoderm formation as well (Weigel et al. 1989; Hoch and Pankratz 1996; Kaufmann and Knöchel 1996). Some of these genes may be under direct regulation by the Nodal pathway. For instance, *Xenopus foxA2* (*HNF3β*) is a direct target of Smad2 (Howell and Hill 1997).

In zebrafish, a family of nine *fox* genes has been described (Strähle et al. 1993; Dirksen and Jamrich 1995; Odenthal and Nüsslein-Volhard 1998), and three of them appear to be expressed in the endodermal cells in a sequential, and partially overlapping, pattern. The first expressed is *foxA3*, formerly known as *forkhead domain2* (*fkd2*), and the homologue of mouse *HNF3γ*. Its expression begins at dome stage in a subset of dorsal cells, which comprise both the deep cells and the yolk syncytial layer (Odenthal and Nüsslein-Volhard 1998; Warga and Nüsslein-Volhard 1999). Like *cas*, yolk cell expression of *foxA3/fkd2* progresses circumferentially to encompass the entire marginal zone by the onset of gastrulation. Unlike *cas*, deep cell expression of *foxA3/fkd2* remains restricted to the dorsal region in both the endodermal and mesodermal territories. Upon gastrulation, *foxA3/fkd2* appears in cells of all three germ layers of the axis up to and including the prechordal plate region. Outside the axis, *foxA3/fkd2* now appears in more lateral cells that have been identified as endodermal precursors as demonstrated by in vivo cell labeling and protein co-localization studies (Warga and Nüsslein-Volhard 1999).

The second *fox* gene to be expressed is *foxA2*, formerly known as *axial* and the homologue of mouse *HNF3β*. Its expression begins at shield stage and closely mirrors that of *foxA3/fkd2*; except that it is not seen in the anterior prechordal plate or in the yolk syncytial layer (Strähle et al. 1993; Schier et al. 1997; Odenthal and Nüsslein-Volhard 1998). Like *cas* and *sox17* it is expressed all around the margin. The third *fox* gene *foxA1*, formerly *forkhead domain7* (*fkd7*) and the homologue of mouse *HNF3α*, becomes expressed in endodermal precursors after the end of gastrulation (Odenthal and Nüsslein-Volhard 1998). While the function of these three *fox* genes remains to be elucidated in zebrafish, chimeric studies in mice demonstrate that FoxA2/Axial/HNF3β is required cell-autonomously for formation of the foregut and midgut endoderm (Dufort et al. 1998).

2.3.3 Other Genes Expressed in the Endoderm

Pitx2 is a member of the bicoid-homeodomain class of transcription factors and experiments in *Xenopus* indicate that it mediates Nodal signaling and plays an important role in endodermal differentiation (Faucourt et al. 2001). In zebrafish, less is known about its function, but *pitx2* expression appears at dome stage in marginal cells of both the endoderm and mesoderm, and during gastrulation it becomes localized to all three germ layers within the prechordal plate region (Essner et al. 2000; Faucourt et al. 2001).

Her5 is a hairy-related/Enhancer of split gene, which belongs to a family of basic helix-loop-helix transcription factors acting as transcriptional repressors. Hairy-related proteins have been implicated in mediating cell fate decisions (Fisher and Caudy 1998) and, in zebrafish, misexpression experiments suggest that *her5* may help regulate cell fate at the anterior- and posterior-most regions of the dorsal hypoblast (Bally-Cuif et al. 2000). *her5* transcripts are ubiquitous in the early zebrafish blastula, but by 30% epiboly become localized to a scattered population of dorsal deep cells. Later, in the gastrula, *her5* transcripts are localized in what appear to be endodermal precursors of the prechordal plate region (Bally-Cuif et al. 2000).

2.4 Regional Expression in the Endoderm at the End of Gastrulation

Although less is known about the function of the endodermal genes expressed in the gastrula, these constitute valuable markers of anterior-posterior patterning of the endoderm, particularly at the end of gastrulation, as we will now discuss and is illustrated in Fig. 5. While *foxA3/fkd2* and *fau* are both expressed in more than one germ layer, their endodermal-specific expression patterns are similar: both appear to be expressed in all dorsal endoderm including the cells anteriorly in what is often referred to as the prechordal plate region. In addition, both are expressed in the more lateral endoderm. They differ, however, in the spatial extent of their ventral endoderm expression: *foxA3/fkd2*

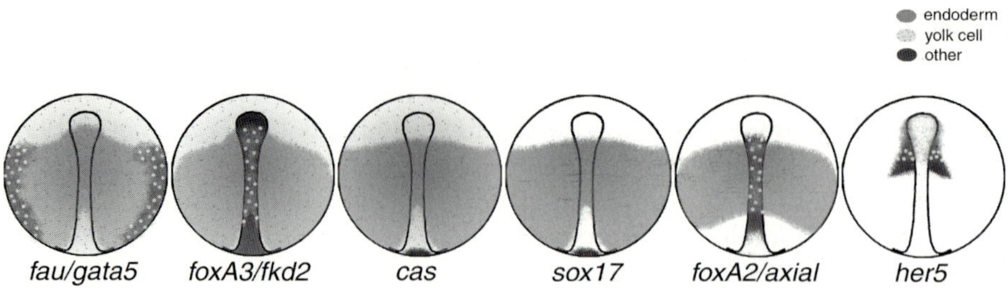

Fig. 5. Endodermal-specific gene expression patterns at the end of gastrulation (100% epiboly). The embryo is presented dorsal up and an outline of the axis is portrayed. *Gray tones* indicate which cell type the gene is expressed in and *white stippled regions* indicate where expression in endodermal cells overlaps with expression in other germ layers

transcripts do not appear in ventral endoderm cells, whereas *fau* transcripts do.

cas and *sox17*, on the other hand, seem only expressed by endodermal cells. Like *fau*, *cas* and *sox17* appear to be expressed by all lateral and ventral endoderm. *cas* and *sox17* are also both expressed in dorsal endoderm though neither appears to be expressed in the more anterior region (Alexander and Stainier 1999; Dickmeis et al. 2001b; Kikuchi et al. 2001). Thus, *foxA3/fkd2* appears to be expressed in all dorsal (including dorsoanterior) and lateral populations, *cas* and *sox17* appear to be expressed in all populations except the dorsoanterior one, and *fau* may be expressed in all populations. Superficially, this may seem unrelated to anterior-posterior position, but, as discussed earlier (see Sect. 2.2), where a cell is located along the margin dorsoventrally ultimately determines its location along the anteroposterior axis.

The expression of *foxA2/axial* resembles that of *foxA3/fkd2*. However, *foxA2/axial* is not expressed in anterior dorsal endoderm and its spatial expression in lateral endoderm appears to extend further ventral than *foxA3/fkd2* (Odenthal and Nüsslein-Volhard 1998). The most marked difference, however, is that unlike all the above genes, *foxA2/axial* is never expressed in the posterior-most endodermal cells (Alexander and Stainier 1999). Thus *foxA2/axial* transcripts appear to be absent from the more anterior and posterior domains of the endoderm at late-gastrula stages. Perhaps the most restricted endodermal-specific expression pattern is that of *her5*. While it is also expressed in neuroectoderm, in the endoderm its transcripts appear to be localized to the dorsoanterior endoderm (Bally-Cuif et al. 2000).

Genes that are expressed simultaneously in both endodermal and mesodermal germ layers also contribute to anterior-posterior patterning of the endoderm. Examples include the T-box gene *spadetail*, which in the gastrula is expressed laterally and ventrally (Griffin et al. 1998) and when mutated leads to similar defects in ventroposterior-derived endoderm and mesoderm (Kimmel et al. 1989; Warga and Nüsslein-Volhard 1998). Another example is the Bmp antagonist Dino, which in the gastrula is expressed dorsally (Schulte-Merker et al. 1997), and when mutated leads to apparently similar decreases in dorsoanterior-derived endoderm and mesoderm as well as similar increases in ventroposterior-derived endoderm and mesoderm (Hammerschmidt et al. 1996b; J. Alexander and D.Y.R. Stainier, unpubl.).

2.5 Further Thoughts on Endodermal Patterning . . .

Once endodermal cell fate has been specified, further differentiation leads the cell to become irreversibly committed and acquire regional identity. Cell transplantation experiments demonstrate that, in zebrafish, cells are not committed to either epiblast- or hypoblast-derived cell fate until mid-gastrulation (~80% epiboly; Ho and Kimmel 1993). It is not known when germ layer fate becomes committed in zebrafish, although in *Xenopus* similar experiments indicate that endodermal cell fate may be committed before gastrulation (Wylie et al. 1987), but regional specification to fate along the anterior-posterior axis does not occur until after the end of gastrulation (Horb and Slack 2001). In this regard, if marginal zebrafish cells from the early gastrula (~60% epiboly) are placed at the animal pole, they never form endoderm, but instead transfate into ectoderm (Ho and Kimmel 1993; Warga and Kane, unpubl.). Thus the zebrafish data appear to be in direct conflict with the *Xenopus* data because they suggest that cells become committed to an endodermal fate quite late. In another example, lateral endodermal precursors in *Zoep* mutants express normal levels of FoxA3/Fkd2 protein throughout gastrulation, indicating that they have undergone some level of specification, but by early somitogenesis only a small percentage of these cells maintain normal FoxA3/Fkd2 expression (R.M. Warga and D.A. Kane, in prep.). As no increase in the rate of cell death has been detected during gastrulation in *Zoep* mutants (Hammerschmidt et al. 1996a), loss of FoxA3/Fkd2 protein may imply that these cells are transfating to another fate and support the idea that endodermal commitment has not yet occurred.

One source of factors promoting endodermal commitment could be the yolk cell itself. The zebrafish yolk cell is perhaps analogous to the extraembryonic endoderm of the mouse, as both provide a nutritive, supportive and inductive role. In mouse, the extraembryonic endoderm has been shown to be necessary for endoderm development, as exemplified by chimeric studies with genes like *smad2* which is expressed in both extraembryonic and definitive endoderm (Tremblay et al. 2000). In zebrafish, the yolk cell may also be necessary for endodermal development as RNase injections into the yolk cell of the early blastulae abolish all but the dorsal expression of *fau* (Chen and Kimelman 2000). For these reasons it is all the more intriguing that many of the endodermal genes in zebrafish are also expressed in the yolk syncytial layer. It remains to be seen how this patterning could occur as yolk cell specific knockdown experiments indicate that *cas* in the yolk cell is not necessary for endoderm formation (Sakaguchi et al. 2001). However, it is also notable that, while most cells of the hypoblast have transient connections to the yolk cell during involution, only the endoderm appears to remain in continuous contact with the yolk cell throughout gastrulation (Warga and Nüsslein-Volhard 1999). Contact with the yolk cell might help promote and maintain endodermal identity once a cell has been specified.

3 Pharyngeal Pouch Endoderm Versus Digestive Tract Endoderm

It will be important to better characterize the origin and migration of the different populations of endodermal cells, as well as the genes these cells express from the late-blastula stage until their terminal differentiation. At this point, lineage data as well as careful examination of *foxA3/fkd2* expression indicate that endoderm which originates from the dorsoanterior region of the hypoblast largely gives rise to pharyngeal endoderm (Warga and Nüsslein-Volhard 1999). This is the endoderm that lines the oral cavity and which will later subdivide into two different populations: one contiguous with the lumen of the gut, and thus part of the digestive tract, and the other giving rise to specialized vertebrate features in the head, namely the pharyngeal pouches. The pharyngeal pouch endoderm is believed to grow out from the wall of the pharynx and these cells that become isolated from the remainder of the pharynx give rise to the segmented gill clefts, thymus, thyroid, parathyroid and parts of the inner ear (Patten 1951; Nelsen 1953; Harder 1975). Current data suggest that the pharyngeal pouch endoderm and the digestive tract endoderm (which includes cells derived from the pharynx) exhibit distinct genetic requirements. For example, in the *van gogh* (*vgo*) mutant, differentiation of the pharyngeal pouches appears to be blocked, whereas development of the digestive tract appears normal, as assessed both histologically and by the pattern of gene expression (Piotrowski and Nüsslein-Volhard 2000). Specifically, the expression domains of *nkx2.3* and *nkx2.7* are strongly reduced or absent in the pharyngeal pouch precursors of the *vgo* mutant, but expression of *foxA1/fkd7* and *sonic hedgehog* in nearby more medial, digestive tract precursors appears mostly normal (Piotrowski and Nüsslein-Volhard 2000). By contrast, in the *cas* mutant, all digestive tract endoderm is lacking, as assessed histologically (Alexander et al. 1999), but pharyngeal pouch precursors appear to be present and their development initiates normally, as assessed by *nkx2.3* expression (J. Alexander and D.Y.R. Stainier, unpubl.). Further development of the pharyngeal pouches in *cas* mutants, however, is defective, presumably because these cells fail to migrate medially (J. Alexander and D.Y.R. Stainier, unpubl.). Thus understanding more fully the exact relationship between these two endodermal populations, the genes that regulate their differentiation and when and where their lineages separate clearly constitutes an important part of understanding endodermal regionalization.

4 Conclusions and Prospects

We are beginning to understand some of the earliest events in endoderm development, including the cellular and molecular mechanisms that lead to its formation and separation from the mesoderm. One goal for the future will be to turn our attention to a more complete understanding of later events. In particular, we need to focus on the following topics: (1) regional patterning of the endoderm; (2) morphogenesis of the gut including its transformation from a sheet into a tube; and (3) organogenesis of the gut and its associated organs which has only recently begun (Biemar et al. 2001). It is to be hoped that the zebrafish will continue to provide insights into our understanding of endoderm development that are widely applicable to other model systems.

Acknowledgements. We thank members of the Kane laboratory: Don Kane, Sumitra Himangshu, and Karen DelKanic, as well as members of the Stainier laboratory: Jeremy Reiter, Sally Horne, Le Trinh, and Yutaka Kikuchi, for thoughtful reading and comments on early versions of this manuscript. We also thank Bernard and Christine Thisse for generously sharing the *cas* misexpression images used in Fig. 2.

Organizer Formation and Function

Masahiko Hibi[1,3], Toshio Hirano[1], and Igor B. Dawid[2]

1 Introduction

During vertebrate development, the embryonic axis is established through a series of inductive signals that are followed by the concerted movement and differentiation of groups of cells. In this process, the dorsal gastrula or Spemann-Mangold organizer plays a major role.

The work of Spemann and his colleagues (reviewed by Spemann 1938; Hamburger 1988), following the initial experiments carried out by Hilde Mangold (Spemann and Mangold 1924), introduced the organizer concept and provided the first description of the nature of this embryonic domain. In the early experiments, equatorial cells derived from the dorsal lip of an early newt gastrula were grafted to the ventral side of an isochronic host, eliciting the development of a secondary neural tube with a well-formed anterior-posterior axis, as well as formation of a secondary notochord, somites, and, in some cases, gut. In these embryos, the graft differentiated into notochord, prechordal plate, variably floor plate, and somites, but most parts of the secondary axis including the neural plate and somites were derived from host cells that normally would not participate in the formation of these structures. The dorsal equatorial cells capable of these activities were named the Spemann or Spemann-Mangold organizer. The initial experiments showed that the organizer is able to generate a complete axis with well-defined anterior-posterior polarity, in part by self-organization, but primarily by emitting inductive signals that control the differentiation and morphogenesis of axial structures. While the original work was carried out in amphibians, an organizer region has been identified in all vertebrate embryos including zebrafish, which is the primary focus of this chapter.

Several lines of evidence indicate that the Spemann-Mangold organizer is established by the action of an earlier dorsalizing center called the Nieuwkoop

[1] Department of Molecular Oncology, Osaka University Graduate School of Medicine, Osaka, Japan
[2] Laboratory of Molecular Genetics, National Institute of Child Health and Human Development, National Institutes of Health, Bethesda, Maryland 20892, USA
[3] Laboratory for Vertebrate Axis Formation, RIKEN Center for Developmental Biology, Kobe, Japan

Results and Problems in Cell Differentiation, Vol. 40
L. Solnica-Krezel (Ed.): Pattern Formation in Zebrafish

center. In the work pointing to the existence of such an early signaling center, Peter Nieuwkoop first studied the phenomenon of mesoderm or mesendoderm induction, using explants of amphibian embryos. When explanted into culture, the animal region of a *Xenopus* embryo, called an animal cap, normally develops into atypical epidermis while the vegetal region will form partially differentiated endoderm. However, when recombined with vegetal explants, animal caps differentiate into mesoderm and pharyngeal endoderm, attesting to the inductive power of the vegetal tissue (Nieuwkoop 1969a,b; Nieuwkoop and Ubbels 1972). Nieuwkoop and colleagues further showed that explants of dorsal vegetal cells induced dorsal structures such as notochord, muscle and head endoderm, whereas ventral vegetal cells induced blood and mesenchyme (Boterenbrood and Nieuwkoop 1973). This result was interpreted to imply a signal that was graded with a dorsal-to-ventral polarity, with high signal strength inducing dorsal and low signal strength inducing ventral tissues. Smith and Slack (1983) modified this idea to generate the three-signal model of mesodermal patterning in which ventral vegetal cells induce ventral mesoderm while dorsal vegetal cells induce the organizer. The organizer generates the third, dorsalizing signal which is responsible for patterning the mesoderm along a dorsoventral gradient. Recent studies have questioned various aspects of this model, but it retains heuristic value for understanding organizer formation and function.

Recent studies in zebrafish have shown that the mechanisms that control axis formation, including formation of the Nieuwkoop center and the organizer, are largely conserved between amphibians and teleosts (Fig. 1). In contrast to amphibians, blastomere cleavages in zebrafish do not penetrate the vegetal yolk sphere and the blastomeres perch on top of the large yolk mass during the cleavage and blastula stages (Kimmel et al. 1995). Thus the vegetal yolk mass of zebrafish does not contain nuclei as it does in amphibian embryos. Instead, around the mid-blastula stage, marginal cells in a deep layer collapse and release their nuclei into the immediately adjoining cytoplasm of the yolk cell, thus forming the yolk syncytial layer (YSL). As discussed below and in other chapters (see Sakaguchi et al.; Kimel and Schier, this Vol.), the YSL plays an important role in mesendoderm induction and axis formation, similar to the role of vegetal and equatorial blastomeres in amphibian embryos. In the *Xenopus* embryo the positions corresponding to the future dorsal (D) and ventral (V) sides can be distinguished at the two-cell stage by pigmentation differences. In contrast, DV orientation is not morphologically apparent in zebrafish until convergence of cells to the dorsal side leads to a thickening of the blastoderm margin at the beginning of gastrulation, forming the embryonic shield, the fish equivalent of the Spemann organizer. As discussed below, it is nevertheless clear that specification of DV polarity is initiated soon after fertilization in zebrafish as it is in *Xenopus*.

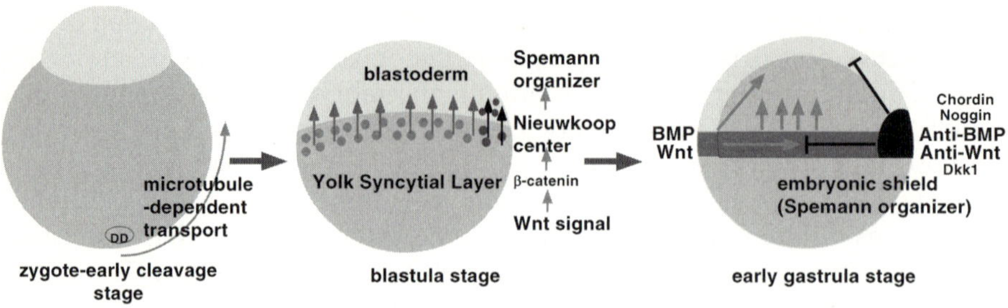

Fig. 1. Model for dorsoventral axis formation in zebrafish. See text for details

2 Dorsal Determinants and the Maternal Wnt Signal

2.1 Dorsal Determinants

What mechanisms determine dorsoventral polarity at an early stage; in particular, what triggers the formation of the Nieuwkoop center? Extensive evidence indicates that dorsal determinants, located at the vegetal pole of amphibian and teleost eggs and transferred to the dorsal equatorial blastomeres shortly after fertilization, lead to the formation of the Nieuwkoop center. In *Xenopus*, the sperm enters anywhere in the animal hemisphere of the egg. During the first cell cycle, the egg cortex rotates about 30° relative to the underlying cytoplasm along a meridian determined by the sperm entry point in a process called cortical rotation. During this process, parallel microtubule tracks are formed below the cortex, with their plus ends oriented toward the future dorsal side (Elinson and Rowning 1988; Houliston and Elinson 1991; Rowning et al. 1997). Although rotational displacement amounts to only 30°, small particles, such as membrane-bound organelles and pigment granules, are transported for about 60° along the microtubule tracks. Disruption of the microtubule array by ultraviolet (UV) irradiation or treatment with low temperature or high pressure leads to ventralization of the embryo (Gerhart et al. 1989), indicating that microtubule-mediated transport of determinants is required for the formation of the Nieuwkoop center and Spemann organizer. Furthermore, removal of vegetal pole cortex before cortical rotation results in embryos that display dorsal deficiencies (Fujisue et al. 1993; Sakai 1996), and the cytoplasm near the vegetal cortex contains transplantable dorsalizing material (Yuge et al. 1990). Transplantation of cortex derived from the vegetal pole of unfertilized egg can generate a secondary axis, and the axis-inducing activity of the cortex shifts from the vegetal to the dorsal region during cortical rotation (Kageura 1997). Collectively, these results indicate that dorsal determinants are transported towards the margin along a microtubule array to initiate the dorsal program in amphibian embryos.

Similar evidence has been obtained for teleost embryos. When the vegetal hemisphere of early cleavage stage goldfish (*Canassius auratus*) is removed, the embryos are deficient in dorsal structures, resembling UV-ventralized *Xenopus* embryos (Tung et al. 1945; Mizuno et al. 1997). In zebrafish, depletion of the vegetal yolk mass at the 1-cell stage resulted in embryos missing dorsal axial structures including neuroectoderm and the anterior-most 14–15 somites (Mizuno et al. 1999b; Ober and Schulte-Merker 1999). Expression of the dorsal organizer-specific genes *goosecoid* and *chordino* was absent, and the ventral gene *eve1* expression domain was expanded to the dorsal side in these embryos; however, expression of *no tail*, marking nascent mesoderm, was unaffected.

The initial signal for DV polarity, which in *Xenopus* is derived from the sperm entry point, is less clear in zebrafish as the sperm enters at the animal pole and no cortical rotation ensues. Nevertheless, a microtubule array arises in the vegetal region approximately 20 min after fertilization, and then disappears; arrayed fibrils were detected again in the equatorial region at the 4-cell stage (Jesuthasan and Stähle 1997). Irradiation of the zygote up to 20 min after fertilization disrupts the formation of the vegetal microtubule array and causes dorsal axis defects. Similarly, exposure to cold (18 °C), or incubation in the microtubule-depolymerizing drug nocodazole prior to the 32-cell stage, prevents axis specification and the expression of *goosecoid*. These treatments also prevent the nuclear accumulation of β-catenin in the dorsal YSL and dorsal blastomeres that is required for Nieuwkoop center formation, as will be described below. Microtubule-dependent transport has been demonstrated by injection of polystyrene beads into the vegetal pole of one-cell to four-cell stage embryos. These data suggest that dorsal determinants are transported to dorsal YSL and blastomeres from the yolk along a microtubule array prior to the 32-cell stage in zebrafish embryogenesis (Jesuthasan and Stähle 1997).

2.2 A Maternally Derived Signal Activating the Wnt Pathway

Although the molecular nature of the dorsal determinants remains unknown, extensive evidence has been adduced for the involvement of the Wnt signal cascade in dorsal axis determination. Many of the multiple effects of Wnt factors in development are executed through a signal transduction cascade now termed the "canonical Wnt pathway" (reviewed in Nusse and Varmus 1992; Dierick and Bejsovec 1999; Sokol 1999; Peifer and Polakis 2000; Seidensticker and Behrens 2000; Fig. 2). Wnt proteins bind to the Frizzled family of seven-pass transmembrane proteins to activate downstream targets, with Dishevelled (Dsh/Dvl) the next known component. In the absence of Wnt ligands, the serine/threonine kinase GSK3 and β-catenin form a complex with the scaffolding protein Axin and the tumor suppressor gene product APC, leading to phosphorylation of the amino-terminal region of β-catenin. Phosphorylated β-catenin is recognized by the F-box-motif-containing protein β-TrCP, a

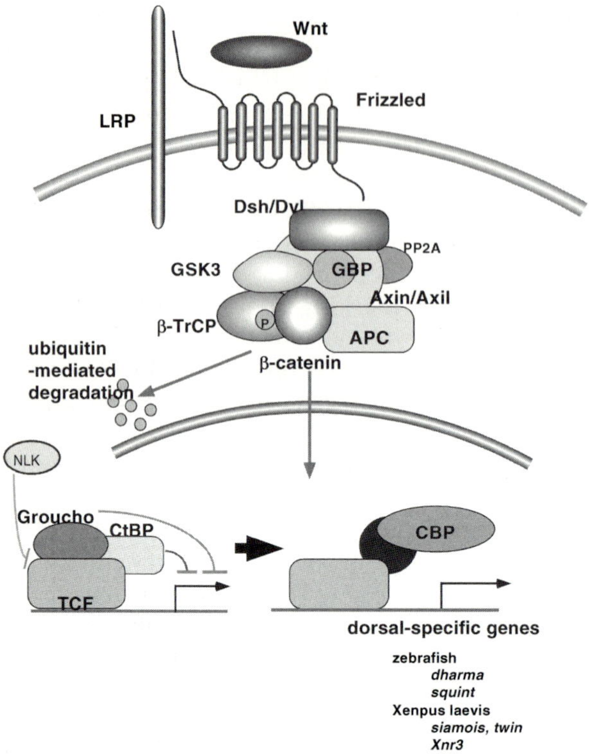

Fig. 2. Canonical Wnt signal transduction pathway. See text for details

component of ubiquitin ligase, and is subjected to ubiquitination-dependent degradation. Upon stimulation by Wnt, Dsh suppresses the phosphorylation of β-catenin by GSK3 by an unknown mechanism, leading to the release and stabilization of β-catenin, which then moves to the nucleus and forms a complex with Tcf or Lef factors. In the absence of β-catenin, the high mobility group (HMG) proteins in the Tcf/Lef family interact with transcriptional co-repressors Groucho and CtBP, and/or with histone deacetylase HDAC1, and repress transcription of target genes (Cavallo et al. 1998; Roose et al. 1998; Brannon et al. 1999; Billin et al. 2000), while the complex between Tcf/Lef and β-catenin activates transcription, possibly through interaction with the co-activator CBP (Takemaru and Moon 2000). In addition to these components, the LDL receptor-related proteins (LRP), the GSK3-binding protein GBP (or its relatives Frat1, 2), casein kinase I (CKI), and PP2A are involved in transmitting the signal positively (Yost et al. 1998; Peters et al. 1999; Sakanaka et al. 1999; Sumoy et al. 1999; Ratcliffe et al. 2000), while Nemo-like kinase (NLK) has a negative role in Wnt signal transduction (Ishitani et al. 1999). Recently, an alternate Wnt-signaling pathway has been discovered which does not result in the nuclear accumulation of β-catenin. Stimulation by certain Wnts, such as

Wnt-4, 5a, and 11, leads to upregulation of cytoplasmic calcium, activation of calmodulin kinase II and protein kinase C (Kuhl et al. 2000), and of the JNK pathway. This so-called PCP pathway is involved in determining planar cell polarity in *Drosophila* (Mlodzik 1999) and cell polarity and movements in *Xenopus* and zebrafish (Heisenberg et al. 2000; Wallingford et al. 2000), as further discussed below.

Activation of the canonical Wnt pathway by overexpression of Wnts, Dsh, dominant negative GSK3, or β-catenin in the ventral marginal zone of *Xenopus* embryos leads to secondary axis formation (Sokol et al. 1991, 1995; Funayama et al. 1995; He et al. 1995; Pierce and Kimelman 1995); overexpression of GBP and β-catenin also leads to second axis formation in zebrafish (Kelly et al. 1995a; Molenaar et al. 1996). Treatment with LiCl before the mid-blastula transition (MBT), which inhibits GSK3 activity (Klein and Melton 1996; Hedgepeth et al. 1997), leads to the ectopic induction of dorsal mesoderm and the organizer in *Xenopus* and zebrafish embryos (Backstrom 1954; Kao and Elinson 1988; Stachel et al. 1993). Inhibition of the canonical Wnt pathway by overexpression of GSK3 (He et al. 1995; Pierce and Kimelman 1995), Axin1 (Zeng et al. 1997; Itoh et al. 1998), Axin2 (Yamamoto et al. 1998b), dominant negative Tcf3 (Molenaar et al. 1996) or the dominant repressor form of β-catenin (Montross et al. 2000), or depletion of GBP or β-catenin (Heasman et al. 1994, 2000) leads to defects in dorsal axis formation and reduction in the dorsal-specific gene expression in *Xenopus* and zebrafish embryos (Pelegri and Maischein 1998; Shimizu et al. 2000).

These results clearly show that the canonical Wnt pathway plays an essential role in dorsal axis formation. However, it is not known at what level the Wnt signal is activated by the dorsal determinants whose molecular nature itself remains unclear. While mutant Dishevelled inhibits Wnt8-dependent secondary axis formation, it has no effect on primary dorsal axis formation in *Xenopus* (Sokol 1996). Thus, Dishevelled may not be involved in the formation of the endogenous dorsal axis; however, it was shown to move along the microtubule array from the vegetal region to the dorsal side during cortical rotation in *Xenopus*, suggesting that Dishevelled is part of the dorsal determinants (Miller et al. 1999a). Overexpression of a dominant negative Frizzled-A in zebrafish and depletion of Xfrizzled 7 protein in *Xenopus* cause defects in the dorsal mesoderm (Nasevicius et al. 1998; Sumanas et al. 2000); as Frizzled is upstream of Dsh in the canonical Wnt pathway, the point of activation of the cascade during dorsal axis determination remains unresolved. Cytoplasmic accumulation of β-catenin on the future dorsal side was observed by the two-cell stage in *Xenopus* and persists through early cleavage stages (Larabell et al. 1997), while nuclear accumulation of β-catenin occurs in both *Xenopus* and zebrafish blastulae (Schneider et al. 1996). In zebrafish, β-catenin-positive nuclei are found within the dorsal YSL at the high stage (3 h postfertilization, hpf), and in the dorsal blastomeres at the sphere stage (4 hpf). Nuclear accumulation of β-catenin is thought to lead to expression of dorsal-specific zygotic genes such as *bozozok/dharma* (*boz/dha*) and *squint* (*sqt*), which are involved in the formation of the Nieuwkoop center and the dorsal organizer (discussed below).

The recessive maternal effect mutant, *ichabod*, displays severe anterior and dorsal defects (Kelly et al. 2000). In *ichabod* mutant embryos, nuclear accumulation of β-catenin in the dorsal YSL is not observed, and *boz/dha* and *sqt* are not expressed. Injection of β-catenin RNA markedly dorsalizes *ichabod* mutants, but injection of RNAs encoding components of the canonical Wnt pathway upstream of β-catenin does not, indicating that the wild-type Ichabod protein acts at the level of nuclear accumulation of β-catenin in axis specification. This work not only provides genetic evidence for the requirement for β-catenin in dorsal axis formation, but also suggests an additional level of regulation of nuclear accumulation of β-catenin in early embryogenesis.

3 The Nieuwkoop Center and Organizer Induction

3.1 Non-cell-autonomous Induction of the Organizer

In addition to Nieuwkoop's studies described above, inducing activities in dorsal blastomeres were revealed by transplantation experiments. Transplantation of dorsal vegetal blastomeres, either into a ventralized embryo or into the ventral side of a normal embryo, leads to axis formation in *Xenopus* (Gimlich and Gerhart 1984; Gimlich 1986). In these experiments, the donor cells do not differentiate into mesoderm but form part of anterior endoderm, revealing their inductive capabilities. Based in part on these experiments, the Nieuwkoop center is generally considered as a region of the early embryo that can induce a complete dorsal axis without contributing directly to axial tissues. Some blurring of this definition derives from the fact that not only dorsal vegetal blastomeres, but also blastomeres in the entire dorsal region were shown to be capable of inducing the organizer (Gimlich and Gerhart 1984; Gimlich 1986; Kageura 1990). The organizer-inducing activities of the dorsal vegetal cells are lost in the late-blastula period (Boterenbrood and Nieuwkoop 1973; Jones and Woodland 1987), suggesting that establishment of the Nieuwkoop center and the organizer-inducing activities relies on maternally supplied factors. It is still an open question, however, whether the maternally derived factors (dorsal determinants) are directly responsible for the activity of the Nieuwkoop center, as zygotic genes appear to be involved in a relay mechanism. It has been demonstrated that blastomeres overexpressing β-catenin induce dorsal mesoderm in a non-cell-autonomous manner and their inductive activity is detected only after MBT, suggesting that β-catenin stimulates the expression of zygotic genes involved in the formation and/or function of the Nieuwkoop center (Wylie et al. 1996).

Recombination experiments in trout, placing largely blastomere-free yolk (containing the YSL) onto the animal pole of another embryo, showed that dorsal structures formed from the host blastoderm contacting the dorsal side of the yolk, revealing that the yolk cell carries positional information for dorsoventral polarity (Long 1983). Similarly, the zebrafish yolk cell can induce mesoderm and the organizer, marked by *no tail* and *goosecoid*, respectively, in host blastoderm cells (Mizuno et al. 1996). When the yolk was prepared from a ventralized embryo that had been deprived of the vegetal yolk mass at the one-cell stage, mesoderm but no organizer was induced, indicating that transport of the dorsal determinants is required for the establishment of the Nieuwkoop center in the dorsal YSL (Mizuno et al. 1999b). These results are consistent with the fact that nuclear accumulation of β-catenin in the dorsal YSL is required for the formation of the Nieuwkoop center (Fig. 1). As we discuss below, *boz/dha* and *nodal*-related genes are candidates as primary targets of β-catenin in Nieuwkoop center function.

3.2 bozozok/dharma

The gene *boz/dha* was first identified by its ability to induce dorsalization of the embryo when misexpressed (Yamanaka et al. 1998; also identified as *nieuwkoid*; Koos and Ho 1998); it encodes a transcription factor that contains a homeodomain in the carboxy-terminal region. *boz/dha* is expressed soon after the MBT at the high stage in dorsal blastomeres, and soon thereafter its transcripts are confined to the dorsal YSL until early gastrula (Koos and Ho 1998; Yamanaka et al. 1998; Fig. 3). Overexpression of *boz/dha* in the YSL is able

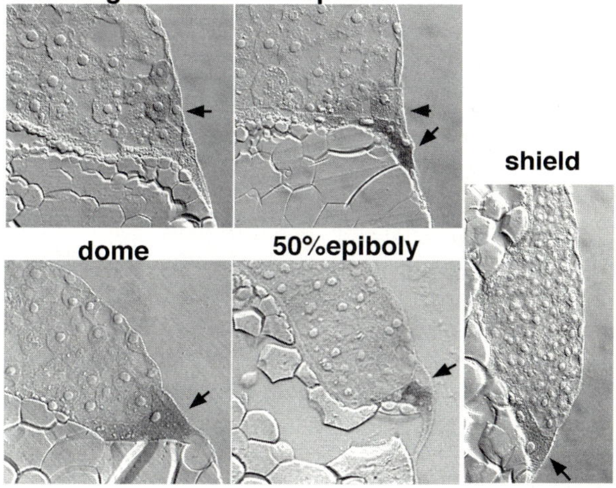

Fig. 3. Expression of *bozozok/dharma*. Expression of *boz/dha* is first detected just after mid-blastula transition (MBT, high stage) in the dorsal blastomeres. At sphere stage, the transcripts are detected in both dorsal blastoderm and dorsal YSL, are confined to the dorsal YSL by dome stage, and disappear after shield stage. After Yamanaka et al. (1998). Expression domains of *boz/dha* are indicated by *arrows*

to confer non-cell-autonomous organizer-inducing activity on the ventrolateral YSL (Yamanaka et al. 1998). This gene is affected in the mutants *boz*^{m168}, and in *momo*, an allele of *boz* (Koos and Ho 1999; Solnica-Krezel and Driever 2001).

Consistent with the overexpression studies, *boz*^{m168} embryos display defects in development of axial structures with variable penetrance and expressivity (Solnica-Krezel et al. 1996; Fekany et al. 1999; Fig. 6). *boz* mutants fail to form the embryonic shield, are defective in the expression of organizer-specific genes, and development of organizer-derived tissues, such as prechordal plate, notochord, hypochord, and floor plate, is severely perturbed (Fekany et al. 1999; Shimizu et al. 2000). Severe *boz*^{m168} mutants display reduction or absence of anterior neuroectoderm, in agreement with the role of the organizer in neural induction and patterning.

The *boz/dha* gene may be a direct target of the canonical Wnt pathway, as activation of this pathway causes ectopic expression of *boz/dha* (Yamanaka et al. 1998; Ryu et al. 2001), whereas inhibition of the pathway or depletion of the vegetal yolk mass abolishes expression (Ryu et al. 2001; Shimizu et al. 2000). Furthermore, a 1.4-kb upstream region of *boz/dha,* which contains nine putative Tcf/Lef-binding sequences, can drive the correct expression of a reporter in the embryo (Fig. 4); mutation of these sites impairs reporter expression

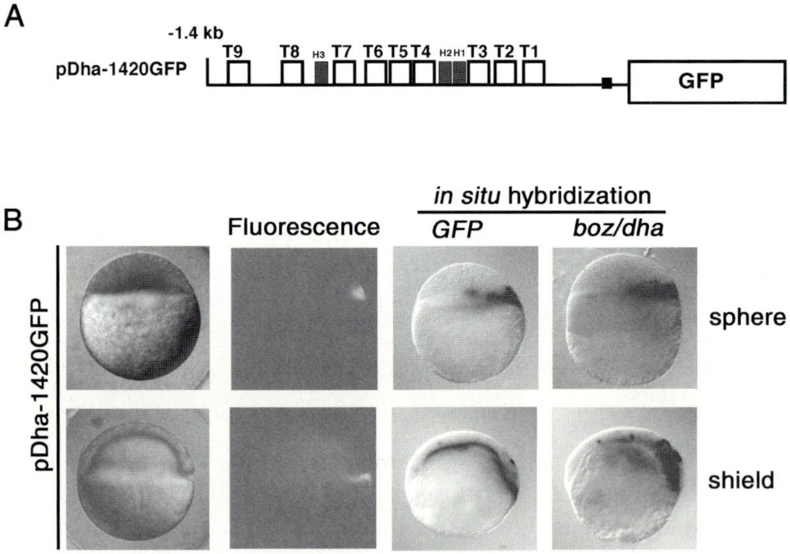

Fig. 4A,B. The *boz/dha* promoter. **A** Schematic presentation of the –1.4-kb promoter region of *boz/dha* attached to the green florescence protein gene (pDha-1420). **B** The –1.4-kb promoter can recapitulate *dha* expression. The reporter gene was injected into one-cell-stage embryos, and expression was detected by fluorescence microscopy and in situ hybridization. Embryos are shown at sphere stage (*upper row*) and shield stage (*lower row*). (After Ryu et al. 2001)

(Ryu et al. 2001). These results imply a direct interaction between the promoter and a β-catenin-Tcf/Lef complex.

3.3 Nodal-Related Genes, *squint* and *cyclops*

Nodal is a member of the TGFβ-family of signaling factors whose effects are transmitted by serine/threonine kinase receptors and the Smad class of signal-transducing molecules (Massague and Chen 2000). In the mouse, only one *nodal* has been reported, and mutant embryos harboring mutations in this gene are defective in the formation of the primitive streak and do not survive gastrulation (Zhou et al. 1993). Overexpression of mouse *nodal* in zebrafish elicits ectopic expression of *goosecoid* and *lim1*, which mark the organizer, suggesting a role of *nodal*-related genes in the organizer formation in zebrafish (Toyama et al. 1995). To date, two and six *nodal*-related genes have been described in zebrafish and *Xenopus*, respectively (Jones et al. 1995; Smith et al. 1995; Joseph and Melton 1997; Erter et al. 1998; Rebagliati et al. 1998a,b; Sampath et al. 1998; Takahashi et al. 2000a). The zebrafish *nodal*-related genes *squint* (*sqt*) and *cyclops* (*cyc*) have different though overlapping expression profiles (Erter et al. 1998; Rebagliati et al. 1998a,b; Sampath et al. 1998). *sqt* (also termed *znr2* or *ndr1*) is expressed at low levels maternally; zygotic transcripts are detected at MBT in the dorsal blastomeres and then in dorsal YSL, expand to the entire marginal blastoderm and YSL, and by shield stage become restricted to the dorsal forerunner cells, a small group of non-involuting superficial cells. *Cyc* (*ndr2* or *znr1*) expression begins at 30% epiboly in the entire blastoderm margin and is progressively confined to the embryonic shield. During mid-gastrula stage, *cyc* is expressed in axial mesendoderm including the prechordal plate and anterior neuroectoderm.

During segmentation, *cyc* is asymmetrically expressed in left lateral plate mesoderm and in a small region of the left diencephalon. Expression of *sqt* and *cyc* in *Xenopus* animal caps revealed that *sqt* has a stronger mesoderm-inducing activity than *cyc*, while *cyc* induces a neural fate (Rebagliati et al. 1998a). Misexpression of β-catenin elicits ectopic expression of *sqt*, while misexpression of Axin, or Axin together with GSK3, inhibits *sqt* expression in the dorsal blastomeres and YSL (Shimizu et al. 2000; Fig. 5). Therefore dorsal expression of *sqt* depends on the canonical Wnt pathway, although it is not clear whether the effect is direct (Shimizu et al. 2000). Intriguingly, expression of *sqt* in a circular domain at the blastoderm margin and YSL is not affected by modulation of the Wnt signal, suggesting different modes of regulation of *sqt* in marginal and dorsal domains. Clearly, maternal factors are involved in the activation of *nodal* genes in the ventrolateral region, as depletion of RNA in the YSL by RNase injection impairs the expression of *sqt* or *cyc* in this domain (Chen and Kimelman 2000). Mutants defective in Nodal signaling such as *one-eyed pinhead* and *schmalspur*, or embryos overexpressing Antivin, a Nodal inhibitor, fail to express *sqt* and *cyc* in the blastoderm margin, suggest-

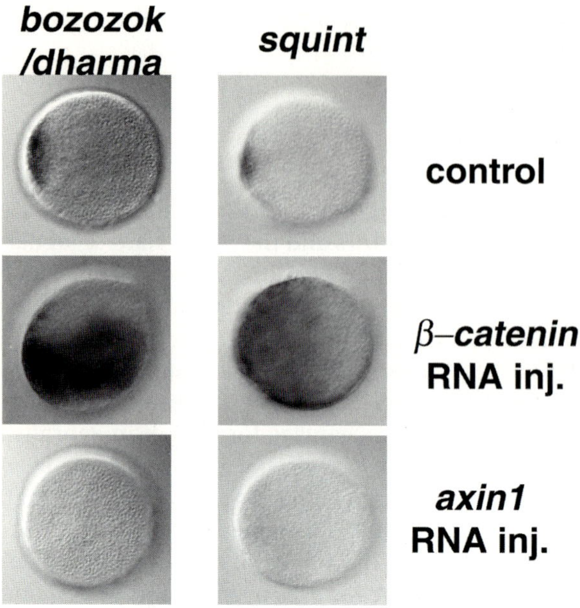

Fig. 5. Expression of *boz/dha* and *sqt* depends on signals from the canonical Wnt pathway. Embryos injected with β-catenin and axin1 RNA were analyzed at the sphere stage. Overexpression of β-catenin elicits ectopic expression of *boz/dha* and *sqt*, whereas *axin1* abolishes expression of these genes. (After Shimizu et al. 2000)

ing that positive feedback regulation of *sqt* and *cyc* has a role in their expression (Meno et al. 1999; Pogoda et al. 2000; Shimizu et al. 2000; Sirotkin et al. 2000b).

The recessive lethal mutation *sqt* leads to reduction of prechordal plate and ventral forebrain and to fusion of the eyes (Heisenberg and Nüsslein-Volhard 1997; Fig. 6). Similarly, *cyc* mutant embryos are cyclopic and show a reduction or absence of prechordal plate, floor plate, and ventral forebrain (Hatta et al. 1991b). Double mutants for *sqt* and *cyc* (*sqt;cyc*) lack endoderm (Fig. 6) and axial mesoderm, and fail to express *goosecoid* (Feldman et al. 1998). In

Fig. 6A–H. Cooperative roles of *bozozok* and *nodal*-related genes. **A** Wild type, **B** *boz*⁻, **C** *sqt*⁻, **D** *cyc*⁻, **E** *boz;sqt* double mutant, **F** *boz;cyc* double mutant, **G** *cyc;sqt* double mutant, and **H** *boz;cyc;sqt* triple mutant embryos at the pharyngula stage (30 hpf). Lateral views for whole embryos, anterior to the left (*leftmost panels*); lateral views of head region, anterior to the left (*second left panels*); lateral views of trunk and tail regions, posterior to the right (*second right panels*); ventral views of the head region (*rightmost panels*) except for *boz;cyc;sqt* shown in dorsal view. The boz^{m168} embryos in this study lacked axial mesendoderm including the prechordal plate, notochord, and floor plate, and were cyclopic; sqt^{cz35} embryos had fused eyes, but normal notochord and floor plate; cyc^{m294} embryos had fused eyes and lacked floor plate. The *boz;sqt* and *boz;cyc* embryos displayed anterior truncations that were more severe in *boz;sqt* embryos. Triple mutant embryos displayed severe abnormalities, essentially lacking mesendoderm and anterior neuroectoderm. (After Shimizu et al. 2000)

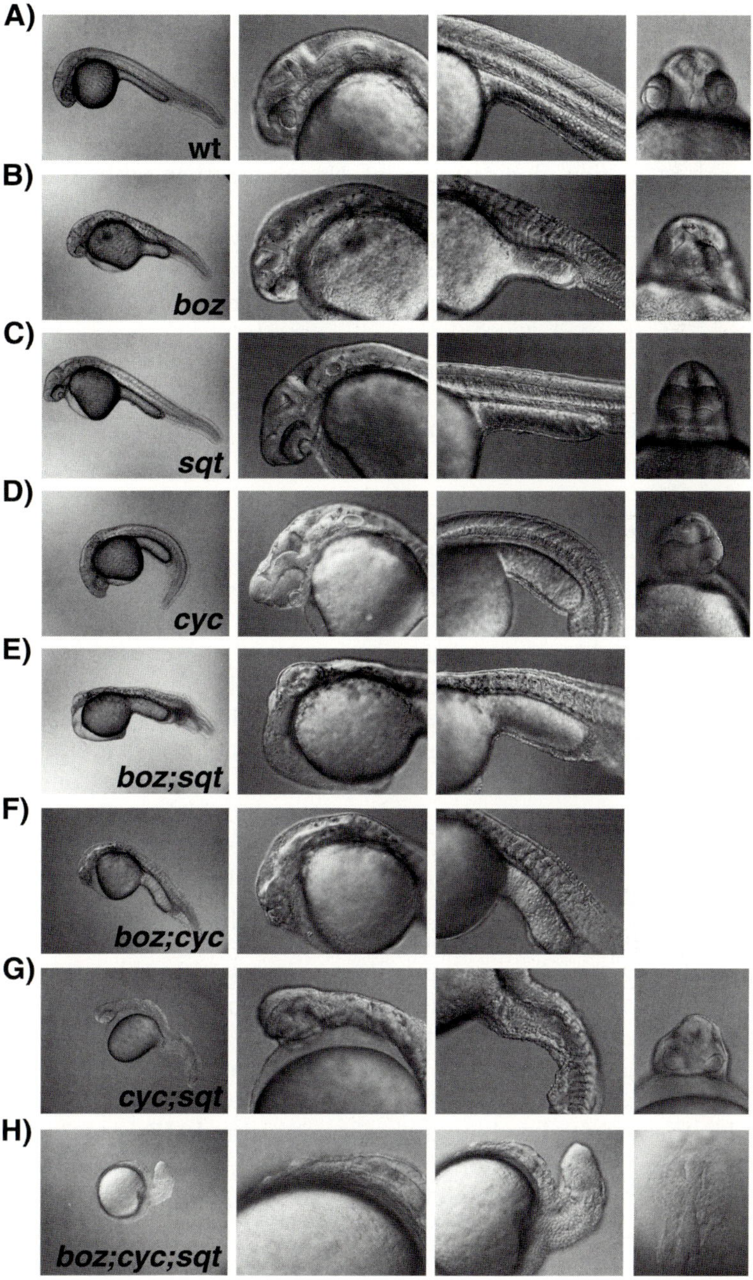
A)
wt
B)
boz
C)
sqt
D)
cyc
E)
boz;sqt
F)
boz;cyc
G)
cyc;sqt
H)
boz;cyc;sqt

addition to *sqt* and *cyc*, two genes, *one-eyed pinhead* (*oep*) and *schmalspur* (*sur*), act in the Nodal-signaling pathway. *oep* and *sur* encode a EGF-CFC-type membrane-anchored protein and the forkhead transcription factor FoxH1/Fast1, respectively (Pogoda et al. 2000; Sirotkin et al. 2000b; Zhang et al. 1998b). Oep is required for Nodal signal transduction, and Sur is involved in transcriptional regulation downstream of Nodal together with Smad2 and 3 (Gritsman et al. 1999; Pogoda et al. 2000; Sirotkin et al. 2000b). Mutants in which *oep* or *sur* are absent both maternally and zygotically (MZ mutants) fail to express organizer-specific genes (Gritsman et al. 1999; Pogoda et al. 2000; Sirotkin et al. 2000b). A Nodal signal is also required for organizer formation in *Xenopus* (Agius et al. 2000).

3.4 Cooperative Roles of *boz/dha* and *sqt* in the Induction of the Organizer

Either *boz/dha* or *sqt* alone is capable of inducing ectopic expression of organizer genes and of a partial secondary axis (Erter et al. 1998; Koos and Ho 1998; Yamanaka et al. 1998), raising two possibilities. First, one of these two genes is sufficient for dorsal axis formation only when it is overexpressed, but the two genes act in parallel in the normal embryo; second, these two genes act within the same pathway. Activation of the Nodal signal by overexpression of either mouse nodal, the active form of zebrafish Vg1, or the activated TGF-β receptor TARAM-A*, rescues the axial mesendoderm defects in *boz* mutants (Fekany et al. 1999), suggesting that *nodal* acts downstream of *boz/dha*. Consistent with this interpretation, *cyc* expression in the embryonic shield is diminished in the *boz* mutants (Sampath et al. 1998). In contrast, *sqt* expression in the dorsal blastoderm and the dorsal YSL is not affected in *boz* mutants, although *sqt* expression in dorsal forerunner cells is absent as these cells are missing in mutant embryos (Shimizu et al. 2000). Conversely, *boz/dha* expression is not affected in *sqt* mutant or Antivin-overexpressing embryos, although maintenance of *boz/dha* depends on an autoregulatory feedback loop at later stages (Shimizu et al. 2000). These data suggest that *boz/dha* and *sqt* are independently activated by a maternal signal mediated by the canonical Wnt pathway.

Although most of the organizer genes are expressed weakly or not at all in *boz* and in the *nodal*-related mutants, *sqt*, *sqt;cyc*, and MZ-*oep* (Feldman et al. 1998; Gritsman et al. 1999; Shimizu et al. 2000), expression of *chordino* is diminished in *boz* but only weakly perturbed in *nodal*-related mutants or in Antivin-overexpressing embryos (Gritsman et al. 1999; Shimizu et al. 2000). Furthermore, overexpression of *boz/dha* elicits ectopic expression of *chordino* in Antivin-overexpressing embryos or in *sqt* or *sqt;cyc* mutants (Shimizu et al. 2000; Sirotkin et al. 2000a). These data indicate the existence of *boz/dha*-dependent and *nodal*-independent processes in organizer formation. Furthermore, *boz;sqt* double mutants display a stronger ventralized phenotype than single mutants, and *boz;sqt;cyc* triple mutants are even more severely affected (Shimizu et al. 2000; Sirotkin et al. 2000a; Fig. 6). Thus *boz/dha* and *sqt* act cooperatively in the zebrafish Nieuwkoop center to mediate organizer formation.

3.5 *vega/vox/vent*

While the mechanisms through which Boz/Dha executes its organizer-inducing effect are not fully understood, an important early event is the repression of a set of two ventral homeobox genes named *vega1/vox* and *vega2/vent* (Kawahara et al. 2000a,b; Melby et al. 2000). Expression of *vega1* is initiated at MBT in all the blastomeres and is downregulated dorsally very quickly, already being excluded from the dorsal domain by dome stage. *vega2* is also expressed ventrally, but somewhat later than *vega1*. *vega* transcripts are expanded dorsally in *boz* mutants, and are reduced or absent in *boz/dha*-overexpressing embryos (Fig. 7). In a complementary manner, overexpression of Vega reduces the expression of *boz/dha* and other organizer genes, and elicits ventralization and a head-less phenotype, similar to severe *boz* mutants. Expression and function of *vega* genes in the early embryo stage are not mediated by BMP as *vega1* expression is not affected in *swirl/bmp2b* mutants or by the injection of a constitutively active BMP receptor (Fig. 7). Recently, Imai and coworkers have reported the isolation of mutants carrying deletions of both *vox/vega1* and *vent/vega2*, as well as a *vox* point mutation (Imai et al. 2001). These authors further used morpholino oligonucleotides to probe the function of these genes. The results show that the two genes function in a redundant manner, with single mutations or injection of a single morpholino having surprisingly little effect. Further, the studies of Imai and coworkers expand the earlier conclusion that the *vega/vox/vent* genes have distinct though partially overlapping functions with the BMP-signaling system.

The *vega/vox/vent* and *boz/dha* genes all act as transcriptional repressors, as engrailed fusions function in overexpression experiments like the wild-type whereas activating VP16 fusions have opposite effects (Kawahara et al. 2000a,b; Solnica-Krezel and Driever 2001; T. Shimizu, Y. Yamanaka, R. Fujii, M. Hibi, T. Hirano, unpubl. results). We suggest that direct mutual inhibition between Boz/Dha and Vega/Vox/Vent immediately after MBT sets up the earliest zygotically determined pattern of DV polarity in the embryo that is independent, and possibly upstream, of the BMP/antiBMP interaction that dominates DV polarity determination during gastrulation.

3.6 Induction by the Nieuwkoop Center Versus Cell-Autonomous Establishment of the Organizer

While much evidence implies an inductive role for the Nieuwkoop center in organizer formation, some experiments indicate that the organizer can form

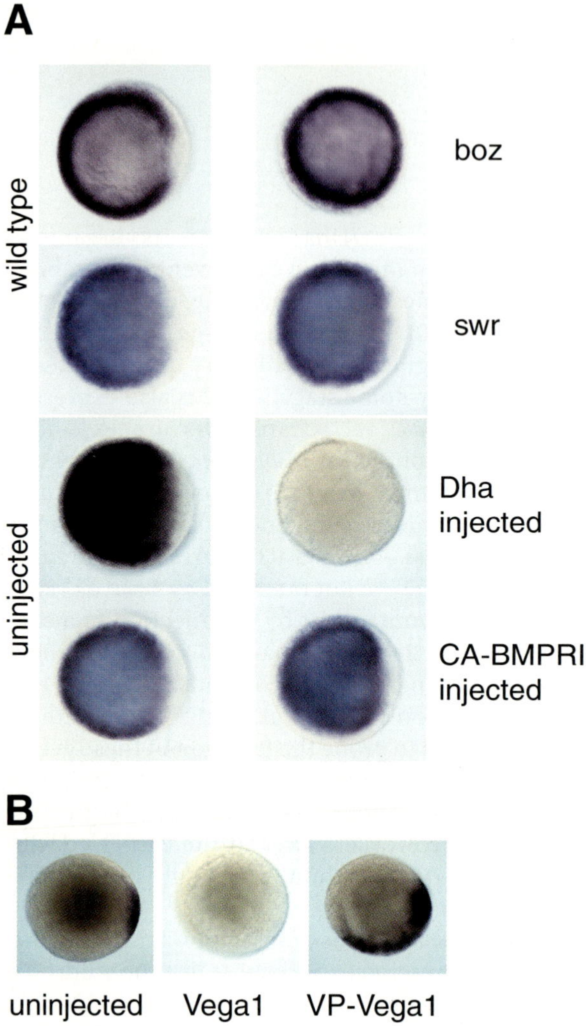

Fig. 7A,B. Interactions of *vega1/vox* with *boz* and the BMP pathway. **A** In situ hybridization with *vega1*, with control embryos in the *left column*. In *boz* mutant embryos at shield stage, *vega1* expression expands into the shield from which it is normally excluded. However, in *swr* mutants at dome stage, dorsal exclusion of *vega1* is unaffected. Injection of *boz/dha* into the embryo completely suppressed *vega1* expression, while injection of a constitutively active BMP receptor had no effect on the *vega1* pattern. **B** In situ hybridization with *boz/dha* at the dome stage. Excess Vega1 suppresses, but VP16-Vega1 stimulates, the expression of *boz/dha*. All animal views. (After Kawahara et al. 2000a)

cell-autonomously in a dorsal marginal domain. In *Xenopus*, dissociation of blastomeres at the 32-cell stage interrupting subsequent inductive interactions does not inhibit expression of *goosecoid*, suggesting that dorsalizing information is already present in marginal blastomeres at that stage (Lemaire and Gurdon 1994). The *Xenopus* homeobox genes *siamois* and *twin* are direct targets of the maternally derived Wnt signal (Lemaire et al. 1995; Brannon et al. 1997; Laurent et al. 1997), and can directly regulate the expression of *goosecoid*; thus, no intercellular signaling is indicated between β-catenin action and organizer gene activation. When the goldfish blastoderm was dissected from the yolk cell at mid-blastula stage, rotated 180°, and recombined with the yolk, it frequently developed two axes (Yamaha et al. 1998), suggesting that the dorsal blastoderm cells had already acquired the ability to differentiate into the organizer. RNase injection into the YSL at the 1K-cell stage of zebrafish, which eliminates transcripts including those encoding *boz/dha* and *sqt* in the YSL, does not affect formation of dorsal mesoderm (Chen and Kimelman 2000), indicating that signals from the YSL are not necessary after this stage.

A modified view of mesoderm and organizer formation has emerged in *Xenopus* based largely on the work of Heasman, Wylie and colleagues (Wylie et al. 1996; Zhang et al. 1998a; Kofron et al. 1999) on the action of β-catenin, which is also critical in the early zebrafish, and VegT, which is not. While the maternal T-box transcription factor VegT plays an essential role in the pre-gastrula *Xenopus* embryo (Zhang et al. 1998a; Clements et al. 1999; Hyde and Old 2000; Xanthos et al. 2001), the apparent zebrafish homologue, *spadetail*, is not expressed maternally and *spadetail* mutants show normal mesoderm induction and organizer formation (Griffin et al. 1998; see also Kimelman and Schier, this Vol.). The work in *Xenopus* can be summarized as showing that β-catenin-expressing cells can induce dorsal structures in a non-autonomous manner, but will only participate in mesoderm formation when they are derived from the marginal zone or receive additional signals. Further, the β-catenin-evoked signal passes to recipient cells only after MBT, indicating that it requires zygotic transcription (Wylie et al. 1996). VegT is required independently of β-catenin for endoderm formation and for the generation of mesoderm-inducing signals. Genes encoding Nodal and other signaling factors depend on VegT for their activation, and injection of *nodal* mRNAs can rescue VegT-depleted embryos (Zhang et al. 1998a; Kofron et al. 1999). Further, at least one *nodal* gene, *Xnr1*, appears to be a direct target gene of VegT. These data again indicate that zygotic transcription is required for the generation of mesoderm-inducing signals. In *Xenopus* it appears, therefore, that mesoderm and organizer formation depends on transcription factors intrinsic to both the vegetal inducing region, traditionally described as the Nieuwkoop center, and the marginal region that forms the organizer. Inducing signals that are required in this process are then generated by post-MBT transcription. In the zebrafish, where the apparent VegT ortholog plays no role in these processes, the relationships between intrinsic factors and intercellular signaling in organizer formation remain to be clarified.

4 The Organizer

4.1 Is the Embryonic Shield the Fish Organizer?

Transplantation of the embryonic shield in teleosts can generate a secondary axis (Oppenheimer 1936; Shih and Fraser 1996; Saúde et al. 2000), showing an activity comparable to that of the prototypical Spemann-Mangold organizer.

Cells within the embryonic shield contribute to the hatching gland, head mesoderm, notochord, somatic mesoderm, endoderm and ventral part of the neuraxis, and the cells close to the midline are biased towards prechordal and chordal mesoderm and ventral neural tissue (Shih and Fraser 1995, 1996). Distinct populations with different gene expression patterns occur within the prospective shield at the early gastrula (Gritsman et al. 2000). At 50% epiboly, an apparent gradient of Nodal signaling specifies prechordal progenitors expressing *goosecoid* (*gsc*) close to the margin, and notochord progenitors expressing *floating head* (*flh*) at a distance towards the animal pole (Gritsman et al. 2000).

While earlier transplantation experiments suggested that the shield is capable of only partial axis induction (Shih and Fraser 1996), more recent results show formation of a complete secondary axis by grafts that include both *gsc*-expressing deep hypoblast and *flh*-expressing epiblast cells (Saúde et al. 2000). Grafts containing the deep region of the shield preferentially induced anterior tissues, and grafts from the superficial region induced posterior structures. The data suggest that there are distinct organizer populations, traditionally identified in amphibians as the head and trunk organizer (Slack and Tannahill 1992).

Although transplantation experiments underscore the organizer activity of the embryonic shield, results of shield ablation are more ambiguous. Removal of the embryonic shield and adjacent marginal tissue leads to a loss of axial mesendoderm and floor plate and to defects in the formation of ventral neuroectoderm, but does not significantly affect anterior-posterior (AP) patterning of the neuroectoderm. The phenotype of such shield-deprived embryos is milder than that of severe *boz* mutants that lack axial mesendoderm and anterior neuroectoderm. Two non-exclusive possibilities can be considered. First, the functional organizer is wider than the morphologically apparent embryonic shield. Consistent with this view, deletion of the entire *chordino*-expressing domain, which is wider than the *gsc*- and *flh*-expressing domains, abolishes the expression of anterior-neuroectoderm-specific gene *opl* (Stachel et al. 1993; Talbot et al. 1995; Miller-Bertoglio et al. 1997; Grinblat et al. 1998). The second explanation suggests that a functional organizer forms prior to the visible initiation of gastrulation. Consistently, most organizer genes, such as *gsc*, *lim1*, *chordino*, and *dickkopf1*, are expressed before gastrulation begins shortly after the MBT (Stachel et al. 1993; Toyama et al. 1995; Miller-Bertoglio et al. 1997; Hashimoto et al. 2000; Shinya et al. 2000). Thus, the organizer forms earlier and in a wider region than the morphologically apparent embryonic shield.

4.2 Organizer Genes

Many genes have been identified that are expressed and function in the organizer (Harland and Gerhart 1997). The majority of these genes encode proteins of two major types, transcription factors such as *goosecoid, lim1, axil/hnf3β/foxa2*, and *floating head*, and various signaling molecules. The latter can be divided into different subgroups by functional characteristics, including BMP inhibitors *chordino, noggin1*, and *twisted gastrulation*, the Wnt inhibitor *dickkopf1*, the *FGFs*, the nodal-related *cyclops, sonic hedgehog*, and others. Among them, the BMP and Wnt inhibitors and the FGFs participate in the initial DV and AP patterning of the embryo.

4.3 *chordino, noggin1, twisted gastrulation*, and *ogon*

Chordin and Noggin prevent interaction of BMPs with their receptors by direct ligand binding, and both factors play essential roles in DV patterning and neural induction (reviewed by De Robertis et al. 2000; Sasai 2001). One *chordin* (also named *chordino*) and three *noggin* genes are known in zebrafish, but only *chordino* and *noggin1* are expressed in the organizer (Miller-Bertoglio et al. 1997; Fürthauer et al. 1999). Overexpression of these genes in zebrafish leads to strong dorsalization, as also observed in *Xenopus* (Miller-Bertoglio et al. 1997; Fürthauer et al. 1999). The phenotypes resemble those of the dorsalized mutants *swirl/bmp2b, somitabun/smad5*, and *snail house/bmp7*, which are defective in different steps of the BMP signal transduction cascade (Kishimoto et al. 1997; Hild et al. 1999; Dick et al. 2000; Schmid et al. 2000). The recessive mutation *chordino*, originally named *dino*, leads to strong ventralization of the embryo (Hammerschmidt et al. 1996a,b; Schulte-Merker et al. 1997); *captain hook* and *turntail* are allelic to *chordino* (Fisher and Halpern 1999). *chordino* is expressed in the dorsal quarter of the blastoderm margin 1 h after MBT, and the transcripts are detected in axial mesendoderm and in bilateral domains in all germ layers during gastrulation, and later in the neuroectoderm (Miller-Bertoglio et al. 1997). The dorsal expression of *chordino* is only weakly affected in embryos lacking Nodal signaling (Gritsman et al. 1999; Shimizu et al. 2000), whereas it is strongly reduced in *boz* mutants during gastrulation (Koos and Ho 1999; Fekany-Lee et al. 2000; Shimizu et al. 2000). However, the initial expression of *chordino* is not affected in *boz* mutants, suggesting that initiation of *chordino* expression is regulated by maternal factors (Fekany-Lee et al. 2000; Shimizu et al. 2000). Consistent with this possibility, the promoter of *chordino* contains a consensus sequence for Tcf/Lef binding, suggesting direct regulation by maternal β-catenin (R. Fujii et al., unpubl. results). Furthermore, double mutants for *boz* and *chordino* (*boz/din*) display a synergistic loss of head and trunk (Gonzalez et al. 2000), confirming the view that *chordino* expression does not depend on *boz/dha* alone (Sirotkin et al. 2000a), and that *boz/dha* suppresses BMP signaling through other mediators in addition to *chordino*. One of these factors might be *noggin1* whose expression depends on Boz/Dha and a Nodal signal (Fürthauer et al. 1999; Sirotkin et al. 2000a).

In addition to *chordino* and *noggin*, several molecules conserved between invertebrates and vertebrates are involved in BMP antagonism. Recently, the *twisted gastrulation* (*tsg*) gene product was shown to bind and antagonize BMP and its *Drosophila* homologue Dpp. Misexpression of *tsg* causes dorsalization in zebrafish in a synergistic manner with Chordin (Chang et al. 2001; Ross et al. 2001; Scott et al. 2001), while injection of an anti-sense morpholino oligo for one of the two *tsg* genes causes ventralization, indicating a role for *tsg* in the embryo.

An additional ventralized mutant *ogon*, and its alleles *mercedes* and *short tail*, have been described (Solnica-Krezel et al. 1996; Miller-Bertoglio et al. 1999). Although the molecular identity of *ogon* is unknown, double mutant and epistatic analysis suggest that the *ogon* gene product cooperates with Chordino in antagonizing BMP signaling (Miller-Bertoglio et al. 1999).

4.4 *dickkopf1*

Inhibition of both Wnt and BMP signaling is required for the formation of anterior neuroectoderm in *Xenopus* (Bouwmeester et al. 1996; Glinka et al. 1997, 1998; Niehrs 1999; Piccolo et al. 1999). In zebrafish, misexpression of *chordino* expands the neuroectoderm but does not suppress the forebrain-defect phenotype in *boz* mutants. However, overexpression of a dominant negative Wnt8 rescues these defects (Fekany-Lee et al. 2000). Loss of the *wnt8* locus results in expansion of the organizer and forebrain (Lekven et al. 2001). Thus, a requirement for Wnt inhibitory factors is indicated. *dickkopf1* (*dkk1*), a homologue of a *Xenopus* gene encoding a Wnt inhibitor, is expressed in the dorsal blastoderm and dorsal YSL after MBT and in the anterior edge of axial mesendoderm during gastrulation, and this expression depends on Boz/Dha and Nodal function (Hashimoto et al. 2000; Shinya et al. 2000). Furthermore, the *dkk1* promoter contains a Tcf/Lef-binding sequence (Shinya et al. 2000), suggesting that initiation of *dkk1* expression is controlled by β-catenin function. Misexpression of *dkk1* in wild-type embryos expands the forebrain and notochord, and rescues forebrain and notochord defects in *boz* mutants (Hashimoto et al. 2000; Shinya et al. 2000; Fig. 8). These data indicate that *dkk1* acts downstream of Boz/Dha and is required for formation of forebrain. In contrast to the effect of Dkk1 on *boz* mutants, overexpression of *dkk1* does not suppress prechordal plate and notochord defects in embryos lacking Nodal signaling but can still anteriorize the neuroectoderm, suggesting that it directly affects its AP patterning (Hashimoto et al. 2000). Recently, Dickkopf was shown to bind to the Wnt coreceptor LRP6, affecting signaling in a manner distinct from that of other known Wnt inhibitory factors (Bafico et al. 2001; Mao et al. 2001; Semenov et al. 2001).

4.5 Fibroblast Growth Factors

Fibroblast growth factors (FGFs) are involved in many aspects of development and have a role in mesoderm induction and patterning of the nervous system

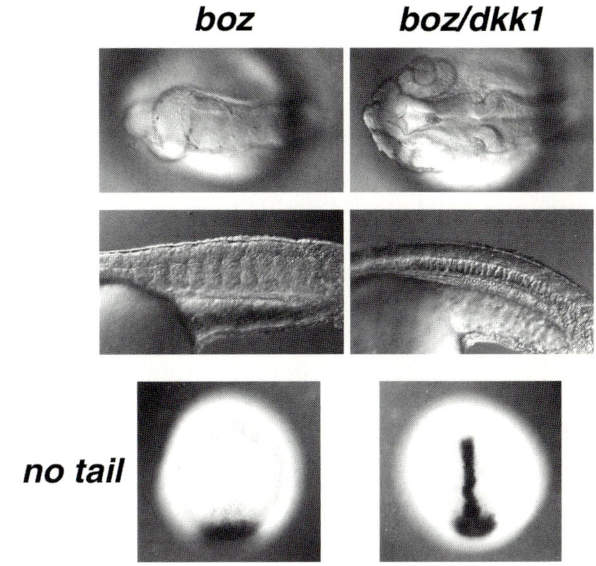

Fig. 8. The Wnt antagonist Dkk1 suppresses the *boz* phenotype. The *boz*^{m168} homozygous embryos used in this study displayed defects in the formation of notochord, floor plate, and anterior neuroectoderm; overexpression of Dkk1 restored development of these structures. Animal pole views for head region (*upper panels*) and lateral views for trunk (*lower panels*). *No tail* (*ntl*) expression, marking the notochord, was not detected in *boz*^{m168} homozygous tail bud stage embryo, but was restored by Dkk1 overexpression. (After Hashimoto et al. 2000)

(see Kimelman and Schier, this Vol.). In the zebrafish, FGFs also have a role in DV patterning. Zebrafish *fgf8* is expressed at the dorsal margin after MBT, and at the early-gastrula stage is expressed in a characteristic dorsoventral gradient within the margin (Fürthauer et al. 1997). Misexpression of FGF8 suppresses *bmp* expression, leads to the expansion of dorsolateral tissues, and can induce a partial secondary axis, suggesting that FGFs can elicit organizer activity in zebrafish. Mutation in *fgf8* (*acerebellar*; *ace*) does not have a major effect on DV patterning or organizer function, but leads to defects in the mid-hindbrain boundary (Reifers et al. 1998). This may be due to the fact that other FGFs complement the function of *fgf8/ace* in DV patterning, as, e.g., *fgf3* is expressed in a similar pattern in the early embryo as *fgf8* (T. Kudoh et al., unpubl. observ.). Consistent with this, overexpression of zebrafish Sprouty4, which can inhibit both Fgf3 and Fgf8, leads to a ventralization of the embryos (Fürthauer et al. 2001).

4.6 Role of the Organizer in AP Patterning

Transplantation experiments suggest the existence of distinct head and trunk organizer domains, but at the same time there is considerable evidence to show that AP patterning in the neuroectoderm does not depend on organizer activity. Implantation of organizer tissue or of Noggin- or Chordin-expressing COS cells into the ventral part of the zebrafish blastula led to the formation of neuroectoderm with well-established AP patterning (Koshida et al. 1998), suggesting that positional values along the AP axis are determined by preexisting information. However, transplantation of mesoderm from the lateral margin to the animal pole transformed forebrain into hindbrain (Woo and Fraser 1997) and transplantation of yolk to the animal pole of the blastoderm reversed AP polarity in the neuroectoderm (Koshida et al. 1998). These reports suggest that posteriorization of the neuroectoderm can be elicited by signals from non-axial mesoderm or from the yolk cell, consistent with Nieuwkoop's two-signal model which holds that ectoderm is induced to anterior neuroectoderm by an "activator" derived from the organizer and is subsequently or simultaneously posteriorized by a "transforming" signal released from paraxial mesoderm (reviewed in Sasai and De Robertis 1997; Niehrs 1999).

If AP positional information is established independently of the organizer, why is it possible to dissect head and trunk organizer regions, and why do *boz* or *boz;sqt* mutants display head defects? Evidence for distinct head organizer activity also comes from gene targeting in the mouse, where deletion of the organizer genes *Lim-1* or *Otx-2* leads to ablation of the head anterior to the otic vesicles but largely normal initial development of trunk and tail (Acampora et al. 1995; Matsuo et al. 1995; Shawlot and Behringer 1995; Ang et al. 1996). It is plausible that the head organizer, represented in zebrafish by deep hypoblast cells of the shield and prechordal plate, not only expresses neural inducers but also factors that antagonize the posteriorizing influence of the trunk organizer, which is localized in the superficial layer of the shield and later in chordamesoderm. In this view, the head organizer reinforces the preexisting AP positional information in the ectoderm.

It appears that the activator function in Nieuwkoop's model is most directly represented by BMP antogonists, primarily Chordin and Noggin. Wnt antagonists also have a role in primary neural induction, but the situation is complicated by the fact that Wnts are among the candidate posteriorizing factors, together with FGFs, retinoic acid (RA), and the Nodals. These factors have been shown to be able to effect posteriorization of the neuroectoderm when applied in excess, and certain antagonists of these factors, such as Dkk1 and the Nodal inhibitor Antivin, are expressed in the head organizer and appear to play a role in head formation (Thisse and Thisse 1999; Hashimoto et al. 2000). This subject is discussed further elsewhere (see Wilson et al., this Vol.).

5 Cell Movements and the Organizer

The organizer is distinguished by several properties, among which self-differentiation and the inductive influences on neighboring tissues have been discussed above. Another aspect of organizer function is the control of cell movements that are critical for gastrulation. Four types of cell movements occur in zebrafish gastrulation: epiboly, involution/ingression, convergence/extension, and anterior migration (for a more extensive discussion of these movements, see Kane and Adams, this Vol.; Solnica-Krezel and Cooper, this

Vol.). In embryos completely ventralized by yolk depletion, epiboly and involution/ingression still take place, suggesting that these cell movements are not obligatorily linked to DV patterning (Mizuno et al. 1999b; Ober and Schulte-Merker 1999). In contrast, convergence/extension and anterior migration are linked to DV patterning, and misexpression of organizer genes not only induces dorsal cell fates, but also recruits additional cells into the dorsal axis. Expression of *gsc* in ventral blastomeres of *Xenopus* embryos induces anterior migration both in *gsc*-expressing cells and neighboring non-expressing cells, indicating that *gsc* acts non-cell-autonomously on the adjacent cells to regulate migration (Niehrs et al. 1993). Misexpression of *fgf8/ace* in zebrafish embryos also induces convergence and extension movements outside the injected domain (Fürthauer et al. 1997), while, in ventralized mutants, such movements are perturbed (Solnica-Krezel et al. 1996).

Some information is available on the genes and molecules involved in cell movements during gastrulation. Specifically, two pathways have been implicated, the first involving a protocadherin, the second involving the alternate Wnt pathway; both of these pathways are regulated by T-box transcription factors. In *spadetail* (*spt*) mutant embryos, the paraxial mesoderm fails to undergo convergence and extension, and formation of somites is disturbed (Ho and Kane 1990). *spt* encodes a T-box transcription factor homologous to *Xenopus* VegT that is expressed in the ventrolateral blastoderm margin and functions cell-autonomously (Griffin et al. 1998). The *paraxial protocadherin* (*papc*) gene is downstream of *spt*, and its function is required for convergence-extension movements (Yamamoto et al. 1998a). PAPC is first expressed at the dorsal margin but is excluded from axial mesoderm during gastrulation by a process that requires *flh*. In *Xenopus*, PAPC likewise is implicated in the regulation of gastrulation movements (Kim et al. 1998b).

Silberblick, *knypek*, and *trilobite* are mutants with reduced convergence-extension movements but no essential cell fate abnormality (Heisenberg and Nüsslein-Volhard 1997; Marlow et al. 1998; Heisenberg et al. 2000; Henry et al. 2000). In these mutants, the prechordal plate is not located properly, resulting in incomplete separation of eye primordia (Heisenberg and Nüsslein-Volhard 1997; Marlow et al. 1998). *silberblick* (*slb*) encodes Wnt11, which is known to activate the alternate, or PCP, Wnt pathway, and mutants of Dishevelled that affect the PCP pathway can phenocopy *slb* (Heisenberg et al. 2000). Transplantation experiments show that *slb/wnt11* acts in the paraxial but not the axial mesendoderm (Heisenberg et al. 2000). In *Xenopus, wnt11* is located downstream of the T-box transcription factor Brachyury whose zebrafish homologue is Ntl (Tada and Smith 2000). The PCP pathway, initiated by certain Wnt molecules such as Wnt11 and Wnt5a and involving calcium waves as well as Dishevelled, has a critical role in convergence-extension movements during gastrulation (Wallingford et al. 2000, 2001b).

Convergence/extension movements must result in appropriate cell arrangements along the DV and AP axes for the successful completion of gastrulation and axis formation. In addition to factors promoting such movements, such as

PAPC and Wnt11, factors exist that limit the domains in which these movements are propagated. One such factor is the anterior gene *otx2*, which is expressed in the organizer at early-gastrula stages and subsequently in anterior neuroectoderm. *otx2* inhibits convergence-extension movements in its expression domain, and it does so by activating the expression of a calponin, an actin-binding protein that can block filament-dependent movements (Morgan et al. 1999). It remains to be elucidated how the different pathways that affect movements are integrated during gastrulation of the vertebrate embryo.

6 Summary and Prospects

The organizer plays a central role in axis formation during vertebrate embryogenesis. Establishment of the organizer requires the microtubule-mediated transport of maternal dorsal determinants to dorsal blastomeres. These determinants activate the canonical Wnt pathway upstream of β-catenin accumulation at a point yet to be elucidated, leading to the expression of a set of genes that identify the Nieuwkoop center. The Nieuwkoop center induces the organizer in an adjacent region, but cell-autonomous organizer formation in a region receiving dorsal determinants appears to contribute to the process. Analysis of zebrafish mutants with axis formation defects and of genes whose overexpression affects DV and AP patterning has led to a great deal of information about the factors that control organizer formation and function. Two genes, *boz/dha* and *sqt*, act in the Nieuwkoop center downstream of the β-catenin-mediated signal, and are required for the expression of organizer genes that control DV and AP axis formation. This linear gene cascade is undoubtedly oversimplified as many organizer genes are activated right after MBT even in *boz*- or *nodal*-related mutants, presumably directly by Nieuwkoop center signals. Similarly, organizer gene expression and function are not independent of each other. Frequently, misexpression of one organizer gene elicits the overexpression of many other organizer genes, although it is not clear which of these effects are direct and which are mediated by other factors. Likewise, organizer genes and ventral genes counteract each other through both direct and indirect pathways.

Many questions remain to be answered. The nature of the primary dorsal determinants remains uncertain although it is clear that nuclear accumulation of β-catenin is a critical consequence of their action. The relationship between mesoderm induction and organizer formation likewise remains an intriguing question that might be represented in molecular terms by the question of the relationship between early dorsalization through β-catenin and Boz/Dha and germ layer induction through Nodal signaling. However, the involvement of additional components in early patterning remains an open issue. The function of the organizer in antagonizing the ventralizing influences of BMP and Wnt factors is quite well understood, but the mechanisms controlling convergence/extension movements in gastrulation have been investigated less thoroughly. The T-box gene-dependent PAPC and Wnt11 mechanisms that clearly affect gastrulation movements appear to affect primarily the formation of trunk and tail structures. Head formation is known to depend on the inhibition of BMP, Wnt and Nodal signals, but the regulation of cell movements involved in head formation is less well understood. These and additional questions continue to make study of the Spemann-Mangold organizer a fascinating subject more than 75 years after its initial discovery.

Dorsoventral Patterning in the Zebrafish: Bone Morphogenetic Proteins and Beyond

Matthias Hammerschmidt[1] and Mary C. Mullins[2]

1 Dorsoventral (DV) Patterning in Frog, Fish, and Fly

Lineage tracing studies carried out in amphibian, fish, mouse and chicken embryos have shown that the fate of cells in the older animal can be traced to the positions of their precursors at pre- or early-gastrula stages of development, which allowed the design of fate maps of vertebrate embryos (Dale and Slack 1987; Kimmel et al. 1990; Hatada and Stern 1994; Lawson 1999). The fate map of the zebrafish gastrula is shown in Fig. 3A. Despite the different architectures of the embryos, the fate maps of the different vertebrate species show high similarities, pointing to conserved mechanisms of early embryonic patterning, which defines positional values within the early embryo. One of the primary systems providing positional information is the genetic network controlling early dorsoventral (DV) patterning, most intensely studied in frog and fish.

Embryological studies carried out in amphibian embryos had initially suggested that DV patterning of the early embryo is strictly linked to mesoderm induction. Mesoderm is induced in the marginal zone of blastula-stage embryos by inductive signals emanating from vegetal (prospective endodermal) cells. Recombinations of vegetal and animal (prospective ectodermal) explants revealed that dorsal vegetal blastomeres preferentially induce dorsal mesoderm like notochord and muscle, while ventral vegetal blastomeres induce ventral mesoderm like mesenchyme and blood. Thus, it was assumed that different mesoderm-inducing signals from vegetal cells lay down DV pattern in the mesoderm. Indeed, factors with such differential mesoderm-inducing activities could be isolated: Activin, a member of the Tgfb superfamily of growth factors, which induces notochord at high and muscle at lower concentrations, and basic fibroblast growth factor (Fgf), which induces ventral mesodermal cell types.

Based on recent findings in *Xenopus laevis* and zebrafish described below and in an accompanying review (see Kimelman and Schier, this Vol.), this

[1] Hans-Spemann Laboratories, Max-Planck Institute for Immunobiology, Stuebeweg 51, 79108 Freiburg, Germany

[2] University of Pennsylvania, Dept. of Cell and Developmental Biology, 421 Curie Blvd., Philadelphia, Pennsylvania 19104-6058, USA

Results and Problems in Cell Differentiation, Vol. 40
L. Solnica-Krezel (Ed.): Pattern Formation in Zebrafish

model of differential mesoderm induction has been revised. According to the current view, mesoderm is induced by a common process that does not discriminate between DV positions, whereas dorsal is defined by a process independent of mesoderm induction that does not discriminate between different germ layers. Mesoderm induction is driven by Nodal proteins, other members of the Tgfβ superfamily (see Kimelman and Schier, this Vol.), whereas dorsal is determined by a Wnt-signaling pathway (see Hibi et al., this Vol.).

In *Xenopus*, the dorsalizing activity was found to reside in dorsal vegetal blastomeres, named the Nieuwkoop center. When transplanted to the ventral side, these dorsal vegetal blastomeres induce an entire secondary body axis, in which even notochord and prechordal plate, derivatives of the dorsal mesoderm, are made from host cells (Gimlich and Gerhart 1984). In fish embryos, in which the vegetal hemisphere does not cleave, the site corresponding to the amphibian Nieuwkoop center is the dorsal yolk syncytial layer and parts of the dorsal blastomeres (Mizuno et al. 1997; see Sakaguchi et al., this Vol.). The primary effect of Nieuwkoop center signaling is the induction of a second inductive center in dorsal regions located animal of the Nieuwkoop center, which – as we will argue below – is characterized by the expression of genes encoding proteins like Noggin, Chordin and Follistatin (Smith and Harland 1992; Hemmati-Brivanlou et al. 1994; Sasai et al. 1994).

In amphibia, this second inductive center has initially been confined to the presumptive axial mesoderm located in the dorsal blastopore lip, called the Spemann-Mangold organizer. According to more recent analyses from zebrafish, chicken and *Xenopus laevis*, the organizer is more generally confined to dorsal cells of both mesodermal and neuroectodermal character (Miller-Bertoglio et al. 1997; Streit et al. 1998; Wessely et al. 2001).

When transplanted to the ventral side, the Spemann-Mangold organizer also induces a complete secondary body axis (Spemann and Mangold 1924). In contrast to the Nieuwkoop center transplantations, however, prechordal plate and notochord of this secondary axis are made from the transplant, whereas the remaining tissues (muscle and neural tube) derive from the host. Chordin, Noggin and Follistatin can mimic the effects of the Spemann-Mangold organizer when ectopically expressed, affecting both the mesoderm and the ectoderm. They convert initially ventrally specified mesoderm in marginal regions of the embryo to intermediate fates such as muscle. In addition, they promote neural specification in dorsal animal cells, which would otherwise give rise to epidermal fates (for Chordin, see Sasai et al. 1994, 1995). Chordin, Noggin and Follistatin are not instructive agents themselves. Thus, target cells do not have receptors for these three proteins. Rather, they fulfill their dorsalizing activity indirectly by inhibiting bone morphogenetic proteins (Bmps), members of the Tgfβ superfamily, which act as dose-dependent ventralizing factors (see for zebrafish, Kishimoto et al. 1997; Neave et al. 1997; Nikaido et al. 1997; Nguyen et al. 1998b).

In target cells, Bmp signaling is mediated by a complex of type I and type II serine/threonine kinase transmembrane receptors. Upon Bmp binding, the type I receptors phosphorylate and activate Smad1/5/8 proteins, which form complexes with Smad4 and translocate to the nucleus where they participate in transcriptional complexes (for review, see Massague and Chen 2000). Thus, by binding Bmps in the extracellular space (Piccolo et al. 1996; Zimmerman and Harland 1996; Fainsod et al. 1997), Chordin, Noggin and Follistatin prevent binding of Bmps to their receptors and attenuate Bmp signaling, thereby determining the local level of Bmp activity, which acts as a morphogen to induce differential cell specification in both mesoderm and ectoderm (Dosch et al. 1997; Knecht and Harland 1997; Neave et al. 1997; Nguyen et al. 1998b, 2000; Barth et al. 1999). In ventral/posterior-most regions of fish and frog embryos, Chordin itself is inhibited via proteolytic cleavage through the metalloprotease Tolloid, which thereby promotes Bmp signaling and ventral specification (Blader et al. 1997b; Piccolo et al. 1997; Connors et al. 1999).

The general importance of Bmps and their inhibitor Chordin during embryonic DV patterning is also revealed by the striking conservation of their roles during evolution, as has become apparent by comparative analyses between vertebrates and the fruitfly *Drosophila melanogaster*. Chordin shares structural and functional homology with *Drosophila* Short gastrulation (Sog; Francois et al. 1994), which acts as an antagonist of the *Drosophila* Bmp proteins Decapentaplegic (Dpp; Biehs et al. 1996), a Bmp2/4 ortholog, and Screw (Arora et al. 1994). These Bmps play a morphogenetic function during patterning of the DV axis of the *Drosophila* embryo (Ferguson and Anderson 1992; Wharton et al. 1993; Francois et al. 1994; Holley et al. 1995; Schmidt et al. 1995). The opposing distributions and activities of Chordin and Bmp4 in *Xenopus* and Sog and the Bmps in *Drosophila* suggest a common mechanism for DV patterning in bilateral metazoa (François and Bier 1995; Jones and Smith 1995; DeRobertis and Sasai 1996), assuming an inversion in the designation of the DV axis in protostomes and deuterostomes, as already suggested by the French naturalist E. Geoffroy Saint-Hilaire in 1822. Other essential components of the genetic system regulating DV patterning of the early fly embryo are the Dpp and Scw receptors Thick veins and Saxophone (Brummel et al. 1994; Neul and Ferguson 1998; Nguyen et al. 1998a), two serine/threonine type I transmembrane receptors with high similarity to vertebrate Bmp receptors; the type II receptor Punt (Ruberte et al. 1995); the intracellular Bmp signaling transducers Mothers against Dpp (Mad; Newfeld et al. 1996) and Medea (Das et al. 1998), which are homologous to the vertebrate Smad proteins Smad1/5/8 and Smad4, respectively; and the metalloprotease Tolloid, which acts as an inhibitor of Sog/Chordin and shows high homology to the vertebrate members of the BMP1/Tolloid family (for review, see Mullins 1998).

2 Mutant Analyses of Vertebrate DV Regulators

The relevance of the fly genes mentioned above for normal DV pattern formation is demonstrated in loss-of-function mutants. In vertebrates, mutant analyses of the homologous genes have been carried out in mice, using gene targeting. However, results obtained in such experiments have left unclear the roles of these genes in DV patterning (for review, see Beddington and Robertson 1999). For Bmp4, a type II and two Bmp type I receptors, Alk2 (ActRIA) and Alk3 (BmpRI), a general requirement prior to, or early in, gastrulation has precluded analysis of the roles of these genes in the later process of DV patterning (Mishina et al. 1995, 1999; Winnier et al. 1995; Gu et al. 1999; Beppu et al. 2000). A small fraction of Bmp4 mutant embryos and chimeric Alk2 mutants survive beyond this stage and display truncations of the posterior axis, which are difficult to interpret but may reflect a dorsalization of the embryo (Winnier et al. 1995; Gu et al. 1999; Mishina et al. 1999). Mutations in Bmp2 (Zhang and Bradley 1996), Bmp7 (Dudley et al. 1995; Luo et al. 1995), the intracellular Bmp signal transducers Smad5 (Chang et al. 1999; Yang et al. 1999a) and Smad1 (Trembley et al. 2001), the Bmp antagonist Chordin (Bachiller et al. 2000), and the putative Chordin inhibitors Bmp1 and Tll1 (Suzuki et al. 1996; Clark et al. 1999) lead to various extraembryonic or later mutant phenotypes in the mouse, whereas early DV patterning appears largely unaffected.

In contrast, mutants with distinct defects in DV patterning have been identified in the zebrafish (Hammerschmidt et al. 1996a; Mullins et al. 1996; Solnica-Krezel et al. 1996). During the past 5 years, the molecular nature of most of these mutations has been resolved, demonstrating the essential role of Bmp signaling during zebrafish DV pattern formation. In the following, we will briefly introduce the various zebrafish mutants, followed by a more detailed description of crucial findings, which helped formulate the current model of DV pattern formation.

Of more than 5000 zebrafish mutants identified during various large- and small-scale mutant screens, approximately 2% display defects during early embryonic patterning. In contrast to the reverse genetics of the mouse, the molecular nature of those ethylnitrosourea-induced zebrafish point mutations is a priori unknown and must be determined in subsequent steps. Thanks to the enormous knowledge that had been acquired in parallel to the genetic approaches, many of the genes affected in the early pattern formation mutants have been identified using the candidate gene testing approach.

3 Zebrafish DV Mutants

Zebrafish DV mutants fall into ten complementation groups. They all display normal mesoderm induction, whereas patterning of the mesoderm and neuroectoderm is defective. In wild-type embryos, their phenotypes can be almost perfectly phenocopied by reducing or elevating the endogenous Bmp activity via injecting variable amounts of mRNA encoding BMPs or BMP inhibitors (Hammerschmidt et al. 1996b; Kishimoto et al. 1997; Nguyen et al. 1998b). The dose-dependence of these effects, together with the range of phenotypes observed in the various mutants, lead to the classification of the strengths of the phenotypes from weak (V1) to strong (V4) ventralization and from weak

(C1) to strong (C5) dorsalization (see Mullins et al. 1996; Hammerschmidt et al. 1996a; Kishimoto et al. 1997; Table 1; Fig. 1). These phenocopies and subsequent mRNA injection experiments with Bmp pathway components that rescued the mutants (Nguyen et al. 1998b) indicated that the DV mutant phenotypes were caused by mutations in genes encoding bone morphogenetic proteins or proteins involved in the transmission or modification of BMP signaling. This notion was confirmed by the subsequent identification of the molecular nature of the various mutants.

The five complementation groups with dorsalized phenotypes isolated during the large-scale screens in Tübingen and Boston (Mullins et al. 1996; Solnica-Krezel et al. 1996; Hammerschmidt et al. 1996a) resulted in the identification of five genes whose products are required for ventral development: the bone morphogenetic protein Bmp2b is mutated in *swirl* mutants (C5; Mullins et al. 1996; Kishimoto et al. 1997; Martínez-Barberá et al. 1997; Nikaido et al. 1997; Nguyen et al. 1998b); Bmp7 is defective in *snailhouse* (C5; Mullins et al. 1996; Dick et al. 2000; Schmid et al. 2000); the Bmp-activated transcription factor Smad5 is mutated in *somitabun* (C4; Mullins et al. 1996; Hild et al. 1999; Müller et al. 1999), the Bmp transmembrane receptor Alk8 is altered in *lost-a-fin* (C2; Mullins et al. 1996; Solnica-Krezel et al. 1996; Yelick et al. 1998; Bauer et al. 2001; Mintzer et al. 2001), and the metalloprotease Tolloid is encoded by *minifin* (C1; Connors et al. 1999). The *somitabun* mutation interacts genetically with both *swirl* and *piggytail* (Mullins et al. 1996), a sixth complementation group of dorsalizing mutations that was tentatively given an independent gene assignment. However, all *piggytail* mutations molecularly investigated thus far are alleles of *somitabun* /*smad5* (Connors et al. 1999; M.C. Mullins and M. Hammerschmidt, unpubl. obs.). Talbot and colleagues have recently isolated chromosomal deficiencies causing a very strong dorsalization (C5 or stronger), which is due to the loss of *vox* and *vent* (*vega1* and *vega2*), two homeobox-containing transcriptional repressors, which negatively regulate the expression of *chordin* (Kawahara et al. 2000a,b; Melby et al. 2000; Imai et al. 2001).

On the other hand, four complementation groups with ventralized mutant phenotypes define four genes required for dorsal organizer formation or function. The strongest ventralization (V4) is caused by the recently isolated, recessive maternal effect mutation *ichabod*, which leads to defects in dorsal organizer formation (Kelly et al. 2000). Although the molecular nature is not yet known, misexpression studies indicate that *ichabod* acts in a Wnt-signaling pathway to regulate β-catenin nuclear import (Kelly et al. 2000). A more downstream interference with organizer formation was revealed for the *bozozok* mutation (Solnica-Krezel et al. 1996). Bozozok could be assigned to the gene encoding the homeodomain protein Dharma/Nieuwkoid, which is expressed in response to β-catenin activation in a Wnt signal transduction pathway (Yamanaka et al. 1998; Fekany et al. 1999; Koos and Ho 1999). It appears to act as a transcriptional repressor, and may directly repress *bmp2b* transcription, in addition to other genes, in dorsal regions (Koos and Ho 1999; Fekany-Lee et al. 2000; W. Driever, pers. comm.). Two other ventralized mutants

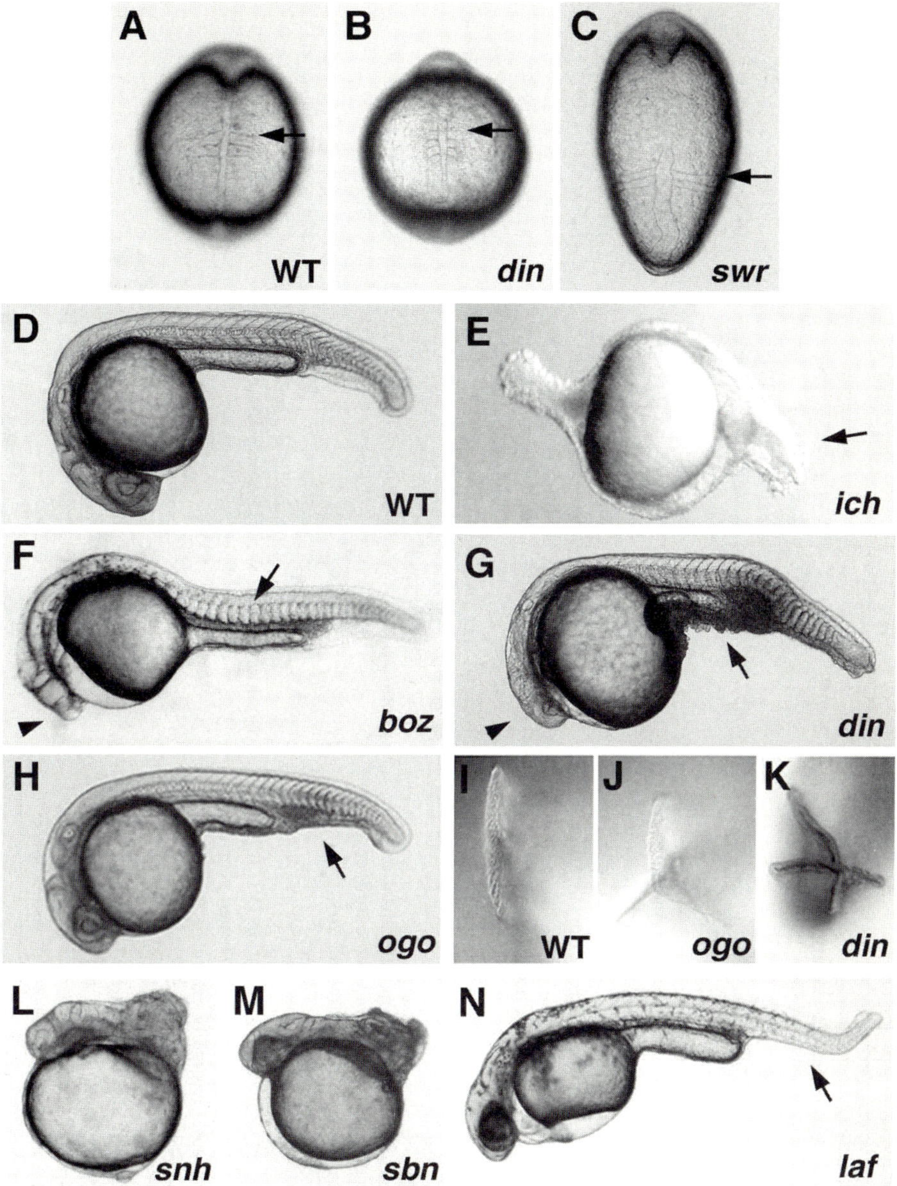

Fig. 1A–N. Phenotypes of ventralized and dorsalized zebrafish mutants. **A–C** 3-somite stage, dorsal view; **A** wild type; **B** din^{tt250}; **C** $swirl/swr^{ta72}$. The first somite is indicated by *an arrow*. **D–H**, **L–N** 28 h postfertilization (hpf), lateral view, anterior to the left; **I–K** 48 hpf, posterior view on tail. **D** Wild type. **E** $ichabod/ich^{p1}$. *Arrow* points to tail. **F** $bozozok/boz^{m168}$; *arrow* points to missing notochord, *arrowhead* to missing anterior head. **G** din^{tt250}; *arrow* points to larger tail with enlarged blood island, *arrowhead* to smaller head and eyes. **H** $ogon/ogo^{tm350}$. **I** Wild type. **J** ogo^{tm350}. **K** din^{tt250}. **L** $snailhouse/snh^{ty68a}$. **M** $somitabun/sbn^{dtc24}$, dominant maternal effect (offspring of heterozygous female). **N** $lost\text{-}a\text{-}fin/laf^{tm110b}$, zygotic effect

Table 1. Zebrafish mutants with defective DV patterning

Mutant	Phenotype	Gene	Reference mutant	Reference gene(s)
	Dorsalized			
swirl	Zygotic effect: strong dorsalization (C5); no maternal effect	*bmp2b*	Mullins et al. (1996) Kishimoto et al. (1997) Nguyen et al. (1998b)	Nikaido et al. (1996) Martínez-Barberá et al. (1997)
snailhouse	Zygotic effect: strong dorsalization (C5); no maternal effect	*bmp7*	Mullins et al. (1996) Dick et al. (2000) Schmid et al. (2000)	Dick et al. (2000) Schmid et al. (2000)
somitabun	Dominant zygotic effect: weak dorsalization (C1) Maternal-zygotic effect: strong dorsalization (C4)	*smad5*	Mullins et al. (1996) Hild et al. (1999)	Hild et al. (1999) Muller et al. (1999)
lost-a-fin	Zygotic effect: weak dorsalization (C1–C2) Maternal-zygotic effect: strong dorsalization (C5)	*alk8*	Mullins et al. (1996) Bauer et al. (2001) Mintzer et al. (2001)	Yelick et al. (1998) Bauer et al. (2001) Mintzer et al. (2001)
mini fin	Weak dorsalization (C1); no maternal effect	*tolloid*	Mullins et al. (1996) Connors et al. (1999)	Blader et al. (1997) Connors et al. (1999)
Df^{st7}, Df^{st8}	Strong dorsalization (C5 or stronger)	*vox* and *vent* (*vega1*, *vega2*)	Imai et al. (2001)	Kawahara et al. (2000a,b) Melby et al. (2000)
	Ventralized			
dino	Zygotic effect: intermediate ventralization (V3; reduced size of head and notochord; enlarged blood islands and ventral tail fin)	*chordin*	Hammerschmidt et al. (1996a)	Schulte-Merker et al. (1997)
	No or very weak maternal effect		Fisher et al. (1997) Schulte-Merker et al. (1997)	Miller-Bertoglio et al. (1997)
ogon (formerly also *mercedes*)	Zygotic effect: weak ventralization (V1; slightly enlarged blood islands; duplicated ventral tail fin; head only slightly smaller)	Not known	Hammerschmidt et al. (1996a)	
	Maternal-zygotic effect (V2)		Solnica-Krezel et al. (1996) Miller-Bertoglio et al. (1999)	
ichabod	Maternal effect: ventralization of variable strength, up to V4 (complete loss of head and notochord)	Not known	Kelly et al. (2000)	
bozozok	Zygotic effect: variable; strong reduction of head and notochord, while ventral tissues only moderately enlarged	*dharma*/*nieuwkoid*	Solnica-Krezel et al. (1996)	Yamanaka et al. (1998)
	Maternal-zygotic effect slightly stronger (Fekany-Lee et al. 2000)		Fekany et al. (1999) Koos and Ho (1999)	Koos and Ho (1998)

do not exhibit defects in the formation, but rather in the function of the organizer: *dino*, which corresponds to zebrafish Chordin, a Bmp inhibitor (Hammerschmidt et al. 1996a; Fisher et al. 1997; Miller-Bertoglio et al. 1997; Schulte-Merker et al. 1997), and the more weakly ventralizing *mercedes/ogon* mutations (V1) whose molecular nature remains unknown (Hammerschmidt et al. 1996a; Solnica-Krezel et al. 1996; Miller-Bertoglio et al. 1999).

A summary of the phenotypes of the various mutants is given in Table 1, mutant embryos are shown in Figs. 1 and 5.

4 Different Phases of DV Pattern Formation

Previous models of DV patterning had proposed the existence of different phases of DV pattern formation. Although these models were partly revised, taking into consideration novel findings in zebrafish and *Xenopus*, the basic principles of the mechanism still hold true. Thus, a coarse DV pattern is set up during Spemann-Mangold organizer formation, induced by factors generated by the mother and deposited into the egg (maternal control). In a second phase, and primarily through zygotic control, this initial pattern is refined. This involves ventralizing signals promoting the maintenance of the initial ventral specification and antagonizing dorsalizing factors generated by the organizer that establish dorsal cell fates. An interaction between the ventralizing Bmp signals and the dorsalizing Bmp antagonistic factors generates a Bmp morphogen gradient that ultimately (in a third phase) leads to the specification of a diverse set of cell types dependent on the interpretation of the graded Bmp signal. In the following, we will discuss the different phases of DV patterning in more detail, focusing on the various zebrafish studies that led to the current model of DV pattern formation.

4.1 Phase 1: Establishment of the Spemann-Mangold Organizer

Let us start this section with a reference to mutants described in another chapter (see Kimelman and Schier, this Vol.), which do not lack Spemann's organizer, although this might have been expected based on older models of organizer formation (see below). These mutants are deficient in the Nodal-signaling proteins Ndr1/Squint and Ndr2/Cyclops or the Nodal-coreceptor One-eyed pinhead and lack nearly all mesoderm, including dorsal mesoderm (as indicated by absence of *goosecoid* transcripts). Nevertheless, they show expression of the dorsal marker *chordin*, and develop relatively normal neural tissue – which was thought to be induced by signals from the organizer. From these findings, two conclusions can be drawn. First, dorsal and ventral mesoderm are induced by the same signals, and, secondly, a dorsal axis can be induced largely independently of mesoderm and mesoderm-inducing factors.

Thus far, two zebrafish mutants have been described in which organizer formation is affected, *ichabod* and *bozozok*. Both mutants point to a requirement of the canonical Wnt signal transduction pathway for organizer formation (see also Hibi et al., this Vol.). Forced ventral expression of Wnt signals or components of the canonical Wnt signal transduction pathway in zebrafish and *Xenopus* embryos induces secondary body axes (e.g., Kelly et al. 1995a,b). However, none of the *wnt* genes cloned thus far displays an expression pattern consistent with such a role. Thus, it has been suggested that the Wnt signal transduction pathway may be activated independently of a Wnt ligand. The key cytoplasmic protein mediating Wnt signaling is β-catenin. Free cytoplasmic β-catenin is normally phosphorylated by the kinase GSK3β, marking it for ubiquitinylation and degradation by proteasomes (Aberle et al. 1997). Activation of the pathway by Wnt signaling, however, inhibits GSK3, thereby stabilizing β-catenin, which forms complexes with LEF/TCF HMG transcription factors and translocates to the nucleus to regulate gene expression. In wild-type zebrafish embryos, nuclear β-catenin protein can be observed in dorsal blastomeres from mid-blastula stages onwards (Schneider et al. 1996). In *ichabod* mutants, nuclear translocation of β-catenin does not occur (Kelly et al. 2000). Using forced expression of various components of a Wnt-signaling pathway, it was shown that *ichabod* mutants still respond to β-catenin, but not to any of the upstream components of the Wnt-signaling pathway (Kelly et al. 2000). The β-catenin gene is not mutated in *ichabod*, indicating that *ichabod* encodes a (thus far) unidentified positive component of the Wnt signal transduction pathway required for the nuclear transport of β-catenin. The effect of the *ichabod* mutation is strictly maternal, consistent with the notion that the first phase of DV patterning is under maternal control. Mutant embryos fail to form the Spemann-Mangold organizer, as indicated by the absence of organizer markers like *chordin* and *goosecoid*. During later stages of development, they are strongly ventralized, lacking axial mesoderm (prechordal plate and notochord) and anterior neural tissues, whereas ventral-posterior structures like blood and tail are strongly enlarged. A weaker ventralization, anticipated by the formation of a smaller organizer, has been obtained upon overexpression of a dominant negative version of FzA, a putative Wnt receptor of the Frizzled family (Nasevicius et al. 1998). In contrast, enlarged or ectopic organizers, followed by a general dorsalization of the embryos, were achieved by antisense morpholino oligonucleotide (MO)-mediated inactivation of Axin or Diversin, two maternally supplied negative components of the canonical Wnt signal transduction pathway (Heisenberg et al. 2001; Schwarz-Ramond et al. 2002). Altogether, these data strongly suggest that formation of the dorsal organizer in zebrafish embryos is governed under maternal control and by canonical Wnt signaling. It has not been determined whether the positive effect of Wnt signaling on *chordin* expression is direct or indirect. Promoter analyses and transactivation assays will be necessary to address this question. Such analyses, however, have revealed a direct Wnt/Lef dependence for the expression of Bozozok/Dharma/Nieuwkoid (Koos and Ho 1998; Yamanaka et al. 1998) in the organizer and the underlying dorsal yolk syncytial layer (Ryu et al. 2001).

In contrast to *ichabod*, *bozozok* mutations have a predominantly zygotic effect. Analyses with Bozozok fusion proteins have revealed that it acts as a transcriptional repressor. One of its targets is the *bmp2b* gene, which contains Bozozok binding sites (W. Driever, pers. comm.). While in wild-type pregastrula embryos *bmp2b* and *bmp7* expression is excluded from the organizer region, *boz* mutants display uniform *bmp2b/7* expression throughout the entire blastoderm (Koos and Ho 1999; Fekany-Lee et al. 2000; W. Driever, pers. comm.). In parallel, *boz* has a positive effect on the expression of organizer-specific genes like *chordin*. Thus, *chordin* transcripts are strongly reduced in *boz* mutants at blastula stages, and completely absent in *boz* offspring of *boz* mutant females (Fekany-Lee et al. 2000). Given that Bozozok is a transcriptional repressor, its positive effect on *chordin* expression cannot be direct. Rather, it is more likely that *boz* represses the expression of other factors such as Wnt8 or Vox/Vent, which may negatively regulate *chordin* transcription (Fekany-Lee et al. 2000; Imai et al. 2001).

In addition to *bozozok*, other organizer-specific genes may be activated by Wnt signaling, such as *admp*, encoding Anti-dorsalizing morphogenetic protein, a Bmp-like signaling protein most similar to Bmp7. In contrast to *bozozok* mutants, MO-mediated *admp* knockdown leads to enlarged organizers, indicating that within the organizer, negative feedback loops are established to restrict the size of the organizer after it has been induced (Lele et al. 2001b; Willot et al. 2001).

Using embryological manipulation experiments, Mangold and Spemann defined the organizer of an amphibian embryo to the upper dorsal blastopore lip, which gives rise to axial mesoderm (notochord and prechordal plate; Spemann and Mangold 1924). Loss of this amphibian organizer region results in loss of dorsal axis formation and dorsoventral patterning. In zebrafish, transplantation of the blastopore lip equivalent, the dorsal embryonic shield, can also induce a secondary axis (Shih and Fraser 1996; Saúde et al. 2000). However, in contrast to the amphibian embryo, surgical or genetical removal of the shield in the zebrafish is characterized by only a loss of dorsal midline mesoderm, with relatively minor defects in dorsoventral pattern formation (Shih and Fraser 1996; Saúde et al. 2000). These differences between the zebrafish and amphibian may be accounted for by the more extensive expression of the Bmp antagonist *chordin* beyond the shield/organizer, within paraxial mesodermal and neuroectodermal tissues in the zebrafish (Fig. 2J; Miller-Bertoglio et al. 1997). Support for this notion comes from double mutant analyses. The *boz* mutation results in specific loss of the dorsal embryonic shield/organizer and causes a phenotype identical to that caused by shield extirpation (Shih and Fraser 1996; Saúde et al. 2000; Fekany-Lee et al. 2000). In contrast, *boz;din* double mutants display a much stronger ventralization (V4; Gonzalez et al. 2000), similar to that observed by blastopore lip removal in *Xenopus*. Thus loss of *boz* or the organizer, in addition to loss of the non-axial dorsal expression of *chordin* in the zebrafish, results in a similar effect as organizer removal in amphibians. The function of Chordin protein in axial and non-axial regions appears to be identical: to antagonize Bmps present in dorsal and lateral regions of mesoderm and ectoderm, as also revealed in *din*/wild-type chimeric analyses (Hammerschmidt et al. 1996b). Hence it is tempting to suggest that the functions attributed to the amphibian organizer extend beyond the dorsal embryonic shield in the zebrafish and encompass non-axial dorsal regions.

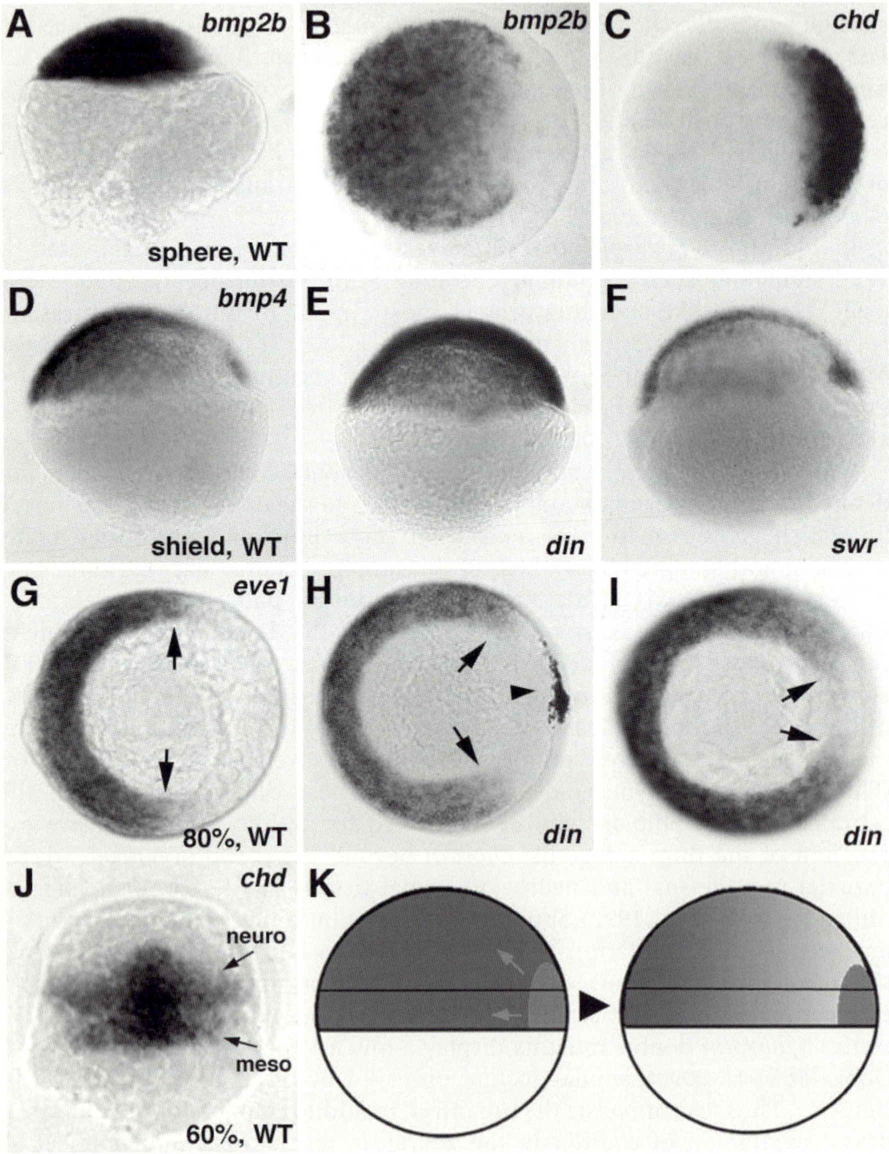

4.2 Phase 2: Establishment of the Morphogenetic Bmp Gradient

In addition to the described DV mutants and morphants with altered organizer formation, DV mutants were isolated that display a rather normal development of the organizer and its derivatives, whereas regions outside the organizer are affected, pointing to specific defects in organizer function rather than in organizer formation. These are the ventralized mutants *dino* (*din*) and *ogon/mercedes* (*ogo*; Hammerschmidt et al. 1996a; Solnica-Krezel et al. 1996; Fisher et al. 1997; Miller-Bertoglio et al. 1999), and the dorsalized mutants *swirl* (*swr*), *snailhouse* (*snh*), *somitabun* (*sbn*), *lost-a-fin* (*laf*) and *mini fin* (*mfn*; Mullins et al. 1996; Solnica-Krezel et al. 1996). All these mutants show normal expression of early organizer markers such as *goosecoid* and *chordin*. Early expression of *bmp2b/7* in ventral and lateral regions of the embryo is also normal, indicating that the early events establishing the DV axis and the organizer are normal in these mutants (for wild type, see Fig. 2A–C).

However, progressive alterations along the DV axis are observed during the further course of development in these dorsalized and ventralized mutants. In wild-type embryos, *bmp2b/7* expression is progressively lost in dorsal regions, leading to a DV gradient of *bmp2b/7* transcripts, which is most prominent at the onset of gastrulation and comprises both the presumptive mesoderm in the marginal region and the presumptive ectoderm in the animal region of the embryo (Fig. 2D). In the ventralized *din/chordin* mutants, this progressive retraction of the *bmp2b/7* expression domain is affected, resulting in the maintenance of high levels of *bmp2b/7* transcripts in dorsal-lateral regions (Fig. 2E; Hammerschmidt et al. 1996b; M. Hammerschmidt, unpubl. observ.). The dorsalized *swr/bmp2b* and *snh/bmp7* mutants, on the other hand, display a general and uniform loss of *bmp2b/7* transcripts along the entire DV circumference

Fig. 2A–K. Establishment of the morphogenetic Bmp gradient. **A–C** *bmp2b* and *chordin* expression at sphere stage (late blastula) in wild-type embryos; **A** *bmp2b*, lateral view, dorsal right; **B** *bmp2b*, animal view, dorsal right; **C** *chordin*, animal view, dorsal right. **D–F** *bmp4* expression at shield stage (early gastrula); lateral views, dorsal right; **D** wild type; **E** *din*^{tm84}; **F** *swr*^{ta72}. *bmp2b* and *bmp7* show a very similar graded pattern in wild type, and a corresponding increase and decrease in *dino* and *swirl* embryos, respectively. **G–I** Chordin has a long-range signaling effect. Labeled wild-type cells were transplanted into the shield of a *dino* mutant embryo; embryo was fixed 2 h after transplantation and stained for *eve1* mRNA. Embryos are shown at 80% epiboly, animal view, dorsal right. Dorsal borders of *eve1* expression domain are indicated by *arrows*. **G** Wild-type control; **H** wild-type/*dino* chimerae, wild-type cells are indicated by *arrowhead*; **I** *din*^{tm84} control. **J** *chordin* is expressed both in the dorsal mesoderm (*meso*) and in the neuroectoderm (*neuro*); 60% epiboly, dorsal view. **K** Scheme of Bmp gradient formation. Lateral view on sphere-stage (*left*) and shield-stage (*right*) embryo; dorsal right. *Arrows* indicate long-range signaling by Chordin and other Bmp antagonists. It is currently unclear whether the Bmp gradient in a late -blastula/early-gastrula embryo does indeed cover the entire DV circumference, or whether it is confined to the dorsal half of the embryo. In the latter case, ventral regions might become patterned during later-gastrula or post-gastrula stages, when the distance to the Bmp antagonists generated in the dorsal axis has become smaller

(Fig. 2F; Kishimoto et al. 1997; Nguyen et al. 1998b; Dick et al. 2000; Schmid et al. 2000). This indicates that Bmp2b and Bmp7 are required for maintenance of their own transcription. Since the *swr/snh* double mutant is not stronger than *swr* or *snh* single mutants, and since *bmp2b* and *bmp7* show a strong synergistic effect upon overexpression, they appear to act together, perhaps as Bmp2b/Bmp7 heterodimers (Dick et al. 2000; Schmid et al. 2000). Their function requires signaling via the novel type I serine/threonine kinase transmembrane receptor Laf/Alk8 and the receptor-activated transcription factor Sbn/Smad5. Thus, loss of *bmp2b/7* expression is observed in *sbn/smad5* mutants (Nguyen et al. 1998b; Schmid et al. 2000) and forced *bmp* expression in *sbn/smad5* mutants cannot restore the DV gradient of *bmp2b/7* transcripts (Hild et al. 1999). Similarly, *laf/alk8* mutants do not respond to forced *bmp2b/7* expression, whereas they do respond to *smad5* (Bauer et al. 2001; Mintzer et al. 2001). Furthermore, *laf* mutants display a strong hypophosphorylation of Smad1/5 (Mintzer et al. 2001). These findings, together with the observation that constitutively active Alk8 can rescue *bmp2b* and *bmp7*, but not *smad5* mutants (Bauer et al. 2001), shows that Alk8 does indeed act as a Bmp receptor. This was not necessarily expected, since Alk8 is quite distinct from the Bmp receptors identified thus far, Alk3/BmpRIa and Alk6/BmpRIb, and more similar, but not orthologous to, Alk1 and Alk2, which appear to function as Tgfβ and Bmp receptors, respectively.

The Chordin protein generated from the organizer can antagonize Bmp signaling at a distance. Transplantation of wild-type cells into the organizer of *din/chordin* mutant embryos leads to a normalization of DV patterning even in lateral regions approximately 50 cell diameters away from the Chordin-generating cells, indicating that Chordin acts in a non-cell-autonomous fashion with a long-range signaling effect (Fig. 2G–I; Hammerschmidt et al. 1996b). Bmps themselves may have a comparably short-range signaling effect, affecting only the Bmp-generating cell itself and direct neighbors (Kishimoto et al. 1997; Nguyen et al. 1998b; Nikaido et al. 1999). Although immunostainings will be necessary to confirm this notion, we propose that the Bmp gradient is set up by the antagonistic interplay of Chordin and Bmps in the following way (Fig. 2K). Chordin, generated in a distinct domain in the dorsal corner of the pre-gastrula embryos, diffuses into dorsolateral regions where it binds Bmp2b and Bmp7, initially uniformly present in the embryo (with the exception of the organizer region itself). Thus, a DV gradient of Bmp activity is generated, which is then further consolidated by the hypothesized Bmp autoregulation on *bmp2/7* transcription, leading to the observed DV gradient of *bmp2/7* mRNA levels during early gastrulation. The role of Chordin in patterning the DV axis through inhibition of Bmp2b/7 is supported by epistasis analyses. Thus, *swirl/dino* and *snailhouse/dino* mutants exhibit the same phenotype as *swirl* or *snailhouse* single mutants, respectively, showing that *bmp2b* and *bmp7* are epistatic to *chordin*, consistent with Chordin affecting DV patterning indirectly via Bmp2b/7 (Hammerschmidt et al. 1996b; Dick et al. 2000). A similar function to prevent positive Bmp/Dpp autoregulation has been described for the

Chordin homologue Short gastrulation during DV patterning of the *Drosophila* embryo (Biehs et al. 1996).

It is known that other factors in addition to Chordin are involved in Bmp gradient formation during zebrafish DV patterning. Thus, there is substantial evidence that Chordin acts in concert with other Bmp inhibitors. The fact that *dino* mutants still form anterior somites and neuroectoderm indicates that inhibition of Bmps is not completely abolished by the loss of Chordin. Thus, it appears that other Bmp inhibitors act in concert with Chordin during DV patterning, including possibly Noggin1 and 2, both of which are present in the organizer (Fürthauer et al. 1999). Another inhibitor is encoded by the *ogon* gene, whose molecular nature is unknown. The *swirl/snailhouse,* and *lost-a-fin* mutations are epistatic to ogon, like *dino/chordin* mutations (Miller-Bertoglio et al. 1999; Wagner and Mullins 2002), consistent with Ogon acting upstream of Bmp2b/7, possibly as a Bmp antagonist. Further support for such a role is found in the ability of forced Noggin and Chordin expression to rescue the *ogon* mutant phenotype (Miller-Bertoglio et al. 1999). Moreover, *ogon/dino* double mutants are more strongly ventralized than *dino* or *ogon* single mutants (Hammerschmidt et al. 1996a). However, a still stronger ventralization is observed in *dino/bozozok* double mutants, likely through the indirect regulation of *boz* on a third Bmp inhibitor (Gonzalez et al. 2000). A dual role may be played by the secreted factor Twisted gastrulation (Tsg), in antagonizing and promoting Bmp signaling through complex formation with Chordin and Bmp ligands (Larrain et al. 2000; Oelgeschläger et al. 2000; Chang et al. 2001; Scott et al. 2001). MO-mediated *tsg* knockdown leads to a mild ventralization, and synergistically enhances the ventralization caused by hypomorphic *chordin* inactivation (Ross et al. 2001).

Further regulation of Bmp signaling is mediated by the dorsolaterally expressed *vox* and *vent* genes, which encode transcriptional repressors negatively regulating expression of *chordin* and other organizer genes (Kawahara et al. 2000a,b; Melby et al. 2000). *vox/vent* double mutants differ from *bmp2b* and *bmp7* mutants in that they display a strong ventral expansion of the *chordin* expression domain already at pre-gastrula stages (Imai et al. 2001), indicating that *vox* and *vent* negatively regulate *chordin* expression directly, independent of Bmp2b/7 signaling. Bmp2b and Bmp7 appear to be required only later for the maintenance of *vox/vent* expression during gastrulation to keep Chordin and other putative Bmp antagonists restricted to dorsal regions of the embryo.

Additional factors regulating *bmp2b/7* expression are the Fibroblast growth factors Fgf8 and Fgf3 and their cytoplasmic antagonist Sprouty 4 (Fürthauer et al. 1997, 2001). All three genes are transiently expressed with specific DV patterns in the marginal zone of pre- and early-gastrula embryos. *fgf3* and *fgf8* knockdown and *spry4* overexpression cause reduced *bmp2b* expression and a general dorsalization, whereas *spry4* knockdown and *fgf3/8* overexpression lead to enhanced *bmp2b* expression and a general ventralization (Fürthauer et al. 1997, 2001). This indicates that Fgf3 and Fgf8 act as negative regulators of *bmp2b/7* expression before gastrulation, thereby regulating DV patterning of the zebrafish embryo (rather than being involved in mesoderm induction, as previously believed from results obtained in amphibia). How early *fgf3* and *fgf8* expression itself is regulated is currently unclear. Finally, inhibitory *smads* (*smad6*, *smad7*) are expressed during blastula and gastrula stages, and might antagonize Bmp signaling in target cells (M. Hammerschmidt, unpubl.).

4.3 Phase 3: Morphogenetic Interpretation of the Gradient by Target Cells

The question arises as to when the Bmp signal begins to be interpreted and how this is regulated. RT-PCR and in situ hybridization analyses indicate that *bmp2b*, *bmp7*, *smad5*, *alk8* and all other known Bmp type I receptors are expressed before *chordin*, and therefore probably before the Bmp gradient is established. Thus, premature interpretation of Bmp signaling at blastula stages could lead to ventralized embryos. Several mechanisms are conceivable for how the onset of morphogenetic action could be controlled. For example, it may be necessary that a maternally supplied signaling inhibitor be diluted out prior to interpretation of the Bmp signal. Such a mechanism could occur through the presence in zebrafish blastulas of maternally supplied *smad6*, an inhibitory Smad (M. Hammerschmidt, unpubl.). Alternatively, or in addition, positive cofactors of Bmp signaling could be recruited after the Bmp gradient has been set up. For example, Smad1, a Bmp-activated transcription factor similar to Smad5, is expressed shortly after the onset of gastrulation. Unlike *smad5*, which shows uniform maternal and zygotic expression, *smad1* expression is restricted to the ventral side of the embryo, and depends on Bmp2b, Bmp7 and Smad5 activity (Dick et al. 1999). Interestingly, recent DNA-binding studies have revealed crucial differences in the binding specificities of Smad1 and Smad5 (Li et al. 2001). Thus, by recruiting Smad1, additional target genes might become accessible to Bmp signaling during gastrula stages.

Mutant analysis provides strong evidence for a morphogenetic action of the Bmp gradient both in the mesoderm and the ectoderm of the zebrafish embryo. According to the fate map of the early gastrula embryo (Fig. 3A; Kimmel et al. 1990), ventral mesodermal cells give rise to blood, ventrolateral

Fig. 3A–H. Interpretation of the morphogenetic Bmp gradient in mesoderm. **A** Fate map of early gastrula embryo. Modified from Kimmel et al. (1990). **B–G** *dino* mutants have more blood precursor cells and fewer muscle cells forming anterior somites; *swirl* mutants lack blood precursors, while anterior somites are laterally expanded and fused on the ventral side of the embryo. Embryos of 8-somite stage, dorsal view; **B–D** *myoD* expression in somites; **E–G** *gata1* expression in blood precursors; **B, E** wild type; **C, F** *dintm84*; **D, G** *swrta72*. **H** Diagram linking shifts in Bmp2b/7-signaling levels in *dino* and *swirl* mutants to shifts in patterning of the mesoderm. *x*-axis shows positions along DV axis, *y*-axis Bmp concentrations and corresponding cell fates

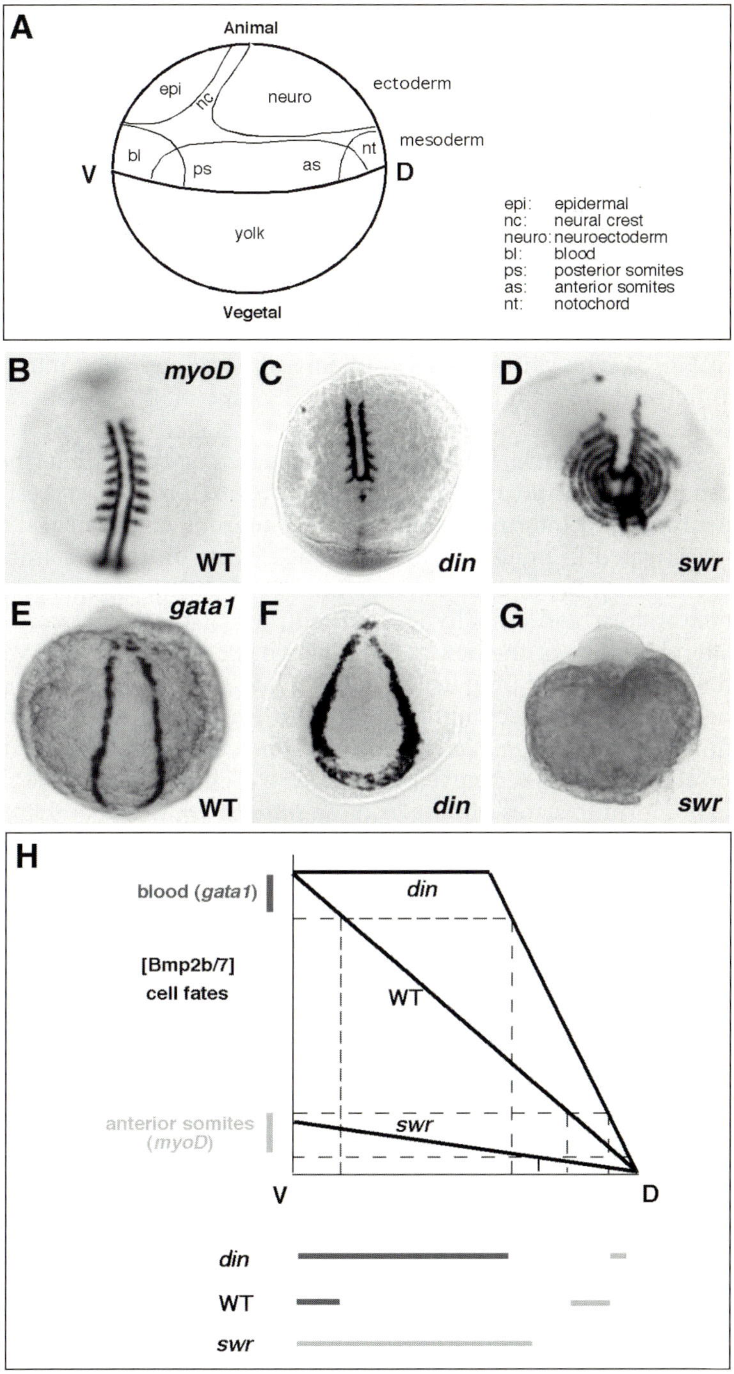
A
Animal
epi
nc
neuro
ectoderm
mesoderm
bl
ps
as
nt
V
D
yolk
Vegetal
epi: epidermal
nc: neural crest
neuro: neuroectoderm
bl: blood
ps: posterior somites
as: anterior somites
nt: notochord
B myoD
WT
C
din
D
swr
E gata1
WT
F
din
G
swr
H
blood (gata1)
[Bmp2b/7]
cell fates
anterior somites
(myoD)
din
WT
swr
V
D
din
WT
swr

cells to pronephros and posterior somites, dorsolateral cells to anterior somites, and dorsal-most cells to notochord and prechordal plate. Dorsalized mutants show a general enlargement of tissues derived from dorsolateral regions of the fate map, whereas tissues derived from ventral cells are reduced or absent. Thus, *swirl/bmp2b* mutants lack blood and pronephric precursors, while anterior somites are laterally enlarged and fused on the ventral side of the embryo (Fig. 3D,G). On the other hand, *dino/chordin* mutants have a significantly enlarged blood and pronephric precursor population, while anterior somites are reduced (Fig. 3B,C,E,F).

The link between the shifts in the Bmp gradient at early-gastrula stages, and the later shifts in the organization of mesodermal derivatives, are illustrated in Fig. 3J. The shifted Bmp gradients of *swirl* and *dino* mutants (Fig. 2D–F) anticipate the quantitative shifts in the size of DV-specific cell populations (Fig. 3B–G).

A similar conclusion about a morphogenetic function of the Bmp gradient can be drawn from analyses of the ectoderm. Here, neural crest, dorsal sensory neurons, and intermediate-positioned neurons of the neural tube have been investigated in mutants with different extents of dorsalization (Nguyen et al. 1998b, 2000; Barth et al. 1999; Schmid et al. 2000). The progenitor cells of the neural crest and dorsally located neurons of the spinal cord originate from lateral regions of the neural plate, which are adjacent to the ventral ectoderm that expresses the Bmp ligand genes. Thus, these cells are in a region that may be under the direct influence of the Bmp gradient. In complete loss-of-function states of *bmp2b/swirl* or *bmp7/snh*, the neural crest is strongly reduced or absent (Fig. 4B; Nguyen et al. 1998b; Schmid et al. 2000). In contrast, in hypomorphic states of *bmp7/snh* or *smad5/sbn*, the neural crest is expanded, up to 5-fold compared to wild type (Fig. 4C; Nguyen et al. 1998b; M.C. Mullins, unpubl.). The opposite phenotypes (expansion versus loss) observed in a hypomorphic versus null state can be explained if a Bmp gradient acts in the specification of the neural crest, as illustrated in Fig. 4J. If a particular, low Bmp-signaling activity is required for neural crest specification, then complete

Fig. 4A–J. Interpretation of the Bmp morphogenetic gradient in ectoderm. **A–I** Shifts in expression patterns of *fkd6* (neural crest), *tlx3* (trigeminal placode and dorsal neurons) and *lim1* (interneurons) in *swirl* and *somitabun* mutants. **A, D, G** Wild type; **B, E, H** swr^{ta72}; **C, F, I** sbn^{dtc24}, maternal effect. **A–C** *fkd6*, 5-somite stage, lateral view, anterior up; **D–F** *tlx3*, 7-somite stage, lateral view, anterior up; note the large trigeminal placode (*TGP*) expansion in the *somitabun* mutant (**F**), in contrast to their absence in the *swirl* mutant (**E**). **G–I** *lim1*, 7-somite stage, lateral view, anterior left, except (**G**) anterior up. **J** Diagram linking shifts in Bmp-signaling levels in *swirl* and *somitabun* mutants and in *swirl* mutants injected with *chordin* mRNA to shifts in ectodermal patterning of the neural crest, trigeminal placode and interneurons. *x*-axis shows positions along DV axis, *y*-axis Bmp activity levels and corresponding cell fates. Both the shifts in the size of the different cell populations and the shifts to ectopic locations can be explained by a response of the different cell types to the morphogenetic Bmp gradient. *TGP* Trigeminal placode; *DSN* dorsal sensory neurons; *IN* interneurons; *N* notochord; *P* pronephros

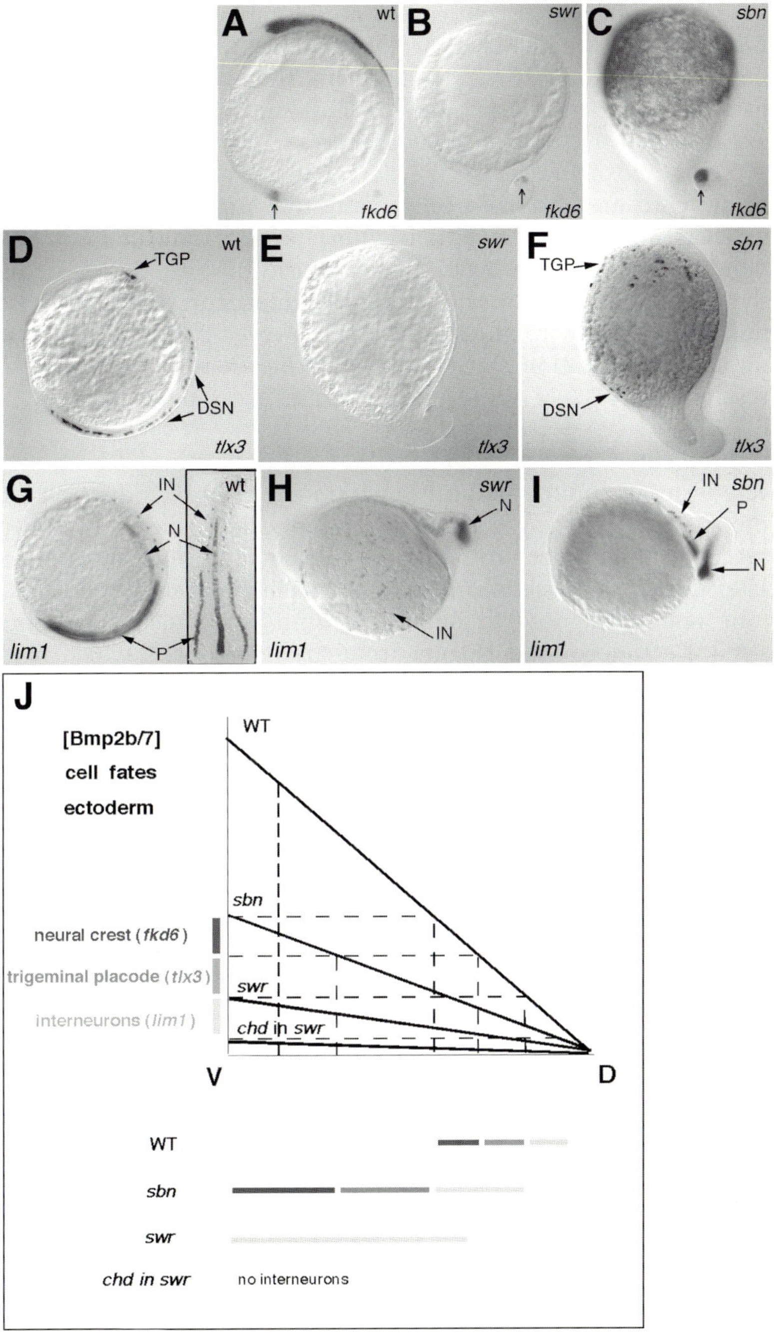
A
wt
fkd6
B
swr
fkd6
C
sbn
fkd6
D
wt
TGP
DSN
tlx3
E
swr
tlx3
F
sbn
TGP
DSN
tlx3
G
IN
wt
N
P
lim1
H
swr
N
IN
lim1
I
IN
sbn
P
N
lim1
J
[Bmp2b/7]
cell fates
ectoderm
WT
sbn
swr
chd in swr
neural crest (fkd6)
trigeminal placode (tlx3)
interneurons (lim1)
V
D
WT
sbn
swr
chd in swr
no interneurons

absence of the signal will lead to no neural crest formation. However, when Bmp activity is reduced uniformly in ventral regions due to the hypomorphic nature of a Bmp pathway component, then a large ventral domain of the gastrula may be subject to the low Bmp-signaling activity, leading to a large expansion in this case of the neural crest.

Similar to neural crest, the dorsal sensory Rohon Beard neurons of the spinal cord are absent in Bmp pathway component null mutants (Fig. 4E), whereas intermediate-positioned lim1-positive interneurons are greatly expanded (Nguyen et al. 2000). In contrast to the situation described above for the neural crest and dorsal neurons, the interneuron expansion is observed in null mutants of *swirl/bmp2b* (Fig. 4H). However, BMP signaling is also required for their specification, since injection of *chordin* mRNA into *swirl* null mutants results in a loss of these interneurons (Nguyen et al. 2000; illustrated in Fig. 4J).

Altogether, these data suggest that the patterning of both mesoderm and ectoderm, including patterning within the neuroectoderm, is driven by the Bmp gradient during gastrulation, which likely occurs by a general planar mechanism. This of course does not rule out later vertical patterning processes after the axis has been formed. In particular, neural crest cells and the neural tube remain exposed to Bmps and Bmp antagonists (e.g., Bmps from the surface ectoderm and Noggin from the notochord), which might refine or maintain the initial patterns set up during gastrulation.

It remains unclear how the Bmp morphogenetic signaling gradient is interpreted. We know, as described above, that the type I receptor Alk8/Laf and Phospho-Smad5/Sbn play major roles. Target cells within the Bmp gradient receive different levels of Bmp signals depending on their DV position, leading to differential activation of the Bmp receptor and consequently differential amounts of Phospho-Smad protein, which then acts to regulate transcription in a dose-dependent manner. According to the morphogen concept, different target promoters are expected to require different levels of a phospho-Smad complex for their transcriptional activation (see also Sect. 7).

5 Implications of DV Patterning on the Anteroposterior Axis

There are aspects of DV and AP patterning that are both independent and interdependent. In the Bmp pathway mutants just described, AP patterning is unaffected in rostral regions. Thus, although ventral regions of the embryo are absent in strongly affected mutants, AP patterning of dorsal tissues is maintained strikingly well (Barth et al. 1999; Nguyen et al. 2000). Hence, rings of neural tissue with distinct AP characters form from anterior to posterior in these mutants.

An interdependence of the two axes is also observed. Thus, strongly dorsalized mutants like *swirl* display a general enlargement of the anterior region at the expense of the posterior region of the embryo, while the ventralized *dino* mutant shows the opposite shift (Fig. 1A–C). These defects might result from a general linkage of DV patterning and dorsal convergence and extension cell movements that form the axis. It has been shown that the ventral-most cells do not converge dorsally, but remain on the ventral side and migrate to the tailbud, giving rise to posterior structures (Myers et al. 2002; see also below). Thus, ventralized mutants develop an enlarged tailbud, whereas, in dorsalized mutants, the tailbud is strongly reduced in size.

A link between the DV and AP systems is observed in the analysis of *wnt8* function. The bicistronic *wnt8* locus is expressed in a complex manner in marginal regions of the gastrula. At mid-gastrula stages (70% epiboly), it is preferentially expressed in the ventral and lateral margin, and is excluded from the dorsal midline. Both gain- and loss-of-function experiments have shown that *wnt8* has a posteriorizing effect on neuroectoderm. *wnt8-1,2* mutants and *wnt8* morphants display a significantly enlarged forebrain region, while hindbrain and spinal cord formation is inhibited (Erter et al. 2001; Lekven et al. 2001). Moreover, they exhibit a dorsalization of tissues as evident by an expansion of dorsal and dorsolateral tissues. The relationship between the Wnt8- and Bmp-signaling pathways is unclear. Bmp signaling does not appear to positively regulate *wnt8* expression (Mullins 1999). However, Wnt8 appears to promote Bmp signaling, as Wnt8 inhibition leads to increased *chordin* expression in wild-type and in *bozozok* mutant embryos (Fekany-Lee et al. 2000). In addition, Bozozok negatively regulates expression of both *wnt8* and the *bmp* genes in dorsal regions (Fekany-Lee et al. 2000). These regulatory relationships can explains why *wnt8* and *bozozok* mutants have much more pronounced head defects than *bmp* and *chordin* mutants (Fekany-Lee et al. 2000; Lekven et al. 2001).

6 Role of Chordin and Tolloid During Ventral Tail Development

The Tolloid metalloprotease encoded by the *mini fin* (*mfn*) gene plays a specific role in DV patterning of the tail (Connors et al. 1999). The *mfn* mutant displays a loss of ventral tail tissues, which includes the ventral tail fin, vein, and somitic mesoderm (Fig. 5D). The phenotype is first apparent within the tail bud at the end of gastrulation, when the organizer becomes confined to the tail bud region. Although *tolloid/mfn* is expressed at earlier stages, it is first required at the end of gastrulation in ventral regions of the tail bud (S.A. Connors and M.C. Mullins, unpubl.). At this stage, the DV extent of *bmp* expression is relatively small compared to mid-gastrula stages and Chordin, in the absence of Tolloid/Mfn, may extend into the most ventral *bmp*-expressing domain and inhibit most Bmp signaling (Connors et al. 1999). Thus, a role has been proposed for Tolloid/Mfn in specifically inhibiting Chordin during post-gastrula, tail bud stages to generate the high Bmp-signaling activity that is necessary for ventral tail development. However, Tolloid/Mfn has been found

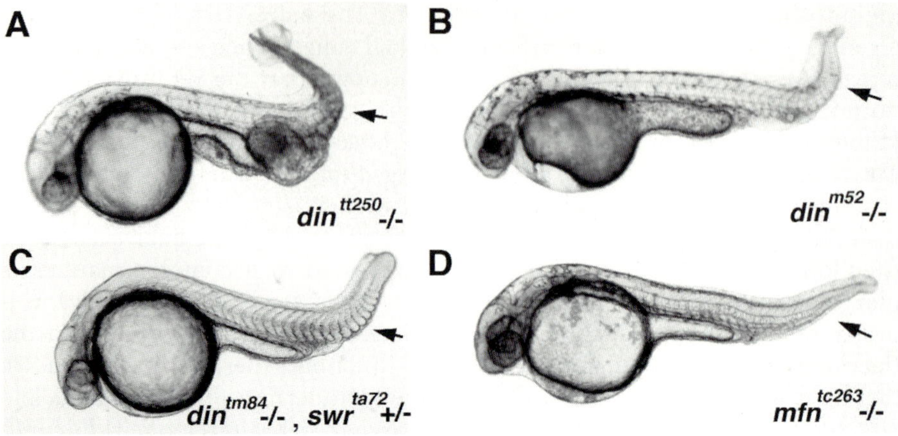

Fig. 5A–D. Chordin enhances Bmp function during ventral tail fin formation. All embryos are shown at 36 hpf, lateral view. *Arrows* point to the missing ventral tail fin. **A** *din*^{tt250}; in some lines, these phenotypes are seen next to mutants without ventral tail fin deficiencies (cf. Fig. 1G). **B** *cpt*^{m52}, according to non-complementation testing, it is another *dino* allele (M. Hammerschmidt and M.C. Mullins, unpubl.). **C** *swr*^{ta72}/+; *din*^{tm84}/*din*^{tm84}; the *swr*/+ siblings did not show any sign of ventral tail fin deficiencies. Similar phentypes have been observed for the *swr*^{tc300} and *din*^{tt250} alleles (Wagner and Mullins 2002). **D** *minifin*/*mfn*^{tc263}

to be dispensable when Ogon antagonism is also reduced, as revealed in the mutual suppression of the tail phenotypes in *mfn/ogon* double mutants (Wagner and Mullins 2002). This mutual suppression likely occurs through increased Chordin antagonism due to loss of inhibition by Tolloid proteolysis, which compensates for the reduction in Ogon antagonism.

While the role of Chordin described thus far is concerned with an inhibition of Bmps, there is also evidence for a Bmp-enhancing function of Chordin during ventral tail development. As discussed above for *mfn*, the ventral tail fin is absent in weakly dorsalized mutants, whereas it is duplicated in weakly ventralized mutants (Mullins et al. 1996; Solnica-Krezel et al. 1996; Hammerschmidt et al. 1996a). However, in some genetic backgrounds, *dino/chordin* mutants display ventral tail fin deficiencies, in addition to their general ventralized phenotype. This can be observed in the *tt250* allele (Fig. 5A), and is more pronounced in the *m52* allele isolated in the Boston screen and formerly named *captain hook* (*cpt;* Fig. 5B; Solnica-Krezel et al. 1996). Most strikingly, *dino* homozygosity causes a significant enhancement of the dominant *swirl/bmp2b* ventral tail fin deficiency (Fig. 5C; Wagner and Mullins 2002). Even in backgrounds with an extremely low penetrance or complete absence of the dominant *swirl* effect, *dino/dino*; *swirl*/+ embryos display a complete absence of the ventral tail fin, while the rest of the embryo has rather wild-type morphology – likely due to the loss of Chordin, which compensates for the lower dose of Bmp2b (Hammerschmidt et al. 1996b). Interestingly, the ventral tail fin phenotype of *dino/dino*; *swirl/+* embryos is indistinguishable from that of *minifin* mutants lacking the Chordin protease Tolloid (Fig. 5D). Thus, it is tempting to speculate that Chordin and Tolloid closely cooperate to enhance Bmp activity during ventral tail formation. Chordin, by binding Bmps, might act as a vehicle to transport Bmps against the gradient of free Bmp proteins, bringing them into the ventral-most region where Tolloid is present to cleave Chordin and free Bmps. A similar role in cooperation with Tolloid has recently been demonstrated for the *Drosophila* Chordin homologue Sog in a long-range enhancement of Bmp signals during amnioserosa formation (Ashe and Levine 1999).

The extent of Chordin's role as a Bmp agonist in zebrafish DV patterning remains unclear. A Bmp-enhancing effect of Chordin has not yet been detected during gastrulation stages. Thus, during gastrulation, *din/din*; *swr/+* mutants still appear slightly ventralized, whereas the dorsalized traits (indicated by ventral tail fin deficiencies) first become apparent following tail bud formation (Hammerschmidt et al. 1996b; Wagner and Mullins 2002). Furthermore, the studies on Tolloid/Mfn function reveal that Tolloid proteolysis of Chordin is critical during post-gastrula stages within the tail bud (S.A. Connors and M.C. Mullins, unpubl.). Thus, we currently favor a role for Chordin acting in conjuction with Tolloid to enhance Bmp activity specifically during post-gastrula stages of development.

7 Perspectives

Despite the rapid progress in the past years, many questions concerning the mechanisms of DV patterning and the role of Bmp signaling during these processes remain open. In addition, recent findings have suggested new roles for Bmp signaling, which require further studies. In the following, we will briefly discuss some very important aspects in our minds of future research.

1. Despite the massive amount of data obtained in embryological studies, there is little direct genetic evidence for the requirement of the canonical Wnt signaling pathway during organizer formation. Therefore, it will be important to determine the molecular nature of the *ichabod* gene. In addition, ongoing screens for maternal effect mutations (Pelegri and Schulte-Merker 1999; M.C. Mullins, unpubl.) will hopefully provide the basis to identify additional essential regulators of organizer formation. In addition, such approaches might help to elucidate how the broad *bmp2b/7* expression is initiated during blastula stages.
2. According to the morphogen concept, specification of different cell types along the DV axis is governed by different local levels of Bmp signaling. Thus, genes involved in DV-specific cell specification processes are expected targets of Bmp-activated Smad proteins and are expected to have different activation thresholds. Strong Bmp signaling/high Phospho-Smad levels

should activate the transcription of ventral specifiers, while moderate Bmp signaling/intermediate Phospho-Smad levels are expected to activate transcription of lateral specifiers. To date, experimental evidence for this notion is largely lacking. While both in *Xenopus* (*Xvent*; Henningfeld et al. 2000) and in zebrafish (*eve1*; M. Hammerschmidt, unpubl.), some ventral marker genes have been shown to be direct transcriptional targets of Bmp-activated Smads, no such dependence has thus far been revealed for more dorsal marker genes. The aforementioned differential response of neural crest and neuroectodermal markers in the various dorsalized zebrafish mutants suggests that such genes might respond to low levels of BMP signaling, whereas progressively higher Bmp activity levels may activate the ventral ectodermal marker *dlx3*, followed by *gata2* (Nguyen et al. 1998b). Promoter analyses will be necessary to give final answers to these questions. Most importantly, it will be necessary to identify differentially regulated, direct targets of the gradient to decipher the transcriptional mechanism used that discerns different positional values and ultimately gives rise to a diversity of cell types along the dorsoventral axis.

3. Does Bmp signaling affect cell movement? Cells along the DV axis do not only differ in their final cell fates, but also in their migratory behavior during gastrulation. This has been nicely investigated by Myers et al. (2002), who defined three different zones of cell movements along the DV axis: a ventral zone (high Bmp levels), where cells neither converge nor engage in extension movements; a lateral zone, where both convergence and extension movements occur at increasing speed; and a dorsal zone, where strong extension movements with little convergence take place. In *dino* and in *somitabun* and *swirl* mutants, the size of the zones shift according to the shift of the Bmp gradient (for further discussion, see Solnica-Krezel and Cooper, this Vol.). Thus, it will be interesting to study whether and how Bmp signaling might regulate the activity of molecules driving such cell movement processes. Interestingly, we have isolated such a regulator of convergence movements, *knypek/glypican 6* (Topczewski et al. 2001c), in an enhancer screen for the *chordin* mutation *dintt250* (M. Hammerschmidt, unpubl.). Thus, it remains to be determined whether Bmp signaling regulates the activity of, or acts in concert with, such proteoglycans and other cell surface molecules driving cell movements and/or regulating cell shape. Interestingly, it has recently been demonstrated that Bmps negatively regulate the expression of *wnt5/pipetail* and *wnt11/silberblick* (Myers et al. 2002), driving extension movements in paraxial regions of the gastrula embryo (Heisenberg et al. 2000). Most intriguingly, *din* mutant embryos display a complete loss of *wnt5* expression in paraxial mesoderm, while *myoD* is still expressed in this region, strongly suggesting that Bmp might impact cell movements in parallel rather than downstream of cell fate specification (Myers et al. 2002).
4. Does Bmp signaling regulate the general time at which cell specification occurs? Cell transplantation and explantation studies have revealed sub-

stantial differences in the timing of cell commitment along the DV axis of gastrulating embryos. Thus, for the ectoderm, it was shown that forebrain cells become irreversibly committed to their neural fate at the beginning of gastrulation, while hindbrain cells become committed at mid-gastrula stages, and epidermal cells after gastrulation (Woo and Fraser 1997; Grinblat et al. 1998). For epidermal fates, this delay in cell specification might be regulated by deltaNp63. deltaNp63, an N-terminally truncated isoform of the p53 tumor suppressor-like p63 protein, is expressed in a Bmp-dependent fashion in the non-neural, epithelial ectoderm from mid-gastrula stages onwards (M. Hammerschmidt, unpubl.), and has been shown to be involved in the maintenance of stem cell character of keratinocytes at later stages of mouse skin development (Yang et al. 1999b). Thus, it will be interesting to investigate whether Bmps, via deltaNp63 or other factors, might generally block cell specification processes, therefore promoting the maintenance of stem cell characters during early zebrafish development. Furthermore, the temporal requirement for Bmp signaling in the various cell specification processes it is implicated in remains to be determined. As cells converge dorsally during gastrulation, they are exposed to different Bmp-signaling levels. Further studies will be necessary to determine at what time point a cell becomes specified to assume a particular cell fate and what additional factors may be involved in regulating the response of a cell to the Bmp signal.

Specification of Left-Right Asymmetry

Christopher V.E. Wright[1] and Marnie E. Halpern[2]

1 Introduction

There has been considerable progress in understanding how restricted patterns of gene expression and differentiation are set up along the anterior-posterior (A-P) and dorsal-ventral (D-V) axes of the vertebrate embryo, and how these provide the fundamental framework for regionalized visceral, neural and cardiovascular organogenesis. Intimately linked, and in many ways equally important, is the morphogenesis that occurs orthogonal to these axes and leads to the left-right (L-R) asymmetric body plan. This plan is evident in all vertebrates as a stereotypical looping of the heart, gut, and the positioning of associated organs. In humans, complete inversion of the normal L-R body plan, called situs inversus, is not deleterious to health, but situations in which some tissues adopt the normal orientation while others are inverted (heterotaxia or situs ambiguus), have profound medical consequences, such as improper connections of the large blood vessels and abnormally chambered hearts (see Kathiriya and Srivastava 2000). Up to 1 in 8500 of human births have congenital defects associated with laterality defects, but there are likely to be many more losses that are unaccounted for in utero. Although the L-R asymmetry program appears to be quite sensitive to a variety of environmental and genetic disturbances, the overall uniformity of L-R patterning both within and between vertebrate species implies a strong, and to some degree conserved, underlying genetic regulation.

L-R asymmetric morphogenesis involves coordinated growth, migration, and remodeling, amongst groups of cells that are, in contrast to A-P and D-V differentiation, essentially similar at the biochemical or physiological level. As in the complex folding and twisting of the gut or heart tubes, these processes also involve communication over long distances. In some cases, formation of asymmetry is linked to the unilateral regression of initially bilaterally paired primordia, such as occurs during the formation of the vascular system in

[1] Department of Embryology, Carnegie Institution of Washington, 115 W. University Parkway, Baltimore, Maryland 21210, USA

[2] Cell Biology Department, Vanderbilt University Medical School, 1161 21st Avenue South, Nashville, Tennessee 37232-2175, USA

Results and Problems in Cell Differentiation, Vol. 40
L. Solnica-Krezel (Ed.): Pattern Formation in Zebrafish

mammals. In addition to revealing the cause of congenital syndromes in humans, studies of L-R asymmetry embody a central goal of developmental biology: to reach a complete molecular description of the genetic and cell-cell communication events that underlie the process of morphogenesis in large tissue sheets or tubes composed of cells from different embryonic germ layers.

Elucidation of the genetic pathways is proceeding rapidly, with input from studies on mouse, frog, and chick. Experimentation in the zebrafish system has not only substantially strengthened our understanding but has provided some unique insights. In this chapter, we will review recent progress in dissecting the L-R asymmetry pathway, emphasizing discoveries in zebrafish that have informed us about the conservation of mechanism or established new principles, set against the background of information obtained from other vertebrates. We will also describe specific areas that need to be addressed in more detail, and outline potential new avenues for exploration that take advantage of the special features of the zebrafish system.

2 Mechanisms Underlying Left-Right Patterning

As has been reviewed previously (Ramsdell and Yost 1998; Capdevila et al. 2000; Schier and Shen 2000; Mercola and Levin 2001; Wright 2001), the development of L-R asymmetry can be conceptually divided into several phases. First, there is the breaking of global embryonic symmetry, followed by the stabilization and/or propagation of this information, involving the establishment of L-R differences in gene expression patterns, the recognition of this information by tissue and organ primordia, and finally its full-scale interpretation by the elaboration of asymmetric tissue morphologies. A diagram of the consensus view of the L-R pathway is presented in Fig. 1.

3 Breaking Symmetry

Two principal unsolved problems are when and how symmetry is initially broken in any vertebrate embryo, and if L-R specification in all tissues and organs occurs as a result of the same initiating event. In general, and although our knowledge is limited, current thinking is that the first symmetry-breaking event occurs quite early in embryogenesis, with L-R information already present in key signaling centers such as the node (the equivalent of the zebrafish embryonic shield) at around gastrulation stages. Laterality signals are subsequently propagated into organ precursor tissues during early somitogenesis, with asymmetric morphogenesis beginning soon afterwards manipulations. In chick, where the most work on the early steps of the pathway has been conducted, led to evidence for a wide-ranging L-R influence within the pre-gastrulation-stage embryonic blastodisc, which is imposed on the forming node and used to set up the asymmetric gene expression patterns diagrammed in Fig. 1. Then, L-R information flows back out of the node to adjacent embryonic tissues, as indicated by the finding that transplanted or rotated nodes can impose asymmetry on surrounding tissues (Pagan-Westphal and Tabin 1998; Capdevila et al. 2000).

Mathematical modeling suggests that large-scale breaks in the initial symmetry of the embryo could be caused by tiny molecular or physiological L-R biases (as shown in cartoon form at http://www.eb.tuebingen.mpg.de/dept4/meinhardt/web_org/organize.html). Unexpected mechanisms may generate this spatial information. For example, a role for organized gap junctional communication (GJC) has emerged from studies in frog and chick (Levin and Mercola 1998a, 1999; Mercola and Levin 2001), where it is posited to be linked to the circumferential displacement of a small "L-R specifier", enriching it in cells on one side of the early embryo. Work is in progress to determine the nature of the transported molecules, and to try to connect this bias to the reproducible orientation of the L-R axis with respect to the primary body axes. Certain genes encoding gap junction components are expressed in regions outside of the primitive streak and node in the early chick embryo, in keeping with the idea that GJC occurs primarily peripheral to the primary body axis. The putative role for GJC in mammalian embryos has been addressed elsewhere (refer to Mercola and Levin 2001). A long-range system of intercellular communication seems appropriate as a mechanism for breaking symmetry, because results from a variety of embryonic manipulations suggest that the

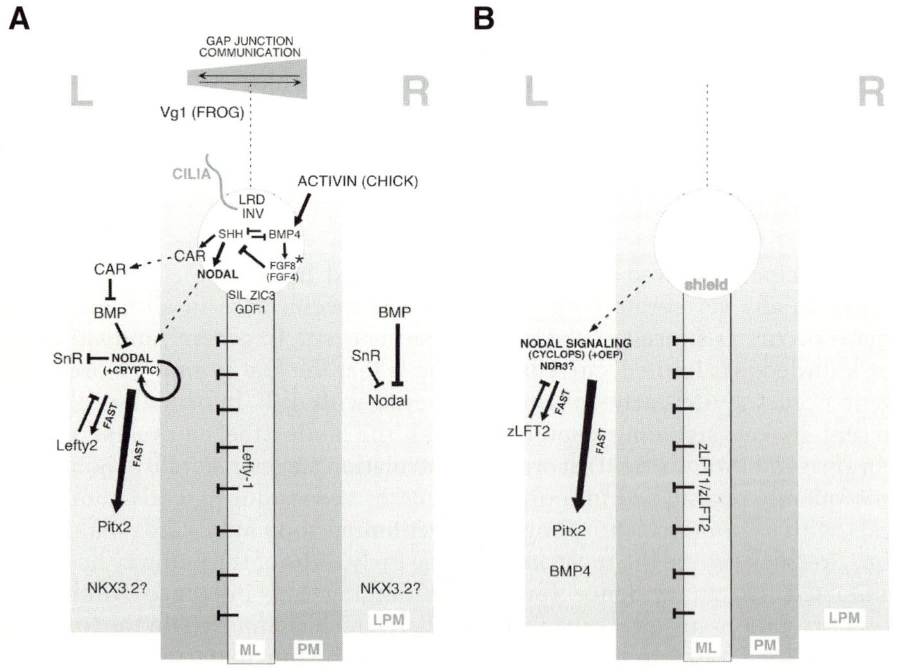

process should be regulative rather than fixed. Such communication could also explain, for example, the apparent interchange of L-R information between side-by-side twinned embryos (Capdevila et al. 2000).

In the mouse embryo, another method of intercellular communication may have evolved that is specifically coupled to L-R axis specification. The concerted rotation of primary cilia that are located on the ventral surface of cells at the node is thought to generate an overall leftward movement of the extracellular fluid – "nodal flow" – within which a left-specifying morphogen is carried and concentrated to one side of the notochordal plate (Nonaka et al. 1998; Okada et al. 1999; Takeda et al. 1999; Capdevila et al. 2000; Juan and Hamada 2001;

Fig. 1. **A** Working draft of the L-R axis specification pathway in vertebrates. Data from several model organisms are depicted. Modified from Wright (2001). Developmental age increases downward. The *shading* of the axial midline (*ML*), paraxial mesoderm (*PM*), and lateral plate mesoderm (*LPM*) blocks suggests the gradual appearance and differentiation of these tissues; before the formation of the node (*central circular structure*), the incipient midline is indicated by the *dashed line*. Not all details are incorporated for the sake of clarity, and the dynamics of the pathway are not fully apparent (see, e.g., Capdevila et al. 2000 for more details). At the head of the cascade, intercellular gap junctional communication may be involved in the transport of a left-right specifier molecule, enriching it on one side of the embryo (here drawn arbitrarily to the right). Propagation of this bias, which may be of small magnitude, leads to asymmetric gene expression patterns. A number of TGFβ-related factors have been implicated in the L-R genetic pathway (Vg1, Activin, BMP, and most notably Nodal). A left-right coordinator, *Vg1*, may be involved at early stages in frog embryos, and right-sided Activin signaling occurs very early in chick. Activin-like signaling may induce right-sided BMP4 expression, which engages in a mutually inhibitory relationship with Shh expressed in the node: BMP4 inhibits Shh directly, and/or via FGF signaling; Shh inhibits BMP4. Right-sided FGF8 inhibits Shh in chick, but the *asterisk* indicates that its function in L or R specification may vary dramatically between species. The dotted line for Caronte (*CAR*) indicates its putative transfer function in moving left cues from the node through the paraxial mesoderm to the lateral plate mesoderm, where it inhibits BMP4 extracellularly and alleviates a repression of Nodal expression. The additional possibility that left-side-enhanced Nodal travels from the node to the LPM is indicated by a *broken line*. In mouse, the left-specifying morphogen(s) may be carried leftward by nodal flow, a net movement of extracellular fluid caused by the concerted rotation of primary cilia located on node cells (only one indicated here); Inversin (*INV*) and left-right dynein (*LRD*) may be involved in this process. The induction of *nodal* expression by the left specifier in the left LPM requires an EGF-CFC family member (Cryptic in mouse); as well as acting as a feed-forward autoactivator of *nodal* expression, Nodal directly induces *lefty-2*, a feedback inhibitor. The *Pitx2* homeobox gene encodes one of the effector transcription factors that initiate programs of asymmetric morphogenesis. Other essential genes in the pathway are *sil*, *gdf1*, and *zic* (see Capdevila et al. 2000), which may be required for the integrity of the midline and node tissues, and for the generation of molecular asymmetries in the node. The chick *SnR* *snail*-related gene is one of the few genes that are expressed specifically on the right. *Nkx3.2* is apparently expressed on opposite sides in mouse and chick (indicated by the *question mark*). *lefty-1* expression on the left side of the axial midline (*ML*) somehow prevents the inappropriate contralateral crossing of "leftness" signals to the right side of the embryo. **B** Conservation of components of the L-R pathway in zebrafish. Not surprisingly, the pathway is lagging behind the combined knowledge from the other vertebrate model systems. The expression of the lefty-related genes *zlft1* and *zlft2* in the axial tissue varies down the A-P axis. The FAST transcription factor is the product of the *schmalspur* gene

Wright 2001). From this location, a "leftness" signal(s) is then transferred unilaterally, either directly or through an intermediary, to the left side of the embryo, where conserved and spatiotemporally dynamic gene expression patterns are induced in the lateral plate mesoderm (LPM), the tissue of origin for the heart and visceral organ precursors (Fig. 1; Capdevila et al. 2000).

This nodal flow model is intrinsically attractive as a way of aligning L-R asymmetry information with respect to the A-P and D-V axes. The fixed direction of ciliary rotation, in synergy with other features such as the arrangement of cells and the shape of the node, the tilt of the ciliary rotational axis, or in-phase variations in angular velocity, could provide a relatively straightforward coupling between the inherent molecular chirality of the monocilia, and a right-to-left nodal flow. These ideas have been formulated in greater detail elsewhere (Capdevila et al. 2000; Mercola and Levin 2001; Wright 2001).

The possible involvement of cilia can be linked to several mouse mutants that show L-R abnormalities, and to an association of L-R anomalies with dysfunctional cilia syndromes in humans (reviewed in Capdevila et al. 2000; Mercola and Levin 2001). The *inversus viscerum* (*iv*) and *inversion of turning* (*inv*) mutations in mouse identify early-acting components of the L-R pathway: *iv* encodes LRD (left-right dynein), a putative structural component of node monocilia (Supp et al. 1997), and *inv* encodes a large Ankyrin-repeat-containing protein of uncertain function (Mochizuki et al. 1998; Capdevila et al. 2000). *iv/iv* embryos essentially display heterotaxia, while *inv/inv* embryos are almost all mirror-image reversed, or situs inversus. Embryos with mutations in the genes encoding the motor proteins Kif3a and Kif3B also have L-R defects (Nonaka et al. 1998; Takeda et al. 1999). The node monocilia fail to form in the motor protein mutants, are immotile in *iv/iv* embryos, and exert a turbulent and slower leftward nodal flow in *inv/inv* mutants (Okada et al. 1999). These observations seem to support the nodal flow model, although we still lack a completely satisfying explanation of how slow but normally oriented flow produces situs inversus in *inv/inv* mice (Okada et al. 1999; Brueckner 2001; Juan and Hamada 2001). Several other issues remain unresolved, including the identity of the asymmetrically-accumulating morphogen, whether nodal flow is an asymmetry setting or propagating event, and whether the *lrd* and *inv* expression patterns might indicate an alternative function, perhaps in some form of intracellular transport of a L-R biasing molecule.

There is little information on the early phase of L-R axis specification in zebrafish. The cellular mechanisms are unknown, and the molecular pathway is more rudimentary in comparison to the one generated from the combined studies in other vertebrates (Fig. 1B). This situation should change rapidly as the transparency and accessibility of the zebrafish embryo, together with the genetic and cell biological attributes of the system, are brought to bear and lead to data that can be generally applied to the pathway, as well as reveal zebrafish-specific mechanisms. Based on the structural similarity of their embryos, one might imagine a greater level of conservation in the early steps of the pathway between frog and fish, and less with mouse and chick. Despite this assumption, the proposed early role for asymmetric signaling via Vg1, which can act as a "left-right coordinator" in *Xenopus* (Hyatt and Yost 1998), has yet to be extended to the zebrafish ortholog. A potential technical problem here is that the frog misexpression experiments may be hard to mimic in zebrafish because of the inability to distinguish, at early cleavage stages, the future left and right sides of the embryo, limiting the capacity for spatially defined overexpression or inhibition studies. It should be fairly easy, however, to test whether GJC is required for formation of the L-R axis in zebrafish, initially by the pharmacological and gene expression approaches used in frog and chick.

The substantially larger size of the embryo, or structural characteristics of the node equivalent – the embryonic shield in zebrafish – might preclude a central role for cilia-mediated nodal flow in breaking or propagating asymmetry in large aquatic embryos. Recently, representatives of *lrd* (J. Yost, University of Utah, pers. comm.) and *inv* (Essner et al. 2001) have been identified in zebrafish. Future descriptive and loss-of-function studies are likely to address whether the embryonic shield carries motile cilia at an appropriate stage, and reveal the identity of putative morphogens. If cilia do not seem to be involved, it will be useful to perform comparative studies of the requirement for LRD, Inv, and Kif3A/Kif3B proteins, by examining whether they are expressed uniformly or in localized groups of cells, and whether their role in L-R specification involves intracellular trafficking. Again, the transparency of the embryo may allow particularly incisive studies of the intracellular location and function of these proteins; preliminarily, Inv-green fluorescent protein (GFP) fusion proteins have been reported to enter the centriolar structures upon expression in live fish (Yost et al. 2001). Multispecies approaches are likely to yield the fastest results, and in mouse, in which an Inv-GFP transgene has been found to rescue the *inv* mutation, the fluorescent protein is localized specifically to 9+0 cilia as well as to the centriole (H. Hamada, Osaka University, pers. comm.).

An intriguing group of cells that have been indirectly linked to L-R patterning in zebrafish are the dorsal forerunners, a wedge-shaped cluster of cells positioned at the leading edge of the dorsal margin of the gastrula-stage embryo. Interestingly, in light of the role for Nodal signaling in L-R patterning described below, dorsal forerunners express the *nodal* ortholog *squint* (Erter et al. 1998; Feldman et al. 1998; Rebagliati et al. 1998a). These cells are also highly endocytic (Cooper and D'Amico 1996; Melby et al. 1996), raising the possibility that they transfer putative inducing factors to overlying embryonic cells from the yolk syncytial layer (YSL), a cellular syncytium that is directly apposed to the embryo proper and whose formation precedes the appearance of the shield. The YSL appears to be a source of inducing signals, and is another site of expression of *squint* (reviewed in Schier and Talbot 2001), and of the Sox-related transcription factor encoded by *casanova* (*cas*; Dickmeis et al. 2001b; Kikuchi et al. 2001). In addition to a severe endoderm deficiency, *cas* mutants lack dorsal forerunners and at later stages exhibit randomization of asymmetric gene expression in the LPM and forebrain (Liang et al. 2000).

Distinguishing the tissue-specific roles of *cas* in the various cellular compartments of the gastrula could provide useful insights into the establishment or maintenance of L-R determining signals. Extending these studies to other mutants that differentially affect the dorsal forerunners and other tissues may yield additional pointers. For example, the maintenance of dorsal forerunners and normal situs in endoderm-deficient *bonnie-and-clyde* mutants potentially defines the forerunners as an essential signaling center for L-R specification (Kikuchi et al. 2001). Whether their cellular composition is fixed or dynamic in nature, the zebrafish dorsal forerunners and the mouse node may share a conserved function as L-R organizers. Both structures may function as a source of L-R information, perhaps arising via the process of nodal flow, which is seeded to cells that transiently move into their neighborhood during gastrulation.

4 Stabilizing, Propagating and Reinforcing Left-Right Asymmetry

The first molecular inroads into L-R specification were afforded by the discovery of left-sided *nodal* expression that is induced by asymmetric Sonic hedgehog (Shh) signaling from the chick node (Levin et al. 1995), and the conservation of left-sided *nodal* expression in mouse and frog (Collignon et al. 1996; Lowe et al. 1996; Lustig et al. 1996; Sampath et al. 1997; Capdevila et al. 2000). This molecular landmark provided a point of reference against which there has been a rapid expansion of the number of players in the pathway (Fig. 1). A standing problem in the field is to determine the degree to which the steps upstream of the Nodal-signaling cascade, such as the role of Shh, are conserved between vertebrates. No asymmetric expression has been noted for *hedgehog* orthologs in any vertebrate other than the chick, although asymmetric signaling could be subtle or transient, or occur in a post-transcriptional fashion. The situs defects observed in mouse *Shh* mutants seem related to the failure of the defective midline to keep left-sided genes expressed on the correct side (Izraeli et al. 1999; Capdevila et al. 2000). Defects in situs are caused in zebrafish by misexpressing Shh or one of its antagonists, protein kinase A (Schilling et al. 1999), similar to results from *Xenopus* (Sampath et al. 1997). No asymmetry defects have been reported, however, in zebrafish *Shh* (*sonic-u*; *syu*) mutants (Chen et al. 1997), although it is possible that parallel L-R signaling pathways may involve the other zebrafish *hedgehog* orthologs as suggested for D-V patterning of the neural tube (Lewis and Eisen 2001). Examining L-R morphology in zebrafish mutants of the transmembrane hedgehog signal transducer Smoothened (Chen et al. 2001a; Varga et al. 2001) and the transcription factor Gli (Karlstrom et al. 1999) may be informative.

On the other hand, it is possible that species-specific differences in embryonic architecture are obligately linked to species-specific differences in the timing and way in which L-R information is generated and/or transferred

(Capdevila et al. 2000). While there is much to be learned about conservation or diversity in these early steps, an over-riding theme is that they apparently all have the same goal: the activation in the left LPM of the Nodal-signaling pathway, involving the Nodal/Oep/FAST/Smad/Lefty/Pitx2 genetic cassette, aspects of which are described elsewhere (see Kimelman and Schier, this Vol.).

How does L-R information spread from axial regions to more outlying tissues? In chick, the *Cerberus*-related gene *Caronte*, which encodes an extracellular BMP-binding antagonist, may act as an intermediate between Shh and Nodal. It is asymmetrically expressed on the left, first in the paraxial mesoderm and then in the LPM. Caronte has been postulated to be fairly diffusible within embryonic tissue, and its asymmetric expression may transfer L-R asymmetries from the node to the left LPM, where it then activates *nodal* expression by relieving a BMP-mediated repression (Capdevila et al. 2000). Additional possibilities cannot be ruled out (Fig. 1): the Nodal ligand itself may move to the LPM by direct diffusion, facilitated transport by a chaperone-like partner (as in the Dpp/Sog interaction proposed in Drosophila), or by deposition onto adjacent tissue layers that grow disproportionately or stretch relative to the organizer/midline. In practice, all of these mechanisms could contribute to a net movement of a "left-inducer" to the organ primordia. In zebrafish, it should be possible to test for conservation of the mechanism for propagating L-R asymmetry from medial regions of the zebrafish embryo to the LPM, by determining whether BMP-inhibition via Caronte-like factors is involved in activating the *nodal* ortholog *cyclops* in the left LPM.

Studies in *Xenopus* indicate that the asymmetric patterns of gene expression and consequent organ morphologies remain plastic until surprisingly late in embryogenesis (see Levin and Mercola 1998b; Lohr et al. 1998), suggesting that much of L-R "determination" consists of maintaining and reinforcing an initial L-R bias. One might presume that a specific point must be reached when L-R information is registered irreversibly by the organ precursors, with effector programs being set into motion that elicit asymmetric morphogenesis. Conversely, it is possible that right up until or even during the morphogenetic process, the tissues are still engaged in a dialogue, among themselves and with other embryonic structures, that affects their overall asymmetry. This proposal has the corollary that even quite late inconsistencies in L-R programming between different parts of organ systems could lead to dysmorphologies.

In light of the need to make a distinction between the left and right side differentiation programs, it seems logical that partitioning mechanisms would arise that prevent the crossing over of L-R signaling molecules from one side of the embryo to the other. A large body of work has repeatedly implicated the midline axial tissues, i.e., the notochord and neural tube floor plate, in performing such a barrier function, because the integrity of these tissues is required for normal situs in mouse, frog and fish. Extirpation experiments in *Xenopus*, for example, show that the left-sided genes are inappropriately expressed on the right when this barrier is interfered with or removed (Danos and Yost 1996; Lohr et al. 1997), and similar conclusions have been reached from analyses of zebrafish and mouse mutants that perturb normal patterning of the midline (for review, see Capdevila et al. 2000). While the mechanism is still unclear, these observations could reflect the ability of the notochord or floor plate to secrete or induce secretion in other tissues of intercellular positive-acting signals or inhibitory factors such as the *lefty* gene products, as described below.

5 Transferring Left-Right Information to the Organ Precursors

The Nodal intercellular signaling pathway has been recruited into several critical patterning events during vertebrate embryogenesis. In the mouse embryo, *nodal* expression is widespread at the egg cylinder stage but around gastrulation becomes restricted to the extraembryonic anterior visceral endoderm, the future posterior region where it marks the incipient primitive streak, and to cells at the node periphery. There is a transient phase of subtly asymmetric *nodal* expression at the node, being more intense and wider in extent on the left side (in the chick, this is linked to asymmetric Shh activity; Fig. 1), which is followed by a short-lived phase of broader expression in the left LPM (Beddington and Robertson 1999). Inactivation of *nodal* causes mouse embryos to arrest development too early to appreciate the effect on the left LPM, but many lines of evidence suggest that severe situs defects would be caused by Nodal depletion specifically within this tissue. Embryos homozygous for a hypomorphic *nodal* allele can overcome the early developmental block but subsequently exhibit situs defects, strongly suggesting the requirement in L-R specification for a fairly high level of Nodal activity (Lowe et al. 2001). Furthermore, L-R defects are found in a number of mutants of the Nodal-signaling pathway, including cofactors, transmembrane receptors and signal transduction components (Schier and Shen 2000).

Two *nodal*-related genes have received detailed attention in zebrafish: *cyclops* and *squint*. Both are expressed during gastrulation and are essential for mesendoderm specification (Feldman et al. 1998; see also Kimelman and Schier, this Vol.). Neither shows any asymmetric expression at shield stages that might be considered equivalent to that seen in the mouse or chick node (Erter et al. 1998; Rebagliati et al. 1998a,b; Sampath et al. 1998). The restriction of *squint* expression to early embryogenesis (Erter et al. 1998; Feldman et al. 1998) rules out a direct function in asymmetric morphogenesis, although it is involved in inducing tissues that play specific roles in later L-R signaling events. During somitogenesis, *cyclops* is transiently expressed in the left LPM and forebrain (see below). Defining a requirement for Cyclops signaling in L-R specification is complicated, however, because the L-R defects in *cyc* null mutants seem too weak to be interpreted as a complete loss of asymmetric Nodal signaling. It is possible that additional zebrafish *nodal* orthologs, such as *ndr3* (Rebagliati et al. 1998a), provide asymmetry cues in parallel to *cyclops*; gene expression and loss-of-function studies will help resolve this issue (in addition, see Sect. 7). That Nodal signaling is essential for L-R specification is supported by the strong laterality defects observed in *schmalspur* mutants (Chen et al. 1997; Bisgrove et al. 2000; Concha et al. 2000), which carry a deficiency in the zebrafish homologue of the FAST/FoxH1 forkhead/winged-helix transcription factor (Kimelman and Schier, this Vol.; Pogoda et al. 2000; Sirotkin et al. 2000b; Whitman 2001), and in embryos in which Nodal signaling is abrogated by loss of an obligatory cofactor, as described next.

A key discovery made first in zebrafish was that the extracellular membrane-bound EGF-CFC factors, represented by One-eyed pinhead, function as indispensable cofactors for reception of Nodal signaling (Kimelman and Schier, this Vol.; Shen and Schier 2000). The dual roles of zebrafish Oep as a Nodal cofactor, first in mesendoderm specification and then in L-R asymmetry, are essentially divided in mouse between the *cripto* and *cryptic* orthologs, respectively. The laterality defects of mouse *cryptic* mutants, which include right thoracic isomerism, are associated with a loss of *nodal* expression in the left LPM, and thus an absence of signals directing the left-specific pattern of morphogenesis (Yan et al. 1999). The current interpretation is that the absence of the Cryptic cofactor interrupts the Nodal autoactivation loop (Fig. 1). In zebrafish, removing maternal and zygotic *oep* function, as in MZ*oep* embryos, produces a gastrula-stage mesendoderm deficiency similar to that caused by downregulating Nodal signaling in *cyclops/squint* double mutant embryos (Feldman et al. 1998; Gritsman et al. 1999; Kimelman and Schier, this Vol.). At the tailbud stage, *oep* is expressed bilaterally in the LPM, where molecular genetic manipulations suggest that it enables receipt and/or propagation of the leftness signal(s). Injecting *oep* RNA into MZ*oep* embryos allows a partial rescue of the early mutant phenotype, such that gastrulation and formation of mesendoderm and prechordal plate all occur. Rescued embryos fail to develop the asymmetric expression domains of *cyclops* and other components of the Nodal pathway in the LPM, and display heterotaxia between the direction of heart looping and positioning of the gut-associated organs (Yan et al. 1999). Thus, the molecular and anatomical laterality defects caused by late-stage reductions in Oep activity support the view that asymmetric activation of the Nodal-signaling pathway in the LPM is essential for L-R patterning in zebrafish. As described below, Nodal/Oep signaling also plays a prominent role in establishing L-R asymmetry in the zebrafish forebrain.

Genetic analyses of the Nodal pathway in mouse and fish illustrate a recurring challenge in the study of L-R determination. Because many of the genes involved are reiteratively used in various aspects of embryonic patterning, it is difficult to unravel the details as to the timing and tissue of action of specific signaling events. Reciprocally, it is important to determine how embryonic cells alter their responsiveness to, for example, Nodal signaling, so that they undergo a massive change in fate during mesendoderm specification at early stages, but later interpret essentially the same signal as a trigger for asymmetric morphogenesis.

Once an asymmetric signal is activated, it must spread rapidly to the appropriate tissues during their period of competence, but the intensity and duration of the signal should also be restricted to allow the proper engagement of effector programs. The *lefty*-related genes, which encode highly diverged members of the TGFβ/BMP superfamily, fulfill at least part of this requirement. The *lefty* genes are direct targets of Nodal signaling and act as feedback inhibitors by breaking the autoregulatory feedforward loop for Nodal, presumably by blocking access to its transmembrane receptor (Fig. 1; see Kimelman and Schier, this Vol.). Extensive analyses of the *lefty* genes began in the mouse. The bilateral expression of *lefty-2* is involved in regulating the level of Nodal-induced mesoderm induction during gastrulation. Conventional *lefty-2* mutants undergo excessive mesoderm specification, and embryogenesis is too aberrant to discern a specific function associated with its asymmetric expression (Meno et al. 1999). Removing an enhancer that is required for left LPM expression, however, reveals that Lefty-2 normally restricts the duration and spatial extent of asymmetric Nodal signaling, and leads to situs abnormalities that include left isomerisms (i.e., bilateral "left" specification). In the absence of feedback inhibition, *nodal* expression in *lefty-2* mutants remains left-sided but is increased and temporally extended. This allows its long-range contralateral diffusion and/or transport to the right side, where it activates several Nodal target genes (Meno et al. 2001). In this vein, an important finding in zebrafish has been that Nodal ligands may act as long-range diffusible inducers, in contradiction to earlier views (Schier and Kimelman, this Vol.; Chen and Schier 2001). In mice, *lefty-1* performs a somewhat overlapping function with *lefty-2* in L-R axis formation. It is expressed in the anterior visceral endoderm of the egg cylinder, and during early somitogenesis in the left side of the ventral neural tube floorplate. The severe asymmetry defects associated with *lefty-1* inactivation, including left thoracic isomerism, and bilateral expression of *nodal* and *lefty-2*, strongly suggest that Lefty-1 acts as part of the midline barrier to block the contralateral diffusion of "leftness" signals (Meno et al. 1998). The biochemical mechanism is currently mysterious, although the possibility has been raised that Lefty-1 may oppose the rightward spread of "leftness" signals, whether involving Nodal or related inducers, by interfering with the autoactivation loop that inherently expands expression through embryonic tissues. Additional insight, however, is likely to come from analyzing more selective temporal and tissue-specific knockouts, in order to rule out a role in earlier parts of the L-R specification process.

The zebrafish *lefty* genes, *antivin/zlft1* and *zlft2* (Bisgrove et al. 1999; Thisse and Thisse 1999), are more similar to each other than to any specific gene in other species, and orthologies to the mouse are difficult to establish. Expression of *zlft1* in the left dorsal forebrain and *zlft2* in the left LPM is, however, reminiscent of the asymmetric expression of mouse *lefty-1* in the floorplate and *lefty-2* in the LPM, respectively. Both zebrafish genes are expressed in overlapping domains in the midline in late gastrulation stage embryos. So far, functional data on zebrafish Lefty factors have been limited, but the general principle that Lefty/Antivin factors function as feedback inhibitors of Nodal signaling has been confirmed. A study using antisense oligonucleotide-based inhibition further supports this notion (Agathon et al. 2001). It remains to be determined unequivocally if Lefty-like proteins, in some circumstances, inhibit other members of the TGFβ superfamily, as suggested by Branford et al. (2000).

The relationship between Nodal and Lefty proteins resembles the interaction of a self-activator and a feedback inhibitor in a hypothetical "reaction-diffusion system", as first proposed by Turing (Juan and Hamada 2001). Recent studies of Nodal and Lefty GFP-fusion proteins in the right LPM of chick embryos (H. Hamada, Osaka University, pers. comm.) suggest that these proteins fulfill a classic tenet of such models: that the feedback inhibitor diffuses faster than the autoactivator. Analyses of *lefty* orthologs in zebrafish should open the door for more complex genetic and molecular epistasis studies, and deeper cell biological analyses, to help tease apart the cellular interactions responsible for generating and elaborating L-R differences. A general aim is to understand how a balance of repressive and activating influences serves to maintain distinct left- and right-sided programs of gene expression, and how these programs are spatiotemporally coordinated along the A-P axis.

6 Effector Programs of Left-Right Asymmetric Morphogenesis

Downstream components of the Nodal-signaling pathway, that show L-R differences during organ morphogenesis, are starting to be identified. The Bicoid-related transcription factor Pitx2 is a case in point. In several vertebrates, Pitx2 is expressed in a left-predominant pattern in the organ anlagen as they undergo asymmetric morphogenesis, leading to the hypothesis that it regulates batteries of genes that mediate the cell biological readout of L-R information (Capdevila et al. 2000). Complex gene modification studies in mouse have revealed that the level of *Pitx2* expression required for correct L-R morphogenesis varies in an organ-dependent context (Liu et al. 2001). Zebrafish *Pitx2* is expressed early during gastrulation and, at later stages, bilaterally in some tissues but asymmetrically in others, very similar to the patterns found in other vertebrates. In mammals, differential promoter usage and mRNA splicing yields three *Pitx2* isoforms – *Pitx2a*, *Pitx2b* and *Pitx2c* – that have identical homeodomains but differences in their amino-terminal regions that are conserved between species (cDNAs representing a fourth isoform, *Pitx2d*, have been detected in *Xenopus*; Essner et al. 2000). Zebrafish cDNAs representing *Pitx2a* and *Pitx2c* isoforms have been isolated, with evidence that they have different activities (Essner et al. 2000). *Pitx2a* is asymmetrically expressed in the left LPM in the region including the heart precursors, while *Pitx2c* shows asymmetric expression in the dorsal forebrain.

An emerging theme is that the genetic regulation in the L-R signaling pathway is intricate. In some instances, related genes (e.g., *zlft1/antivin* and *zlft2*) are expressed on the left side, but in different tissues, or within different domains of the same tissue. In other cases, such as *Pitx2*, transcriptional and/or post-transcriptional regulation of the same gene leads to the translation of functionally divergent protein isoforms in specific regions or tissues. A detailed mechanistic understanding of such processes will likely be aided by inter-species comparisons of *cis*-regulatory regions, as reported for amphibian and mammalian *Pitx2*. A major mediator of the initiation of left-sided *Pitx2* expression is a conserved asymmetrically activated enhancer (ASE) that requires several FAST/FoxH1 sites, a finding consistent with *Pitx2* representing a transcriptional target of Nodal-like signaling. The maintenance of asymmetric *Pitx2* expression also requires a conserved binding site for Nkx-type transcription factors (Shiratori et al. 2001), members of which have been linked to the L-R pathways in chick and mouse (Fig. 1; Schneider et al. 1999).

7 Cardiac Left-Right Asymmetry

The proper interpretation of L-R asymmetry cues is particularly important in mammals for the formation of the multichambered heart and connections to the pulmonary and vascular systems (see Kathiriya and Srivastava 2000; Yelon et al., this Vol.). Several congenital, sometimes fatal, structural defects found in human syndromes can be traced back to abnormalities in this process. The developing heart is easily viewed in zebrafish, and the long-term goal of creating a collection of mutant lines covering most of the genome has yielded information regarding the molecular control of this organ's asymmetric placement and chiral looping. Heart formation begins during the segmentation period when bilateral sets of myocardial precursors migrate to envelope medially located endocardial precursors to form a cylinder at the midline. The first anatomical readout of L-R vectorial information in zebrafish embryos is the jogging of the atrial end of this tube to the left side, away from its initial midline position (Chen et al. 1997). Subsequently, the heart tube loops to the right. Surveys of mutant lines showed that defects in the integrity of midline tissues such as the notochord cause heterotaxia and are associated with defects in heart jogging and looping, as well as disturbances in the pattern of left-side expressed genes (*cyclops*, *antivin/zlft2*, *Pitx2*; Chen et al. 1997; Bisgrove et al. 2000; Chin et al. 2000).

It has been suggested that the direction of heart tube jogging is an excellent predictor of the chirality of heart looping, and a better indicator of initial cardiac laterality because of its independence from secondary defects that might be related to defective myocardiocyte differentiation (Chen et al. 1997;

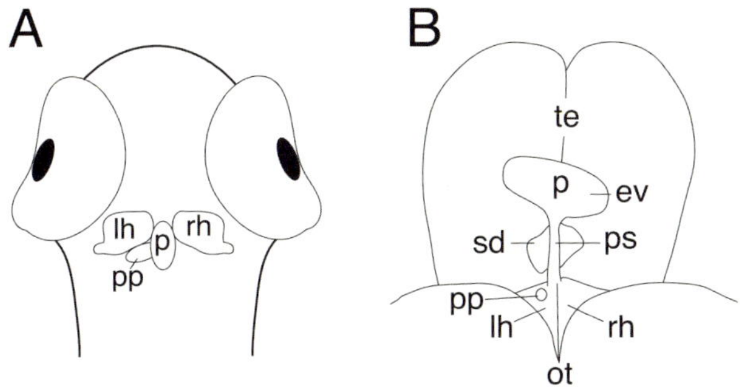

Fig. 2A,B. Asymmetry of the zebrafish epithalamus. **A** Schematic view of the dorsal diencephalic region of a 4-day-old larva. Anterior is to the top. The left-sided parapineal organ (*pp*) is detected anterior to the epiphysis or pineal anlage (*p*) at 48 h (Gamse et al. 2002) but by 4 days comes to lie in a more ventrocaudal position relative to the developing pineal. At this stage, the dorsal habenular nuclei are visibly asymmetric, with the left habenula (*lh*) being slightly larger in size and denser in neuropil than the right (*rh*; Concha and Wilson 2001). **B** Schematic dorsal view of the epithalamic region of the adult zebrafish brain (adapted from Liang et al. 2000), which is situated between the caudal telencephalon (*te*) and the optic tectum (*ot*). The mature pineal organ consists of the photoreceptive end vesicle (*ev*) and the pineal stalk (*ps*) that connects to the diencephalic roof at a slightly left to medial position. The membranous saccus dorsalis (*sd*) is closely associated with the pineal. The rudimentary parapineal remains at the base of the pineal, closely apposed to the left habenula (J. Gamse and M.E. Halpern, unpub. results). The gene expression patterns associated with the transient asymmetric activation of the Nodal-signaling pathway are restricted to segmentation stages, before 24 h

Goldstein et al. 1998; Yelon 2001). In some mutants that alter L-R laterality, notably those affecting notochord development, the direction of jogging can be genetically uncoupled from that of looping (Chin et al. 2000). This finding is consistent with the idea mentioned above that L-R information is continually monitored and acted upon within the heart anlage, even during the morphogenetic process. A surprising corollary from these studies, however, was the low incidence of heart looping defects observed in ENU-induced *cyclops* null mutants (Chen et al. 1997; Bisgrove et al. 2000; Chin et al. 2000) suggesting, as described above, that additional Nodal-related L-R specifiers exist in zebrafish. A much higher incidence of situs defects occurs in *cyc*^{b16} deficient mutants (Chen et al. 1997; Chin et al. 2000), which may reflect the deletion of additional relevant genes. One candidate is *Hex*, which is expressed in the YSL and encodes a transcription factor that may be connected to the generation of early Nodal signals from this extraembryonic tissue (Ho et al. 1999; Liao et al. 2000).

As in other species (e.g. frog; Breckenridge et al. 2001), the TGFβ-related factor BMP4 may serve a bridging function in translating extracardiac L-R asymmetry signals into molecular asymmetry and morphogenesis within the zebrafish heart tube. When the heart tube first forms, *BMP4* expression is initially symmetric, but then becomes predominantly left-sided just before the leftward jogging event. Two groups of zebrafish mutants that all affect midline structures (notochord or neural tube floorplate) alter *BMP4* expression. In one group, symmetric *BMP4* expression is associated with a failure of the heart tube to jog and a consequent randomization of looping direction. In the other, *BMP4* expression is left-sided, right-sided or symmetric, and the proportions of these patterns are correlated with the incidences of leftward, rightward, or unlooped heart tubes (Chen et al. 1997). It has been reported that *BMP4* misexpression on the right side of the embryo, in a non-tissue restricted manner, inverts heart looping but leaves visceral organ asymmetry undisturbed. While this result may suggest a local effect for BMP4 in the cardiac region (Schilling et al. 1999), higher resolution misexpression and inactivation experiments must be carried out. In addition to *BMP4*, early molecular asymmetries at the gene expression level in the zebrafish heart tube that may directly regulate the processes of asymmetric morphogenesis include the rtk2 receptor tyrosine kinase, DM-GRASP adhesion protein, and transcription factors such as fli-1 and nkx2.5 (Schilling et al. 1999).

8 Asymmetry of the Zebrafish Forebrain

One area of research where the zebrafish is providing a unique contribution is in revealing the molecular basis for L-R specialization of the brain. It has long been known that primitive and teleost fish possess characteristic anatomical asymmetries in the dorsal diencephalon or epithalamus, and asymmetries in the same region have also been well documented in amphibia and reptiles (for a comprehensive review, see Concha and Wilson 2001). The epithalamus consists of paired habenular nuclei, connected via a prominent dorsal commissure, and the pineal complex, which includes the epiphysis cerebri or pineal organ (associated with the non-neural saccus dorsalis) and, often in fish, an asymmetrically positioned parapineal organ. Depending on the species, the dorsal (or medial) habenular nuclei can show L-R differences in their size, compartmentalization, neuronal distribution and structure, neuropilar density, efferent projections and biochemical properties. Although there is considerable diversity in the organization and asymmetry of the epithalamus amongst vertebrates, the habenulae lie at a central relay point of an evolutionarily conserved neuronal circuit that extends from the telencephalon to the interpeduncular nucleus of the midbrain (see Sutherland 1982). Variability in the presence or directionality of the lateralized features of the epithalamus suggests that, while the circuitry is largely preserved, there is an inherent flexibility in utilizing or processing L-R cues in this part of the brain.

Work in zebrafish provided the first evidence for asymmetric expression of the Nodal-signaling cassette in the embryonic epithalamus. The *cyclops* (Rebagliati et al. 1998a,b; Sampath et al. 1998), *zlft1/antivin* (Bisgrove et al. 1999; Thisse and Thisse 1999) and *Pitx2* (Concha et al. 2000; Essner et al. 2000; Liang et al. 2000) genes all show transient asymmetric expression in the left side of the forebrain during segmentation stages. The expression domain was initially designated as the left dorsal habenula, but careful double labeling studies revealed that all three genes are transcribed within a region of the left dorsal diencephalon that corresponds to the epiphyseal anlage. The precise boundaries of *cyc*, *atv/zlft1*, and *pitx2* expression within the left epiphysis differ, identifying another site where information about the transport properties of inducers and their antagonists will be useful. Furthermore, although components of the Nodal-signaling cassette are transcribed on the left side, the cofactors of the pathway, *oep* and *sur*, are expressed bilaterally in the paired epiphyseal precursors (Zhang et al. 1998b; Pogoda et al. 2000; Sirotkin et al. 2000b). By mid-segmentation, a strip of cells lying rostrally within the pineal anlage strongly expresses these genes, consistent with their selective ability to respond to Nodal signals (Concha et al. 2000; Liang et al. 2000).

The process whereby asymmetric gene expression in this region of the forebrain is regulated is under intense scrutiny and several key findings have emerged. First, some of the same mechanisms that modulate LPM asymmetry also influence the molecular asymmetry of the epiphysis. For example, when midline development is compromised as in notochord mutants, gene expression in the brain is later bilateral. Second, early patterning events, such as those regulated by *cas* or *sqt* before or during gastrulation, affect asymmetric gene expression in the brain detected almost 15h later. Third, no single positive or negative regulatory interaction can account for left-sided gene expression in the brain. While normal levels of *pitx2* expression seem to require signals provided by both the *sqt* and *cyc* Nodal-related genes, *atv* expression appears largely dependent on *cyc* activity. Moreover, an uncoupling of regulation can be infrequently observed, where one gene retains left-sided epiphyseal expression while another is bilateral (Liang et al. 2000). Finally, the analysis of a diverse array of zebrafish mutants demonstrates that laterality of gene expression in the brain can be completely uncoupled from molecular or anatomical laterality of the viscera (i.e., Sampath et al. 1998; Bisgrove et al. 2000). For instance, *cyc* is expressed bilaterally in the forebrain of notochord-less mutants, yet remains left-sided in the LPM. Such observations challenge the existence of a globally acting process that initiates and maintains L-R polarity throughout the A-P axis of the developing embryo. Rather, early events could trigger an overall asymmetry in the organism that requires some shared regulatory steps for its maintenance, but also relies on local, tissue-specific events for reinforcement and elaboration of L-R differences.

The significance of asymmetric Nodal signaling in the embryonic forebrain was not immediately obvious, because the pineal organ is most often described as originating from the midpoint of the dorsal diencephalic roof. As with the LPM (Yan et al. 1999), RNA-mediated rescue of early Nodal signaling in MZoep mutants has provided some insight. Although rescued embryos recover many aspects of Nodal-dependent tissue patterning, they lack the transient asymmetric gene expression in the forebrain (Concha et al. 2000; Liang et al. 2000). At adulthood, the brains of these rescued fish appear relatively normal, but the position where the stalk of the pineal organ connects to the diencephalic roof can be displaced significantly from its usual location just left of the midline (Liang et al. 2000).

A conspicuous asymmetry of the epithalamus in many fish species is the parapineal organ, which in zebrafish typically arises to the anterior and left of the pineal as a distinct cluster of cells that contains opsin-positive photoreceptors, Islet-immunoreactive neurons (Concha et al. 2000), and expresses otx5 (Gamse et al. 2002). In mutants defective for Nodal signaling, or in midline mutants that show bilateral forebrain expression, the position of the parapineal becomes L-R randomized (Concha et al. 2000; Gamse et al. 2002). The directionality of parapineal sidedness relative to the displacement of the pineal stalk needs to be examined further in rescued MZ*oep* larvae, because the interdependence of their development is unknown. Overall, the data lead to the conclusion that the function of asymmetric activation of the Nodal pathway in the brain is to ensure the reproducible laterality of the pineal complex throughout the population by biasing an otherwise stochastic L-R positioning to the left side of the epithalamus.

In larval zebrafish, the anatomical asymmetry of the dorsal habenulae is less dramatic than in some other types of fish or amphibia, but is reproducibly characterized by the larger size and denser neuropil of the left side (Concha et al. 2000). Recently, some genes have been found that are expressed in the left dorsal habenula, in both an expanded region of cells and at significantly higher levels than on the right side (J. Gamse, B. Thisse, C. Thisse and M.E. Halpern, unpubl. results). Molecular asymmetry is evident in 2-day-old larvae, but it is not yet known how it correlates with the appearance and organization of the afferent inputs to the dorsal habenulae. In a number of species, the parapineal preferentially innervates the left habenula (Bisazza et al. 1998; Concha and Wilson 2001). Examination of zebrafish mutations that alter the asymmetry of the diencephalon has revealed a strong correspondence between the L-R position of the parapineal and the habenula nucleus that develops the "left" identity, suggesting that either a shared mechanism confers laterality on both structures, or that the asymmetry of one depends on the other (Concha et al. 2000; J. Gamse and M. E. Halpern, unpubl. results). Recent experiments support the idea that the parapineal has an instructive influence on the habenula because L-R differences in habenular gene expression and neuropil density are not observed after the parapineal is laser-ablated earlier in development (J. Gamse and M.E. Halpern, unpubl. results). The picture that is emerging for epithalamic asymmetry, therefore, is that Nodal signaling sets up a L-R bias that is further propagated through local tissue interactions. The molecular nature of the interaction between the parapineal and left habenula, and whether the dorsal habenular asymmetry might influence L-R differences elsewhere in the brain, are issues that are being eagerly pursued.

There is a large gap in knowledge between the structural asymmetries of the diencephalon and their physiological or behavioral relevance. The zebrafish parapineal is a light-responsive organ that exhibits endogenous, circadian transcriptional activation similar to the pineal (Gamse et al. 2002), and could potentially modulate activity of the left habenula in a cyclical manner. Selective blockage of circadian-regulated gene expression in the parapineal (and pineal) organ (as in Gamse et al. 2002) does not affect habenular asymmetry (J. Gamse and M. E. Halpern, unpubl. results) but may have functional consequences. The possible link between circadian-regulated and lateralized brain functions merits further exploration.

Functions attributed to the dorsal habenulae are ill-defined in lower vertebrates (Sutherland 1982) and the purpose of habenular asymmetry is unclear. In contrast, some progress has been made in the general characterization of lateralized behaviors that fish display. Many teleost species, including zebrafish, show a consistent preference for left eye use when viewing their mirror image (Miklósi et al. 1998; Sorvano et al. 1999) but favor the right eye for viewing predators or foreign objects (Miklósi and Andrew 1999). Such visual preferences have been correlated with schooling behavior and with the directionality of the escape response, a rapid turning of the body mediated by the giant Mauthner hindbrain neurons (see Bisazza et al. 1998). Preferential eye use, with corresponding visual input through contralateral projections to the optic tectum, implies an underlying L-R difference in processing visual stimuli in the brain. The stereotypic swimming behavior that is elicited in response to visual cues also requires L-R differences in neuronal firing and neuromuscular activation (Hatta and Korn 1998). At present, there is no evidence that the circuitry for these lateralized neural behaviors intersects with or relies on the asymmetry of the habenulointerpeduncular system.

9 Summary and Future Perspectives

The ultimate goal of studies in L-R asymmetry is to draw a complete molecular picture of how an initial L-R bias is transformed into asymmetric morphogenesis, and to understand how these pathways arose during evolution. Despite progress, the current L-R pathway remains a rather stark framework with large gaps that need to be filled at the genetic and cell biological levels, and uncertainty about the conservation of key steps between vertebrate species.

Continued studies in zebrafish will lead to the identification of new L-R regulators, as well as to the isolation and functional characterization of orthologs of genes placed into the pathway in other vertebrates. For example, studying the zebrafish orthologs of *SIL*, *Zic3* and *gdf1* (Fig. 1) may permit a greater appreciation of their cell biological function, using genetic and molecular loss- and gain-of-function approaches. There is also currently a bias with respect to left-sided genes, and more signaling molecules and transcription factors will surely be linked to the active specification of the right side. An advantage of zebrafish may be that several functions of a single mammalian gene are assorted between several orthologs, allowing easier uncovering of specific functions. Once genome and expressed sequence databases become extensive enough, it should be possible to use "virtual cloning" to identify immediately the zebrafish ortholog(s) of any gene implicated in the pathway from other species. Moreover, genomic information will help resolve questions about the number of orthologs, clarifying issues such as the possibility of potential parallel-acting players in L-R specification.

Undoubtedly, huge gains will continue to come from forward genetic screens (e.g., see Patton and Zon 2001) that, as well as relying on anatomical criteria, use RNA or antibody probes to detect subtle alterations in the asymmetric expression patterns of key genes or proteins (e.g., Beattie et al. 1999; Moens and Fritz 1999). A technical advance that could make a dramatic impact is the production of transgenic lines in which tissue-specific labeling by GFP can be followed in real time in live fish (e.g., Huang et al. 2001). Transgenic Pitx2c-GFP strains have been reported that appear to express GFP in a pattern similar to the endogenous gene (Yost et al. 2001). Such strains would enhance screening for mutants in L-R specification, aid in maintaining mutant lines, and provide useful tools for a deeper analysis of situs defects in visceral and neural tissues.

New players will also come from molecularly oriented screens (Crosier et al. 2001; Kudoh et al. 2001; B. Thisse and C. Thisse, IGBMC Strasbourg, pers. comm.) that reveal differences in left versus right gene expression. Subtractive or microarray screens, using dissected embryonic tissues, mutants or manipulated embryos, may well yield downstream targets as reported for Nodal signaling during mesendoderm induction (Dickmeis et al. 2001b). These studies may be augmented by the use of fluorescently tagged genes to guide tissue microdissection or enable fluorescence-activated cell sorting of distinct cell populations. In lieu of mutants, functional testing will utilize direct methods for gene inactivation: homologous recombination may become applicable in zebrafish, but other gene knockdown methods are available, such as the specific protein translational block achieved by morpholino antisense oligonucleotides or by activation of caged RNA/DNA compounds (Nasevicius and Ekker 2000; Ando et al. 2001).

Studies in zebrafish will not only inform us on the specializations of the L-R pathway in this species, but will also be directly pertinent to mammals and, ultimately, humans. For example, mutations in the orthologs of genes placed into the L-R asymmetry from extensive studies in model systems, notably zebrafish *oep*, have been linked to laterality defects in humans (see Goldstein et al. 1998; Bamford et al. 2000; Casey 2001). It may also be possible to test whether sensitive steps in the L-R determination pathway can be debilitated by environmental toxicological agents, thereby explaining some of the spontaneous human congenital situs defects. Furthermore, whether human L-R defects arise in a polygenic fashion is feasible to investigate systematically in a genetically tractable vertebrate such as the zebrafish, extending findings in mouse, where situs defects were discovered in compound heterozygous *nodal*$^{+/-}$;*HNF3β*$^{+/-}$ embryos (Beddington and Robertson 1999). The simplicity of the fish body plan may preclude it from offering insight into the asymmetries seen in more elaborate mammalian organs, such as the cardiovascular system and the differentially lobated left and right lungs. It will be important, however, to continue comparative studies on multiple vertebrate model systems in order to determine if the early steps of the L-R pathway are really as diverse as is currently believed and, if so, how very different initiating events universally unfold into the activation of the Nodal-signaling pathway in the left LPM.

In all species, it is essential to place the developing genetic network into a cell biological context. Techniques will need to be developed for detecting protein products involved in L-R specification in living tissues. This is particularly true for the diffusible inducers and inhibitors, such as Nodal and Lefty, because their range of action may extend well beyond the region in which their genes are transcribed. Being able to determine accurately how the effective local concentration of inducers varies with space and time would be fundamental to cell biological models that attempt to explain the long-range coordination of asymmetric tissue morphogenesis. In considering the downstream programs that control cellular morphology, migration, adhesion, proliferation and survival, the target genes of Pitx2, and other transcription factors such as the Nkx proteins, must be defined, as these will likely include modulators of cytoskeletal structure and dynamics, cell-cell and cell-matrix adhesion. For instance, *N-cadherin*, which encodes an adherens junction component, is asymmetrically expressed in the early chick embryo, and inhibition experiments alter heart looping in addition to disturbing the lateralized expression of *Pitx2* (but not *lefty* or *nodal*), thus implying a role parallel to, or downstream of, asymmetric Nodal signaling. Studies in *Xenopus* have implicated the tight junction component Claudin in L-R asymmetry (reviewed in Mercola and Levin 2001). Tagging such structural genes with GFP to produce chimeric proteins in transgenic zebrafish embryos might allow detailed studies of the modulations in cell morphology and adhesion that occur during asymmetric morphogenesis.

The functional ramifications of lateralized anatomy in vertebrates are, in many ways, still not fully appreciated. The asymmetric patterning of the abdominal region is probably related to the need to fit a long gastrointestinal tract and the associated organ systems into a compact body plan without compromising their physiological activity. On the other hand, a compelling explanation still needs to be found for the different lobation observed in the left and right lungs of mammals. Perhaps nowhere is the functional implication of L-R specification more exciting than in the brain, where hemispheric specializations of the cortex have long fascinated neurobiologists and psychologists alike. Structural asymmetries continue to be sought that correlate with the lateralized neural processing associated with language, cognition, behavior and handedness, and that are disrupted in neurological disorders. A tempting speculation is that the mechanisms gleaned from molecular genetic approaches in the zebrafish will provide a new entry point for exploring the basis of L-R differences that regulate higher cortical functions. It has been previously argued that L-R specialization of the neocortex arose from a preexisting asymmetric template, and there is good reason to believe that the diencephalon provided such an evolutionary template (i.e., Harris et al. 1996; Vallortigara et al. 1999). On the other hand, overt asymmetry of the epithalamic region is not found in the mammalian brain, nor is there as yet evidence for lateralized expression of the Nodal pathway. A challenge for the zebrafish field, therefore, will be to decipher how the anatomical asymmetry of the forebrain leads to L-R differences in functional output, and, in combination with studies in other species, to explore the evolutionary preservation as well as elaboration of such asymmetries in the brains of higher vertebrates.

Acknowledgements. We thank Josh Gamse for Figure 2 and for allowing us to cite his recent studies. Thanks also to Christine Thisse and Bernard Thisse for sharing unpublished results. CVEW acknowledges support from the NIH and MEH from the NSF.

II. Gastrulation Movements

Life at the Edge: Epiboly and Involution in the Zebrafish

Don Kane[1] and Richard Adams[2]

1 Introduction

In the developing zebrafish, the mid-blastula stage embryo finds itself organized as an abashed mass of blastomeres on top of a large yolk cell. Over the next few hours, these cells begin a beautiful sweep over the yolk cell as they form the germ layers. These dramatic and exquisite movements form the zebrafish gastrula, and the changes in the embryo are so obvious that they form the basis for staging the embryo for much of the first day of development (Fig. 1). What physical forces underlie these movements? What is the genetic basis for these movements? And, what is the immediate consequence of cell movement for the embryo?

The early stereotyped cell movements of the embryo lead to the emergence of the fate map in early epiboly. In zebrafish, the fate map in the early blastula is non-determinative; the progeny from any blastomere of the blastoderm contributes to all tissues of the forming embryo. Cell movements during this period are not directed or coherent. However, in the late blastula, shortly after the beginning of epiboly, the fate map becomes restricted at the tissue level. That is, cells near the margin of the blastoderm become mesodermal and endodermal tissues and cells at the animal pole become ectodermal tissues. It is at this time that the nature of cell movement changes to a predictable, cohesive pattern. This early epiboly fate map is not the result of commitment to cell fates, for when cells are moved from one position to another, the cells take the fate of their new position. Rather, it seems that the blastula fate is a consequence of cells moving in a stereotypical pattern from their position in the blastula to a position in the forming embryo.

Here, we discuss the early cell movements in zebrafish. First, we focus on the beginning of cell motility in the zebrafish, a somewhat non-directed early motility. Then, we discuss the spreading of the blastoderm to engulf the yolk cell, a process termed epiboly. Finally, we discuss how the most marginal cells of the blastoderm move to the inside of the blastoderm to form the inner layer

[1] Department of Biology, University of Rochester, 213 Hutchison Hall, Rochester, New York 14627, USA

[2] Department of Biology and Biochemistry, University of Bath, Bath BA2 7AY, UK

Results and Problems in Cell Differentiation, Vol. 40
L. Solnica-Krezel (Ed.): Pattern Formation in Zebrafish

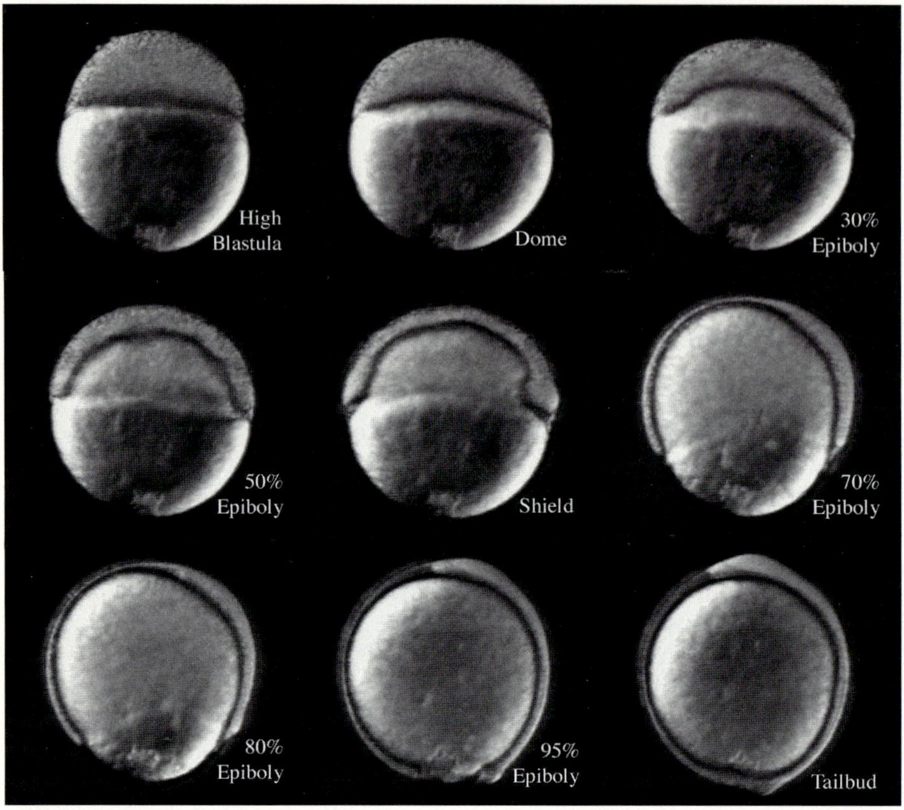

Fig. 1. Staging series during late blastula and epiboly periods. *High-blastula* stage, about 3.5 h after fertilization: The embryo is in the 11th cell cycle, and cells are still partially in synchrony. At this magnification the deep cells of the blastoderm dominate the view; it is not possible to discern the enveloping layer or the syncytial layer of the yolk cell. Note how the blastoderm perches on top of a spherical yolk cell. *Dome* stage, 4.3 h: The yolk is beginning to dimple into the middle of the blastoderm. The embryo as a whole is now spherical. *30% Epiboly*, 4.6 h: Staging of epiboly is based on the apparent percentage of the yolk covered by the blastoderm. *50% Epiboly*, 5.5 h: The blastoderm is thinner and more uniform than at 30%. Because of the delay as epiboly pauses at the equator of the yolk cell, this 40-min time period is subdivided into 50% epiboly, germ ring, and shield. At the germ ring, cells at the edge of the blastoderm dive deep to form the hypoblast, which appears as a thickening around the margin of the blastoderm. *Shield* stage, 6 h: Dorsal is to right. Dorsal can be distinguished in the live embryo as a thickening in the germ ring. *70% Epiboly*, 7.5 h, and *80% Epiboly*, 8.5 h: The dorsal side of the blastoderm is thicker, and two layers can be discerned. *95% Epiboly*, 9.5 h: Note the obvious 'Polster' of the hypoblast at the animal pole. *Tailbud* stage, 10 h: Epiboly is complete, and in about 20–30 min, the first segment will form

of cells, the hypoblast, a process which has historically been described as involution (Pasteels 1936), but here we will term the process "hypoblast formation". Epiboly and hypoblast formation use different cellular mechanisms, so we discuss them separately. The feature that unifies them is that they are seen to effect changes at the margin of the blastoderm, the edge where the blastoderm meets the yolk cell. Here the three earliest recognizable tissues of the zebrafish meet: the yolk cell, a large multinucleated cell, the enveloping layer, an epithelial layer covering the blastoderm, and the deep cells of the blastoderm, the cells that form the embryo. This trinity of tissues joins at a ring around the embryo, forming not only a focus for early cell movement in the fish embryo, but a focus for transcriptional activation of many early expressed genes.

2 Mid-Blastula Transition and the Beginning of Cell Motility

The earliest cell rearrangements in the zebrafish blastula are the simple result of cell divisions. In the cleavage-stage embryo, cells divide synchronously and rapidly, every 15 min. In the early divisions, the cleavage pattern is very stereotypic and daughter cells tend to fill the space of the mother blastomere. In fact, many of the cells are physically constrained by their cytoplasmic attachment to the future yolk cell (Kimmel and Law 1985c). Thus there is little net movement. However, in later stages, after about the 64-cell stage, division planes are more stochastic and cells are not constrained by the yolk, and cell divisions often spread daughter cells more than a cell diameter apart. This causes a mild dispersion of progeny and intermingling of distant cell lineages (Kimmel and Law 1985b).

The first motile cells appear at division 10 as the cell cycle slows, signaling the zebrafish mid-blastula transition (Kane and Kimmel 1993). During the next two divisions, the average cell cycle length doubles and, because some cells have a longer cycle than others, synchrony is rapidly lost. This fish mid-blastula transition is similar in many respects to that in the frog, which begins in cycle 12, and that in the fly, which begins in cycle 9 or 10. For example, the timing of the fish mid-blastula transition is controlled by the ratio of nuclei to cytoplasm, as in frogs and flies, and more notably in terms of any consideration of zygotic gene expression, the embryo first becomes transcriptionally active at this time.

Despite the apparent uniformity of the blastomeres of the mid-blastula stage, careful inspection reveals differences amongst the cells in terms of cell division, cell adhesion and motility (Kane et al. 1992). In the early cleavage stages, divisions are incomplete, leaving the yolk mass undivided and the cells cytoplasmically continuous with the yolk. Despite the horizontal divisions that separate the majority of the blastomeres from the yolk, a ring of blastomeres always remains at the yolk-blastoderm margin (Wilson 1889; Kimmel and Law 1985a). These so-called Wilson cells remain cytoplasmically continuous with the yolk. At division 10, as the cell cycle begins to slow, the Wilson cells collapse into the large yolk cell to form a syncytial layer. These nuclei continue to divide rapidly for several more rounds and then arrest at division 13 or 14.

Cells on the outside surface of the blastoderm adhere to each another and prevent intrusion of the extraembryonic medium into the blastoderm. This barrier is highly protective of the embryo such that even small molecules such as alpha-amanitin or ethanol do not readily pass into the embryo when added to the bathing solution. At divisions 10 and 11 these "outside" cells begin to form a thin epithelium. Shortly thereafter, the cells acquire a long asynchronous cell cycle and, when cell divisions occur, daughters remain within the plane of the epithelium (Kimmel et al. 1990). This lineage-restricted epithelium, termed the enveloping layer, ultimately becomes the embryo's first skin, the periderm.

Inside the blastoderm, covered by the enveloping layer and underlain by the yolk cell, are the deep cells of the blastoderm. After the mid-blastula transition, the deep cells acquire a rhythm of divisions intermediate between that of the enveloping layer and the yolk cell in terms of both synchrony and length. Cycle 12 is reached during the high-blastula stage and cycle 13 is reached about the beginning of epiboly. Although as divisions progress in the zebrafish the cells become very asynchronous, cells, on average, tend to enter cycle 14 when the embryo is in early epiboly and gastrulation, a situation remarkably parallel to the relationship of cycle 14 coinciding with gastrulation in flies.

At cycle 10, the deep cells of blastoderm become highly motile, extruding large blebs and moving in seemingly random directions, producing a fine mixing effect at the cellular level. Clones produced by labeling cells at cycles 9 or 10 mix readily with non-clonal neighbors, even before more extreme mixing due to the radial intercalations of early epiboly. On close examination of the deep cells in the blastula stages, there is an impression of a loose aggregation of cells with ample intracellular space, and thus it is no surprise that the stages between the mid-blastula and gastrulation are a favorite time to perform cell transplantations in the zebrafish.

3 Epiboly

Epiboly is the first morphogenetic movement of the embryo, the spreading of the blastoderm to completely enclose the yolk cell. This process is analogous to that of many creatures that produce teleolecithal eggs, where smaller, quickly dividing animal pole derived cells internalize large slowly dividing vegetally derived yolky cells, a strategy seen in distantly related groups thoroughout the animal kingdom.

Epiboly begins in the late-blastula stage, at about 5 h after fertilization, and continues until about 10 h, at which time the yolk cell is covered (Fig. 1). Just before the beginning of epiboly, at high-blastula stage, the blastoderm is perched on the top of a spherical yolk cell as a mound of cells that is some 6–8 cells thick in the middle. At sphere stage, these cells move into a lenticular shape as the embryo as a whole become spherical. At dome stage, the formal beginning of epiboly, the cells begin to spread over the yolk and the yolk cell pokes, or "domes", upward into the blastoderm. The blastoderm then begins a slow vegetal-ward advance, about 100 μm per hour, towards the vegetal pole, hesitating only for a short time at the equator of the yolk cell, coinciding with the formation of the germ ring.

The mechanism of epiboly seems to require an interplay amongst all three tissue types, although the role for each is not completely known (Fig. 2). Before the beginning of epiboly, the yolk cell itself is a sphere, an energetically low conformation for a droplet, and, perched on top of the yolk cell, is the ovoid (but cellularized) blastoderm, perhaps also in an optimum shape when bounded by the yolk cell on one side. If released from the yolk cell at this stage, the blastoderm rounds up. As epiboly begins, the entire embryo becomes spherical. The animal surface of the yolk cell now conforms to the vegetal surface of the blastoderm. The cell membrane of the vegetal three-quarters of the yolk cell forms the equatorial and vegetal surface of the sphere and the cells of the enveloping layer form the animal surface of the sphere. This arrangement is consistent with the view that epiboly involves an interplay of forces transmitted tangentially between the vegetal cortex of the yolk cell and the enveloping layer of the blastoderm.

The function of the zebrafish yolk cell during epiboly seems similar to that of the more thoroughly examined *Fundulus* yolk cell. In *Fundulus*, elaborate connections develop between the yolk cell and the enveloping layer of the blastoderm, suggesting that the yolk cell is towing the enveloping layer (Betchaku and Trinkaus 1978). Before epiboly begins, a portion of the yolk nuclei move out from under the edge of the blastoderm, and, as epiboly begins, "lead" the blastoderm to the vegetal pole. In a now classic experiment conducted by John Trinkaus, where he removed the blastoderm in the mid-blastula, the vegetal movement of the yolk syncytial nuclei continued, demonstrating that the movement of the yolk syncytial layer (YSL) is independent of the blastoderm and autonomous to the yolk cell (Trinkaus 1951). Consistent with a function as a towing motor, the zebrafish yolk cell has microtubules and microfilaments in the proper orientation to effect such movement, and treatments that interfere with these structures interfere with epiboly (Strähle and Jesuthasan 1993; Solnica-Krezel and Driever 1994). Such a model predicts that vast amounts of cell membrane must be recovered from the vegetal side of the yolk cell as the margin of the blastoderm sweeps vegetal-wards, and, indeed, endocytosis occurs vegetal to the blastoderm margin in both *Fundulus* and zebrafish (Betchaku and Trinkaus 1978; Solnica-Krezel and Driever 1994). In fact, this endocytosis may be part of the motor itself, for removal of cell membrane vegetal of the blastoderm would draw the cortex of the yolk cell vegetally.

After it forms, the syncytial layer of the yolk divides for several rapid divisions in both *Fundulus* and zebrafish. These divisions tend to distribute the nuclei evenly over the animal cap of cytoplasm of the yolk cell. In both species, the mitotic arrest of these nuclei occurs just before the beginning of epiboly (Kane et al. 1992; Trinkaus 1992, 1993). In zebrafish, this last mitotic division when counted from fertilization can be either 13, 14 (normally) or 15. Because the syncytium forms at cycle 9, 10, or 11, the last mitotic division when counted from formation of the syncytium can be either 2, 3, 4, or 5. Hence, there seems to be no precise counting mechanism for the number of cell cycles, either after fertilization or after formation of the YSL. One might imagine that the arrest of the cell cycle is caused by the beginning of cell movement of epiboly, as the use of microtubules in mitosis and motility might be expected to be mutually exclusive. Regardless – and formally – we are seeing the terminal mitosis of a differentiated cell.

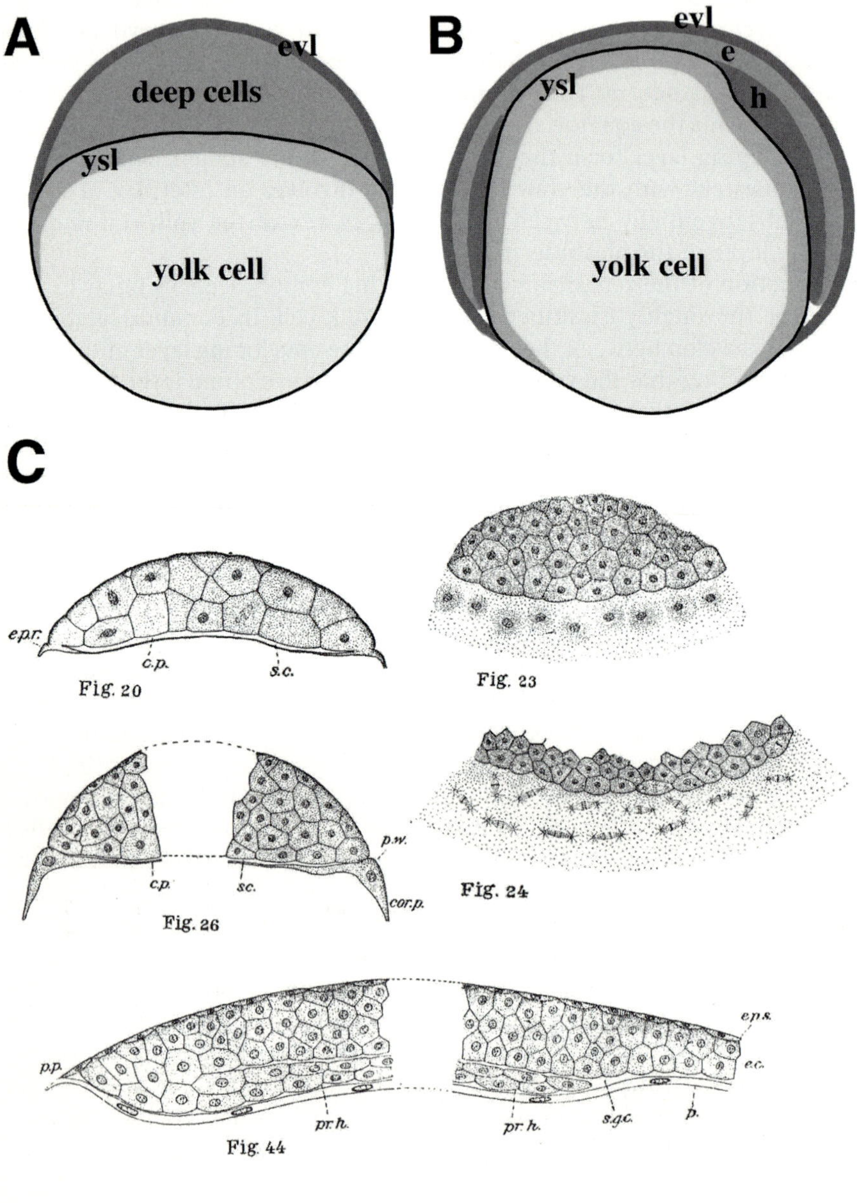

The enveloping layer of the blastoderm seems to be a passive component of epiboly given its tight attachment to the yolk cell and its stretched-out appearance late in epiboly. This epithelium becomes lineage-restricted by division 13, a timing that coincides with the rounding of the embryo from high-blastula to sphere stage. The lineage restriction may reflect the increasing tension on the enveloping layer cells, as we imagine that the mitotic apparatus can be aligned by forces external to the cell, forcing tangential divisions that give rise to daughters within the enveloping layer. While it may be a passive partner in epiboly, the enveloping layer maintains an active process of cell rearrangement within the epithelium. As the margin narrows over the *Fundulus* yolk cell, many enveloping layer cells leave the margin and intercalate amongst more animal-ward cells (Keller and Trinkaus 1987; Fink and Cooper 1996), while constantly maintaining a barrier against the outside environment.

No experiment has demonstrated the role of the deep layers of the blastoderm in epiboly. Before epiboly begins, the cells of the deep layer are motile, but this motility correlates with no future aspect of epiboly. The deep cells seem to merely fill the space between the yolk cell and the enveloping layer. As epiboly proceeds, the yolk cell domes towards the animal pole, thinning the blastoderm and forcing a dramatic radial intercalation of cells as deeply placed deep cells mix with those more shallow (Kimmel and Warga 1987a; Wilson et al. 1995). One interesting aspect of this process is that the blastoderm thins quite uniformly, from about five cells thick at 30% epiboly to about two to three cells thick at 50% epiboly. It is unclear what mechanism could explain this uniformity.

Fig. 2A–C. Schematic of epiboly. **A** Structure of the embryo at high-blastula stage. The enveloping layer (*dark*) covers the deep cells (*light*); both structures comprise the blastoderm. The yolk cytoplasmic layer (*light shading in yolk cell*), exaggerated in thickness, covers the animal pole hemisphere of the yolk cell. The cell membrane of the yolk cell is shown in *black*. **B** Structure of the embryo at 70% epiboly. The dorsal side is to the right. The cytoplasmic layer (and YSL) moves toward the vegetal pole. The enveloping layer follows the cytoplasmic layer of the yolk, remaining on the surface of the embryo. At the margin of the blastoderm, the deep cells turn inward to form the hypoblast (*dark*), the presumptive mesoderm and endoderm; the non-involuting deep cells become the epiblast (*light*), presumptive ectoderm. *ysl* Yolk syncytial layer; *evl* enveloping layer; *e* epiblast; *h* hypoblast. **C** Selected plate figures from Wilson (1889). Most of these figures were based on fixed and sectioned material from samples collected in the early-summer seasons of 1889 and 1890 at the Wood's Holl (sic) station. *Fig. 20*: Section through mid blastula stage. Note the loss of cell synchrony, and the cytoplasmic connections of the marginal cells to the yolk. *Fig. 23*: Surface view of blastula after formation of the syncytial layer of the yolk cell. *Fig. 24*: The syncytial layer of the yolk cell dividing in synchrony. *Fig. 26*: Section through late-blastula stage. Note the hint of the epithelial nature of the outer most cell layer of the blastoderm. *Fig 44*: Section through shield of gastrula -stage embryo, with dorsal up and the animal pole to the right. The enveloping layer (labeled the epidermic stratum), syncytial layer of the yolk cell (labeled the periblast), epiblast and hypoblast can easily be distinguished. *e.p.r.* Early periblastic ridge; *c.p.* central periblast; *s.c.* segmentation cavity (a hopeful homology to the amphibian blastocoel); *p.w.* periblast wall; *cor.p.* cortical periblast; *p.p.* posterior pole of blastoderm; *pr.h.* primitive hypoblast; *s.g.c.* subgerminal cavity (late stage of segmentation cavity); *p.* periblast; *e.c.* ectoderm (epiblast); *ep.s.* epidermic stratum

Studies of cell behaviors in live embryos show that as epiboly begins the movements of the deep cells become ordered and predictable. In an analysis of cell movements, where hundreds of cells were followed simultaneously, a picture emerges showing the paths taken by cells at the margin of the deep cell domain as well as the YSL itself. All these cells (and nuclei) move at about equal and constant velocity in the vegetal-ward direction when epiboly first begins. These movements generate a velocity gradient in the plane of the blastoderm that diminishes animal-ward away from the margin. These movements precede the formation of the hypoblast.

When the germ ring begins to form, the margin of the blastoderm slows and pauses its vegetal-ward movement, beginning a 40 min delay in epiboly. During this delay, there is a substantial reorganization of the nuclei of the yolk cell. Many nuclei slip animal-ward under the blastoderm while others remain beneath the external cell membrane of the yolk cell, similar to that seen in *Fundulus* (Trinkaus 1984b). While this is happening, cells of the epiblast continue to move forward towards the margin, and their vegetal-ward velocity towards the margin even increases, despite the margin itself slowing or stopping.

Could the delay in epiboly be caused by the initial formation of the hypoblast? In mutants that form little or no hypoblast, we still observe these delays in epiboly. One such mutant lacks both the maternal and zygotic function of the gene *one-eyed pinhead*, a gene that is necessary for Nodal function in the embryo. In the mutant, the progress of the margin pauses at the time that the germ ring is forming in wild-type siblings. Furthermore, the nuclei of the yolk cell reorganize, many moving animal-ward under the blastoderm, in a manner indistinguishable from the movements seen in wild-type embryos. The only difference in the mutants is that movement of the deep cell domain towards the margin ceases. This begins close to the margin and spreads like a wave of relaxation animal-ward through the cell sheet. Thus, in the absence of germ ring formation there is still a pause in the process of epiboly indicating that the movements of the yolk may themselves be quite sophisticated and independently timed to coincide with the normal germ ring formation at the margin. How this is achieved is not yet understood.

Later, the dependence of the deep cells of the blastoderm on a yolk cell motor becomes less clear, especially given the transient and changing connections between the deep cells and the yolk cell. For instance, at any one time, individual deep cells are moving relative to the enveloping layer, relative to the margin of the blastoderm, relative to the yolk cell and relative to other deep cells. When epiboly restarts, the vegetal-ward movement of the yolk nuclei is followed by the cells closest to the margin and spreads animal-ward in a renewed wave of movement. Again this reactivation of movement generates a gradient of velocity, consistent with a model of the blastoderm being pulled across the yolk by a force generated close to the margin, quite possibly by the movement of the yolk cell, in a way analogous to the movement of cells of the enveloping layer. It is also possible that the process of internalization itself produces a tension within the plane of the blastoderm from the point of cell reorganization at the margin in the same way that the epiboly of the yolk cell generates tension. Although clearly this movement is not necessary, for, in the maternal-zygotic mutant of *one-eyed pinhead* mentioned above, there is no change in the rate of epiboly.

It remains possible that other types of forces could aid or assist epiboly. In *Xenopus*, which has no enveloping layer and no yolk syncytial layer, epiboly occurs albeit to a lesser extent; it is thought that this movement is mediated initially by radial intercalation (Keller 1980), and later by a little understood pursestring-like movement of the ventral and lateral edges of the lip. Cellular morphology and microtubule staining show that at the vegetal pole enveloping cells exhibit characteristic shape changes typical of pursestring mechanism at the end of epiboly. But do the deep cells themselves accomplish some component of epiboly autonomously? We do not yet know.

4 The Epiboly Mutants

The earliest phenotypes identified in both the Tübingen and Boston screens for new zebrafish mutants were in fact those that affect epiboly. The type phenotype for this class is characterized by a mutation named *half baked* (*hab*), which, when homozygous, slows and arrests epiboly by 60 or 70% (Fig. 3). Four mutants were isolated in the Tübingen screen (Haffter et al. 1996). Three of these mutants segregated with the dominant effect phenotype that we discuss below, hence complementation testing was not possible (Kane et al. 1996). Now we know that all of these mutants map to a single locus, and they are considered alleles. A fifth mutant that affects epiboly, named *volacano*, was isolated in the Boston screen (Driever et al. 1996; Solnica-Krezel et al. 1996), but mapping and complementation of this mutant has not been yet carried out. Based on its phenotype, it is likely to be allelic to *hab*, but this has not yet been confirmed.

After epiboly arrests, the yolk cell begins to lyse near the vegetal pole, and the mutation is lethal by 1–2 days. Mutant embryos that typically die in the first day of development raise the possibility that *hab* could be generally required for cellular metabolism. Such mutants are usually autonomous cell lethals. Fortunately, cells mutant for *hab*, when transplanted into wild-type hosts, survive irrespective of their final position in the host and contribute to a variety of tissues, formally demonstrating that *hab* is not a cell lethal (Kane et al. 1996).

While it is certain that the zygotic activity of *hab* is necessary for epiboly, *hab* activity is also essential maternally. This mutation produces a phenotype described as a zygotic-maternal dominant effect, a novel genetic effect. Females heterozygous for *hab*, when mated to homozygous wild-type males, produce clutches in which a portion of the embryos display slowed epiboly and neural

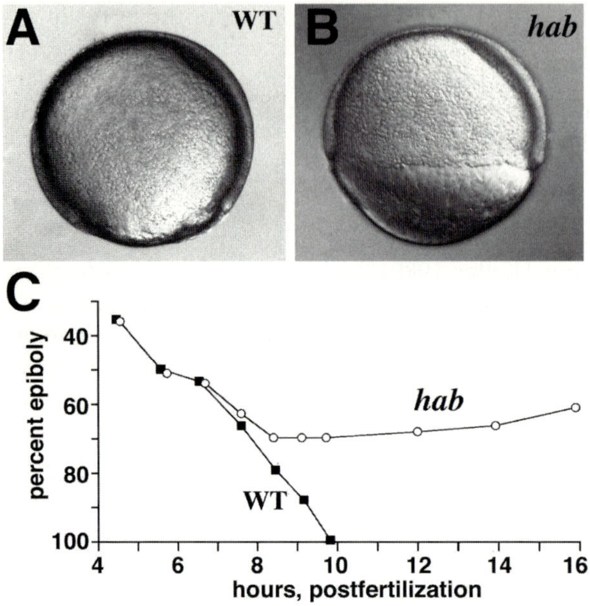

Fig. 3A–C. Arrest of the blastoderm in *half-baked* mutants. **A** Surface view of wild type at 8h and **B** surface view of *hab*-recessive phenotype at 8h. Note that the extent of the movement of the prechordal plate towards the animal pole is the same in both wild type and mutant. **C** Relationship between progression of the deep cells of the blastoderm in normal embryos (*filled squares*) and *hab* embryos (*open circles*)

tube defects, another defect in morphogenesis of the forming neural tube. This dominant phenotype segregates with heterozygous carriers. Males that are heterozygous for *hab*, when mated to homozygous wild-type females, produce normal embryos except for yet another phenotype, a dominant phenotype that enlarges the hatching gland of *hab* heterozygotes. Thus, these mutations produce a dominant maternal phenotype that is only revealed when both the mother and the embryo are heterozygous. The zygotic-maternal dominant phenotypes of those alleles that possess them express weak attributes of the recessive phenotypes. Epiboly is not arrested but slowed, proceeding at about 90% the normal rate. Because of the lethality of *hab* homozygotes, it is not known if *hab* has a recessive maternal phenotype. One interpretation of this complex phenotype is that it is the result of dosage effects adding a maternal and zygotic *hab* gene product. If so, the maternal recessive phenotype should be more severe, potentially revealing the complete – and currently unknown – null phenotype of *hab*.

Closer examination of the *hab* phenotype reveals that mutations in *hab*, when homozygous, do not arrest all aspects of epiboly. Actually, only the epiboly of the deep cell domain is affected. The yolk syncytial layer and the enveloping layer undergo epiboly normally. In wild-type embryos, the yolk syncytial layer normally migrates toward the vegetal pole at about 100 μm/h and the deep cell domain trails typically 50–100 μm behind. In *hab* embryos this gap between the yolk syncytial layer and the deep cells begins to widen at 60% epiboly. Close examination of the live mutant embryo demonstrates that the enveloping layer completes epiboly along with the yolk cell.

The so-called forerunner cells of the dorsal lip are also unaffected in the *hab* mutant (Solnica-Krezel et al. 1996). This group of cells resides at the margin of the blastoderm vegetal of the shield region, tucked between the enveloping layer and the yolk cell. These cells do not enter the hypoblast during epiboly, and lead the shield vegetally during epiboly (Cooper and D'Amico 1996). Later, the progeny of the forerunners are found in the tail, where they are often transiently associated with the epithelial lining of Kupffer's vesicle (Melby et al. 1996). The origin of the forerunners is not completely clear, as it seems some of the cells may be derived from the enveloping layer and some may be derived from the deep cells of the blastoderm. Also, it is difficult to distinguish the forerunner cells from the enveloping cells, to which they are so tightly associated. True to this attraction, these cells are pulled down to the vegetal region of the yolk in *hab* mutants, where they form a small island of cells, and still express many of the same markers, such as NoTail, that they normally express in wild-type embryos.

The zygotic-maternal dominant and the dominant phenotypes of *hab* may give some insight to the function of the *hab* locus. In the case of the zygotic-maternal dominant phenotype, a neural tube defect, the morphogenesis of the neural tube, is affected. This is an ectodermal structure that is traditionally thought to use mechanisms of morphogenesis that act within an epithelium. One such example would be changes in cell shape, but there are many others. Little is known of the mechanisms of morphogenesis of the zebrafish neural keel, but we might expect to find some mechanisms that are related to those used in the formation of the neural tubes in higher vertebrates. If so, then perhaps the basis for *hab* function in the blastoderm may be related to morphogenesis of epithelial structures in the blastoderm, i.e., the epiblast of the blastoderm. Related to this is the *hab* dominant phenotype, the enlarged hatching gland. Here we see no increase in the number of hatching gland cells, which are of mesendodermal origin, but rather an increase in the hatching gland support cells, which are of ectodermal origin. Again, we see changes in a structure that has an epithelial type of morphogenesis.

In any case, it seems that the epiboly of the deep layers of the blastoderm may be genetically separable from the epiboly of both the yolk cell and the enveloping layer: *hab* is necessary only for the movement of the deep cell domain. However, it is not known where *hab* acts. It could act in the yolk cell or enveloping layer, e.g., as a ligand necessary for the moment of the blastoderm, or it could act in the deep cell domain, e.g., as an autonomous or non-autonomous factor necessary for deep cell movement.

5 Towards a Unification of Vertebrate Epiboly

Within the teleosts, epiboly seems well conserved. In other fish model systems, such as *Fundulus* and *Medeka*, and even the sea bass that Wilson described over a century ago, the functions of the enveloping layer and the yolk cell in epiboly seem identical to that of the zebrafish. Although the teleost blastoderm has a longer journey, both zebrafish and *Xenopus* epiboly envelope the yolk and cover the yolk plug. Embryos of the sturgeon, related to teleost ancestors (Bemis 1992; Bolker 1993, 1994), share many morphogenetic similarities with embryos of *Xenopus*, suggesting that the teleost egg is derived from a frog-like ancestral blastula. The extreme teleost epiboly may have evolved after the separation of the fish and tetrapod lineages. As we mention above, the amphibian example of epiboly is thought to be a result of the thinning of the *Xenopus* animal blastocoel roof (Keller 1980). In this example, the roof of the blastocoel intercalates into a single cell layered epithelium, which, in principle, causes the expansion of the animal hemisphere, and is thought to be partially responsible for driving the equatorial regions of the amphibian embryo (containing, e.g., the future dorsal lip) toward the vegetal pole. Where is the teleost yolk cell? The yolky vegetal blastomeres of *Xenopus* have recently been shown to rotate and thus aid in the internalization of material at the dorsal lip (Winklbauer and Schurfeld 1999); this movement, like the movement of the syncytial layer in teleosts, is autonomous. This internalization process could help pull the yolk plug itself into the embryo.

What would an intermediate between teleosts and amphibians look like? Unfortunately, the bones of epiboly have not been well preserved in the fossil record, but perhaps we can imagine what it was like based on the partial characterization of epiboly in *Amia*, a so-called living fossil (Ballard 1986). The embryo of this fish has a telolecithal egg like teleosts, but a holoblastic cleavage pattern like in sturgeons and amphibians. At the border of the blastoderm are large cells termed "hypoblast cells", a name destined to confuse the unwary. These cells are never internalized and remain on the outer surface of the large yolky vegetal cells, in a manner reminiscent of the zebrafish forerunner cells. Although only sectioned material has been examined, these hypoblast cells seem to lead the *Amia* blastoderm vegetally, perhaps as a cellular version of the syncytial layer of the teleost yolk cell.

In the amniotes, epiboly occurs at such a huge scale that it is usually completely overlooked. In the chick, the "epiboly" of the yolk mass is accomplished by the expansion of a ring of cells known as the *zone of juncture* that reside initially at the edge of the area opaca. The zone of juncture is the lateral wall of the blastocoele, connecting the endodermal epithelium with the ectodermal epithelium. With these epithelia following – or in tow, no one knows – this ring transits the entire yolk mass, a process that takes several days. Thus, the zone of juncture moves in a manner analogous to the *Amia* hypoblast cells and the syncytial layer of the zebrafish yolk cell. Moreover, the associated ectodermal epithelium is homologous to the mammalian trophectoderm, both contributing to the somatopleure in their respective embryo. Could the expansion of the trophectoderm be homologous to teleost epiboly? And could the mechanisms – and genes – used to internalize the yolk of lower vertebrates be homologous to those used to implant the mammal oocyte? As we learn which genes do what in these disparate groups of creatures, we may be able to reconstruct part of our fishy past, an exciting prospect.

6 Hypoblast Formation

The process of germ ring formation in zebrafish begins after approximately 5 h of development when the epiboly of the blastoderm has reached the equator of the embryo (Kimmel et al. 1995). The onset of internalization is marked also by a pause in epiboly that lasts for about 40 min in zebrafish but may not be evident in all teleosts (J. Trinkaus, pers. comm.). When the margin pauses we can now first begin to see the most marginal cells start to move deep down the inner face of the YSL to form the germ ring (Fig. 4). A great deal of current interest lies in identifying the signals that orchestrate these events and the cell behaviors that enact them.

Recent studies of how cells internalize to form the hypoblast have concentrated on three species, *Fundulus* (Trinkaus 1996), the rosy barb (Wood and Timmermans 1988) and zebrafish (Warga and Kimmel 1990; Shih and Fraser 1995; Melby et al. 1996; Adams et al. 2002). Similarities and differences have been found between these species, which has generated some debate about whether these are caused by variation on a common theme or if completely different mechanisms are at play. The growing impression is that the signaling mechanisms that pattern the early embryo are highly conserved between species, which would imply that perhaps the responses of cells to those signals might have changed during evolution. How cells reorganize to form the early hypoblast has been postulated to involve one of two possible mechanisms:

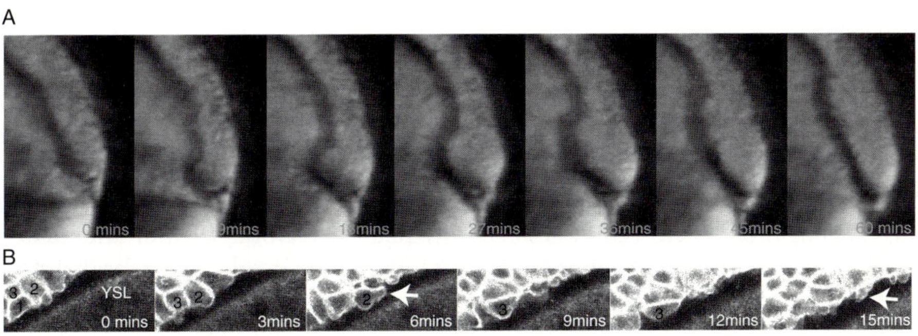

Fig. 4A,B. Cells at the margin internalize when gastrulation begins. **A** This series depicts the reorganization that takes place in the shield, taken at 9-min intervals beginning at 50% epiboly and ending at shield stage. At this low magnification it can be seen that cells move inside then move up towards the animal pole. **B** At higher magnification the movements of individual cells can be resolved. Three marked cells can be followed as they arrive at the margin and ingress individually into the cleft adjacent to the yolk cell (*YSL*). In many instances active cell protrusions can be seen on those cells that are crawling into the hypoblast (*arrows*)

involution – the flowing of a sheet of cells around an edge; or *ingression* – the sinking of individual cells from the epiblast into the hypoblast (Trinkaus 1984b). In both the rosy barb (Wood and Timmermans 1988) and the zebrafish embryo (Warga and Kimmel 1990), cell internalization appears to be restricted to the very edge of the embryo and involves the entire margin of the blastoderm sheet. In these cases, it appears that the whole tissue is folding over the blastoderm margin and tucking in underneath to make the germ ring. Time-lapse analysis of cell movements at the margin of the *Fundulus* embryo shows cells leaving the surface, descending individually into the hypoblast (Trinkaus 1996). Further, these cells often move away from the marginal edge of the blastoderm before moving deep a few cell diameters away from the edge. This behavior would appear to be a very clear example of ingression and counter to our perceptions of the mechanisms of involution. These two extreme mechanisms suggest that quite different cell behaviors might underlie germ ring formation in these diverse species of teleosts. Close examination of cell behavior near the margin during germ ring formation in zebrafish now suggests that the differences between the ingression versus involution models of internalization may not be so vast (Adams et al. 2002). This study confirms that in zebrafish the passage of cells into deeper locations to contribute to the germs does indeed seem to be restricted to those cells at the very margin – within one or two cell diameters (Fig. 5). It also shows that virtually all cells are moving deep together but, as the cells are beginning to descend from the surface of the blastoderm, they appear to be moving as individuals. In many instances, cells can be seen changing places with their neighbors and are extending active protrusions consistent with (but not proof of) autonomous motility. This behavior has components of both models: neighboring cells are moving quite synchronously so the blastoderm sheet at the margin appears to be moving inside but the movements that constitute the actual journey into the hypoblast are as individuals with very strong resemblance to the ingression seen in *Fundulus* (Trinkaus 1996), albeit at a higher density and restricted to the very margin of the embryo.

The specification of endodermal and mesodermal cell fate is dependent on a number of signaling mechanisms. One important signaling pathway without which there is no anterior and trunk mesoderm or endoderm in zebrafish involves ligands of the TGF-β superfamily related to Nodal (Schier and Talbot 2001). Two *nodal*-related genes have been found to be involved in zebrafish fate specification during gastrulation and these are named after the originating mutant lines, *cyclops* (Hatta et al. 1991b) and *squint* (Feldman et al. 1998). The phenotype of the $cyc^{-/-}/sqt^{-/-}$ double mutant is much more severe than with either single mutation alone; these embryos lack axial head and trunk mesoderm and endoderm (Feldman et al. 1998). Furthermore, at the earliest stages of gastrulation, the germ ring and shield both fail to form and early markers of these structures such as *no tail*, *goosecoid* and *floating head* are absent. The cells usually fated to form these structures become part of an expanded ectoderm. (Note, however, there is some mesoderm in the embryo that forms in

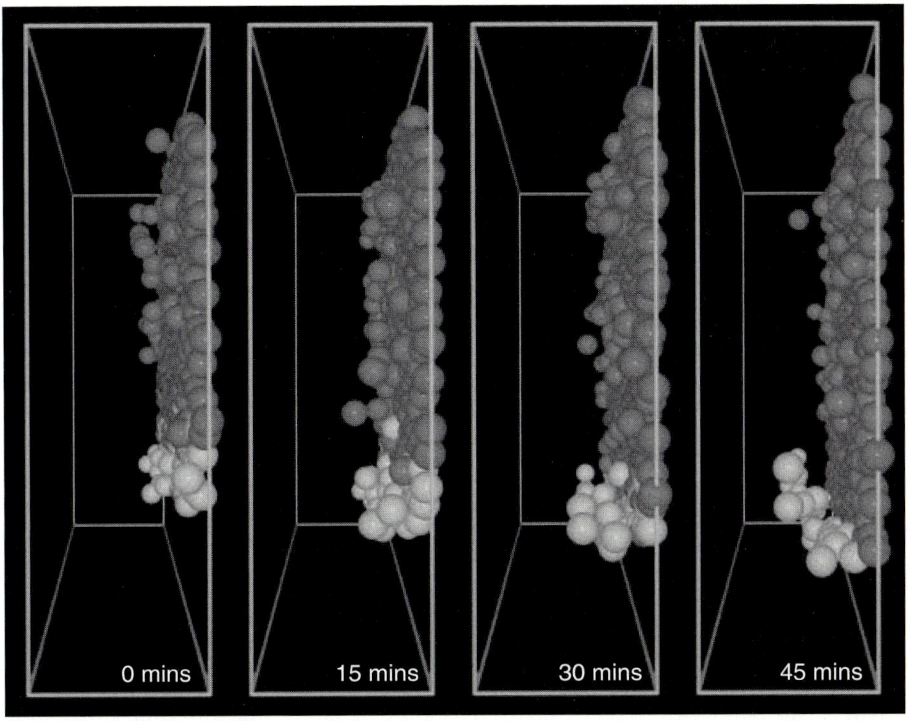

Fig. 5. Tracking the motions of all cells in the surface layer of the epiblast permits us to visualize the ordered internalization of cells at the margin. The cells that will internalize within the early period of germ ring formation are colored *light gray*. Images taken at 15-min intervals

the ventral regions.) This striking phenotype is closely mimicked in another mutant line, *oep*, when both maternal and zygotic transcripts (MZ*oep*$^{-/-}$) are mutated (Gritsman et al. 1999). The gene encoded by the *oep* locus was found to be a member of the EGF-CFC family of extracellular proteins that have also been associated with the MAPK-signaling pathway (Zhang et al. 1998b). The similarity of the phenotype of MZ*oep*$^{-/-}$ and *cyc*$^{-/-}$/*sqt*$^{-/-}$ double mutants suggested a link between Oep protein function and Nodal signaling. This was further strengthened with evidence that neither the *cyc* nor *sqt* gene products could rescue the MZ*oep*$^{-/-}$ phenotype but other downstream factors in the Nodal-signaling pathway could. Together it appears that the *oep* gene product plays an essential role in the reception of Nodal signals and the specification of mesendodermal fates as well as the morphogenetic movements associated with the early reorganization of these cells in the germ ring and shield.

Several groups have taken advantage of the increasing understanding of the role of Nodal signaling in specifying endoderm and mesoderm in the zebrafish to ask how these events relate to the cell behavior of hypoblast formation. MZ*oep*$^{-/-}$ mutants provide a useful model system in which to investigate the cell behaviors that underlie these transformations. Several groups have now used this approach (Carmany-Rampey and Schier 2001; David and Rosa 2001; Adams et al. 2002). The activity of the Oep protein is shown to be cell-autonomous, i.e., it appears to be necessary on the surface of the cell receiving the Nodal signal. Transplantation studies, both MZ*oep*$^{-/-}$ in wild-type hosts and wild-type cells into MZ*oep*$^{-/-}$ hosts, were performed (Carmany-Rampey and Schier 2001) to ask if individual cells with the normal response to Nodal signaling could internalize when their neighboring (mutant) cells could not. Conversely, if mutant cells – unresponsive to Nodal signals – would internalize and join the hypoblast if their wild-type neighbors were all doing so. The result was that wild-type cells could individually move into the hypoblast and express genes appropriate to the mesendodermal fate if transplanted from the margin of the wild-type embryo into the margin of an MZ*oep*$^{-/-}$ embryo where they would continue to be exposed to the appropriate Nodal signals. In the opposite direction it was found that mutant cells could also move deep onto the yolk when the neighboring cells were also doing the same but that, in this case, this did not display other hypoblast-specific morphogenetic movements or express markers associated with a mesendodermal fate, suggesting that they may be taken non-specifically into a deeper location in association with cells undergoing normal morphogenesis but that was not sufficient to cause them to adopt that fate. Further, wild-type cells from the margin region (where they have presumably already been exposed to Nodal signals) moved to a region of blastoderm away from the margin were still able to enter the hypoblast but now presumably by a direct route circumventing the need to move to the margin. This is shown most elegantly in experiments in which Nodal signaling was bypassed entirely by the use of a constitutively active presumptive receptor for Nodal signals, Taram-A (David and Rosa 2001). In this case, even single cells transplanted into the animal pole of normal embryos, far from the known sources of a Nodal signal, moved directly to the yolk surface. This result has important implications for understanding the mechanism of normal hypoblast formation at the margin.

In a complementary comparison of cell behaviors in wild-type and MZ*oep*$^{-/-}$ embryos, an analysis of cell movements and the displacement of the whole epiblast were investigated by following the movements of hundreds of cells in 3D time-lapse movies. This method allows the construction of a picture of the paths cells have taken and the relative movements of cells within the tissue of the margin as morphogenesis proceeds. When internalization begins in normal embryos, the margin of the blastoderm slows and pauses its vegetal-ward movement and there is a substantial reorganization of the nuclei of the yolk cell. Many nuclei move animal-ward under the blastoderm while others remain on the surface (Trinkaus 1993). Coincident with this pause in epiboly of the yolk nuclei, the cells of the blastoderm begin to internalize to form the germ ring.

Analysis of the behaviors of individual cells as this takes place shows cells producing large cytoplasmic protrusions and beginning to move independently of their neighbors. This is in contrast to the ordered and coherent motion of the marginal cells of the epiblast. Internalization is restricted to the cells at the very margin of the blastoderm; in fact, just the first one or two rows of cells move deep into the germ ring at any time. While this is happening, cells of the epiblast continue to move forward towards the margin so in fact their vegetal-ward velocity towards the margin even increases (relative to the preceding period of epiboly) despite the margin itself slowing or stopping. These behaviors are in contrast to those seen in MZ*oep*$^{-/-}$ mutants at this time. In the mutant, the progress of the margin also pauses at about 50% epiboly and nuclei of the yolk cell reorganize in a manner indistinguishable from the movements seen in the wild-type. The significant difference in behavior is seen in the movements of the deep cells of the blastoderm. Whereas in wild-type the rate of movement persists or even speeds-up during the phase of rapid internalization and paused epiboly, cells in MZ*oep*$^{-/-}$ embryos are no longer seen to move deep at the blastoderm edge to form the hypoblast. Movement of the blastoderm towards the margin ceases. This begins close to the margin and spreads like a wave of relaxation animal-ward through the cell sheet. Rather than a wholesale internalization of a large band of marginal cells to form the germ ring, only the closest row of cells moves slightly deeper as the yolk nuclei rearrange but even these cells move as part of the epiblast when epiboly resumes. When epiboly restarts, the vegetal-ward movement of the yolk nuclei is followed by the cells closest to the margin and spreads animal-ward in a renewed wave of movement. Again this reactivation of movement generates a gradient of velocity, consistent with a model of the blastoderm being pulled across the yolk by a force generated close to the margin, quite possibly by the movement of the yolk cell, in a way analogous to the movement of cells of the enveloping layer. It appears that the process of internalization itself produces a tension within the plane of the blastoderm from the point of cell reorganization at the margin in the same way that the epiboly of the yolk cell seems to generate a tension during epiboly. Both of these forces serve to pull cells towards the blastoderm margin. Further, in the absence of Nodal signaling, there is still a pause in the process of epiboly indicating that the movements of the yolk may themselves be quite sophisticated and independently timed to coincide with the normal Nodal-dependent internalization at the margin. How this is achieved is not yet understood.

7 How Do Cells Internalize at the Margin: Involution or Ingression?

There remains some debate as to whether the manner in which cells at the blastoderm margin internalize to form the hypoblast is by the *involution* of the blastoderm sheet or *ingression* of marginal cells. These two models, and their variations in the literature, lead to two different images of the cellular mechanism of reorganization at the edge. We will briefly explore the evidence for and against each model and ask if either of these terms adequately expresses what we currently know of germ ring formation in zebrafish and teleosts in general. We feel that what is most important in any discussion of these events is to elucidate the manner in which the forces that drive reorganization are generated and therein lies the crucial components of the mechanism. We do not yet have all of the information we might like to fully answer this question. We do have a number of studies, as outlined above, that provide various detailed descriptions of the process. From these we must try to intuit what cell behaviors might cause these rearrangements and so the mechanics of the process. In zebrafish (Warga and Kimmel 1990; Concha and Adams 1998; Adams et al. 2002) and another small-egged teleost, the rosy barb (Wood and Timmermans 1988), the reorganization of cells at the margin involves a dense layer of cells in which there is little free space between individual cells. Movement of cells in the epiblast of zebrafish has every appearance of the cells being organized into a cohesive sheet (Concha and Adams 1998; Adams et al. 2002). The majority of movements into the depth of the embryo to form the hypoblast seem to be restricted to cells within one or two cell diameters of the margin and it appears that all cells of local cohorts are moving together – either in the plane of the epiblast or in turning the edge to move deep into the germ ring and hence the hypoblast. However, it is not the case that only cells within touching distance of the margin are able to leave the epiblast surface; cells within the whole plane of the epiblast are able to move within its depth both deep and shallow to the surface; however, these movements change only their location within the epiblast and do not apparently move cells into the hypoblast. It would appear that the vicinity of the margin holds special properties or signals that cause the reorganization that leads to germ ring formation. The transplantation studies described above suggest that it is the local signals rather than any particular physical property of the margin that might be most important. Similar conclusions are drawn for *Fundulus*, where cell movements to form the germ ring are clearly as individual ingressions. The results of imaging cell movements in MZ*oep*$^{-/-}$ mutant embryos suggest that this process is active, generating a force, quite likely autonomous to those cells moving inside. The act of internalization generates sufficient force to pull the blastoderm surface towards the margin just as it does during epiboly. Tracing movements of cohorts of cells at the edge as they move inside suggests that, although adjacent cells move together, they are able to change their relative positions within the confines of the margin, consistent with an autonomous component of motion. The finding that single transplanted cells specified (and probably committed) to a mesendodermal fate are able to move deep to the YSL surface wherever they are located in the embryo suggests that what is special about the margin region are the signals it provides to initiate these cell responses rather than its physical properties.

An alternative view might be that the margin is either pulled inside by the action of cells deeper in the germ ring, perhaps beginning their migration animal-ward, or cells could be adhering to the surface of the YSL as it changes shape beneath the margin. We can see that many of the nuclei of the YSL move underneath the blastoderm during the pause in epiboly (Adams et al. 2002). We do not yet have evidence that the surface of the yolk itself also moves but it does remain a possibility that at least some of the force needed to internalize cells could be provided by a reorganization of the nuclei of the yolk cell. Clearly, the entire surface of the yolk cell cannot change shape since the attachment sites of the enveloping layer just external to where these movements are taking place remain essentially stationary throughout this period of the pause but then move when epiboly resumes. The most salient features of the process, as they appear at present, are that the entire marginal region of the epiblast will move deep to form the germ ring, individual cells change their relative locations while this is happening and that these movements are restricted to the very edge of the epiblast, along the internal face of the yolk cell.

It could then be the case that in the embryos of zebrafish, and perhaps other small-egged teleosts where cell density is relatively high at the onset of gastrulation, cell movements to form the hypoblast do also involve the internalization of individual cells by a process that is mechanically similar to ingression but that the close proximity of so many cells all undergoing the same reorganization gives the appearance of a flow of cells into the hypoblast. Whether in such a situation the process is called ingression or involution is unclear and perhaps not useful to pursue. The formal definition of involution – the flowing of a sheet of cells – implies that the organization of cells into a sheet is an integral part of the reorganization. This might not need to be the case here but this force does seem to be propagated within the plane of the epiblast. A question remains as to whether cell-cell connectivity is needed to achieve germ ring formation. It is likely that cell adhesion is needed to transmit the force generated by this rearrangement within the plane of the epiblast to generate the observed velocity gradient within the plane of the epiblast.

8 Conclusions and Prospects?

We have shown that even these earliest behaviors of cells in zebrafish embryo are both varied and sophisticated. Within a few hours of the onset of development, gastrulation has begun, the early germ layers become segregated and the embryo is completing its epiboly over the surface of the yolk. A number of cellular mechanisms appear to be involved in these events and we are just beginning to unravel their individual roles. The power of genetic analysis, high-resolution time-lapse imaging and the ease of experimental manipulation hold great promise for the answering of many of the remaining questions in the coming years. As we and others are starting to identify the genes and molecules that are used in these processes, not only will these tools aid us in identifying the mechanisms that drive epiboly and formation of the hypoblast, but they will also help us understand the relationship of the earliest stages of development and morphogenesis across the animal kingdom.

Cellular and Genetic Mechanisms of Convergence and Extension

Lilianna Solnica-Krezel[1] and Mark S. Cooper[2]

1 Introduction

The zebrafish blastula consists of a mound of blastomeres that lies atop a large syncytial yolk cell (Fig. 1). Around the mid-blastula stage, the initial dorsoventral polarity of the embryo has already been established (see Sakaguchi et al.; Hibi et al., this Vol.) and the induction of mesendoderm has begun (see Kimelman and Schier, this Vol.; Warga and Stainier, this Vol.). During the course of the next 10 h of development, cells within the blastoderm will acquire specific fates (see Hammerschmidt and Mullins, this Vol.), as well as engage in morphogenetic movements to secure appropriate positions in the nascent embryo. These cell migrations and rearrangements transform the mound of blastomeres into an embryo that features a typical vertebrate body plan. During gastrulation, epibolic movements spread the blastoderm and the yolk syncytial layer towards the vegetal pole. Germ layer formation begins when mesendodermal cells move beneath ectoderm (see Kane and Adams, this Vol.). Epiboly and mesendoderm internalization take place in a relatively uniform manner along the entire dorsoventral axis, except for a small region in the gastrula organizer. This chapter focuses on the third gastrulation movement, convergence and extension, through which dorsoventral and anteroposterior embryonic polarity become morphologically distinct. The process of convergence and extension fashions the dorsoventral axis of the nascent embryo by narrowing its tissues from back to belly. Through a simultaneous elongation of tissues from head to tail, anteroposterior polarity is elaborated.

Convergent extension can be broadly defined as all morphogenetic processes that result in a *simultaneous* mediolateral narrowing and anteroposterior elongation of tissues or cell populations in vertebrate gastrula (Fig. 1; for review see Keller et al. 2000). Convergence, the mediolateral narrowing of tissue, can be associated with its anteroposterior elongation (extension), as shown in Fig. 1C. However, extension is not the only possible tissue transformation

[1] Department of Biological Sciences, Vanderbilt University, Nashville, Tennessee, USA
[2] Department of Zoology and Center for Developmental Biology, University of Washington, Seattle, Washington, USA

Results and Problems in Cell Differentiation, Vol. 40
L. Solnica-Krezel (Ed.): Pattern Formation in Zebrafish

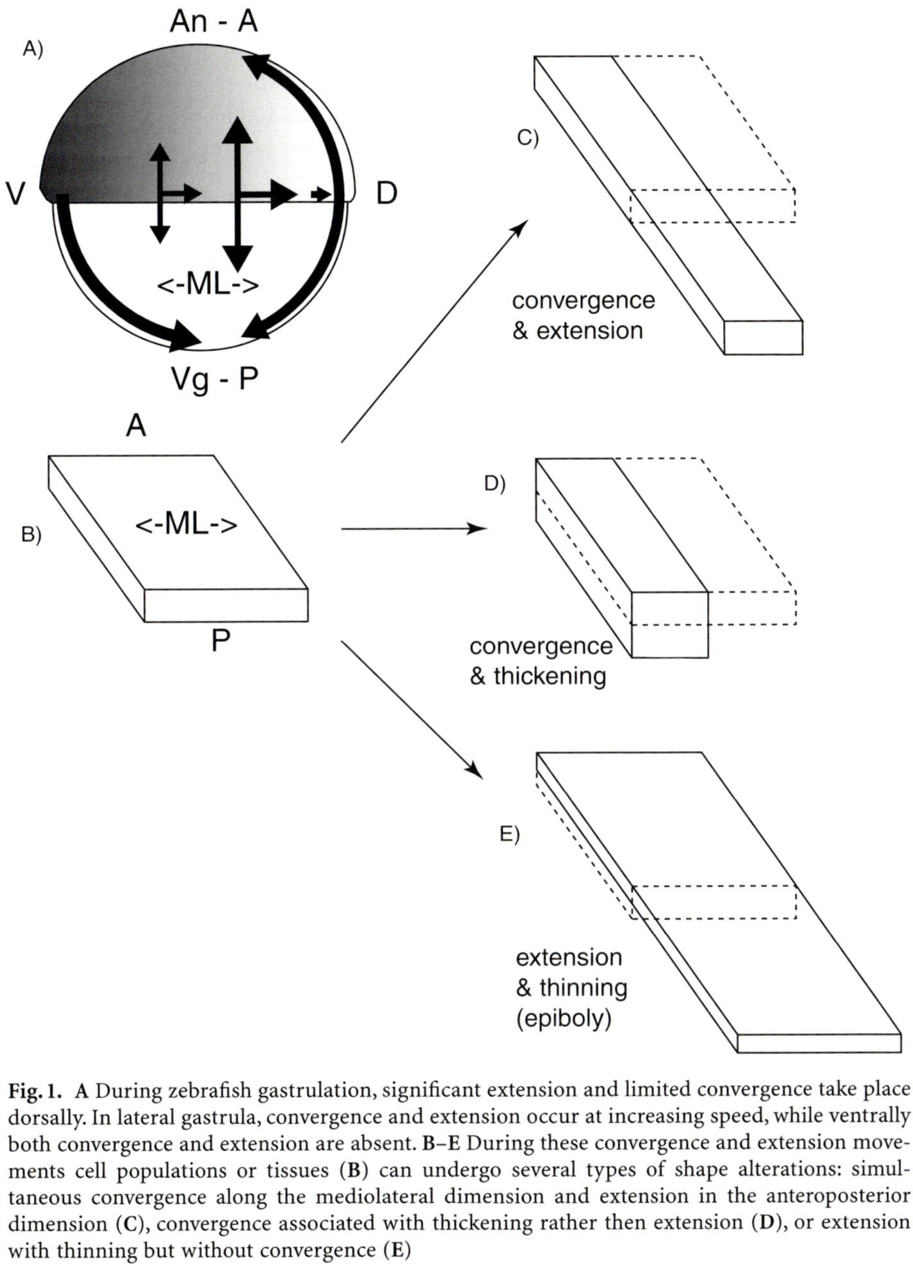

Fig. 1. **A** During zebrafish gastrulation, significant extension and limited convergence take place dorsally. In lateral gastrula, convergence and extension occur at increasing speed, while ventrally both convergence and extension are absent. **B–E** During these convergence and extension movements cell populations or tissues (**B**) can undergo several types of shape alterations: simultaneous convergence along the mediolateral dimension and extension in the anteroposterior dimension (**C**), convergence associated with thickening rather then extension (**D**), or extension with thinning but without convergence (**E**)

associated with convergence. During gastrulation, tissue convergence can lead to its thickening rather than extension (Fig. 1D; e.g., somitic mesoderm, Wood and Thorogood 1994). Alternatively, an embryonic tissue might extend when decreasing its thickness. Such radial thinning has been observed during epiboly (Fig. 1E), a process that does not have to be associated with tissue convergence (Warga and Kimmel 1990).

This chapter summarizes the progress achieved towards understanding the cellular and molecular mechanisms underlying convergence and extension in zebrafish. It has been demonstrated that the speed and magnitude of convergence and extension movements are regulated spatiotemporally in the course of zebrafish gastrulation: at least three distinct domains of convergence and extension movements along the dorsoventral gastrula axis are observed. Advances have been made towards identifying the migratory cell behaviors that drive these distinct convergence and extension movements. We will next discuss the relationship between convergence and extension movements and the concurrent movements of epiboly and migration of mesendoderm. We also discuss mutant analyses, which suggest that the processes of convergence and extension are genetically separable during zebrafish gastrulation. Hence, we will use the term "*convergence and extension*" to refer to the processes that narrow and elongate tissues in the zebrafish gastrula. Furthermore, mutational analyses and overexpression experiments have implicated several signaling pathways and cell-cell interaction molecules in regulation and execution of normal convergence and extension movements. Possible mechanisms, which act to coordinate convergence and extension movements with ongoing cell fate specification events, will also be considered.

2 Compaction at Blastula Stages

Whereas convergence and extension takes place predominantly during the gastrula period, recent studies have revealed that dorsal narrowing of tissues is already initiated at the blastula stages (Warga and Nüsslein-Volhard 1998). In these studies, single marginal blastomeres, which were labeled randomly with lineage tracer at the mid-blastula stage, gave rise to small clones. These clones were unexpectedly concentrated in the dorsal half of the late blastula, compared to lateral and ventral clones. Following the onset of epiboly, cells in the dorsal, but not ventrolateral region of the blastula, flatten and increase the number of intercellular contacts, suggestive of strengthened cell adhesion. Together, these observations indicate that in the dorsal region of the blastula a process of cellular compaction leads to a displacement of marginal cells toward dorsal and simultaneous displacement of dorsal midline cells toward the margin (Fig. 2A). Hence, before the onset of gastrulation, an early process of dorsal convergence and animal-vegetal (anteroposterior) shortening of the dorsal blastoderm is already taking place in the zebrafish embryo.

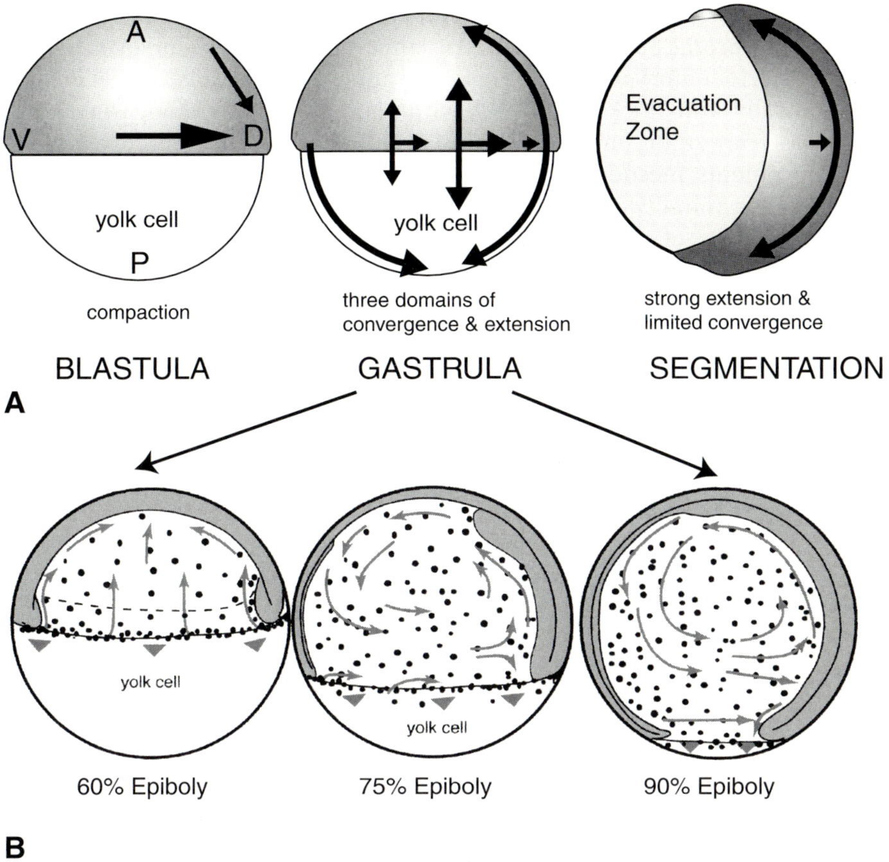

Fig. 2A,B. Spatiotemporal regulation of convergence and extension movements in zebrafish. **A** Convergence and extension movements in blastoderm. Compaction at blastula stages. Three domains of distinct convergence and extension movements during the gastrula period: ventral domain of inhibited convergence and extension, lateral domain of increasing convergence and extension movements and dorsal domain of strong extension and limited convergence movements. Strong extension and slow convergence movements at segmentation stages. **B** Diagram of YSL nuclear movements during gastrulation. *Arrows* summarize the regional trajectories of YSL nuclei (not necessarily the displacements of individual nuclei). Nuclei begin converging toward the dorsal midline during mid-gastrulation (75% epiboly). Concurrently, nuclei located in the axial YSL domain extend along the AP axis (**B** adapted from D'Amico and Cooper 2001)

3 Distinct Domains of Convergence and Extension Movements in the Zebrafish Gastrula

The majority of convergence and extension movements occur at the gastrula stages which is clearly manifest by morphological changes in the embryo. At the onset of gastrulation, the blastoderm in the form of an inverted cup covers the animal half of the spherical, giant yolk cell (Fig. 2A). Owing to convergence and extension movements, the entire embryo, the yolk cell and the blastoderm covering it become highly asymmetric, thus revealing the dorsoventral and anteroposterior embryonic axes (Figs. 2A,B and 5). The prospective germ layers thin ventrally and thicken dorsally, while simultaneously elongating in the anteroposterior embryonic axis. Eventually, due to both convergence and epibolic movements, mesoderm- and endoderm-depleted evacuation zone forms ventrally. The initially spherical embryo and the yolk cell acquire an ovoid shape at the end of the gastrula period (9.5h postfertilization, hpf). This is apparent by the completion of epiboly when the blastoderm encloses the vegetal pole of the yolk cell. At the same time, the elongating embryonic axis becomes clearly visible in the dorsal midline, with the head rudiment reaching the animal pole and tailbud residing at the vegetal pole. Whereas epiboly and internalization of mesendoderm are largely complete by this stage, convergence and extension movements continue into segmentation stages. This is evidenced by the steady elongation of the anteroposterior embryonic axis and mediolateral narrowing of organ rudiments such as neural plate, notochord and somites (Kimmel et al. 1995; Sepich et al. 2000).

Convergence and extension movements occur in all germ layers, with some interesting differences (Fig. 3). It appears that the entire mesoderm and endoderm, except the most ventral mesodermal cells (see below), engage in convergence and extension (Warga and Nüsslein-Volhard 1999; Myers et al. 2002). Both mesoderm and endoderm undergo dramatic anteroposterior elongation and narrow mediolaterally. However, in the dorsal region, extension of mesendodermal cell populations exceeds that of ectodermal cell groups during the gastrula period (Myers et al. 2002). Effectively, most of the prospective mesodermal and endodermal cells gather dorsally, leaving a zone depleted of cells in the ventral gastrula regions (Fig. 3B,D). In contrast, convergence and extension of nascent neuroectoderm are accompanied by mediolateral expansion of non-neural ectoderm (Fig. 3A,C; Woo and Fraser 1995; Concha and Adams 1998). Consequently, there is no region devoid of cells in the ventral ectoderm.

The speed and magnitude of convergence and extension movements of different tissues and cell populations are both temporally and spatially controlled in zebrafish gastrulae. We have just begun to understand this dynamic process, by tracing behavior of individual ectodermal cells with Nomarski optics (Concha and Adams 1998), tracing clonal progeny of labeled neural progenitors (Kimmel et al. 1994), and measuring movements of labeled mesendodermal or ectodermal cell groups (Heisenberg et al. 2000; Sepich et al. 2000). These analyses demonstrate that both convergence and extension movements are slow during the first part and accelerate during the second part of gastrulation.

Measurements of labeled cell populations in gastrulating embryos revealed that speed and magnitude of convergence and extension movements also vary along the dorsoventral axis of the gastrula (Fig. 2A). In the dorsal region of both the hypoblast and epiblast, cell populations undergo strong extension, but

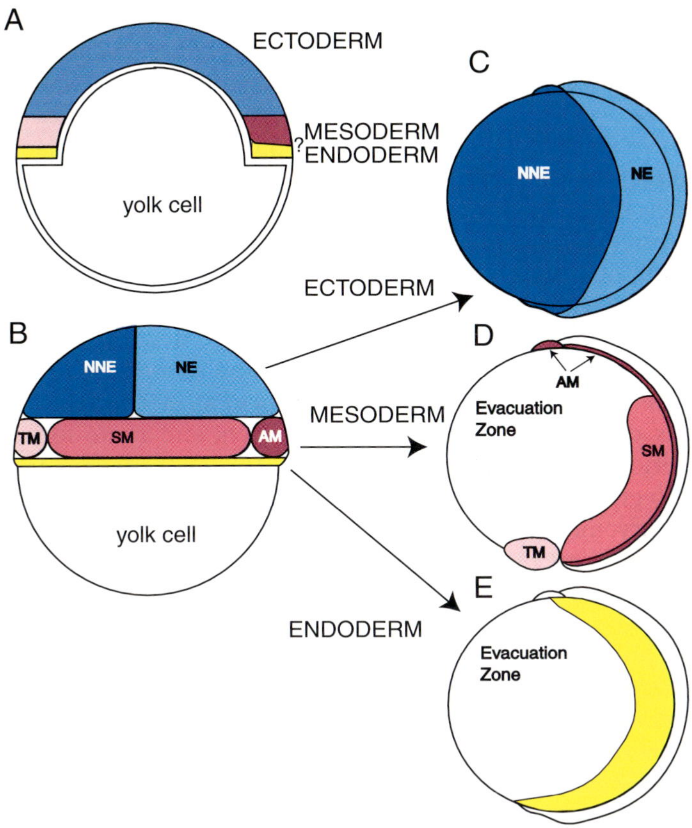

Fig. 3A–E. All germ layers of the zebrafish gastrula engage in convergence and extension movements. **A** At the onset of gastrulation, three germ layers can be distinguished within the blastoderm covering the animal hemisphere of the yolk cell. Prospective ectoderm forms the most animally positioned layer, future mesodermal cells form the middle layers, whereas a thin layer of endoderm is positioned at the vegetal margin of the blastoderm. However, the depth and distribution of presumptive endodermal cells within the marginal zone are largely unknown (*question mark*). **B** Within ectoderm, prospective neural tissues reside dorsally, whereas non-neural ectoderm occupies the ventral region. Within mesoderm, axial mesoderm resides dorsally, somitic mesoderm occupies a broad dorsolateral region, and tailbud tissues are positioned ventrally. **C** Owing to epibolic movements ectoderm expands to cover the entire embryo. Dorsal convergence and extension movements narrow prospective neuroectoderm mediolaterally and elongate it anteroposteriorly, while non-neural ectoderm concomitantly expands in the mediolateral direction. **D** Upon internalization, dorsolateral mesoderm narrows mediolaterally and extends anteroposteriorly due to the combination of anterior migration, epiboly and convergence and extension movements. By contrast, the ventralmost mesoderm migrates to the tailbud. **E** Internalized endoderm undergoes anteriorward migration, epiboly and convergence and extension to gather dorsally along the nascent anteroposterior embryonic axis only limited convergence. Cell populations originating in the lateral region of the gastrula (90° from the dorsal midline) initially converge and extend slowly. These cell groups increase their speed of convergence and extension movements as they move dorsally. Extension of the lateral mesendodermal cell populations is smaller than extension of equivalent cell populations in the dorsal midline (Sepich et al. 2000). Finally, convergence and extension movements of mesodermal cells are largely absent in a ventral region of blastoderm margin, referred to as the "*no convergence no extension zone*" (Myers et al. 2002). Cell groups occupying this small 20–30° arc of the ventral gastrula margin move beneath ectoderm, where they disperse. Instead of engaging in convergence or extension movements as the more dorsally located cells do, these cells migrate towards the vegetal pole and contribute to the forming tailbud at the end of epiboly (Fig. 3B,D; Myers et al. 2002). Next these ventral tailbud cells undergo the process of subduction, and move to underlie the dorsal (anterior) tailbud (Kanki and Ho 1997). During the segmentation period the tailbud region engages in convergence and extension, contributing significantly to embryonic axis elongation. Collectively, domains of convergence and extension movements reside along the dorsoventral axis of the zebrafish blastoderm margin: (1) the *dorsal domain* of strong extension and limited convergence, (2) the *lateral domain* of increasing convergence and extension movements, and (3) the *ventral domain* straddling 20–30° arc of the ventral gastrula margin in which both convergence and extension are absent (Fig. 2A). Whether there is a sharp or rather a gradual transition between the lateral domain of increasing convergence and extension movements and the dorsal domain of slow convergence and strong extension, and where this transition occurs in the gastrula, remains to be determined. Based on the available data, we predict that the dorsal domain occupies ca. 40–50° of the dorsal margin (Sepich et al. 2000; Myers et al. 2002; D.S. Sepich and L. Solnica-Krezel, unpubl. observ.).

It is well recognized that the yolk syncytial layer (YSL) endowed with a set of nuclei engages in epiboly (see Sakaguchi et al.; Kane and Adams, this Vol.). Yolk syncytial nuclei initially forming a ring within the external YSL just below the blastoderm migrate animally to populate the internal yolk syncytial layer below the blastoderm. A subset of the yolk syncytial nuclei also moves vegetally in a process that requires an intact microtubule network (Solnica-Krezel and Driever 1994). Recent 4D confocal imaging studies have revealed that yolk syncytial nuclei undergo long-range convergence and extension movements during gastrulation and early segmentation (D'Amico and Cooper 2001). During mid-gastrulation, YSL nuclei located in ventral and lateral positions in the embryo begin to converge toward the dorsal side where they engage in extension movements along the anteroposterior axis. YSL nuclei intercalate along the dorsal axis, and form a spatially distinct axial domain of dorsal YSL nuclei that underlies the notochordal rudiment (Fig. 2B). Those nuclei already present near the dorsal midline undergo axial extension, and move longitudinally, either toward the animal or vegetal pole depending on their latitudinal positions. As the axial YSL domain undergoes extension, YSL nuclei located initially near the animal pole become displaced ventrally and laterally. Hence, the overall convergence and extension movements of nuclei in the YSL parallel the morphogenetic movements of cells within the overlying blastoderm (Fig. 2). It is a fascinating question to consider to what degree the mechanisms underlying the regulation and execution of these morphogenetic movements are similar.

4 Cellular Behaviors Effecting Convergence and Extension Movements

Whereas all germ layers undergo convergence and extension movements, the cellular organization of these tissues differs. Prospective ectodermal cells congregate and increase intercellular contacts as gastrulation begins, and thereafter move as a coordinated sheet (Concha and Adams 1998). By contrast, mesodermal cells move as individuals or small groups, while large endodermal cells move exclusively as individuals (Trinkaus et al. 1992; Warga and Nüsslein-Volhard 1999). However, as discussed above, small or large cell groups labeled in the equatorial region of ectoderm or mesendoderm at the onset of gastrulation undergo very similar convergence and extension movements depending on their position along the dorsoventral axis (Sepich et al. 2000; Topczewski et al. 2001). Whereas dorsal cell populations undergo dramatic anteroposterior extension with little convergence, the lateral cell populations converge dorsally and significantly extend anteroposteriorly at an increasing speed. Hence, all dorsolateral clones or polyclones labeled at the beginning of gastrulation are transformed into "*cellular strings*" concentrated dorsally and elongated anteroposteriorly by the end of the gastrula period (Kimmel et al. 1994; Sepich et al. 2000; Myers et al. 2002).

Current evidence indicates that convergence and extension of zebrafish cell populations are driven by a suite of cellular behaviors whose diversity we have just begun to elucidate. Employment of distinct sets of migratory behaviors likely accounts for spatiotemporal differences in the speed of convergence and extension movements described above. In this regard, the challenge is to identify all cell behaviors that contribute to convergent and extension movements expressed in cell populations at specific times and regions of the gastrula and link them to specific genetic pathways or networks.

Domingo and Keller (1995) have suggested that embryonic cells can be viewed as expressing both *behavioral* and *structural phenotypes*. Mediolateral intercalation behavior (MIB) is an example of a behavioral phenotype that is expressed in notochordal cells, a morphogenetic cell behavior that drives the dramatic convergence and extension of the notochordal rudiment (Domingo and Keller 1995). These same notochordal cells will later form intracellular vacuoles (i.e., a structural phenotype), as the notochord straightens during later stages of notogenesis.

We now know from mutagenesis studies in zebrafish that a number of late-acting genes are required for notochord vacuolation (Odenthal et al. 1996a; Stemple et al. 1996). Through similar mutant analysis, a number of early-acting genes have been linked to the expression of MIB in the zebrafish embryo. In the next sections of this chapter, we discuss how these genes modify the *behavioral phenotypes* of deep cells, as they undergo their stereotyped morphogenetic activities during zebrafish gastrulation.

4.1 Epiboly and Anteriorward Mesendoderm Migration Contribute to Convergence and Extension

When considering cell behaviors underlying the mediolateral narrowing and anteroposterior elongation of tissues, it is important to note that, from the onset of the gastrula period (6 hpf) until the end of epiboly (9.5 hpf), convergence, extension, internalization and anterior migration of mesendoderm and epibolic movements occur simultaneously. In fact, all of these movements variably contribute to mediolateral narrowing and/or anteroposterior elongation of germ layers (Fig. 4). Because epibolic movements thin and spread all germ layers along the animal vegetal axis (Fig. 1E), which anticipates the anteroposterior embryonic axis, they contribute to tissue extension during this time period (Fig. 4B). The significance of epiboly to convergence and extension is underscored by mutations and impairment of microtubules with taxol that appear to interfere specifically with epibolic movements. Several mutations have been found such as *half baked* and *volcano* that block epibolic movements of germ layers without affecting epiboly of the enveloping layer or YSL (Kane et al. 1996; Solnica-Krezel et al. 1996). At the end of the gastrula period, axial mesoderm in these mutants reaches the normal anterior position but is shortened in the posterior region compared to wild type. Furthermore, axial mesoderm is also broadened in these mutants, suggesting that dorsal convergence movements are correspondingly reduced. However, formation of kinked and twisted notochords has been observed in the epiboly mutants, suggesting that some aspects of notochord extension, likely driven by cell vacuolation, can occur despite arrested epiboly (Kane et al. 1996). These observations clearly show that epibolic movements are an important component of the convergence and extension process. Intriguingly, the reverse does not appear to be true in zebrafish. Even compound *knypek;trilobite* or *bozozok chordino* mutants, in which convergence and extension movements are dramatically impaired, complete epiboly on schedule (Marlow et al. 1998; Gonzalez et al. 2000). It is thought that epiboly at the blastula stages is driven primarily by radial intercalation whereby cells from deeper cell layers intercalate between their more superficial neighbors, thus thinning and spreading the tissue (Warga and Kimmel 1990). To what degree radial intercalation continues to be the driving force of epiboly during the gastrula period (6–9.5 hpf) remains to be elucidated (see Kane and Adams, this Vol.).

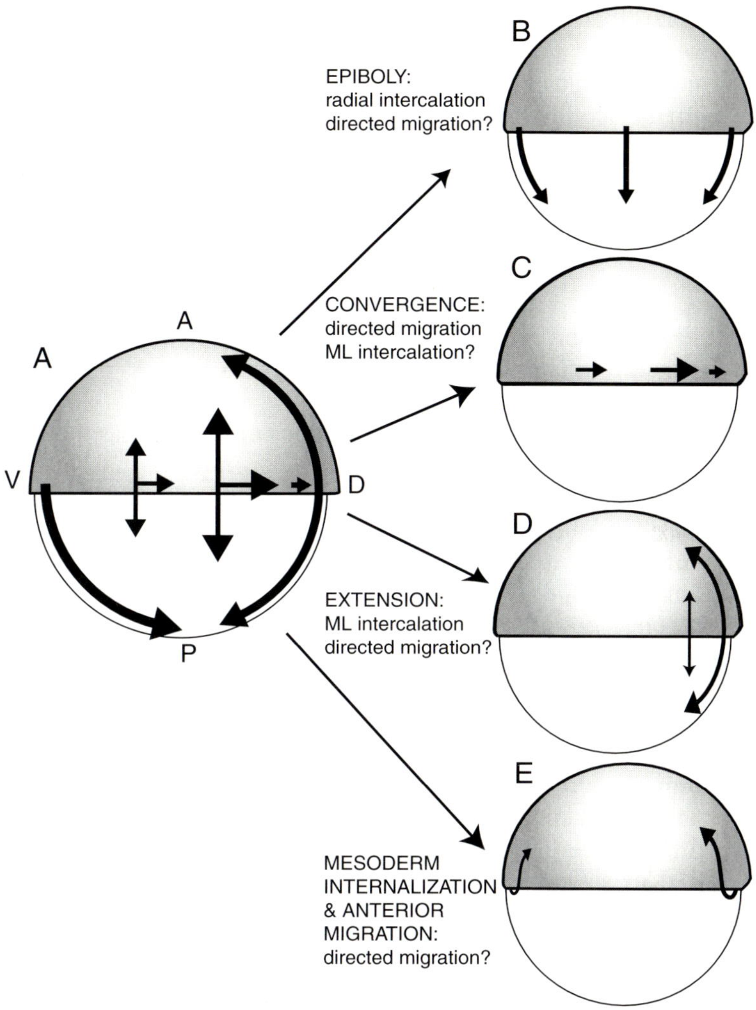

Fig. 4. **A** Convergence and extension movements of ectodermal and mesodermal tissues along the dorsoventral axis of the zebrafish gastrula involve several types of cell movements and rearrangements. **B** Epiboly, vegetal spreading of tissues due to radial cell intercalation and directed migration, contributes to tissue extension. **C** Convergence of tissues is thought to be driven by directed cell migration and mediolateral cell intercalation. **D** Anteroposterior elongation of tissue is also effected by mediolateral cell intercalation and possibly directed cell migration. **E** Upon internalization, anterior migration of mesoderm likely involves directed cell migration

Upon internalization, mesendodermal cell populations not only undergo epiboly towards the vegetal pole, but also spread towards the animal pole, resulting in anteroposterior extension of tissues (Warga and Kimmel 1990). During *Xenopus* gastrulation, convergent extension movements aid in involution of the marginal zone (Shih and Keller 1994). It is noteworthy that in zebrafish even severe inhibition of convergence and extension movements in *knypek;trilobite* or *bozozok chordino* double mutants does not appear to interfere with mesendoderm internalization (Marlow et al. 1998; Gonzalez et al. 2000). However, in *bozozok chordino* mutants, anterior migration and thus extension of mesoderm appears inhibited with most of the mesodermal cells migrating vegetally and accumulating in the tailbud region (Myers et al. 2002).

4.2 Directed Migration is a Key Cell Behavior Underlying Convergence and Extension in Lateral Regions of the Gastrula

What cellular behaviors underlie simultaneous convergence and extension of lateral cell populations (Fig. 4A)? Several studies indicate that, in the lateral region of the gastrula, individual cells are engaged in epiboly, convergence and extension simultaneously. Moreover, mesodermal and endodermal cells are also involved in anteriorward migration. Therefore, a single cell, depending on its position, might be moving towards the dorsal midline (thus contributing to tissue convergence), and also engage in epiboly towards the vegetal pole. The net movement direction of such a cell can be depicted by a vector pointing dorsovegetally as a consequence of dorsally directed convergence movement and vegetally directed epibolic and/or extension movement. Generally, cells residing animally with respect to the gastrula equator exhibit a net dorsoanimal movement. By contrast, cells located below the equator move dorsovegetally (Concha and Adams 1998). Exceptions are cells residing just animally to the equator that converge dorsally without significant displacement along the animal-vegetal axis (Trinkaus et al. 1992; Concha and Adams 1998; Sepich et al. 2000). Thus, a lateral cell population undergoes convergence and extension, as some of the cells move dorsally, other cells move in the dorsovegetal direction, and the remaining cells move dorsoanimally (Fig. 4A).

The speed of convergence movement of lateral ectodermal and mesodermal cells increases in the second half of gastrulation when these cells move closer to the dorsal midline (Concha and Adams 1998; Sepich et al. 2000). Time-lapse analyses revealed that the increased speed of lateral cell populations' convergence can be explained by individual cells taking more direct routes as they move closer to the dorsal midline. Although the lateral mesodermal cells migrate with a relatively constant total speed, they take much more direct routes towards dorsal and meander significantly less in the dorsolateral region, achieving a faster rate of dorsal translocation (Fig. 6; D.S. Sepich and L. Solnica-Krezel, unpubl. observ.). Such an increasingly persistent dorsal migration of cells was observed in pioneering studies of Trinkaus during convergence movements in the germ ring of another teleost fish, *Fundulus* (Trinkaus et al. 1992). These studies raise an intriguing possibility that convergence of lateral cell populations in teleost gastrula may be driven by directed cell migration up a chemoattractant gradient, which peaks dorsally. The migratory behaviors of lateral cells towards the vegetal and animal poles that drive epiboly of all germ layers, and anterior migration of mesoderm and endoderm, respectively, and contribute to anteroposterior elongation of germ layers during the gastrula period remain to be identified.

4.3 Mediolateral Cell Intercalation is a Key Cell Behavior Underlying Convergence and Extension in Dorsal Regions of the Gastrula

As lateral mesendodermal and ectodermal cell populations approach the dorsal midline late during gastrulation, their speed of convergence decreases, but extension movements intensify (Kimmel et al. 1994; Heisenberg et al. 2000; Sepich et al. 2000). Strong extension with slow convergence of tissues in the dorsal region of the gastrula appears to be driven in large part by mediolateral cell intercalation. Mediolateral intercalation cell behavior is best understood in *Xenopus* gastrula, where it is thought to be the main motile behavior underlying convergent extension of mesodermal tissues (Keller et al. 2000). In the course of convergent extension in *Xenopus*, cells become mediolaterally elongated. The mesodermal cells develop bipolar protrusive activity to exert traction on and eventually intercalate between their lateral neighbors, displacing them along the anteroposterior axis (Shih and Keller 1992). A somewhat different cell behavior, characterized by dorsally directed monopolar protrusive activity, has been proposed to underlie convergent extension of neural explants (Elul and Keller 2000).

Similarly to frog convergent extension, late during zebrafish gastrulation dorsal ectodermal and mesodermal cells become highly elongated (10.3 hpf; length-to-width ratio, LWR=2.1 ± 0.7) compared to their ventral counterparts (LWR=1.7 ± 0.5). Furthermore, the majority of dorsal cells align their cell bodies parallel to the mediolateral (ML) embryonic axis (70% of cells ±20° from ML axis); such alignment is less pronounced in the ventral gastrula (Concha and Adams 1998). The significance of this mediolaterally polarized morphology in vivo is underscored by the observation that in *knypek* mutants the lack of mediolateral cell elongation in dorsal tissues is associated with impaired extension movements (Topczewski et al. 2001). Conversely, mediolateral cell elongation is observed ectopically in the ventral region of dorsalized *somitabun* mutants in which convergence and extension movements driven by mediolateral intercalation occur around the embryonic circumference (Fig. 5; Myers et al. 2002).

Within the ectodermal layer this polarized cell morphology is thought to reflect mediolaterally biased protrusive activity. This enables cells to move between their neighbors preferentially in the mediolateral direction, thus extending embryonic tissues (Kimmel et al. 1994; Concha and Adams 1998). Collectively, these studies point to mediolateral cell intercalation as an important mediator of convergence and extension in late zebrafish gastrula. However,

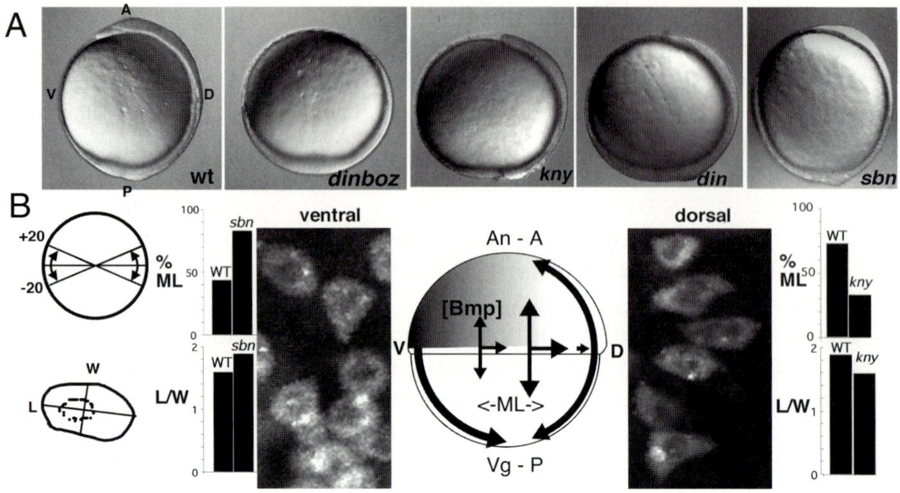

Fig. 5. **A** Convergence and extension is differentially affected in several zebrafish mutant strains. Lateral images of live embryos at the end of the gastrula period (YPC-tailbud). **B** Mediolateral cell elongation is associated with convergence and extension movements in vivo. In the dorsal region of the wild-type gastrula strong extension and limited convergence are associated with mediolateral cell elongation. Such cell polarization is less pronounced in the ventral region where convergence and extension movements are absent. In *kny* mutants, reduction of mediolateral cell elongation is linked to impaired convergence and extension movements, whereas ectopic mediolateral cell elongation of ventral cells is associated with ectopic extension movements in the ventral region of *sbn* mutants

protrusive activity during zebrafish gastrulation is yet to be characterized and quantified. It also remains to be determined whether zebrafish cells develop unipolar protrusive activity as observed for frog neuroectoderm and/or the bipolar protrusive activity characteristic of *Xenopus* mesodermal cells engaged in convergent extension (Shih and Keller 1992; Elul and Keller 2000). Furthermore, cell behaviors driving convergence and extension dorsally in the early gastrula are not well understood. In *Xenopus,* MIB is initiated in a small region within the Spemann-Mangold gastrula organizer, called the vegetal alignment zone, and subsequently spreads laterally (Lane and Keller 1997). Whether a similar region forms in the embryonic shield of zebrafish embryo remains to be elucidated.

4.4 Cellular Segregation, Directed Migration, and Mediolateral Intercalation Underlie Dorsal Hypoblast Formation

Is mediolateral cell intercalation the only cell behavior driving convergence and extension during late zebrafish gastrulation, or are other cell behaviors, such as radial intercalation or directed vegetalward and animalward cell migration, involved? In this section we discuss recent studies of convergence and extension movements in the dorsal mesendoderm layer (hypoblast) that underscore the complexity of underlying cellular mechanisms.

Dorsal hypoblast formation is initiated when presumptive mesendodermal marginal cells in the organizer region move from surface layers in the dorsal marginal blastoderm to deeper layers within the blastoderm (Schmitz and Campos-Ortega 1994; Kane and Adams, this Vol.). Once deep cells enter the dorsal hypoblast, they participate in collective flows directed toward the animal pole (Warga and Kimmel 1990; Fig. 7).

Not all dorsal marginal deep cells in the organizer region become internalized during early gastrulation. A population of *non-involuting endocytic marginal (NEM) cells* can be first distinguished through their endocytic activity in the superficial region of the dorsal marginal zone in the late blastula (Cooper and D'Amico 1996). As deep marginal cells behind and beneath the NEM cells become internalized to form the early hypoblast of the embryonic shield, deep NEM cells segregate from the rest of the dorsal blastoderm, and migrate in front of the blastoderm margin to become a cell population known as *forerunner cells* (D'Amico and Cooper 1997). Forerunner cells remain at the leading edge of the blastoderm, and are displaced toward the vegetal pole during the remaining course of epiboly (D'Amico and Cooper 1997). Forerunner cells undergo mediolateral convergence during the course of gastrulation, becoming a compacted wedge-shaped mass that is subducted by the constricting germ ring around 90% epiboly (for an example of local convergence and thickening, see Fig. 1D; D'Amico and Cooper 1997).

In the anterior hypoblast, cells that are located near the embryonic midline do not engage in strong dorsal convergence movements, but rather move *en masse* toward the animal pole region (Fig. 7). These *anterior hypoblast* cells outpace anterior YSL nuclei that are also coherently moving *en masse* in the same direction (D'Amico and Cooper 2001). *her5* expression is progressively excluded from the prospective prechordal plate (*gsc*-expressing cells) during early gastrulation, and becomes localized exclusively in anterior hypoblast cells that lie immediately lateral to the prospective prechordal plate (Bally-Cuif et al. 2000). The central domain of the anterior hypoblast contains *prospective head mesoderm* (future prechordal plate), whereas the lateral domains of the anterior hypoblast contain *prospective head endoderm* (future hatching gland and pharyngeal endoderm). Thus, the anterior-dorsal hypoblast becomes subdivided and morphologically transformed into distinct domains during the earliest periods of gastrulation (Fig. 6).

Posterior to the anterior hypoblast, there are three ongoing morphogenetic processes that are regionally expressed in the forming dorsal hypoblast. These processes are superimposed to different extents during the course of gastrulation: (1) continued internalization of mesendodermal cells in the marginal zone, (2) convergence and extension movements, and (3) epiboly (D'Amico and Cooper 2001). Hypoblast cells located near the dorsal midline exhibit little mediolateral (ML) movement (convergence). Instead, they engage in a

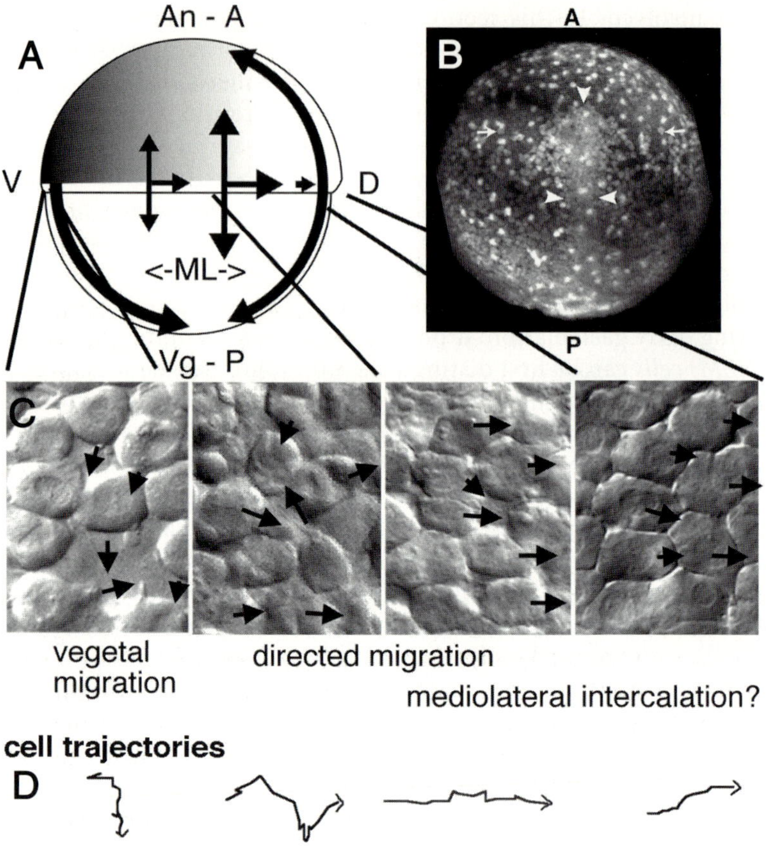

Fig. 6. Spatially distinct domains of convergence and extension movements in zebrafish gastrula (**A**) are characterized by different cell morphology, migratory cell behaviors and movement trajectories. **B** Distribution of hypoblast cells in the late-gastrula (80% epiboly) zebrafish embryo. The embryo was injected into the yolk cell with Alexa-Histone at sphere stage. The YSL nuclei are brightly stained, while hypoblast deep cells have a weaker stain. The anterior border of the embryonic shield (prechordal plate, *arrowhead at top*) is migrating over the yolk cell toward the animal pole. Cells in the axial hypoblast (*arrowheads*) are tightly packed and cannot be individually distinguished. In contrast, deep cells in lateral mesendoderm can migrate individually (*arrows*). **C** Morphology of mesodermal cells in different morphogenetic domains of convergence and extension revealed by Nomarski microscopy. Note rounder shapes and loose packing of ventral and ventrolateral cells, and tight arrangement of mediolaterally elongated cells in dorsolateral and dorsal domains. **D** Typical trajectories of cells shown in **C** determined by time-lapse microscopy. Trajectories of ventral cells are directed vegetally, while trajectories of lateral and dorsal cells point dorsally. Whereas ventrolateral cells migrate dorsally along elaborate trajectories, dorsolateral and dorsal cells move along almost straight paths (**B** courtesy of L.A. D'Amico and **C** courtesy of D.S. Sepich)

coherent cellular flow toward the animal pole (extension), whereas cells in more lateral positions predominantly undergo ML convergence. As cells move closer to the dorsal midline, they become progressively more involved in AP cellular flow (i.e., extension; Fig. 7). The increased AP spreading of hypoblast cells near the dorsal midline results, at least in part, from mediolateral cell intercalation (Glickman et al. 2002).

Dorsal marginal cells that enter the dorsal hypoblast during mid-gastrulation end up in intermediate positions along the AP axis. Dorsal hypoblast cells in this region undergo little net movement with respect to the animal-vegetal axis (see inset, Fig. 7). In contrast, cells that become internalized during late-gastrulation are eventually carried toward the vegetal pole by continued epiboly and extension in the hypoblast. The overall pattern of cellular trajectories in the axial hypoblast (D'Amico and Cooper 2001; Glickman et al. 2002) is similar to the trajectories of deep cells in the dorsal epiblast (Concha and Adams 1998), as well as nuclei contained within the dorsal YSL (D'Amico and Cooper 2001).

During gastrulation, the chordamesoderm (axial mesoderm) and presomitic mesoderm (paraxial mesoderm) both undergo a dramatic and coordinated mediolateral convergence. Around mid-gastrulation, a distinctive notochord-somitic boundary appears (Kimmel et al. 1995). Local morphogenetic cell behaviors are organized around this boundary, including the polarization of adaxial cells that palisade along the length of the notochordal rudiment (axial mesoderm). Cell intercalation takes place across the notochord-somitic boundary, even though the notochordal and presomitic mesoderm continues to undergo a steady ML convergence (Wood and Thorogood 1994).

An elegant new study, utilizing a combination of 4D imaging and computer-assisted visualizations, has examined the detailed movements and intercalations of axial and paraxial mesodermal cells as they undergo convergence and extension during gastrulation and early segmentation (Glickman et al. 2002). In wild-type embryos, axial mesoderm (chordamesoderm) undergoes extension concurrently with ML convergence. However, in *ntl* mutants that fail to form a notochord, extension of axial mesoderm continues to occur, even though ML convergence is severely reduced. Moreover, mediolateral intercalation is also disrupted in the *ntl* mutants. Abnormal cell mixing between the somite-forming paraxial domain, as well as the axial floorplate-forming domain, is also apparent in the *ntl* mutants.

Glickman et al. (2002) conclude that the disorganized local ML convergence of cells towards one another underlies the global disruption in cellular convergence in the axial domain. Their results indicate that convergence is not required for extension. This conclusion agrees with analyses of other mutants suggesting that convergence and extension are genetically separable. Similar to the observations from *ntl* mutants, whereas the dorsal migration (convergence) of lateral cell populations in dorsalized *somitabun* gastrulae is reduced, extension movements are not reduced, but might be increased compared to wild type (Myers et al. 2002). In *trilobite* mutants, both convergence and extension movements of lateral cell populations are compromised. However, comparison of equivalent cell populations that converged to the same distance from the dorsal midline in wild-type and mutant embryos shows reduced anteroposterior elongation of the mutant cell population. This indicates that extension defect in *tri* is not simply a consequence of defective convergence

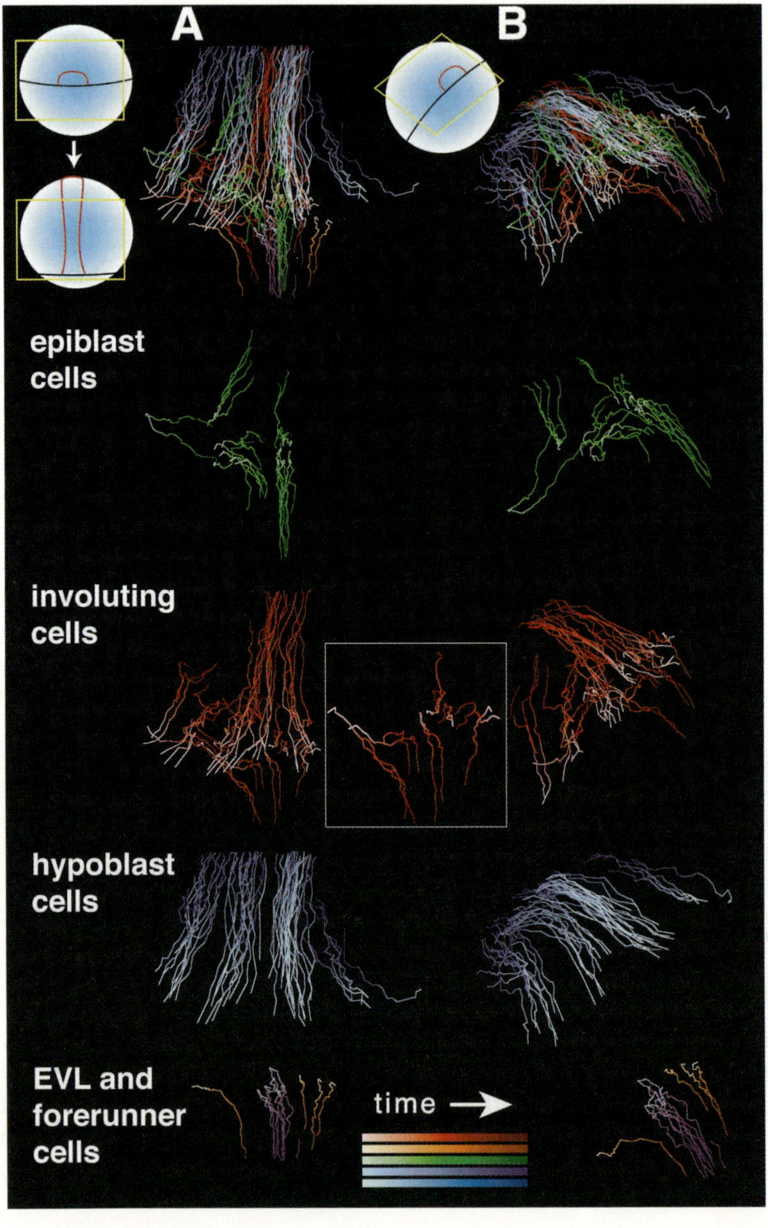

(Sepich et al. 2000). Glickman et al. (2002) also conclude that mediolateral cell intercalation cannot be the only process mediating hypoblast extension in zebrafish, further supporting the notion that blastoderm epiboly may also play an important role.

5 Molecular Genetic Basis of Convergence and Extension Movements

The variety of cell behaviors underlying convergence and extension movements, including directed cell migration or cell intercalation, must entail changes in cell polarity, cytoskeletal organization, and adhesive properties. Accordingly, treatments of sphere-stage embryos with nocodazole to disorganize the microtubule network completely prevents all gastrulation movements, including convergence and extension (Solnica-Krezel and Driever 1994). Disruption of actin microfilaments has an even more dramatic effect inducing dissociation of the enveloping layer and subsequent detachment of deep gastrula cells (Zalik et al. 1999).

Understanding the molecular basis of the convergence and extension process will require elucidation of cytoskeletal organization and cell polarity and adhesive properties of cells exhibiting specific migratory behaviors. Because convergence and extension is driven by movements of cell populations we also need to understand how the motile behaviors of individual cells are coordinated in time and space. Another important question is how cellular movements during gastrulation are connected with cell fate specification events. In recent years, the first inroads into the molecular underpinnings of convergence and extension movements have been made. We will first discuss pathways and molecules which are specifically and directly involved in regulation and/or execution of convergence and extension movements, but appear to be dispensable for cell fate decisions during gastrulation. Then the focus will be on pathways impacting both cell fate specification and movements, which might provide a molecular basis for coordination of inductive and morphogenetic events of gastrulation.

Fig. 7A,B. Morphogenetic movements of cells on the dorsal side of a gastrulating zebrafish embryo (shield stage to 85% epiboly). **A** Dorsal view. The embryonic midline lies near the center of the forerunner cell trajectories (magenta). **B** Oblique dorsolateral view. Deep cells in the epiblast and hypoblast exhibit greater convergence movements when located further away from the dorsal midline. Epiblast cells closer to the blastoderm margin move toward the vegetal pole, while epiblast cells located further away from the margin move toward the animal pole (see also Concha and Adams 1998). Cells already located within the hypoblast at the beginning of the sequence (*blue*) undergo directed movement toward the animal pole. Cells involuting after shield stage (*red*) move away from the margin toward the animal pole, while those cells that become internalized later (75–80% epiboly) move away from the blastoderm margin, as it advances toward the vegetal pole. Cells that involute during intermediate times (70% epiboly) first move toward the animal pole, then switch direction and move toward the vegetal pole (see *inset*). (Reprinted from D'Amico and Cooper 2001)

5.1 Wnt Planar Cell Polarity Pathway

Convergence and extension (C&E) movements in zebrafish and *Xenopus* are dependent on non-canonical Wnt signaling. The Wnt proteins constitute a large family of extracellular ligands that mediate multiple inductive and morphogenetic events during development (Miller et al. 1999b). Although all known Wnt signals are received by seven transmembrane receptors of the Frizzled family, extracellular modulation of their activity as well as intracellular signal transduction pathways are diverse. The Wnt signaling promoting C&E does not operate through the canonical pathway engaging β-catenin, which specifies Nieuwkoop center formation (see Sakaguchi et al.; Hibi et al., this Vol.) and anteroposterior neural patterning (see Wilson et al., this Vol.). Rather, it follows a non-canonical pathway, similar to the planar cell polarity (PCP) signaling pathway described in *Drosophila* (Djiane et al. 2000; Heisenberg et al. 2000; Tada and Smith 2000; Wallingford et al. 2000). In this pathway, which mediates cell polarization along the proximal-distal axis within a plane of epithelium, Frizzled is the most upstream known component activating Dishevelled (Dsh), with more downstream components including small GTPases of the Rho family, Rho kinase and Jun N-terminal kinase (Mlodzik 1999; Winter et al. 2001).

In zebrafish, both overexpression and loss-of-function mutant phenotypes demonstrate the requirement for Wnt11 and Wnt5 function in extension movements of dorsal cell populations (Fig. 8). Such similarities between gain- and loss-of-function phenotypes are typical for components of the PCP pathway in *Drosophila* (Mlodzik 1999). In particular, *slb* mutations inactivating the *wnt11* gene impair cell intercalation underlying extension movements (Heisenberg et al. 2000). Impairment of convergence and extension movements in *slb* and other mutants in this category brings about a set of morphological defects discernable at the end of the gastrula period in live embryos or via altered gene expression patterns. The embryonic axis is shortened, forming a neural plate and notochord that are shorter anteroposteriorly but broadened in the mediolateral direction, and the prechordal mesendoderm is positioned more posteriorly than in wild-type embryos. Intriguingly, at 1 day of development, there is recovery of the axis extension defect, with synopthalmia or cyclopia distinguishing the *slb* mutants from the wild type (Hammerschmidt et al. 1996a; Heisenberg and Nüsslein-Volhard 1997). Inactivation of zygotic *ppt* function becomes morphologically manifest only during segmentation through a shortened tail region and mediolaterally broadened and anteroposteriorly compressed somites (Hammerschmidt et al. 1996a). However, interference with both zygotic and maternal *ppt* (*wnt5*) function using morpholino antisense nucleotides produces a more severe compression of the body axis already at gastrulation (Lele et al. 2001a). Despite severe morphogenic defects, these mutants exhibit normal numbers of somites and form diverse cell types, suggesting that Wnt11/5 signaling is required for convergence and extension movements, but is dispensable for cell fate specification (Hammerschmidt

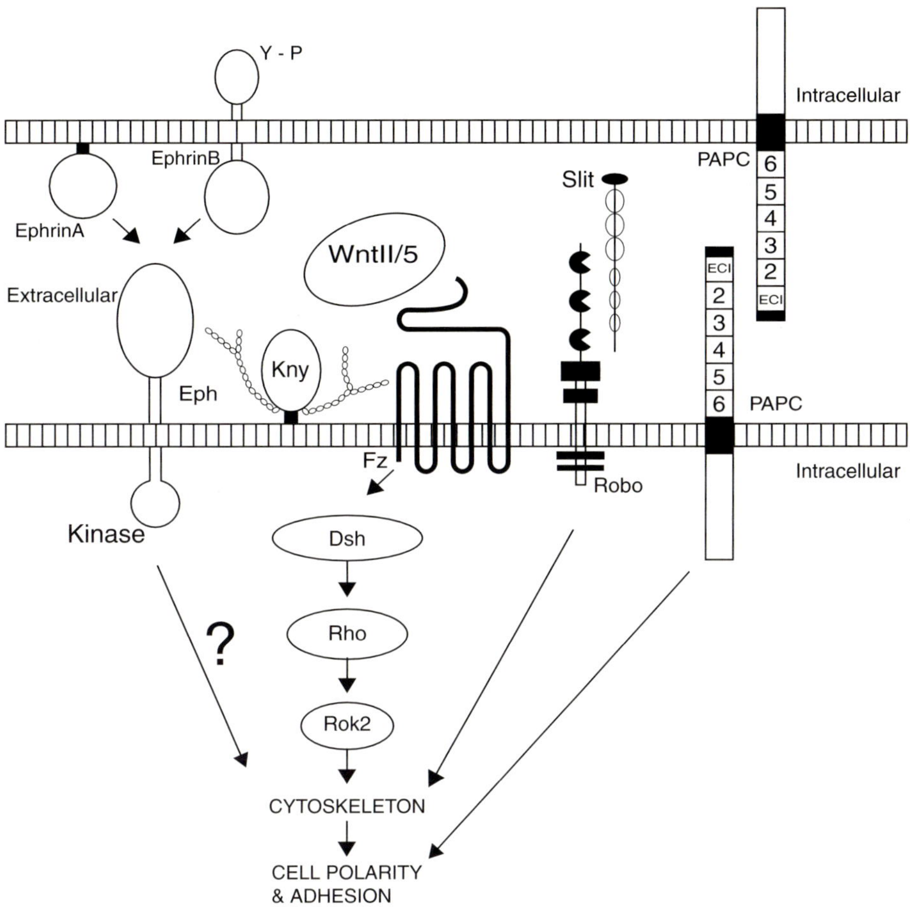

Fig. 8. Molecules and pathways implicated in convergence and extension movements in zebrafish

et al. 1996a). To what degree Wnt11, Wnt5 and possibly other Wnt ligands compensate for each other's function, and the consequence of the complete inactivation of the non-canonical Wnt signaling on cell movement and fate specification, requires further analyses.

Of the many Frizzled receptors expressed during gastrulation, *frizzled2* morpholino knockdown phenocopies the convergence and extension *ppt* mutant, indicating a role for this putative Wnt receptor in convergence and extension (Sumanas et al. 2001). A novel component of this pathway that regulates convergence and extension has been identified by mutations in the *knypek* (*kny*) locus encoding a zebrafish glypican (heparan sulfate proteoglycan; Fig. 5; Topczewski et al. 2001). *kny* mutants are defective in convergence and extension movements of both dorsal and lateral cell populations. Fate mapping experiments reveal that similarly to *slb*, *kny* mutations perturb cell movements without significantly altering cell fate specification events. Genetic and molecular epistasis analyses demonstrate that Knypek promotes Wnt11 (Slb) signaling. Therefore, Knypek might be involved in the reception of the Wnt11 signal during convergence and extension (Fig. 8).

Overexpression of Dishevelled mutant forms specifically blocking or activating PCP signaling implicate this phosphoprotein in zebrafish C&E, as demonstrated for *Xenopus* embryos (Heisenberg et al. 2000; Tada and Smith 2000; Wallingford et al. 2000). Two cloned *dishevelled* genes (*dsh2* and *dsh3*) are expressed and likely contribute to C&E in zebrafish (J.T. Topczewski and L. Solnica-Krezel, unpubl. observ.). While studies in *Xenopus* point to Cdc42 as a downstream component of this pathway, evidence for involvement of the small GTPases in zebrafish C&E is still lacking (Djiane et al. 2000). However, Rho-associated kinase 2 (Rok2), which in mammalian tissue culture and *Drosophila* PCP pathway mediates some aspects of Rho signaling, has been recently implicated in zebrafish C&E. Overexpression of dominant negative Rok2 protein phenocopies *slb* and *kny* axis shortening and synopthalmia, inhibiting convergence and extension without any significant effect on cell fate specification in the gastrula. Conversely, overexpression of wild-type Rok2 ameliorates convergence and extension defects and cyclopia of *slb* mutants, placing Rok2 downstream of Slb in the pathway regulating C&E (Fig. 8; Marlow et al. 2002).

Impairment of dorsal convergence and extension movements associated with inhibition of the Wnt PCP pathway is likely a consequence of defective cell polarization. In *Xenopus*, overexpression of a dominant negative form of Xdsh, corresponding to mutations in *Drosophila* Dsh that selectively impair PCP signaling, links deficient convergent extension of tissue explants to cell polarization defect (Wallingford et al. 2000). Both the mediolateral cell orientation and cell elongation observed during convergent extension of dorsal mesoderm tissue explants (Shih and Keller 1992) are disrupted in explants overexpressing dominant negative and wild-type Xdsh (Wallingford et al. 2000). Furthermore, cells overexpressing dominant negative Xdsh extend and withdraw more protrusions, and form less stable protrusions compared to normal cells or cells overexpressing wild-type Xdsh (Wallingford et al. 2000). The similar failure of *kny* mutant cells to elongate and align along the mediolateral axis links for the first time impaired cell polarization with defective convergence and extension movements in vivo (Topczewski et al. 2001). Hence, it is tempting to hypothesize that Wnt11 and Wnt5 mediate convergence and extension movements by regulating cell polarization, acting as instructive or permissive signals. Defective cell polarity most likely perturbs convergence and extension by inhibiting the process of mediolateral intercalation. However, decreased cell adhesion was associated with the axis extension defect observed in embryos overexpressing Wnt4 (Ungar et al. 1995). Defective cell adhesion could inhibit mediolateral intercalation directly or indirectly via its effect on cell polarization.

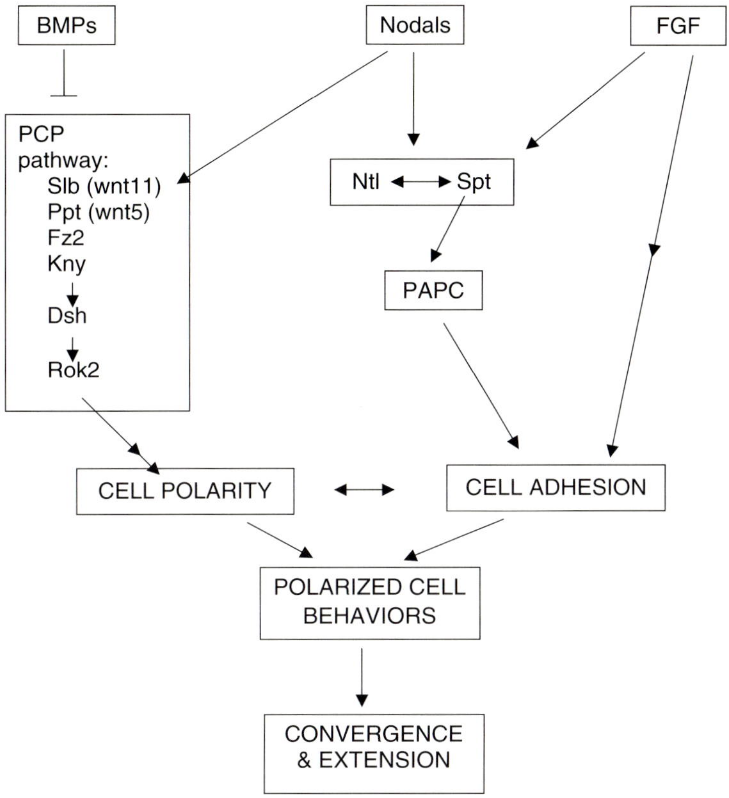

Fig. 9. Model for genetic regulation of convergence and extension

5.2 Cell Adhesion Molecules

The requirement for proper cell adhesion during C&E is further underscored by functional studies of paraxial protocadherin (PAPC) expressed in nonaxial mesoderm during gastrulation (Yamamoto et al. 1998a). Overexpression of a dominant negative secreted form of PAPC leads to mediolateral expansion of somites, but not notochord, where this gene is not expressed. This suggests that specific interference with PAPC function inhibits convergence and extension in nonaxial mesodermal tissues. Similar convergent extension defects have been observed in *Xenopus* embryos overexpressing a dominant negative mutant form of frog PAPC, and were associated with defective cell adhesion (Kim et al. 1998b). Together, these studies suggest that homophilic cell-cell interactions mediated by PAPC are influencing adhesive cell properties required for convergent extension movements (Fig. 9). In *Xenopus*, PAPC overexpression promotes cell shape changes associated with convergent extension (Kim et al. 1998b). It is not clear, however, whether mediolateral cell polarity depends on PAPC function.

5.3 Slit

Current evidence indicates that molecules known to mediate cell interaction and guidance in neural tissue might be involved in the regulation of C&E. Signaling of secreted Slit proteins through their Robo receptors has been initially shown to control axon crossing at the midline in the *Drosophila* central nervous system (Rothberg et al. 1988; Kidd et al. 1999). Since then, short- and long-range Slit signaling have been shown to act in invertebrate and vertebrate embryos as repellents on the axons from the olfactory bulb, the spinal motor neurons, and on neuronal precursors migrating towards the olfactory bulb from the anterior telencephalon subventricular zone (Brose et al. 1999; Wu et al. 1999). Furthermore, Slits can both negatively and positively influence the migration of mesodermal muscle precursors during *Drosophila* development (Kidd et al. 1999; Kramer et al. 2001). Two zebrafish *slit* homologues, *slit2* and *slit3*, have been isolated and shown to be expressed in a dynamic pattern starting at mid-blastula (*slit2*) and gastrula (*slit3*) stages (Yeo et al. 2001). Intriguingly, expression of both genes is enriched in the axial mesoderm during gastrulation. That Slits are involved in the regulation of C&E is suggested by gain-of-function experiments. Embryos overexpressing wild-type Slit2 from the cleavage stages through gastrulation, or transiently during late-gastrulation, exhibit alterations in gene expression domains consistent with impaired convergence and extension. Similar to *slb* and *kny* mutants, Slit2 overexpressing embryos show broader neural anlage, broader but shorter chordamesoderm and posteriorly positioned prechordal mesoderm, suggesting that movements in both ectodermal and mesodermal layers are affected. This apparent inhibition of convergence and extension movements due to excess Slit2 signaling does not discern whether normal Slit function negatively or positively regulates convergence and extension movements (Yeo et al. 2001). It is noteworthy, however, that the speed of convergence movements decreases significantly, as lateral cells approach the dorsal midline expressing high levels of *slit2* and *slit3* (Sepich et al. 2000; Yeo et al. 2001). Hence, Slit at the dorsal midline could exert an inhibitory role on convergence and/or extension movements. Furthermore, high levels of Slit in axial mesoderm cells could prevent their mixing with paraxial mesoderm cells undergoing dorsal convergence.

5.4 Eph Receptors and Ephrins

One mechanism regulating cell-cell interactions during convergence and extension might involve signaling by Eph receptor tyrosine kinases and their ephrin ligands. Ephrins are thought to act as intermediaries between patterning and morphogenic events, during such diverse processes as repulsive axon guidance, cell migration, somitogenesis, topographic mapping and angiogenesis (Holder and Klein 1999; Klein 2001). Ephrin ligands are broadly divided into two classes based on the mode of cell surface attachment and receptor specificity. EphrinA ligands use glycosylphosphatidylinositol (GPI) as a membrane anchor and bind EphA receptors, while the ephrinB ligands feature transmembrane and short cytoplasmic domains and bind to EphB receptors (Fig. 9; Klein 2001). Crucial for their role in intercellular communication is that interaction between a ligand on one cell with a receptor on another cell can trigger bidirectional signaling.

A potential role for Eph receptor/ephrin in gastrulation movements is supported by studies involving several ephrin ligands and their receptors expressed in zebrafish gastrulae (Xu et al. 1994; Cooke et al. 1997; Durbin et al. 1998; Chan et al. 2001). Soluble ligands have been used to assess Eph signaling functions in vivo because they are able to inhibit both components of Eph/ephrin signaling (Davis et al. 1997). When such dominant negative forms of soluble human ephrinA5, EphA3 and of zebrafish ephrinB proteins are expressed in early zebrafish embryos, they result in changes in embryonic morphology and gene expression patterns suggestive of a convergence and extension, rather than a cell fate specification defect (Oates et al. 1999; Chan et al. 2001). The cell behaviors and activities dependent on ephrin function remain to be elucidated. Nevertheless, based on the mechanisms used by Eph/ephrin in other developmental contexts, Eph/ephrin signaling could impact convergence and extension via modulation of cytoskeleton or cell adhesion, or by influencing other signaling pathways (Klein 2001; Lu et al. 2001).

5.5 Calcium

Calcium released into intercellular space might play an important role in convergent extension (Wallingford et al. 2001a). In *Xenopus* explants undergoing convergent extension and in zebrafish gastrulae, waves of intercellular calcium mobilization have been observed (Creton et al. 1998; Gilland et al. 1999). Pharmacological repression of calcium dynamics in *Xenopus* embryos inhibited convergent extension movements without altering cell fate specification. Hence, calcium waves might provide a mechanism for coordination of morphogenetic cell behaviors during convergent extension (Wallingford et al. 2001a).

5.6 Ethanol

Notably, treatment of zebrafish blastulae with 2.4% ethanol impairs migration of prechordal mesendoderm, and possibly extension of the neuroectoderm midline and tail development, resulting in varied degrees of synopthalmia and cyclopia. By contrast, convergence and extension of chordamesoderm is not significantly affected by ethanol treatment (Blader and Strahle 1998). Identification of the molecular target for ethanol and the underlying cellular defect might provide cues into the regional mechanisms of convergence and extension.

6 Molecular Genetic Coordination of Convergence and Extension Movements with Cell Fate Specification

In the course of gastrulation cells participate in several types of mass movements, while simultaneously making decisions about their future fates. How is the precise coordination of cell fate decisions and convergence and extension movements accomplished? This difficult problem might be addressed by functional analysis of genes required for both cell fate specification and gastrulation movements.

6.1 *Spadetail*

Spadetail (*spt*), the first identified mutant gene in this category (Kimmel et al. 1989), was subsequently shown to encode a T-box transcription factor (Griffin et al. 1998). As far as movements are concerned, *spt* acts initially during blastula stages to mediate the process of dorsal cell compaction (Warga and Nüsslein-Volhard 1998). At gastrulation, the mutation impairs convergence movements in a cell-autonomous manner, such that lateral mesodermal cells migrate vegetally and enter the tailbud, instead of converging dorsally (Ho and Kane 1990). As a consequence, *spt* mutants are deficient in trunk somitic mesoderm and blood, but form a relatively normal head and tail with excess cells in the tail (Kimmel et al. 1989). *spt* mutations block *papc* expression, suggesting that decreased cell adhesion might cause movement defects. However, *papc* overexpression in *spt* mutants failed to suppress the convergence and extension defect, indicating that *spt* must regulate other genes required for C&E (Griffin et al. 1998). Ubiquitous expression of *spt* mRNA at blastula stages and the enlargement of notochord in *spt* mutants are indicative of a broad function of *spt* in promoting muscle development (Amacher and Kimmel 1998). It remains to be determined whether the dorsal blastoderm compaction and convergence defects are a secondary consequence of a cell fate specification defect(s), or rather represent distinct functions of this transcription factor.

6.2 Nodal

Nodal signals encoded in zebrafish by the functionally redundant *squint* (*sqt*) and *cyclops* (*cyc*) loci are absolutely required for development of endoderm and dorsolateral mesoderm (see Kimelman and Schier, this Vol.). In Nodal-signaling-deficient embryos, dorsolateral marginal tissues develop as ectoderm (Feldman et al. 2000). By the end of epiboly, these embryos develop a shortened anteroposterior embryonic axis with the forebrain region residing at the equator, suggesting impairment of extension movements (Feldman et al. 1998; Gritsman et al. 1999; Thisse and Thisse 1999). Accordingly, fate mapping of individual and cell populations revealed that dorsal convergence movements occur in Nodal-deficient embryos, but extension movements are strongly impaired (Feldman et al. 2000; Erter et al. 2001). Likely, in the absence of Nodal signaling, convergence and extension driven by mediolateral cell intercalation is blocked, whereas directed cell migration mediating convergence and epiboly are less affected.

Another noteworthy observation is that extension of ectoderm is blocked although *nodal*-related genes are neither expressed nor necessary for ectoderm formation. In frogs, neural tissue does not require mesoderm for convergence and extension in the context of whole embryo (Wallingford and Harland 2001), but mesoderm can influence cell behaviors employed in the overlying ectoderm (Elul and Keller 2000). When explanted alone, deep neural explants undergo mediolateral cell intercalation using bipolar protrusive activity, as observed for mesoderm (Elul et al. 1997). By contrast, the explants of deep neural tissue made with underlying mesoderm exhibit distinct and regionalized cell behaviors, including unipolar dorsally directed protrusive activity and multipolar, randomly oriented protrusive activity at the midline, where the notoplate would form (Elul and Keller 2000). Hence, abnormal extension in the ectoderm of Nodal-deficient embryos might be a consequence of distinct cell intercalation cell behaviors employed in the absence of underlying mesoderm. Alternatively, Nodal signaling may be required either directly or indirectly for the production of a signal mediating mediolateral cell polarization. In frogs, a more divergent Nodal ligand previously implicated in cell movements, Xnr-3, has been proposed to regulate convergence and extension (Smith et al. 1995). Based on overexpression experiments in animal cap explants, Xnr-3 has been proposed to promote C&E dorsally acting downstream of β-catenin, possibly independent of cell fate specification (Kuhl et al. 2001).

6.3 Bone Morphogenetic Proteins

Cell fates within the ectodermal and mesodermal layer are thought to be specified by a ventral-to-dorsal gradient of bone morphogenetic protein (BMP) activity (see Hammerschmidt and Mullins, this Vol.). Analyses of convergence and extension movements in wild-type, ventralized and dorsalized mutants revealed that the same BMP activity gradient plays an instructive role in the regulation of convergence and extension. High BMP activity levels in the ventral region of the gastrula are necessary and sufficient to inhibit both convergence and extension movements without interfering with mesendoderm internalization or epiboly (Myers et al. 2002). Intermediate levels of BMP activity observed in the lateral regions of zebrafish gastrula are required for lateral cell populations to converge and extend at increasing speed as they move dorsally. Finally, in the context of the whole embryo, low BMP activity levels specify dorsally biased cell intercalation that drives strong extension movements with limited convergence. Thus, a single ventral-to-dorsal gradient of BMP activity may coordinate cell fate specification with convergence and extension movements (Myers et al. 2002).

By what mechanism do BMP activity levels impact convergence and extension? BMPs could negatively regulate the production of dorsally derived signals required for C&E and/or negatively regulate a cell's ability to respond to such signals or to execute convergence and extension movements. Evidence exists for both mechanisms. First, the expression of *slb* (*wnt11*) and *ppt* (*wnt5*) genes encoding presumptive PCP signals is downregulated dorsally in ventralized mutants with inhibited convergence and extension movements. Conversely, ectopic mediolateral cell polarization and intercalation driven convergence and extension in dorsalized *sbn* mutants is correlated with ventral expansion of *slb* (*wnt11*) expression. These observations are consistent with BMP negatively regulating convergence and extension by limiting dorsally derived signals required for convergence and extension movements. Secondly, cells overexpressing an activated form of the type I BMP receptor upon transplantation into wild-type embryos fail to participate in convergence and extension and migrate towards the tailbud, as observed for cells residing in the "no convergence no extension zone" (Myers et al. 2002). This result supports the notion that high levels of BMPs inhibit in a cell-autonomous manner the ability of a cell to respond to C&E signals and/or execute specific cell behaviors.

6.4 Fibroblast Growth Factor

Inhibition of fibroblast growth factor (FGF) signaling by overexpression of a dominant negative receptor severely impairs mesoderm development and results in a dramatic truncation of the posterior embryonic structures, tail and trunk (Amaya et al. 1991; Griffin et al. 1995). Specific effects of deficient FGF signaling on convergence and extension movements have not been elucidated. It is also not clear whether FGF is directly involved in specification of cell movement behaviors, or acts indirectly by influencing the expression of other genes affecting cell fate specification. During *Xenopus* and zebrafish development, FGF signaling regulates expression of T-box transcription factors, such as *Xbra* (*ntl*) and *spt* (discussed above). These transcription factors in turn regulate expression of genes required for normal convergence and extension movements, including *wnt11* (in frog) or *papc* (in zebrafish). Furthermore, in zebrafish, overexpression of an FGF-signaling antagonist, Sprouty4, causes ventralization of the embryo (Fürthauer et al. 2001). Hence, inhibition of FGF signaling could impair C&E movements indirectly, by increasing Bmp signaling, thereby limiting convergence and extension dorsally, as discussed above. However, studies in *Xenopus* suggest a direct involvement of FGF signaling in

C&E, possibly via calcium regulation (Nutt et al. 2001). In the mouse gastrula, FGF signaling has also been implicated in mesoderm patterning and migration (Sun et al. 1999). Recent analyses of mice harboring inactive Fgf receptor 1 indicate that FGF signaling impacts gastrulation via interdependent pathways regulating cell adhesion, morphogenesis and cell fate determination (Ciruna and Rossant 2001).

Cumulatively, these studies suggest several possible scenarios for how the major signaling molecules can coordinate cell fate specification and movement. FGF signaling might follow distinct pathways to regulate convergence and extension and cell fate specification, as suggested for *Xenopus* gastrula (Nutt et al. 2001). Alternatively, FGF triggers several interdependent events to coordinate cell fate determination and migration (Ciruna and Rossant 2001). In zebrafish, a morphogenic BMP activity gradient might inhibit convergence and extension movements by limiting expression of genes required for convergence and extension movements, which are dispensable for cell fate specification (Myers et al. 2002). To compound the complexity of this problem, the same molecule might employ different strategies in distinct developmental context.

7 Role of C&E Movements in Generating Embryonic Morphology

Morphometric analyses together with studies of cell population movements in wild-type and mutant gastrulae reveal how spatial regulation of convergence and extension movements during the course of gastrulation shapes the zebrafish body plan. Increasingly faster convergence and extension movements of lateral cell populations, together with strong extension and limited convergence dorsally, provide an explanation for the gradual narrowing and elongation of germ layers, as well as their dramatic anteroposterior extension in the dorsal midline (Fig. 2A). Inhibition of convergence and extension movements in the ventral region ensures contribution of cells to the tailbud, providing a supply of cells for subsequent tail development. Impaired convergence and extension of lateral cell populations, and extension of dorsal cell populations in *tri* and *kny* mutants, can account for the characteristic anteroposterior shortening and mediolateral expansion of embryonic structures and organ rudiments, including neural ectoderm and somitic mesoderm in these mutants (Henry et al. 2000). Similarly, impairment of extension movements without significant inhibition of mesodermal convergence movements in *slb* mutants produces an embryo with a shortened anteroposterior axis, but relatively normal somites (Heisenberg et al. 2000).

Analyses of dorsalized *sbn* and *swr* mutants underscore the significance of limiting convergence and extension to the dorsal region. Expansion of cell intercalation behavior into lateral and ventral regions of the gastrula prevents effective movements of the lateral cells dorsally leading first to thinning and subsequently to impaired anteroposterior elongation of the embryonic axis. Similarly, ventral cells fail to migrate vegetally and contribute to the tailbud, eventually leading to truncations of posterior tail structures, characteristic of this mutant class. Instead, both lateral and ventral cells engage in dorsally directed cell intercalation narrowing the embryonic circumference along the mediolateral axis and elongating it in the animal-vegetal direction, which is reflected by the dramatic tubular shape of these embryos during early segmentation (Myers et al. 2002). Continued narrowing of these tissues, especially prominent in the somitic region, might squeeze the yolk extensively, eventually causing it to rupture (Mullins et al. 1996).

Modulation of convergence and extension movements in ventralized mutants provides an important insight into how a ventral-to-dorsal gradient of BMP activity can impact anteroposterior organization of the embryo. Progressive expansion of the "no convergence no extension" zone in the moderately ventralized *chordino* and strongly ventralized *bozozok chordino* compound mutants leads to formation of an enlarged tailbud characteristic of these mutants (Gonzalez et al. 2000). The excessive migration of lateral and dorsolateral cells towards the tailbud occurs at the expense of their participation in dorsal convergence and extension movements, resulting in formation of a reduced head and trunk in *chordino* mutants and the absence of these structures in *bozozok chordino* double mutants (Fig. 5).

8 Future Directions

Recent decades have witnessed significant progress in our understanding of the cellular and molecular events of convergence and extension during zebrafish gastrulation. The studies summarized here reveal the lateral domain of increasing convergence and extension movements mediated by directed cell migration and a dorsal domain of strong extension and slow convergence driven by intercalation of mediolaterally elongated cells. Similarly to *Xenopus*, cell polarization underlying these convergence and extension movements absolutely requires signaling via the vertebrate equivalent of the *Drosophila* PCP pathway, likely involving Wnt11, Wnt5, Frizzled 2, glypican Knypek, Dishevelled and Rho kinase. Furthermore, several other molecules, including Slit and Ephrins, have been implicated in the regulation of convergence and extension.

Further progress towards understanding the roles of these and yet to be identified genes is critically dependent on studies that define cell behaviors mediating convergence and extension in different regions of zebrafish gastrula. We need to learn whether zebrafish cells utilize bipolar or unipolar protrusive activity during cell intercalation. How is the mediolateral intercalation behavior spatiotemporally controlled during zebrafish gastrulation? To what degree are mediolateral intercalation and directed cell migration involved in the lateral and dorsal domain of convergence and extension? And are any other cell behaviors contributing to tissue narrowing and elongation? New computer-assisted methods for visualizing and analyzing movements of cell groups and individual cells (D'Amico and Cooper 2001; Glickman et al. 2002) will aid in answering these questions.

Having defined significance of mediolateral cell polarization for convergence and extension and several molecules essential for its expression, molecular mechanisms underlying this process need to be addressed. The ability to micromanipulate and observe cells in living zebrafish embryos and micromanipulate zebrafish embryos, along with genetic tools, will facilitate cell biological and biochemical approaches to these problems. With the aid of GFP-tagged reagents, one should be able to address the mechanisms of cell polarization in the context of the gastrulating embryo. Indeed, the zebrafish embryo is a *living laboratory* in which fundamental connections between cell fate specification and embryonic morphogenesis can be experimentally determined. The power of classical and reverse genetics will likely be employed to elucidate pathways that connect morphogens with cell fate decisions and motile behaviors, ensuring that the next decade of research on zebrafish gastrulation will also be an exciting one.

Acknowledgements. We thank Dr. Chuck Kimmel and members of our laboratories for discussions and constructive comments on the manuscript. We thank Piotr Solnica and Dr. Diane Sepich for help with figure preparation. L.S.K. would like to acknowledge support from NIH, March of Dimes Birth Defects Foundation, and Pew Scholars Program in Biomedical Sciences. M.S.C. gratefully acknowledges grant support from the NSF (IBN-9808224).

Primordial Germ-Cell Development in Zebrafish

Erez Raz[1] and Nancy Hopkins[2]

1 Introduction

In sexually reproducing organisms, primordial germ cells (PGCs) give rise to gametes that are responsible for the development of a new organism in the next generation. In many cases, specification of primordial germ cells occurs early in embryogenesis when a small group of cells is set aside and follows a unique pathway of differentiation into gametes (reviewed in Saffman and Lasko 1999; Wylie 1999, 2000).

Several distinctive aspects of germ-cell development made this cell population an appealing research subject in invertebrates and vertebrates. First, the germ cells follow a characteristic differentiation path that is manifested in specialized regulation of basic cell functions. Thus, the cells display unique control over RNA transcription, exhibit a unique cell cycle program and show a characteristic cellular, as well as subcellular morphology that distinguishes them from neighboring somatic cells. Considering that the PGCs carry the genetic information to the next generation, the cells should maintain their totipotency, i.e., the gametes that they form should be able to give rise to a complete new organism. This property of germ cells is at the basis of studies aimed at understanding the molecular mechanisms that allow germ cells to maintain a feature that is gradually lost in somatic cells. The ability of germ cells to maintain totipotency is of particular practical importance for the purpose of modern genetic manipulation of different organisms. Indeed, in several systems, it was possible to isolate or culture germ cells and then re-introduce them into the animal to obtain subsequent differentiation into multiple cell types or gametes (Lin et al. 1992; Matsui et al. 1992). Finally, a common feature of primordial germ cells is the fact that they migrate from the position where they are specified towards the position of the gonad where they differentiate into gametes (Saffman and Lasko 1999; Wylie 1999, 2000). This behavior serves as a general model for long-range cell migration in multicellular organisms with the aim

[1] Germ Cell Development, Max-Planck-Institute for Biophysical Chemistry, Am Fassberg 11, Göttingen 37070, Germany

[2] Department and Center for Cancer Research, Massachusetts Institute of Technology, Cambridge, Massachusetts 02139, USA

Results and Problems in Cell Differentiation, Vol. 40
L. Solnica-Krezel (Ed.): Pattern Formation in Zebrafish

of identifying the underlying molecular mechanisms of this process (e.g., Howard 1998; Moore et al. 1998; van Doren et al. 1998a; Anderson et al. 2000).

In this chapter, we review development of PGCs in zebrafish. We summarize the "classical" knowledge regarding specification of PGCs in fish in general, and specifically in zebrafish. Following this description we turn to more recent analyses of germ-cell specification and the migration processes in zebrafish. These studies have been enabled by the identification of molecular markers for PGCs. In conclusion, we compare the findings obtained in zebrafish to those in other organisms.

2 Specification of Germ Cells in Fish

2.1 The Premolecular Markers Era

PGC specification has been studied by microscope analysis, embryological experiments and, importantly, by analysis of maternal-effect mutations in *Drosophila* and *C. elegans* (reviewed in Williamson and Lehmann 1996; Saffman and Lasko 1999; Wylie 1999, 2000). The main conclusion from the studies in invertebrates and from studies in *Xenopus laevis* was that asymmetrical localization of cytoplasmic determinants – the germ plasm – is responsible for the early specification of the germ line lineage. The germ plasm is characterized by the presence of polar granules, electron-dense structures that are associated with mitochondria and contain RNA molecules and proteins. The specification of the germ line is different in mammals. Both morphological, as well as experimental evidence, argue that mammalian germ cells are induced via cellular interactions during gastrulation rather than by inheritance of cytoplasmic factors (Saffman and Lasko 1999; Wylie 1999, 2000).

Germ cell specification in fish was studied mainly in bony fish (teleosts) where the work relied on observations using light and electron microscopy. The germ cells were recognized by their large size and by the typical electron-dense "nuage" structures that are found in germ cells of different organisms. Using these methodologies, the earliest time point at which germ cells were recognized in teleosts was around early somitogenesis stages (Hamaguchi 1982; Timmermans and Taverne 1989; Gevers et al. 1992a). Thus, on the basis of these morphological criteria, PGCs could be identified only at a rather advanced stage of embryonic development, and the presence of electron-dense material in the germ cells suggested common mechanisms of PGC specification between fish and germ-plasm-containing organisms such as *Xenopus, Drosophila* and *C. elegans*. Some investigators deduced that the position where the PGCs were first identified was the position where they originated (for references, see Nieuwkoop and Sutasurya 1979). However, the question of the time and place at which these cells originate remained open since it was possible that, initially, a small population of PGCs existed before the end of gastrulation and migrated to the position where it was identified morphologically.

In some more revealing experiments, PGC specification was indirectly examined at earlier stages of fish embryogenesis. In one experiment, one blastomere of a 64-cell *Barbus conchonius* embryo was labeled with a lineage tracer and the presence of labeled germ cells was monitored at later stages (Gevers et al. 1992b). Several important conclusions were derived from this study that were later confirmed using direct assays for the number and position of PGCs in fish. First, blastomeres that were positioned in the more animal tier of the blastula never formed germ cells, while cells at the more vegetal tier gave rise to labeled germ cells, suggesting that PGC progenitors were located within the vegetal tier of cells. Second, at this stage (64-cell stage), no cases were found where an injected blastomere gave rise exclusively to germ cells, indicating that, at this early stage, lineage restriction for the germ line did not occur. Finally, when one blastomere was labeled, less than 20% of the PGCs were labeled suggesting that already at this time point more than one PGC progenitor cell existed.

In zebrafish, the number of PGCs at the cleavage stages was indirectly determined from the frequency of γ-ray-induced mutations at these stages (Walker and Streisinger 1983). In this experiment, wild-type embryos at different blastula stages (at the 8-cell, 1000-cell, and the sphere stages) were irradiated and were later test-crossed to homozygous *gol*-1 fish of the opposite sex to screen for induction of mutation in the *gol* locus. Importantly, even at the earliest time point of irradiation, i.e., at the 8-cell stage, the germ line of the irradiated fish was found to be a mosaic of mutated *gol* germ cells and wild-type germ cells. Thus, as early as the 8-cell stage, several chromosomal targets for γ-rays exist that are inherited by PGCs. Interestingly, comparing the *gol* clone size between fish containing 8 and over 4000 cells at the time of irradiation, revealed a similar clone size suggesting that the number of chromosomal targets remained unchanged during this period. Furthermore, based on the number of chromosomal targets, the number of PGCs was estimated to be about 5 during blastula stages. In an independent study, transplantation experiments were performed in which 50 cells were transferred from wild-type embryos at the 1000-cell stage into an *albino* recipient of the same developmental stage (Lin et al. 1992). Assuming that the number of PGCs at this stage is 5, then the chance of transferring one wild-type PGC among the 50 transplanted cells is 25%. Indeed, this was approximately the value that was measured as the frequency of germ-line chimeras obtained in these experiments.

To summarize, fish PGCs could be directly identified only at relatively late stages of development when they exhibited characteristic cellular and subcellular morphology. Indirect genetic experiments suggested that, at early blastula stages, before the PGCs could be morphologically identified, several progenitors for PGCs existed in more vegetal parts of the embryo. At early blastula stages these progenitors were not specified as germ cells and consequently contributed to somatic structures as well. As will be discussed below, these conclusions were confirmed following the identification of the first molecular marker for PGCs in zebrafish, *vasa* RNA.

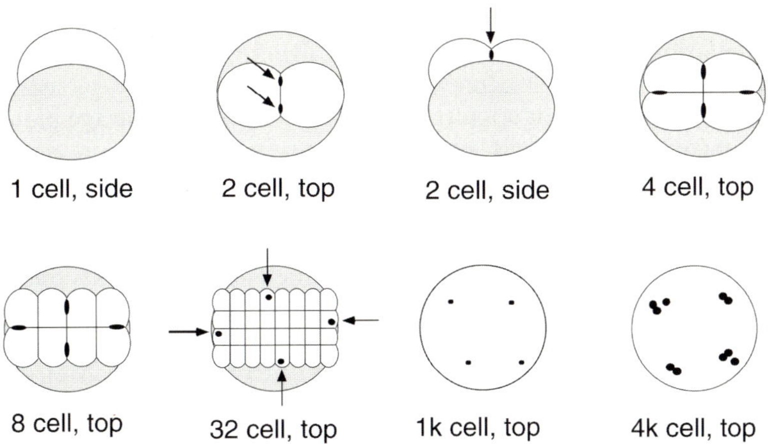

Fig. 1. Schematic drawings of *vasa* RNA expression in early zebrafish embryos as determined by whole-mount in situ hybridization. The stages and views are indicated below each drawing and the yolk is *shaded*. *Arrows* point to the *vasa*-expressing cells in selected stages. *vasa* RNA is first detected along the first two cleavage planes (*arrows*), then condenses into four subcellular clumps by the 32-cell stage (*arrows*). These four clumps are segregated into four cells through the 1000-cell stage. By the 4000-cell stage, the cells that inherited *vasa* RNA, the PGCs, have begun to divide and *vasa* RNA is located throughout the cytoplasm. (Reproduced from Yoon et al. 1997 with permission from the Company of Biologists)

2.2 The Molecular Markers Era

The first molecular marker for PGCs in zebrafish is *vasa* (Olsen et al. 1997; Yoon et al. 1997), an ATP-dependent RNA helicase of the DEAD-box family (Hay et al. 1988; Lasko and Ashburner 1988) that is expressed in the PGCs of many species (reviewed in Raz 2000). *vasa* was originally identified in *Drosophila* as a maternal-effect gene required for the formation of abdominal segments and for germ-cell specification (Schüpbach and Wieschaus 1986).

Similar to many other organisms, zebrafish *vasa* (*vas*) mRNA is maternally supplied and it is detected in PGCs during all stages of germ-line development (Yoon et al. 1997; Braat et al. 1999, 2000; Weidinger et al. 1999; Knaut et al. 2000). The cloning of the zebrafish *vasa* gene revolutionized the study of PGC development in this organism by allowing close examination of these cells from the earliest stages of embryonic development to later stages at which they differentiate into gametes (Yoon et al. 1997; Braat et al. 1999; Knaut et al. 2000). Whole-mount in situ hybridization at the two-cell stage revealed that the *vasa* transcript is enriched at the marginal positions of the first cleavage plane (Yoon et al. 1997; Fig. 1). Similar accumulation of *vasa* transcript is found at the second cleavage plane leading to the formation of four stripes of intense *vasa* RNA staining (Fig. 1). Mutant analysis and the use of microtubule-inhibiting drugs suggested that interaction between microtubules and *vasa* RNA-containing particles is essential for localization of the RNA at the correct position (Pelegri et al. 1999). Interestingly, *vasa* expression is detected not only in four stripes, but also at a lower level along additional cleavage planes and, at later stages, in other cells around the margin as well (Yoon et al. 1997). However, the RNA molecules, which are not localized to the first two cleavage furrows, apparently soon disappear. Additional RNA molecules, e.g., *nanos* mRNA, exhibit similar early distribution of the maternally supplied RNA that is subsequently refined to become PGC-specific (see below). This observation is similar to that described for RNA molecules of some genes important for germ-cell development in *Drosophila* and in *C. elegans* that are degraded outside of the germ cells (Seydoux and Fire 1994; Bashirullah et al. 1999). Analysis of *cis*-acting elements in the zebrafish *vasa* RNA revealed that sequences within the 3′ untranslated region (UTR), as well as within the coding portion of the RNA, are important for the differential stability of the RNA in the soma and the germ cells contributing to PGC-specific distribution of the maternal RNA (Wolke et al. 2002).

Transmission electron microscopy showed that at the 4-cell stage *vasa* RNA is embedded in an electron-dense matrix that resembles nuage, indicating that at this stage the RNA resides within the zebrafish germ plasm (Knaut et al. 2000; Fig. 2A,B). During cleavage stages, up to the 1000-cell stage, no increase in the number of the *vasa*-containing cells is observed despite the dramatic increase in total cell number in the embryo during this period. The basis for this apparent discrepancy is that at these stages the germ plasm consists of a tight structure that is inherited only by one of the two dividing blastomeres, thereby keeping the number of *vasa* RNA-containing cells constant (Knaut et al. 2000; Fig. 2C,D). Interestingly, unlike the findings in *Drosophila*, at these stages, Vasa protein is neither localized, nor confined to, a specific cell population (Knaut et al. 2000).

From sphere stage on, the tight morphology of the germ plasm changes as it spreads into the cytoplasm and is symmetrically distributed between the dividing blastomeres (Fig. 2E), thus resulting in a net increase in the cell number of *vasa*-positive cells; from about 4 at sphere stage (just before the initiation of zygotic vasa transcription) to 25–50 by the end of the first day of development (Yoon et al. 1997; Braat et al. 1999; Weidinger et al. 1999; Knaut et al. 2000). The nature of the signal that directs the transition from asymmetric to symmetric distribution of the germ plasm, marking the time point of the PGC specification, is unknown. This switch in the mode of germ plasm inheritance is independent of zygotic transcription or nucleoplasmic ratio and, therefore, is likely to be controlled by maternally provided factors (Knaut et al. 2000).

Experiments carried out using *vasa* as a molecular marker confirmed the classical finding presented above concerning the position and time of PGC specification, as well as their number. Namely, as suggested for *Barbus conchonius* (Gevers et al. 1992b), from early cleavage stages in zebrafish more than

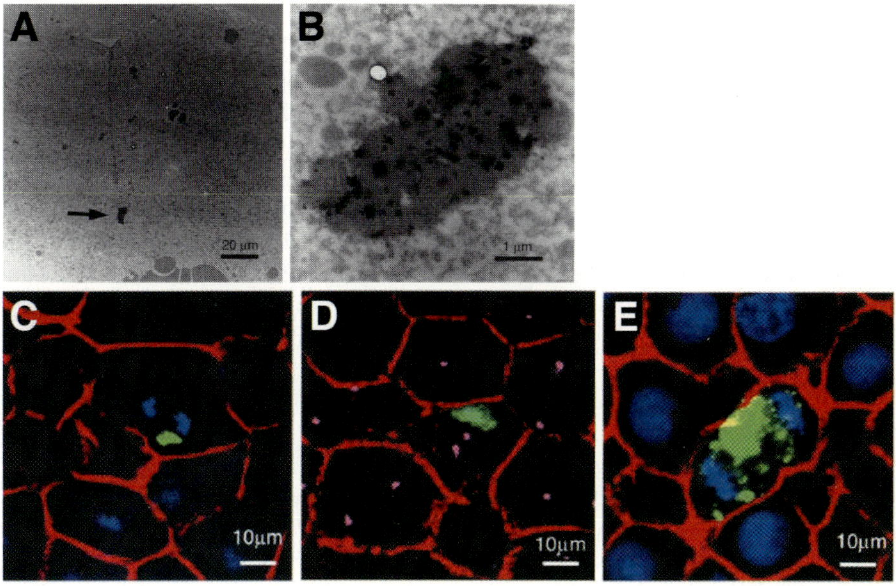

Fig. 2A–E. Germ plasm and *vasa* mRNA distribution in early zebrafish embryos. **A, B** Electron microscopy sections of 4-cell embryos stained for *vasa* RNA along the animal-vegetal axis. *vasa* RNA is confined to a structure vegetal to the indented furrow (*arrow* in **A** and *magnified* in **B**). **C** In the 512-cell stage embryos, dividing cells segregate *vasa* RNA (*green*) to only one daughter cell (DNA is *blue*). **D** During the asymmetric segregation, the *vasa* RNA (*green*) is always colocalizing with one of the two centrioles (γ-tubulin in *pink*). **E** Sphere-stage embryos distribute *vasa* RNA (*green*) equally to both daughter cells (DNA in *blue*). (Reproduced from Knaut et al. 2000 with copyright permission from The Rockefeller University)

one blastomere possesses *vasa*-labeled germ plasm and during later blastula stages they are more often found in more vegetal positions of the embryo. Furthermore, it was shown that the blastomeres are not committed to the germline lineage until late-sphere stage when they start to divide symmetrically. The early estimates of the PGC number in zebrafish at blastula stages (ca. 5 PGCs) that were based on the γ-ray irradiation and the PGC transplantation (Walker and Streisinger 1983; Lin et al. 1992) are in agreement with the findings that about 4 blastomeres harbor germ plasm labeled by *vasa*.

Germ plasm was implicated in germ-cell specification in organisms such as anuran amphibians, *C. elegans* and *Drosophila*. While in *Drosophila* and *C. elegans* many germ plasm components have been identified through genetic screens for maternal effect mutations, screens of this kind are relatively difficult to perform in zebrafish (Pelegri et al. 1999). An alternative approach for the identification of genes that are important for PGC development is screening for genes expressed in these cells. Such an approach brought about the identification of several genes that are expressed in the primordial germ cells during early embryogenesis (Köprunner et al. 2001), some corresponding to genes whose function is important for germ-cell development in other organisms. One example of a gene that was isolated in this way is that of a *nanos*-related gene whose RNA is localized to the region of the germ plasm similar to *vasa* RNA (Köprunner et al. 2001). *nanos*, an RNA-binding protein, had been previously identified in *Drosophila* and in *C. elegans* as a germ-plasm component. The shared morphological characteristics, as well as the identification of similar components (e.g., the gene products of *vasa* and *nanos*) in the germ plasm of invertebrates, anuran amphibians and zebrafish, suggests that the mechanisms controlling early germ-cell development are conserved among these phylogenetically distant organisms.

3 PGC Migration in Zebrafish

Similar to many other organisms, the PGCs in zebrafish are specified prior to the formation of the somatic portion of the gonad and at a different position. As described above, the four groups of PGCs are arranged at positions that are determined by the location of the first two cleavage planes. This mode of positioning the PGCs in zebrafish creates an interesting situation wherein the PGCs initiate their migration from four different directions rather than from one defined point, as described for *Drosophila*, *Xenopus*, and the mouse. A further complication results from the fact that, in zebrafish, the orientation of the first two cleavage planes is random relative to the position of the future dorsal aspect of the embryo (Abdelilah et al. 1994; Helde et al. 1994). Therefore, zebrafish PGCs are initially found and specified at random positions with respect to the dorsoventral axis, yet virtually all of them arrive at the target by the end of the first day of development (Yoon et al. 1997; Weidinger et al. 1999; Fig. 3).

Using *vasa* as a molecular marker for the migrating cells in combination with markers of somatic tissues yielded a detailed description of PGC migration in zebrafish (Weidinger et al. 1999). Before the onset of the migration, *vasa*-labeled blastomeres are found in a marginal or vegetal position and therefore are in contact with the endodermal and mesodermal cells from early gastrulation. The mechanism by which the PGCs reach the deep layers of the blastoderm may involve oriented asymmetrical cell divisions in which the germ-plasm-containing blastomere is positioned vegetally (Knaut et al. 2000).

The analysis of PGC migration in wild-type zebrafish embryos showed that the process can be divided into six steps, which culminate in formation of two clusters by the end of the first day of development (Weidinger et al. 1999; Fig. 4). An important intermediate step during the migration is evident by the end of gastrulation when the PGCs appear to align along the lateral and ventral borders of the mesoderm, as well as along the border between the trunk and head mesoderm (Figs. 4 and 5). Thus, the PGCs that are specified in marginal positions of the blastoderm are found in close proximity to the region where

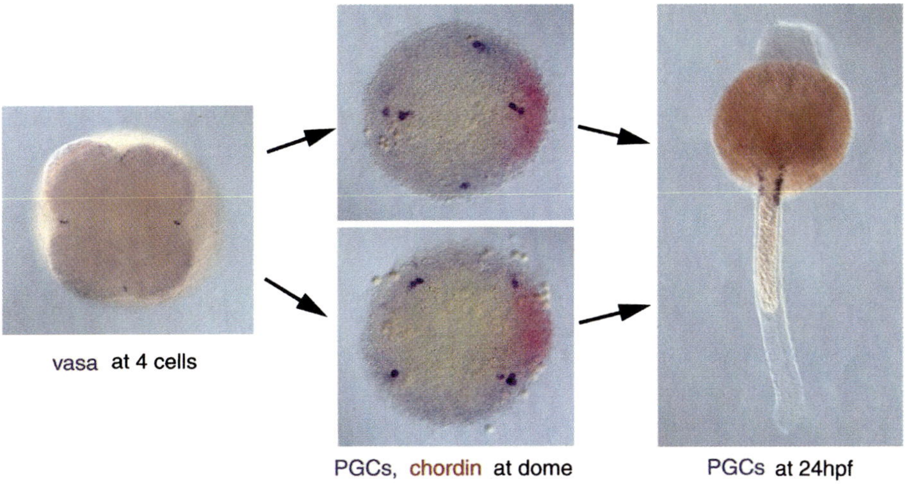

Fig. 3. Random dorsoventral positioning of PGCs before the initiation of migration. At the 4-cell stage *vasa* mRNA and germ-plasm components are enriched in the region of the first two cleavage furrows (*left panel, blue*). Since the orientation of the first cleavage planes is random with respect to the future position of the dorsal aspect of the embryo, before gastrulation PGC clusters are found in different arrangements relative to the dorsal (*middle panel*, PGCs in *blue*, dorsal aspect in *red*). Despite the initial random dorsoventral positioning of the cells, most of the PGCs arrive at two clustering points within the first day of development (*right panel*, PGCs in *blue*)

the mesoderm forms. This allows the cells to arrive at a common intermediate target during gastrulation despite the random dorsoventral starting position (migration steps IIIa and IIIb, Fig. 4). Subsequent to PGC alignment along the mesoderm borders, during early somitogenesis, they migrate towards two lateral positions at the anteroposterior level of the first somites and form two PGC clusters (steps IV and V, Fig. 4). During later segmentation stages, the PGC clusters migrate further posteriorly, to reach the level of the eighth somite where their final target may reside, i.e., cells comprising the somatic portion of gonad.

The general conclusion from the analysis of germ-cell migration in *Xenopus*, *Drosophila* and mouse is that the migration of the cells towards the gonad relies on directional cues provided by the somatic environment. PGC migration in zebrafish is consistent with this notion since the migrating cells appear to align along specific structures in the early embryo, they consistently avoid certain tissues, and they migrate towards attraction points at defined positions within the developing embryo. A direct demonstration of somatic cell control of PGC migration was provided when the migration was analyzed in zebrafish mutants in which patterning and/or differentiation of specific somatic tissues was defective (Weidinger et al. 1999). This analysis showed that PGCs indeed respond to positional cues from somatic cells such that their absence results in aberrant PGC migration.

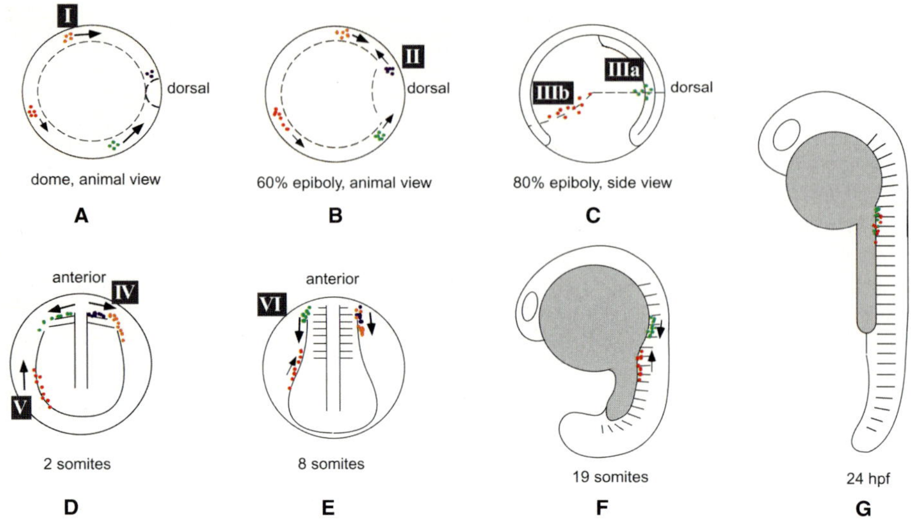

Fig. 4. The six steps of early PGC migration in zebrafish. Schematic drawings of embryos from dome stage to 24 h postfertilization (hpf) showing the positions and movements of the four PGC clusters. At dome stage, four clusters of PGCs are found close to the blastoderm margin in a symmetrical 'square' shape. All possible orientations of the square relative to the dorsal side of the embryo can be observed. **A** Here, an intermediate arrangement is shown with one cluster close to, but not directly at the dorsal side. Beginning before gastrulation, lateral and ventral clusters move towards the dorsal, with ventral clusters migrating more slowly (*step I*, convergence towards the dorsal). This movement is shared with somatic cells and can be attributed to early compaction before gastrulation and dorsal convergence of hypoblast cells during gastrulation. **B** Clusters located very close to the dorsal migrate away from the dorsal midline and are therefore rarely found on the notochord from the 60% epiboly stage on (*step II*, exclusion from the dorsal midline). At the 60% epiboly stage, these movements have resulted in loss of the 'square' arrangement of PGC clusters; convergence continues, but some PGCs can still be found in far ventral positions until the end of gastrulation. **C** Dorsally located PGCs align along the border between the head and trunk paraxial mesoderm marked by the anterior border of the papc expression domain depicted by a *dashed line* (*step IIIa*, alignment along the anterior border of the trunk mesoderm). Ventrally located clusters align at the lateral border of the mesoderm that early on is also marked by the border of papc expression and at late gastrulation starts to express the pronephros marker pax8 (*step IIIb*, alignment along the lateral border of the mesoderm). **D** At the 2-somite stage, most PGCs have arrived in two lines at the level of the first somite. These anterior located PGCs migrate towards the lateral (*step IV*, formation of two lateral PGC clusters). Cells that were initially located ventrally migrate towards the anterior along the anlage of the pronephros (*step V*, anterior migration of trailing PGCs). In this illustration, the positions of the PGCs are drawn relative to the adaxial cells, the somites and the lateral border of the pronephric anlage. **E** At the 8-somite stage, all anterior PGCs are found lateral to the paraxial mesoderm in a cluster extending from the first to the third somite. These clusters start to move towards the posterior (*step VI*, posterior positioning of the PGC clusters), while the trailing cells tightly align on the lateral border of the pronephros and continue to migrate anteriorly. **F** At the 19-somite stage, the main clusters have shifted to more posterior positions and in 60% of embryos some trailing cells are still seen. **G** At 24 hpf, the PGC clusters are located at the anterior end of the yolk extension, which corresponds to the eighth to the tenth somite level. In most embryos, all PGCs have reached this region, only a few trailing cells are found close to the main clusters. (Reproduced from Weidinger et al. 1999 with permission from the Company of Biologists)

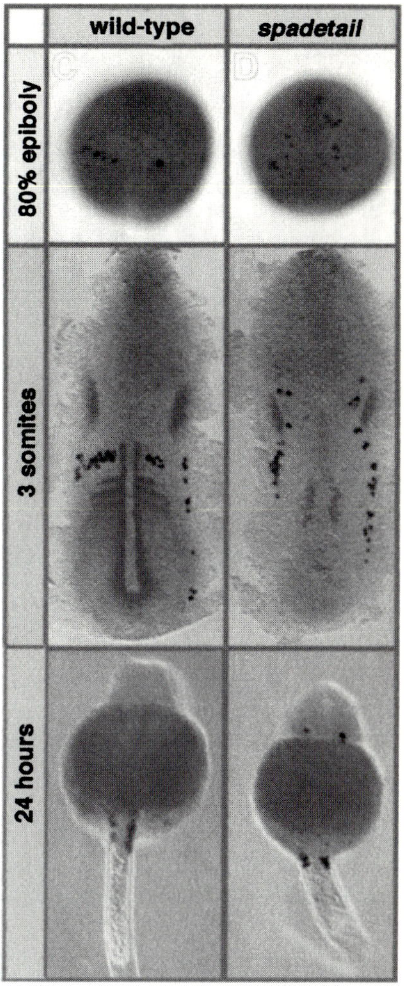

Fig. 5. Ectopic PGCs are found in *spadetail* mutant embryos. Unlike the alignment of PGCs at the anterior border of the trunk mesoderm seen in wild-type embryos during gastrulation and early somitogenesis, in *spadetail* mutants, PGCs are found in more anterior position (*upper* and *middle panels*). During later stages of somitogenesis, the PGCs migrate back towards the clustering position while some migrate anteriorly towards an ectopic clustering point in the head (*lower panels*). (Reproduced from Weidinger et al. 1999 with permission from the Company of Biologists)

A striking example reflecting the importance of cues provided by somatic intermediate targets is the PGC migration phenotype exhibited by *spadetail* mutant embryos (Weidinger et al. 1999). The *spadetail* gene encodes a T-box protein required for proper development of trunk paraxial mesoderm (Griffin et al. 1998). At gastrulation stages, the expression level of various RNA molecules that mark the anterior border of the trunk is severely reduced in *spadetail* mutants. Thus, as judged by gene expression, the anterior border of the trunk paraxial mesoderm, which constitutes an intermediate target for the PGCs, is absent. Indeed, during gastrulation, many PGCs in *spadetail* mutants are not found in their normal intermediate target within the trunk, and as a result end up in ectopic positions in the head (Fig. 5).

Normally, in wild-type embryos, a posterior-to-anterior and medial-to-lateral cell migration along the mesoderm borders towards the clustering positions (steps IV and V, Fig. 4) takes place. Interestingly, ectopic anterior cells in wild-type and more prominently in *spadetail* embryos can migrate back in an anterior-to-posterior direction towards the clustering position (Weidinger et al. 1999, 2002; Fig. 5). Thus, the intermediate target where the PGCs initially cluster during early somitogenesis appears to be able to attract the migrating cells from several different directions. During later stages of somitogenesis, the PGCs leave this intermediate target towards a more posterior position, the region where the gonad will form (Weidinger et al. 2002).

Examining PGC migration in live embryos confirmed that these cells migrate actively towards the clustering position (Weidinger et al. 2002). Throughout their migration the cells exhibit morphological features characteristic of motile cells, including extension of numerous cellular processes and clear change of their position relative to their somatic neighbors. The migration patterns are highly dynamic: individual PGCs frequently alter their speed, direction and position relative to each other in the course of migration. These observations suggest that the PGCs migrate as individual cells rather than as a coordinated group of cells.

4 Maintenance of the Fate of Migrating PGCs

During their migration, PGCs face conflicting requirements regarding their response to cues provided by the somatic environment. On the one hand, the PGCs should respond to signals provided by somatic tissues guiding them towards the gonad; on the other, while migrating, the PGCs have to ignore signals that would lead to their differentiation into other cell types. This issue of PGC competence to respond to somatic differentiation cues was investigated in *Drosophila* by following the fate of transplanted labeled PGCs. In this case, PGCs that ended up in regions outside of the gonad never assumed somatic fates, never incorporated into somatic tissues and eventually died (Underwood et al. 1980; Technau and Campos-Ortega 1986; Kobayashi et al. 1993). Analysis of the fate of ectopic PGCs in zebrafish suggests that the situation is similar to that in *Drosophila*. Specifically, cells that arrive at ectopic positions during early development maintain their PGC fate as judged by ongoing expression of molecular markers and preservation of characteristic PGC morphology (Weidinger et al. 1999). Thus, the PGCs of zebrafish appear to be committed to the germ-line lineage from early stages of development.

In both *Drosophila* and *C. elegans*, the maintenance of the PGC fate, their survival, and their incorporation into the gonad require the function of the *nanos* gene (Lehmann and Nüsslein-Volhard 1991; Wang and Lehmann 1991; Wharton and Struhl 1991; Deshpande et al. 1999; Subramaniam and Seydoux 1999). A *nanos*-related gene that is expressed in PGCs has been identified in zebrafish (*nanos-1*, Köprunner et al. 2001) and, similar to the mRNAs of the

Drosophila and *C. elegans nanos* genes, the zebrafish *nos*1 mRNA is provided maternally. During the first two mitotic divisions, the mRNA is enriched in the cleavage planes of the early embryo, while non-localized mRNA can still be detected in whole-mount in situ hybridization until the late blastula stages. This expression pattern is comparable to that described for the zebrafish *vasa* mRNA which at these stages is found within an electron-dense structure, the putative zebrafish germ plasm (Yoon et al. 1997; Knaut et al. 2000). During later blastula stages, *nos*1 mRNA is incorporated into the PGCs that characteristically migrate to form two PGC clusters, as described above (Weidinger et al. 1999; Köprunner et al. 2001). The mechanisms controlling the distribution of the maternal mRNA of *nanos* were found to be similar to those shown for *vasa*, in that the RNA is stabilized in the germ cells while being rapidly degraded in other cells in a process that depend on the 3'UTR of the gene. The function of Nanos in germ-cell development of the zebrafish was determined by specifically inhibiting the translation of the corresponding RNA using morpholino antisense oligonucleotides (Nasevicius and Ekker 2000; Köprunner et al. 2001). Interestingly, even when a very high amount of *nanos*1 morpholinos was injected into the embryos, no change in PGC numbers at pre-gastrulation stages was observed. However, while reducing Nos1 level did not affect PGC formation, it did deleteriously affect their migration and survival; in morpholino-injected embryos the PGCs failed to migrate normally, reached ectopic positions and exhibited morphological characteristic of cells undergoing programmed cell death (Fig. 6). This phenotype is strikingly reminiscent of the *nanos* loss-of-function phenotype in *Drosophila* and *C. elegans*, which is manifested by cell survival and cell migration defects.

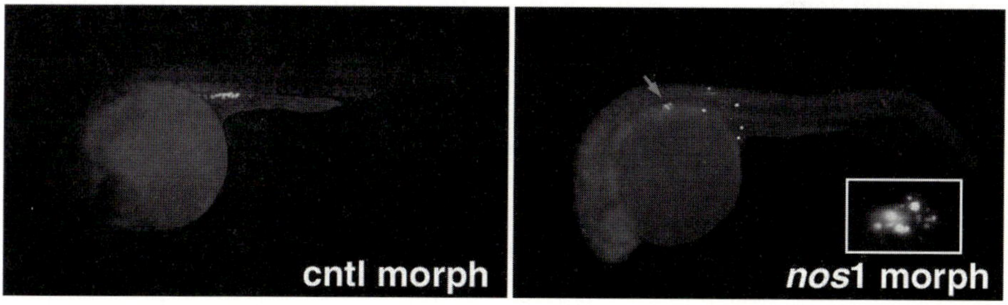

Fig. 6. Reduction of zygotically translated Nanos1 protein results in PGC death and migration defects. In embryos injected with a control morpholino the PGCs are found at the normal position at the end of the first day of development (*left panel*). In embryos injected with morpholino antisense oligonucleotides directed against the mRNA of *nanos*1, a dramatic reduction in PGC number is seen while the remaining cells exhibit migration defects and morphology characteristic of cells undergoing programmed cell death (*right panel, inset*). (Reproduced with permission from Köprunner et al. 2001)

5 PGC Development in Zebrafish as Compared with That in Other Organisms

The current understanding of PGC development in zebrafish suggests that the developmental program of these cells is controlled similarly to that of *Drosophila*, *C. elegans* and *Xenopus* and differs from that of mammals. A common denominator for these organisms is that the PGCs appear to be specified in positions where specific asymmetrically localized molecules are present, as well as cytoplasmic structures that can be visualized by electron microscopy – the germ plasm. Importantly, the identity and the function of germ-plasm determinants appear to be conserved among these organisms. Nevertheless, a closer look at germ-cell development in zebrafish reveals some important differences between this vertebrate model organism and the invertebrates mentioned above. The most striking difference between the function of zebrafish germ plasm and that of *Caenorhabditis* and *Drosophila* concerns the time at which zygotic transcription is initiated in the germ cells. Both in *Caenorhabditis* and *Drosophila* the germ plasm represses early zygotic transcription (Zalokar 1976; Seydoux and Dunn 1997; Asaoka et al. 1998; van Doren et al. 1998b). In contrast, in zebrafish transcription appears to initiate in all nuclei at the same time, irrespective of the presence of germ plasm (Knaut et al. 2000). Additional differences include the uniform Vasa protein distribution in zebrafish prior to PGC specification as opposed to the early PGC-specific expression of the Vasa proteins in *C. elegans* and *D. melanogaster*. Further, unlike in *C. elegans* and *D. melanogaster*, it is *vasa* RNA and not its protein that is part of the germ plasm at early stages of PGC development (Knaut et al. 2000).

The cloning of *vasa* homologues from other teleost species allows one to ask whether the findings regarding PGC specification in zebrafish are applicable to other fish species. Curiously, however, the early subcellular localization of *vasa* RNA at the cleavage planes has not been described for any of the other fish species from which *vasa* was cloned (Kobayashi et al. 2000; Shinomiya et al. 2000; Yoshizaki et al. 2000). The most thorough analysis of *vasa* in other teleosts was performed in medaka – *Oryzias latipes* (Shinomiya et al. 2000). The expression pattern of the medaka *vasa* homologue (*olvas*) was monitored from the 2-cell stage and onwards. Surprisingly, in contrast to the clear subcellular enrichment of the zebrafish *vasa* mRNA in the first two cleavage planes, *olvas* transcripts were found to be evenly distributed in the cytoplasm of the early blastomeres. During blastula stages, the level of the uniformly distributed mRNA gradually declined until late gastrulation stages, when *olvas*-positive cells, scattered in the posterior part of the embryonic shield, become detectable. By early somitogenesis stages, the *olvas*-containing cells are found on either side of the body axis in a line that extends along the first four somites. The finding that early on *olvas* mRNA is uniformly distributed and that specific expression of the gene in the PGCs is observed at a defined somatic region (the posterior part of the shield) raises the possibility that the mechanism of

PGC specification in this fish differs from that in zebrafish. Specifically, the lack of early subcellular localization of *olvas* RNA combined with the fact that the PGCs appear in a specific position relative to the dorsoventral axis suggest that PGC specification in medaka may depend on inductive signals rather than on prelocalization of cytoplasmic determinants. Nevertheless, a series of experiments conducted in medaka suggest that, despite these differences, the PGCs are likely to be specified by a mechanism similar to that of zebrafish (S. Rohr and E.R. unpublished). Consistent with the model stating that PGC formation in medaka does not depend on inductive signals, the number of PGCs was not altered when a dramatic change in the somatic environment was induced, that is, formation of a secondary axis.

6 Conclusions and Future Directions

In recent years our understanding of germ-line development in fish has been enhanced to a large extent by work performed in zebrafish. The most fundamental questions concerning the mode of PGC specification and their origin, questions that until recently had no answer, have been resolved. The cloning of the zebrafish *vasa* gene and the analysis of its expression pattern support the notion that germ plasm is responsible for the early segregation and specification of the PGC lineage in fish. These findings, combined with the recently described function and regulation of the zebrafish *nanos*, revealed a previously unsuspected conservation between PGC development in invertebrates and fish. Considering that germ plasm has been identified in other vertebrate groups such as the anuran amphibians and birds, it is conceivable that the finding regarding PGC development in fish will be relevant for those organisms as well (Rongo and Lehmann 1996; Saffman and Lasko 1999; Houston and King 2000a; Knaut et al. 2000; Tsunekawa et al. 2000). While the mechanisms of PGC specification and early development differ between mammals and the species mentioned above, germ-plasm components are expressed in the germ line of mammals at later stages of development and are important for gametogenesis (e.g., Eberhart et al. 1996; Houston and King 2000b; Tanaka et al. 2000). Therefore, the genes and the mechanisms controlling early PGC development in zebrafish are likely to be relevant for germ-cell development in mammals as well.

III. Neural Development

Patterning the Zebrafish Central Nervous System

Steve W. Wilson[1], Michael Brand[2], and Judith S. Eisen[3]

1 Introduction

One of the hallmarks of the vertebrate nervous system is the enormous diversity of neurons of which it is composed. The ability of an animal to carry out its normal behavioral repertoire requires that all of these neurons develop in the appropriate numbers and at the appropriate times and positions and that they elaborate the appropriate differentiated characteristics. This process begins with patterning of the nascent nervous system, the neural plate, along the dorsoventral (DV) and anteroposterior (AP) axes of the body. In this chapter we consider what is currently known about the mechanisms involved in patterning the zebrafish neural plate. Although we discuss many of the similarities and differences between zebrafish and other vertebrates, we limit our primary focus to zebrafish, as including an exhaustive description of neural patterning in other model vertebrates would exceed the space limitations of this chapter. We apologize for any work we have overlooked and for discussion of work on other model vertebrates we direct the reader to several excellent reviews (Goulding and Lamar 2000; Jessell 2000; Patten and Placzek 2000; Wilson and Rubenstein 2000; Altmann and Brivanlou 2001; Briscoe and Ericson 2001).

2 Nervous System Morphogenesis

As for other vertebrates, the zebrafish central nervous system (CNS) begins as an ectodermal epithelium on the dorsal side of the embryo, the neural plate. A dorsomedial thickening of cells immediately after gastrulation (about 10 h; hours postfertilization at 28.5 °C) is the first morphological manifestation of neural plate formation (Schmitz et al. 1993). Within the next 1–2 hr, two lateral

[1] Department of Anatomy and Developmental Biology, University College London, Gower Street, London WC1E 6BT, UK
[2] Max Planck Institute for Molecular Cell Biology and Genetics, Dresden, Pfotenhauerstrasse 108, 01307 Dresden, Germany
[3] Institute of Neuroscience, University of Oregon, Eugene, Oregon 97403, USA

Results and Problems in Cell Differentiation, Vol. 40
L. Solnica-Krezel (Ed.): Pattern Formation in Zebrafish

thickenings appear. Morphogenetic convergence movements (Kimmel et al. 1994) fuse the lateral and medial thickening into a neural keel around 13 h and into a neural rod around 16 h. Cavitation beginning around 17 h establishes the central canal of the hollow neural tube (Raible et al. 1992; Schmitz et al. 1993) that forms brain anteriorly and spinal cord posteriorly. Thus, the lateral edges of the neural plate form dorsal CNS and the medial neural plate forms ventral CNS, as has been shown directly by labeling studies (Papan and Campos-Ortega 1994). Much of the patterning we will discuss actually begins in the neural plate; thus, when we refer to DV patterning, we mean patterning that begins along the mediolateral neural plate axis and results in cells with specific soma positions along the DV axis of the CNS. We consider patterning of the spinal cord and brain separately, starting with the spinal cord because it is more simply organized.

The entire CNS is underlain by axial mesoderm that internalizes during gastrulation (see Kane and Adams; Kimelman and Schier, this Vol.). Anteriorly, this mesoderm forms the prechordal plate. Posterior to hindbrain rhombomere 4, this mesoderm forms the notochord which extends to the tip of the tail (Hatta and Kimmel 1993). Axial mesoderm is bounded on both sides by paraxial mesoderm that later forms somites and head mesoderm (see Kimelman and Schier, this Vol.). As we describe, signals emanating from axial mesoderm play key roles in nervous system patterning. Signals emanating from somites may also be involved in neural patterning, although considerably less is currently known about such signaling.

3 The Spinal Cord

Although the spinal cord is more simply organized than the brain, it shows both AP and DV patterning. AP patterning is most clearly manifest by the development of limb-innervating motoneurons in spinal segments adjacent to the pectoral fins (Myers 1985). There is also more fine-grained, segmental AP patterning, revealed by the soma positions of early-developing primary motoneurons that innervate myotomal muscles (Myers 1985; Eisen et al. 1986; Westerfield et al. 1986). However, the most obvious spinal cord patterning is along the DV axis such that distinct types of neurons have somata in specific DV locations.

As for other anamniote vertebrates, the zebrafish spinal cord has both early-developing primary neurons and later-developing secondary neurons (Kimmel and Westerfield 1990). Primary neurons are large, few in number, born starting about 9–10 h (Myers et al. 1986; Kimmel and Westerfield 1990), undergo axogenesis between 14–24 h and comprise all modalities, including sensory, inter- and motoneurons. Each type of primary neuron has a specific DV location within the embryonic and larval spinal cord: primary motoneurons are ventrally located, Rohon-Beard (RB) primary sensory neurons are dorsally located, and a variety of primary interneurons are located throughout the

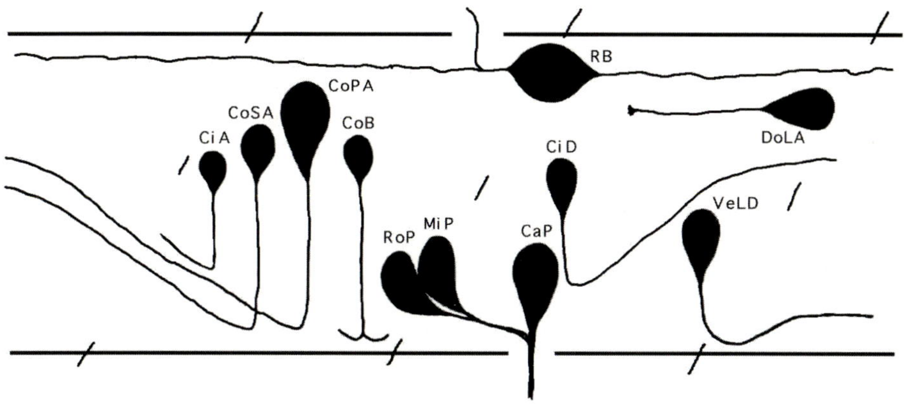

Fig. 1. Several types of primary and secondary neurons have been described in the embryonic and larval zebrafish spinal cord. Drawing shows lateral view of spinal cord, anterior to the *left*. Rohon-Beard (*RB*) sensory neurons, ventral longitudinal descending (*VeLD*) and commissural primary ascending (*CoPA*) interneurons and caudal primary (*CaP*), middle primary (*MiP*) and rostral primary (*RoP*) motoneurons are all primary neurons, whereas commissural secondary ascending (*CoSA*) neurons are secondary interneurons. Dorsal longitudinal ascending (*DoLA*), circumferential descending (*CiD*), circumferential ascending (*CiA*), and commissural bifurcating (*CoB*) neurons are all interneurons, although it is unclear whether they should be considered as primary or secondary; secondary motoneurons are not pictured here. (Adapted from Bernhardt et al. 1990)

midregion of the spinal cord. Bernhardt and colleagues (1990) used axonal morphology to classify several types of primary interneurons (Fig. 1) and Eisen and colleagues (1986) used axonal morphology to classify three types of primary motoneurons. Primary motoneurons have somata located in the medial motor column, innervate fast muscles derived from the myotomes and persist through adulthood (Westerfield et al. 1986). Persistence through adulthood is probably also the case for many primary interneurons, as cell death has not been described in this population. In contrast, RBs are a transient neuronal population. Many RBs die by 3 d (days of development at 28.5 °C); death apparently depends on signaling via TrkC1 receptors (Williams et al. 2000) and sodium-channel-mediated electrical activity (Svoboda et al. 2001).

Secondary neurons are smaller, more numerous, born later (starting about 13–14 h; Myers et al. 1986; Kimmel et al. 1994; Appel et al. 2001) and typically have finer axons than primary neurons (Kimmel and Westerfield 1990). Although secondary neurons are probably more similar to the spinal neurons studied in amniote vertebrates (Kimmel and Westerfield 1990), considerably less is known about these cells than about zebrafish primary neurons. Bernhardt and colleagues (1990) have classified only one type of secondary interneuron by axonal morphology. Myers (1985) and Pike and Eisen (Pike et al. 1992) described several types of secondary motoneurons based on axonal morphology in embryos and larvae and Westerfield and colleagues (1986) identified several types of secondary motoneurons in adults based on innervation fields and axonal morphology. Some secondary motoneurons are located in the lateral motor column and innervate fin muscles. However, many more secondary motoneurons are located in the medial motor column and innervate myotomal muscles (Eisen 1994).

The spinal cord contains several other cell types in addition to neurons. Floor plate cells occupy the spinal cord ventral midline and it is likely that roof plate cells occupy the dorsal midline, although they have not been well described. The floor plate comprises three longitudinal columns of cells; medial floor plate forms a single column in the midline and lateral floor plate forms flanking columns on both sides of the midline column (Bernhardt et al. 1992b; Strähle et al. 1997; Odenthal et al. 2000). Lateral floor plate probably generates Kolmer-Agduhr (KA) neurons (Bernhardt et al. 1992b), a class of GABAergic neurons that contact cerebrospinal fluid and may act as proprioceptive position sensors (Dale et al. 1987a,b), although they may also generate other fates. Radial glia have been described in the zebrafish spinal cord (Appel et al. 2001), although their origin is currently unknown.

3.1 Bmp Signaling Establishes DV Pattern in the Spinal Cord

As described elsewhere (see Hammerschmidt and Mullins, this Vol.), bone morphogenetic proteins (Bmps) are key regulators of embryonic DV patterning. Within the developing ectoderm, Bmp activity has been shown to promote epidermal differentiation and inhibit neural differentiation. Because different levels of Bmp activity specify different cell fates within the mesoderm, Bmps are considered morphogens (Dosch et al. 1997; Neave et al. 1997). However, rather than establishing different levels by diffusion, it seems likely that different Bmp levels are established by at least four distinct proteins, Chordin, Follistatin, Noggin and Cerberus, that emanate from the dorsal side of the embryo and inhibit Bmp function (Dale and Wardle 1999; Piccolo et al. 1999). Of these, *noggin* (Bauer et al. 1998; Fürthauer et al. 1999), *follistatin* (Bauer et al. 1998) and *chordin* (Schulte-Merker et al. 1997) have been described in zebrafish.

Analysis of zebrafish Bmp pathway mutants reveals that Bmps participate in patterning throughout the entire mediolateral extent of the neural plate, thus the entire DV axis of the spinal cord. Studies of avian embryos have implicated Bmps in specification of dorsal neural fates and neural crest (Liem et al. 1995), a cell type that emigrates from the dorsal spinal cord, migrates along well-defined pathways and gives rise to a diverse set of derivatives including peripheral neurons and glia, pigment cells and fin ectomesenchyme (Eisen and Weston 1993; see Kelsh and Raible, this Vol.). Analysis of zebrafish embryos homozygous for mutations in Bmp pathway genes, including *swirl/bmp2b*, *snailhouse/bmp7* and *somitabun/smad5* (Nguyen et al. 1998b, 2000), provides evidence that Bmp signaling is essential to establish neural crest as well as RB neurons that also arise from the lateral neural plate domain that generates neural crest (Cornell and Eisen 2000). However, Bmp effects are not limited to the lateral neural plate/dorsal spinal cord. Characterization of embryos with different levels of Bmp signaling provides evidence for graded effects of Bmp throughout the mediolateral axis of the neural plate (Barth et al. 1999; Fig. 2). Thus, severe depletion of Bmp signaling by overexpression of *noggin* leads to loss of RBs as well as interneurons in the midregion of the spinal cord and additionally results in expansion of ventral spinal cord fates. Slightly less severe Bmp depletion in *swirl/bmp2b* mutants leads to loss of RBs and expansion of interneurons. Mild Bmp depletion in *somitabun/smad5* mutants leads primarily to lateral displacement of RBs and interneurons.

A surprising observation of these studies was that severe Bmp depletion leads to expansion of the floor plate (Barth et al. 1999; Nguyen et al. 2000). Thus Bmp effects extend to the most medial region of the neural plate in zebrafish. Consistent with this observation, mouse embryos with augmented Bmp signaling resulting from a targeted deletion of the *noggin* gene lack floor plate and motoneurons in the caudal spinal cord (McMahon et al. 1998), suggesting that Bmp signaling suppresses formation of these ventral cell types. This effect could be direct, or it could be mediated by interactions between the Bmp-signaling pathway and other signaling pathways, such as the Hedgehog (Hh) pathway, that have been shown to specify ventral spinal cord fates (Goulding and Lamar 2000; Jessell 2000; Patten and Placzek 2000; Briscoe and Ericson 2001). Indeed, recent in vitro studies in chick embryos suggest that specification of these ventral spinal cord fates depends on integration of signals in these two pathways (Liem et al. 2000).

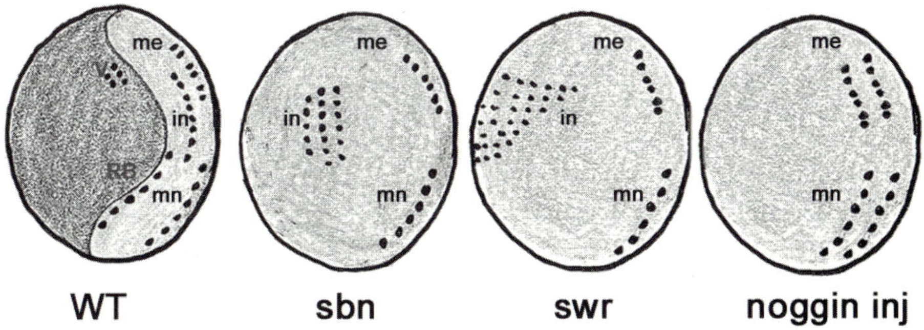

Fig. 2. Bmp signaling affects patterning of the entire neural plate. Series of embryos from wild-type (*WT*) to most severely Bmp-depleted (*noggin inj*) showing the arrangement of motoneurons (*mn*), interneurons (*in*), spinal sensory neurons (*RB*), and medial cranial neurons (*me*). *Darker gray* represents non-neural ectoderm and *lighter gray* represents neural ectoderm. (Adapted from Barth et al. 1999)

3.2 Hedgehog and Nodal Pathways Pattern the Ventral Spinal Cord

Elegant studies in avian embryos demonstrated that underlying notochord patterns the ventral neural plate (Placzek et al. 1990; Yamada et al. 1991, 1993). These studies provided evidence that, without notochord, floor plate and motoneurons do not form. However, two zebrafish mutants lacking notochord, *no tail* (Halpern et al. 1993) and *floating head* (Talbot et al. 1995), still form at least some floor plate and motoneurons, raising the possibility that the mechanisms underlying neural patterning might differ between zebrafish and amniote vertebrates.

Although one of the first studies implicating Sonic hedgehog (Shh) as the primary notochord-derived signaling molecule responsible for ventral neural tube patterning was carried out in zebrafish (Krauss et al. 1993), initial loss-of-function studies reinforced the idea that the role of Shh might be different in zebrafish and other vertebrates. In vitro studies in chick (Roelink et al. 1995) as well as targeted deletion of the mouse *Shh* gene (Chiang et al. 1996) demonstrate that Shh is both sufficient and necessary for specification of ventral spinal cord fates, including both floor plate and motoneurons. In contrast, zebrafish embryos homozygous for a deletion of the *sonic-you* (*syu/shh)* gene have motoneurons and medial floor plate, although lateral floor plate is absent (Schauerte et al. 1998; Odenthal et al. 2000). These results demonstrated that Shh signaling is necessary for specification of lateral floor plate, but also raised the possibility that other aspects of ventral spinal cord patterning in zebrafish do not require Shh signaling.

This controversy has been addressed in part by our current understanding that during evolution, the ancestral bony fishes apparently underwent a genome duplication with the consequence that zebrafish have an additional copy of some genes (Postlethwait et al. 1998). Thus, zebrafish have two copies of the ancestral *shh* gene, one called *shh* and the other called *tiggywinkle hedgehog* (*twhh*; Ekker et al. 1995; Zardoya et al. 1996). Initially, both genes are expressed in the dorsal embryonic shield; however, by 90% epiboly, *shh* is expressed only in presumptive notochord and *twhh* is expressed only in presumptive floor plate (Etheridge et al. 2001). During later segmentation stages, *shh* is expressed in both notochord and floor plate (Krauss et al. 1993), but *twhh* expression remains confined to the floor plate (Ekker et al. 1995). In addition, an *Indian hedgehog* homologue, *echidna hedgehog* (*ehh*; Currie and Ingham 1996; Zardoya et al. 1996), is also expressed in the notochord. Because these Hhs may function redundantly, and because mutations in all three genes are not currently available, it has been difficult to address the role of Hh signaling in zebrafish ventral spinal cord patterning.

One way the question of Hh redundancy during specification of ventral spinal cord cell fates has been addressed is by injection of morpholino antisense oligonucleotides (MOs). MOs prevent translation and thus effectively "knock down" gene function (Summerton 1999). Medial floor plate forms normally in wild-type embryos injected with *shh*-MO plus *twhh*-MO (Nasevicius and Ekker 2000) and *syu* mutants injected with *twhh*-MO (Etheridge et al. 2001) or *twhh*-MO plus *ehh*-MO (Lewis and Eisen 2001), providing strong evidence that Hh signaling is unnecessary for medial floor plate specification.

Given that Hh signaling is unnecessary for specification of medial floor plate but necessary for specification of lateral floor plate, what about its role in specification of motoneurons? Severely reducing function of any of the three *hh* genes alone has a negligible effect on the number of primary motoneurons (Beattie et al. 1997; Schauerte et al. 1998; Lewis and Eisen 2001), suggesting that these three Hhs have redundant functions during primary motoneuron specification. Addressing the importance of each *hh* gene has been difficult because mutations in either *twhh* or *ehh* have not yet been isolated. Thus, this issue has been investigated in two ways: first by characterizing *cyclops* (*cyc*) mutants lacking floor plate, and thus diminished in Shh and Twhh signaling, *flh* mutants lacking notochord, and thus diminished in Shh and Ehh signaling, *cyc;flh* double mutants that should lack most Hh signaling (Beattie et al. 1997) and *cyc;flh;syu* mutants that should lack essentially all Hh signaling (Lewis and Eisen 2001), and second by injection of morpholinos for combinations of the *hh* genes (Lewis and Eisen 2001). These experiments reveal that the number of primary motoneurons is proportional to the level of Hh signaling, such that embryos lacking function of two of these Hhs have fewer primary motoneurons than wild types and embryos lacking function of all three Hhs have even fewer primary motoneurons (Beattie et al. 1997; Lewis and Eisen 2001), occasionally lacking primary motoneurons altogether (Lewis and Eisen 2001). However, a few rogue primary motoneurons typically remain, suggesting either that Hh signaling has not been entirely eliminated or that other signaling pathways may contribute to motoneuron specification.

Isolation of mutations in the *smoothened* (*smoh*) gene (*smooth muscle omitted*; *smu*) has also been extremely important in analysis of the role of Hh signaling in ventral spinal cord patterning (Barresi et al. 2000; Chen et al. 2001a; Varga et al. 2001). Just like *syu* mutants, *smu* mutants retain medial floor plate but lack lateral floor plate, again providing strong evidence that Hh signaling is not involved in specification of medial floor plate. Interestingly, though, medial floor plate begins to degenerate early in the second day of development in *smu* mutants, showing that Hh signaling is required for medial floor plate maintenance. *smu* mutants also form some primary motoneurons anteriorly, but this appears to result from the activity of maternal Smoothened (Chen et al. 2001a; Lewis and Eisen 2001). Taken together, these studies show that, in zebrafish, as in amniotes, Hh signaling is necessary for specification of at least the vast majority of motoneurons.

Because Hh signaling is unnecessary for medial floor plate specification, zebrafish must have another signaling pathway that carries out this function. Medial floor plate is absent from embryos homozygous for mutations in the *cyclops* (*cyc*; Hatta et al. 1991b) and *squint* (*sqt*) genes that encode Nodal-related members of the TGFβ family (Feldman et al. 1998), as well as in mutants for *one-eyed pinhead* (*oep*; Strähle et al. 1997; Odenthal et al. 2000), that encodes an EGF-CFC protein that acts as an extracellular cofactor for Nodal signaling (Gritsman et al. 1999). Thus Nodal signaling is required for medial floor plate specification. Interestingly, however, medial floor plate does form later in *cyc* and *oep* mutants, suggesting that other pathways may be involved as well (Strähle et al. 1997).

Does this mean that floor plate specification in zebrafish differs from that in amniote vertebrates? Although the precise role of Nodal signaling in floor plate specification in mouse is currently unclear, in part because of the severity of the *Nodal* mutant phenotype (Zhou et al. 1993; Conlon et al. 1994), a number of recent studies suggest a link between Nodal and Shh signaling. For example, mouse embryos transheterozygous for mutations in *Nodal* and *Smad2*, a downstream effector of Nodal signaling, have many ventral CNS defects similar to mouse *Shh* mutants (Normura and Li 1998) and mutations in human TGIF, that acts as a Smad2 transcriptional corepressor, also result in defects in the ventral CNS similar to those caused by *Shh* mutations (Gripp et al. 2000). Consistent with these mammalian studies, Nodal signaling has been shown to induce *shh* expression in ventral neural tubes of zebrafish and chick embryos (Müller et al. 2000). Based on these studies, Müller and colleagues (2000) have proposed that zebrafish medial floor plate specification requires Nodal signaling to initiate expression of *shh* and *twhh* and then Shh and Twhh signaling initiate expression of downstream floor plate genes. It will be important to learn whether this model is correct and to resolve whether a similar pathway acts during floor plate specification in amniotes.

3.3 Delta/Notch Signaling Segregates Neural Fates Within Neural Plate Domains

Because different cell types arise at different mediolateral positions within the spinal cord neural plate, the neural plate can be thought of being composed of domains, each of which forms a longitudinal stripe. Several of these stripes can be defined as "proneural" domains, territories that express particular transcription factors of the bHLH class (Sommer et al. 1996; Blader et al. 1997a; Korzh et al. 1998). Two of these domains have been well studied: the lateral domain that generates neural crest and RB neurons, and the more medial domain that generates primary and secondary motoneurons. In each case, the choice of cell fate relies on signaling via the Delta/Notch signaling pathway. Thus, in the lateral neural plate, cells strongly expressing the bHLH gene, *ngn1*, become RB neurons (Blader et al. 1997a). These cells are isolated from one another and the cells between them become neural crest (Cornell and Eisen 2000). Studies in which signaling by the products of "neurogenic" genes, *delta* and/or *notch*, is blocked show that the surrounding cells are capable of becoming RBs, providing evidence that Delta/Notch mediated lateral inhibition prevents them from taking this fate (Appel and Eisen 1998; Haddon et al. 1998; Cornell and Eisen 2000). Similarly, in the medial neural plate domain,

Delta/Notch signaling prevents cells surrounding prospective primary motoneurons from taking this fate and they become secondary motoneurons instead, and possibly other cells types as well (Appel and Eisen 1998; Haddon et al. 1998). The roles of proneural and neurogenic genes are discussed in more detail elsewhere (see Appel and Chitnis, this Vol.).

3.4 Later Signals May Refine Cell Identity

Patterning of the spinal cord neural plate results in the appropriate number of neurons being specified at the right time and place. However, additional signals may be necessary to establish the final identities of these cells. For example, the neural crest cells that arise in the lateral neural plate generate both neural and non-neural derivatives. A variety of studies in zebrafish have begun to unravel both the signals and the transcriptional responses involved in specification of neural crest derivatives. As described in more detail (see Raible and Kelsh, this Vol.; Dorsky et al. 2000a), Wnt signaling is involved in establishing the pigment cell fate by regulation of *nacre*, the zebrafish gene encoding MITF (Dorsky 2000b).

Primary motoneurons also show later patterning. Initially, all of these cells express *islet1* that encodes a transcription factor shown to be required for the motoneuron fate in mouse (Pfaff et al. 1996). Later, specific primary motoneurons downregulate expression of *islet1* and express a related gene, *islet2* (Appel et al. 1995; Tokumoto et al. 1995). This dynamic change in gene expression follows somite formation and occurs in a segmental pattern within the spinal cord that mirrors the segmental pattern of the overlying somites, establishing a fine-grained AP pattern for this cell type. At least one type of ventral interneuron, VeLD, also shows this segmental pattern (Bernhardt et al. 1990; Eisen and Pike 1991). Recent studies have shown that *islet2* is required for normal development of the primary motoneurons that express it and when Islet2 function is knocked-down, these cells develop a VeLD-like morphology and express GABA, a VeLD neurotransmitter, rather than ACh, the normal primary motoneuron neurotransmitter (Segawa et al. 2001). The signals responsible for establishing the patterning of these cells are currently unknown, although circumstantial evidence from mutants with disrupted somites (Eisen and Pike 1991) and heat shock experiments (Kimmel et al. 1988) suggest that the signals could be of paraxial mesodermal origin. This idea fits well with the known role of paraxial mesoderm in determining motoneuron subtypes in chick, as demonstrated by spinal cord and somite reversals (Ensini et al. 1998). It also fits well with the results of transplantation studies (Eisen 1991; Appel et al. 1995; Fig. 3). These studies revealed that the fates of individual primary motoneurons are not fixed until after somitogenesis and that the precise AP position of each motoneuron soma within the spinal cord, and relative to the overlying somite boundaries, determines the cell's identity, as assayed by *islet* gene expression and axonal trajectory.

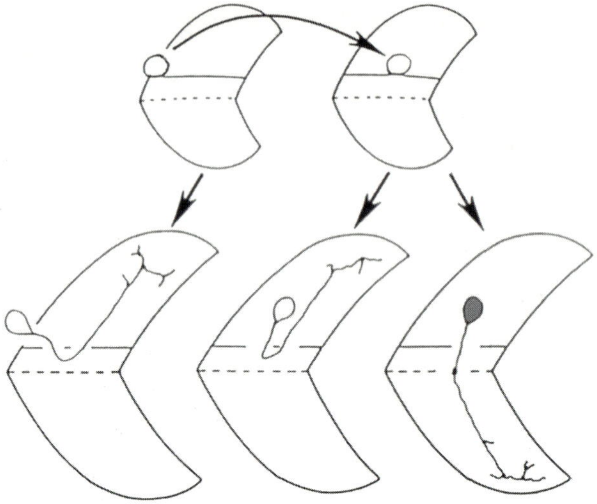

Fig. 3. Position determines the identity of primary motoneurons. An *islet1*-expressing primary motoneuron transplanted from the MiP soma position (*upper left*) to the same position (*lower left*) develops a normal MiP axonal trajectory in the overlying myotome. A similar cell transplanted to the CaP soma position (*upper right*) also develops a MiP axonal trajectory if the transplant is done within 1 hr of axogenesis (*lower middle*), but develops a CaP axonal trajectory if the transplant is done earlier (*lower right*). In this case, the cell also turns on expression of *islet2* (*gray shading*), a CaP-specific gene. (Adapted from Eisen 1991; Appel et al. 1995)

4 The Forebrain

The forebrain is the most complex region of the CNS, both in terms of its connectivity and also in terms of its morphogenesis. As with more caudal regions of the CNS, it derives from the simple neuroepithelial cell sheet of the neural plate. However, subsequent to neural tube formation, the anterior CNS undergoes a series of morphogenetic events that transform the simple neural tube and result in generation of the highly complex derivatives of the telencephalon, optic vesicles and diencephalon. It is fair to say that we know almost nothing about the mechanisms that regulate morphogenesis of the forebrain (although see Loosli et al. 2001). However, we are beginning to understand the genetic pathways that underlie allocation of regional fates within the anterior neural plate. In the following section, we review some of the progress that has been made in understanding how cells acquire their regional identities within the most rostral regions of the CNS.

The major derivatives of the anterior neural plate are the telencephalon, the optic vesicles, the hypothalamus, the ventral and dorsal thalamus and the pretectum (Fig. 4A,B). The telencephalon is the region that in mammals includes the cerebral cortex (what we normally consider to be the "thinking" part of the brain), the olfactory bulb, and the basal ganglia. Although defining the axes in

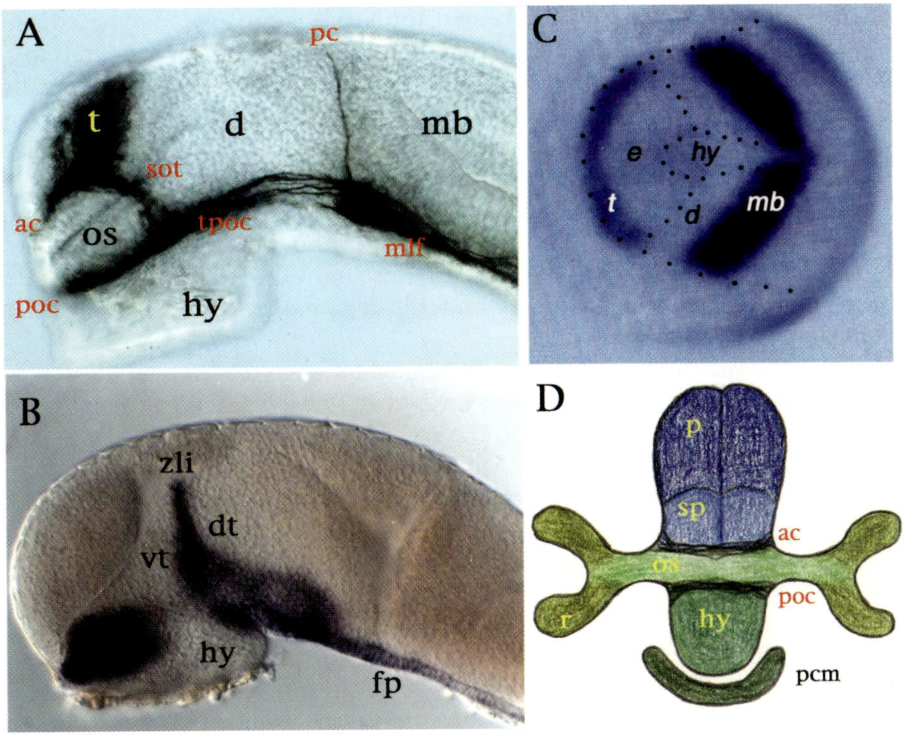

Fig. 4A–D. The embryonic zebrafish forebrain. **A** Lateral view of a 1-day-old zebrafish brain labeled with an antibody to acetylated tubulin (*black axons*). The major regions of the brain are labeled in *black* or *yellow* and the axon pathways in *red*. The eye, which is an evagination of forebrain tissue, has been removed, revealing the optic stalk positioned at the interface between telencephalon and diencephalon. Adapted from Macdonald et al. (1994). **B** Lateral view of a 1-day-old zebrafish showing expression of the *shh* gene (*blue*). *shh* expression extends into the dorsal brain along the *zli*, dividing the dorsal diencephalon into ventral thalamus and dorsal thalamus (plus pretectum – not indicated). Adapted from Barth and Wilson (1995). **C** Dorsal view of the anterior neural plate showing expression (*blue*) of *anf* in the prospective telencephalon and *pax2.1* in the prospective midbrain. The *dots* show the approximate positions of cells that will contribute to forebrain structures (see, e.g., Varga et al. 1999). At this stage, prospective hypothalamic cells are still moving rostrally within the anterior neural plate displacing eye field cells laterally (and hence into left and right eyes). Cells that remain medially in the eye field probably primarily contribute to the optic stalks (Varga et al. 1999). Adapted from unpublished work of D. Delaney and S.W. Wilson. **D** Cartoon of a frontal view of a highly schematized forebrain showing the major domains of the telencephalon (pallium and sub-pallium), the optic vesicles (retina and optic stalks) and the anterior ventral diencephalon (hypothalamus). At earlier stages, the brain is underlain by prechordal mesendoderm (sometimes called the prechordal plate). By the stage cartooned, the prechordal mesendoderm would have dispersed to other sites. *ac* Anterior commissure; *d* diencephalon; *dt* dorsal thalamus; *e* eye field; *fp* floor plate; *hy* hypothalamus; *mb* midbrain; *mlf* medial longitudinal fasciculus; *os* optic stalk; *p* pallium; *pc* posterior commissure; *pcm* prechordal mesendoderm; *poc* postoptic commissure; *r* retina; *sot* supraoptic tract; *sp* subpallium; *t* telencephalon; *tpoc* tract of the post-optic commissure; *vt* ventral thalamus; *zli* zona limitans intrathalamica the forebrain is problematic (see below), the cortex is considered to be a dorsal, pallial telencephalic subdivision while the basal ganglia are considered to be ventral, sub-pallial derivatives (Wilson and Rubenstein 2000). The adult fish telencephalon is very different from the adult mammalian telencephalon and neuroanatomical homologies remain uncertain. However, pallial and sub-pallial telencephalic divisions are evident in all vertebrates and there is no reason to think that early stages of telencephalic development vary greatly between species. Indeed, the extensive similarities in gene expression patterns between mammals and fish during early forebrain development suggest that very similar genetic mechanisms underlie regional patterning of the telencephalon in all vertebrates.

The optic vesicles derive from cells located between telencephalic and diencephalic regions of the neural plate (Varga et al. 1999; Fig. 4C). The distal/lateral regions of the vesicles give rise to the retinal components (neural retina and pigment epithelium) of the eyes, while the proximal portions form the optic stalks that later contribute glial cells to the optic nerves. Other components of the eyes, such as the lens, cornea and sclera, are primarily derived from the surface ectoderm overlying the optic vesicles or from neural crest cells that migrate around and into the eye cups (see Easter and Malicki, this Vol.).

Several diencephalic territories are formed from cells caudal and ventral to the eye-forming regions of the neural plate. In dorsal regions of the forebrain, the diencephalon generates (in anterior to posterior sequence) ventral thalamus, zona limitans intrathalamica (zli), dorsal thalamus and pretectum. The zli is considered to be an important AP boundary within the forebrain separating chordal from epichordal regions of the CNS (Rubenstein et al. 1998; Zeltser et al. 2001) and although various mutations are known to disrupt formation of the zli in fish (e.g., Macdonald et al. 1994), little is known about the function of this prominent boundary. Indeed, with the exception of the epithalamus (see below), little work has been done on dorsal diencephalic development in zebrafish.

The most medial region of the anterior neural plate forms the hypothalamus (Fig. 4C), a brain region involved in regulation of autonomic and endocrine functions. Fate mapping studies (Varga et al. 1999; Mathieu et al. 2002; Woo, Shih and Fraser, pers. comm.) have shown that hypothalamic precursors originate in more posterior regions of the embryo and subsequently move through the neural plate to arrive at their final anterior location. This discrete origin suggests that, at least at early stages, the hypothalamus should be considered a separate compartment from the more dorsal forebrain territories that it eventually comes to underlie. The hypothalamus (or at least parts of it) may be considered an anterior extension of the floor plate and indeed the same genetic pathways are implicated in both floor plate and hypothalamus formation.

To date, studies in zebrafish have contributed very little to our understanding of the later steps of forebrain development where the greatest progress has come from analysis of transgenic and knockout mice. In contrast, the earliest steps in patterning the forebrain are currently better understood in zebrafish than in other vertebrate model systems and it is upon these early stages of forebrain patterning in fish that we will focus.

4.1 DV Patterning of the Zebrafish Forebrain

Unlike the midbrain, hindbrain and spinal cord, the anterior ventral forebrain forms neither floor plate nor motoneurons. Despite this, the Nodal- and Hh-signaling pathways are as crucial for development of ventral tissues in the forebrain as they are elsewhere in the CNS. In embryos carrying mutations affecting these signaling pathways, defects are observed in development of the hypothalamus, patterning of the optic vesicles and establishment of ventral regions of the telencephalon. As is the case in more posterior regions of the CNS, the exact epistatic relationship between Nodal and Hh signals is not entirely clear but some inroads have been made in dissecting the relative contribution of these two pathways to ventral forebrain development.

The text-book view of ventral forebrain development suggests that mesendodermal tissues of the prechordal plate migrate underneath the developing forebrain and send signals that induce the hypothalamus and split the eye field into left and right eyes (Kiecker and Niehrs 2001; Roessler and Muenke 2001). This view is probably largely correct, but is certainly a simplification of the true situation as it has proved to be extremely difficult to dissociate the role of the prechordal plate from that of the overlying neural midline tissue. Both originate in a region of the embryo close to the shield and both express receptors and ligands of the Nodal- and Hh-signaling pathways. Thus, while we can be certain that axial tissues are crucial for patterning ventral cell types in the forebrain, the relative contribution of the axial mesendoderm versus the axial neural ectoderm remains to be determined.

4.1.1 Formation of the Hypothalamus

The embryonic hypothalamus is usually considered to equate to the ventral regions of the forebrain that lie below the tract of the postoptic commissure (Fig. 4). This definition is rather vague and includes some territories, such as posterior tuberculum, that would not classically be defined as hypothalamic. However, until better markers of specific regions and nuclei are developed, this and other working definitions will have to suffice. Defects in early hypothalamic development are usually associated with varying degrees of cyclopia (see below) and, as this is an easy phenotype to distinguish, many mutations affecting the hypothalamus have been isolated. Most of these mutations fall into either the Hh- or the Nodal-signaling pathways.

A role for the Hh-signaling pathway in hypothalamic development is supported by studies in other species which have shown that hypothalamus is missing in mice lacking Shh (Chiang et al. 1996) and that Shh can promote hypothalamic fates in in vitro assays (e.g., Dale et al. 1997). Indeed, it is generally assumed that Hh signaling is essential for hypothalamic induction in amniotes. However, in zebrafish, some degree of hypothalamic development occurs in all known Hh pathway mutants. Most notably, in *smu* mutant embryos, early expression of the hypothalamic marker *nk2.1a* occurs and it is only later that hypothalamic tissue becomes severely reduced (Rohr and Concha 2000; Varga et al. 2001). One interpretation of these observations is that Hh signaling is required for the maintenance of the hypothalamus but not for its early induction. However, a major caveat is that residual Hh signaling is present at early stages in *smu* mutants (Chen et al. 2001a; Varga et al. 2001), and this early Hh activity may be sufficient to mediate early hypothalamic induction.

Although an absolute requirement for Hh signaling in hypothalamic induction is not established, it is clear that Hh activity is required for subsequent patterning of the ventral diencephalon. For instance, the homeobox gene *nk2.2* is expressed in a band of cells along the dorsal boundary of the hypothalamus and expression of this gene is reduced or absent in *you-too* (*yot/gli2*), *syu* and *smu* embryos (Karlstrom et al. 1999; Rohr et al. 2001; Varga et al. 2001). Thus, cell fates lateral (dorsal) to the ventral midline of the brain are more severely affected in Hh pathway mutants than the ventral midline tissue itself. This is highly reminiscent of the situation in more posterior regions of the CNS where medial floor plate is always present in Hh pathway mutants but more lateral fates, such as lateral floor plate and motoneurons, are reduced or absent (Odenthal et al. 2000; Chen et al. 2001a; Etheridge et al. 2001; Lewis and Eisen 2001).

In contrast to the Hh pathway, there is no doubt that the Nodal-signaling pathway is required for establishment of all hypothalamic tissue. *cyc*, *sqt*, *oep* and maternal-zygotic *schmalspur* (*MZsur/fast1*) mutant embryos all exhibit cyclopia and have reduced, or more often an absence of, hypothalamic tissue (Feldman et al. 1998; Rebagliati et al. 1998a,b; Sampath et al. 1998; Gritsman et al. 1999; Pogoda et al. 2000; Sirotkin et al. 2000a,b; Rohr et al. 2001, and references within). However, despite these striking phenotypes, where, when and how Nodal signals are required for establishment of hypothalamus is still unclear. Indeed it seems likely that Nodal signaling may act at different times and in different places to mediate different aspects of hypothalamic development. First, as mentioned above, hypothalamic precursors originate close to the shield and subsequently move anteriorly to their final position in the medial region of the anterior neural plate. This movement is compromised in *cyc* mutants (Varga et al. 1999), and presumably other Nodal pathway mutants. Therefore, one key role for Nodal signaling may be to facilitate movement of prospective hypothalamic tissue to the correct position within the neural plate for the cells to receive the signals that establish their hypothalamic identity.

Nodal signals are likely to be among those that are received by prospective hypothalamic tissue once it is positioned within the anterior neural plate. The

Nodal ligand Cyc is expressed in the prechordal plate underlying the hypothalamus and also within the prospective hypothalamus itself (Rebagliati et al. 1998a,b; Sampath et al. 1998). However, to disentangle the role of Cyc within these tissues from its earlier role in axial development is not easy. The best evidence that Nodal signals must be received by prospective hypothalamic cells has come from recent experiments in which Nodal signaling has been manipulated either in prechordal plate or in ventral brain (Matthieu et al. 2002). From these studies, it appears that cells that cannot receive Nodal signals are unable to contribute to the most ventral hypothalamus but can become part of more dorsal hypothalamic tissue. Similarly, restoration of Nodal activity in the prechordal plate but not the brain leads to recovery of dorsal but not ventral hypothalamus. Overall then, our current hypothesis is that locally acting Nodal signals are required for establishment of ventral hypothalamic fate while Hh signals are required for proper development of more dorsal hypothalamic fates. As Hh expression is lost in anterior regions of Nodal pathway mutants, both Hh activity and Nodal activity are compromised in these mutants. This disruption of both signaling pathways may explain the severity of the hypothalamic phenotypes in this class of mutants.

4.1.2 Establishment of the Optic Stalks

As prospective hypothalamic cells move anteriorly within the neural plate, they displace cells in medial regions of the eye field to more lateral positions where they will subsequently give rise to left and right retinas (Fig. 4C,D). Although fate mapping studies are incomplete, it is likely that those cells of the eye field that remain close to the midline form the optic stalks (Varga et al. 1999) which connect the retinal compartments of the eye to the brain. Mutations that disrupt the specification and/or migration of the hypothalamus or underlying axial tissues also usually affect optic stalk development. In the most severe cases, optic stalk tissue fails to be specified at all and instead medial regions of the eye field form retina resulting in cyclopia (Hatta et al. 1994; Macdonald et al. 1995). Such observations suggest that signals either from prospective hypothalamus or from underlying axial tissues promote optic stalk identity and, once again, the Hh pathway is implicated in these fate determination events.

The first indication that Hh signaling regulates early regional patterning of the optic vesicles came from analysis of embryos in which Hh signals were overexpressed. In the normal optic vesicle, *pax2.1 (no-isthmus; noi)* expression defines the compartment that will form optic stalks whereas *pax6.1* expression defines the compartment that will form retina (Macdonald et al. 1995; Macdonald and Wilson 1997). In embryos with increased Hh activity, *pax2.1* expression expands throughout the optic vesicles while *pax6.1* expression is suppressed (Ekker et al. 1995; Macdonald et al. 1995). This suggests that graded Hh activity contributes to subdivision of the optic vesicles into proximal optic stalk tissue and distal retinal tissue. In support of this hypothesis, mice lacking Shh function lack *pax2* expression and have fused retinas (Chiang et al. 1996). In zebrafish, *shh/syu*) mutants have robust *pax2.1* expression within the optic vesicles (Schauerte et al. 1998) raising doubts as to the requirement for Hh signals in optic stalk induction. However, recent data has shown that *pax2.1* expression is absent in *smu* mutants (Varga et al. 2001) indicating that another Hh protein, probably Twhh, can compensate for the loss of Shh alone.

Altogether, these results are consistent with the possibility that Hh signals derived from midline tissue induce medial eye field cells to form optic stalk and not retina. However, as usual, the story is more complicated than this and several outstanding issues still need to be resolved. First, although *pax2.1* expression is lost in *smu* mutants, these embryos do not exhibit as severe cyclopic defects as Nodal pathway mutants. Therefore, there may be other optic-stalk-promoting signals (perhaps including Nodals themselves) that are lost in Nodal pathway mutants. Second, perhaps most intriguing of all, is the observation that *pax2.1* expression is restored in embryos with very severely compromised Nodal activity (Feldman et al. 2000; Masai et al. 2000). There appears to be no Hh gene expression in the anterior brain in these embryos, and so one is faced with the possibility that optic-stalk-specific gene expression can be induced in the absence of Nodal and Hh signaling. This phenomenon is not yet understood but one possibility is that Hh signaling is required to overcome repression of optic stalk identity. If the repression is alleviated in severely Nodal-depleted embryos, then *pax2.1* expression may recover, even in the absence of Hh activity. This "repression of repression" model again has parallels in more posterior regions of the CNS. In caudal regions of mouse embryos, Gli3 appears to act as a repressor of ventral fates and, if Gli3 activity is removed in embryos lacking Shh activity, then ventral fates, which are lost in *Shh* single mutants, are restored (Litingtung and Chiang 2000). Thus Hh activity is required to overcome Gli3-mediated repression of ventral cell fate identity in the spinal cord. It will be interesting to see if a similar mechanism is operating in the forebrain and eyes.

Mutations that disrupt the anteriorly directed movement of axial cells during gastrulation can also lead to reduced or absent optic stalks and cyclopia (Heisenberg and Nüsslein-Volhard 1997; Marlow et al. 1998). In these mutants, which include *silberblick* (*slb/wnt11*; Heisenberg et al. 2000) and *knypek* (*kny/glypican4/6*; Topczewski et al. 2001), signals required for optic stalk development, such as Shh and Twhh, are still present. However, the disrupted anterior movement of tissues that are the source of these signals (prechordal mesendoderm and perhaps prospective hypothalamic tissue) has the consequence that the axially derived signals fail to act at the correct place and time to properly pattern the eye field. Analysis of these mutants emphasizes the fact that the genes regulating morphogenetic events that bring tissues into proximity with each other are just as crucial to forebrain development as the genes that encode the signals that specify cell fate.

4.1.3 Establishment of Ventral Telencephalic Fates

Ventral sub-pallial telencephalic markers are reduced or absent whereas dorsal pallial markers are unaffected or expanded in zebrafish embryos carrying mutations in the Hh-signaling pathway (Rohr et al. 2001; Varga et al. 2001). Reciprocal results are observed in Hh signaling gain-of-function experiments in fish and other species (reviewed in Wilson and Rubenstein 2000). Recent studies in mice have suggested that the role for Hh signals in the telencephalon may once again be to overcome Gli3-dependent repression of sub-pallial identity (G. Fishell, pers. comm.). Thus Hh signaling appears to play comparable roles in the optic vesicles and in the telencephalon, in both cases affecting the allocation of regional identities within the compartment, without affecting the commitment of cells to a telencephalic or to an optic vesicle fate.

Ventral telencephalic markers are also absent in Nodal pathway mutants; however, this may primarily be due to the secondary loss of Hh signaling in anterior regions of the mutants (Rohr et al. 2001). Thus, if Hh signaling is restored to embryos completely lacking Nodal activity, then telencephalic cells express sub-pallial markers. This indicates that, even in the absence of Nodal activity, Hh activity is sufficient to induce ventral telencephalic identity. However, this does not imply that Hh signals are the only mediators of early DV patterning in the telencephalon. Indeed, Fgf signaling also influences the development of sub-pallial territories. In *acerebellar* (*ace/fgf8*) mutants, expression of the sub-pallial marker *nk2.1b* is reduced (Shanmugalingam et al. 2000) and, in embryos with more severely compromised Fgf signaling, expression of this marker is absent altogether (Shinya et al. 2000). How the Fgf- and Hh-signaling pathways interact to promote subpallial development will be an interesting area of investigation in the coming years.

4.1.4 Specification of Dorsal Forebrain Fates

To date, very few studies in zebrafish have addressed the mechanisms by which dorsal forebrain fates are established. However, the role of the Bmp pathway in regulating DV patterning of neural tissue appears to be conserved at all AP levels of the CNS (Nguyen et al. 1998b, 2000; Barth et al. 1999, and references cited therein). For instance, Bmp signaling appears to define the extent and position of DV subdivisions of the forebrain such as the epithalamus, the dorsal-most region of the diencephalon. *flh* is a homeobox gene required for epithalamic neurogenesis (Masai et al. 1997) and for notochord development (Talbot et al. 1995). Within the brain, *flh* expression is very tightly restricted to the epithalamic region of the dorsal diencephalon. Analysis of embryos with variably increased or decreased levels of Bmp activity has shown the medial and lateral extent of *flh* expression within the prospective epithalamic region of the neural plate is set between thresholds of Bmp activity (Barth et al. 1999). Thus, *flh* is not expressed in medial regions as Bmp activity is too low and it is not expressed beyond the limits of the neural plate where Bmp signaling is too high. Even in embryos lacking all axial tissues, *flh* expression remains absent from medial regions of the neural plate (Concha et al. 2000) indicating that repressive signals from ventral tissue are not required to limit *flh* expression to dorsal regions. Although the Bmp pathway has a profound influence on DV patterning of neural tissue, it has little or no effect on AP patterning of the CNS. Thus, in all Bmp pathway mutants, *flh* expression remains limited to the same AP position within the embryo even though the DV extent of expression can vary enormously.

4.2 Left/Right Patterning in the Brain

Although the CNS is largely symmetrical, there are functionally important differences between left and right sides. All vertebrate, and many invertebrate, animals exhibit lateralized behaviors that are mediated by differences in neuroanatomy between the left and right brain. For instance, in humans it is well established that aspects of language processing are dominant in the left hemisphere of the cortex. Lateralization is not limited to the telencephalon and, in most vertebrates that have been studied, the epithalamic region of the diencephalon is lateralized. While the function of this lateralization is currently unknown, its presence in all classes of vertebrates indicates strong evolutionary conservation (Concha and Wilson 2001).

Neuroanatomical differences between left and right sides of the epithalamus appear relatively early in zebrafish development and are evident as two prominent asymmetries (Concha et al. 2000). The first is a left-side-specific photoreceptive nucleus called the parapineal that is located just anterior to the larger midline photoreceptive epiphysis or pineal organ (Fig. 5). The parapineal projects to a second lateralized nucleus in the epithalamus, termed the habenula. Habenular nuclei are present on both sides of the brain, but the left nucleus is larger, at least in terms of labeling of neuropil. Recent studies have shown that the Nodal-signaling pathway determines the laterality of these two asymmetries (Concha et al. 2000). This pathway also appears to influence the position of the pineal stalk in adult fish (Liang et al. 2000).

An involvement for Nodal signals in regulating brain laterality seemed likely from analysis of the expression of various genes that function in the Nodal pathway, some of which show left-side-restricted expression (Bisgrove et al. 2000; Concha et al. 2000; Liang et al. 2000; see also Wright and Halpern, this Vol.). It is likely that cells on both sides of the epithalamus are able to respond to Nodal signals, but the signals themselves may be restricted to the left side and, consequently, genes expressed in response to Nodal activity are only induced on the left side. In embryos with disrupted Nodal signaling, parapineal and habenular asymmetries are always established and so Nodal activity is not required to specify asymmetry (Concha et al. 2000). However, the laterality of these CNS asymmetries is totally randomized in mutants.

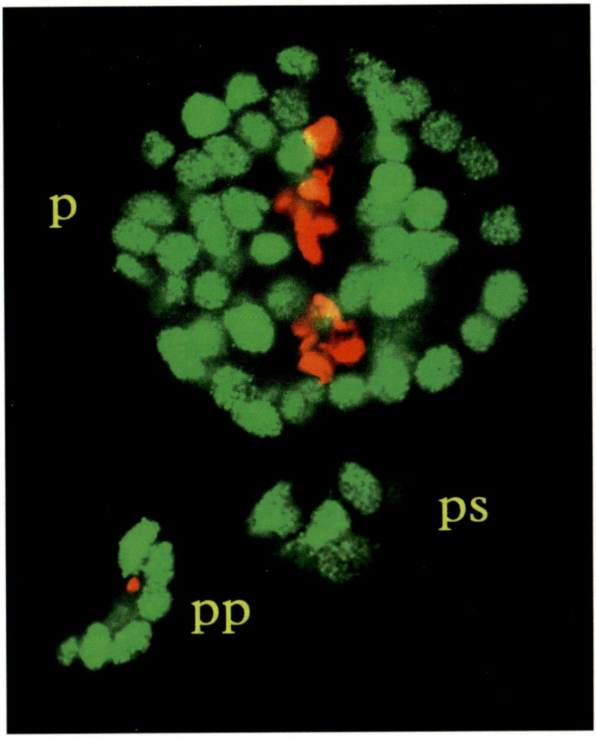

Fig. 5. Asymmetry in the epithalamus. Composite of confocal images of dorsal views looking onto the epithalamic region of the diencephalon. Neurons are labeled with an antibody to Islet1/2 (*green*) and to opsin (*red*). The opsin labeling in the pineal organ (epiphysis) indicates the position of the dorsal midline. Pineal neurons are located to both left and right sides of the midline. However, deeper into the brain, the parapineal organ is located only on the left side of the diencephalon. *p* Pineal; *ps* pineal stalk; *pp* parapineal. (Adapted from Concha et al. 2000 and courtesy of Miguel Concha)

Wheareas over 95% of wild-type fish are "left-brained", fish lacking lateralized Nodal signaling are 50% "left brained" and 50% "right-brained".

It is still early days for studies of laterality and asymmetry in the brain and one hopes that directed genetic screens should reveal many new genes that act in the pathways leading to the establishment of a lateralized CNS. Further discussion of left/right patterning in the brain and in other organ systems can be found elsewhere (see Wright and Halpern, this Vol.).

4.3 AP Patterning of the Prospective Brain

4.3.1 Establishment of Early AP Pattern in the Neural Plate

Several years ago, transplantation studies demonstrated that germ ring cells produce posteriorizing signals that inhibit development of anterior neural fates (Woo and Fraser 1997; Koshida et al. 1998). However, posteriorizing signals appear to be absent, at least at early stages, from the dorsal organizer region (shield), and indeed the organizer probably actively antagonizes such signals. These studies suggested that AP positional values within the neural plate might be established by cells reading their positions with respect to their distance from the germ ring and shield. Our understanding of the signals that arise from the organizer and germ ring and influence AP pattern are still rudimentary. However, several recent studies have provided compelling evidence that Wnt proteins (particularly Wnt8) contribute to the posteriorizing activity and that Wnt antagonists (such as Dkk) at least partially underlie the ability of the organizer to antagonize such signals [see Hibi et al., this Vol., for a more detailed consideration of these issues in zebrafish and Yamaguchi (2001) for a discussion of Wnt activity in AP patterning in other species].

The complex activity of germ ring cells in both producing and inhibiting signals that regulate AP patterning is evident from analysis of Nodal pathway and other mutants that affect the establishment of germ ring fates (reviewed in Wilson and Rubenstein 2000; updated in Erter et al. 2001). Simplistically, most results suggest that, if induction of dorsal germ ring fates is compromised, then the activity of posteriorizing signals from more lateral and ventral germ ring cells will be increased and anterior fates, such as telencephalon, will be reduced. However, if induction of germ ring fates is more widely and severely compromised, then induction of the posteriorizing signals will be compromised and anterior neural fates will expand.

Confirmation that Wnt8 activity in vivo promotes posterior and suppresses anterior development has come from several studies in which increased or decreased Wnt8 activity has been shown to directly correlate with reduced or increased specification of anterior fates (Erter et al. 2001; Levken et al. 2001 and references cited therein). For instance, embryos that lack most or all Wnt8 activity exhibit a hugely expanded forebrain and lack hindbrain and spinal cord fates. Furthermore, suppression of Wnt activity can often restore anterior CNS fates in embryos that normally lack these structures due to mutations that directly or indirectly lead to enhanced Wnt activity in the germ ring (e.g., Fekany-Lee et al. 2000; Hashimoto et al. 2000; Kim et al. 2000a; Shinya et al. 2000; Erter et al. 2001).

Although these various studies implicate Wnts and Wnt antagonists in AP CNS patterning, it is uncertain how Wnt signals that originate in the germ ring, and are thought to diffuse poorly, could directly influence global AP pattern in the CNS. If establishment of global AP positional values is indeed a direct effect of Wnt8, then we have much to learn about the transport of Wnt proteins throughout the embryo. An alternative or additional possibility is that Wnt8 may influence the activity of other genes, possibly including Wnt pathway components that secondarily influence AP pattern. As we describe below, modulation of Wnt activity continues to influence AP patterning throughout gastrulation, and it appears that local activity of Wnt agonists and antagonists within the forming anterior neural plate influences regional fate determination in the forebrain.

4.3.2 Local Induction of the Telencephalon and Eyes

The telencephalon is generated from cells around the anterior margin of the neural plate (Varga et al. 1999; Whitlock and Westerfield 2000) and can therefore be defined both as "anterior", as it forms from cells at the front of the neural plate, and as "dorsal", as it derives from cells at the margin of the neural plate. Because of this origin, mutations that affect either the establishment of dorsal neural cell fates, such as Bmp pathway mutants, or the establishment of AP positional values both affect early development of the telencephalon. With respect to the Bmp pathway, it seems likely that induction of telencephalon can only occur between certain thresholds of Bmp activity (Barth et al. 1999). As described above, if Bmp activity is high then non-neural fates are specified, if Bmp signaling is low, then medial neural plate fates are promoted. Between these two extremes, cells at the anterior margin of the neural plate are competent to form telencephalon, if exposed to the correct inducing signals.

Cells at the anterior margin of the neural plate have the ability to non-autonomously induce telencephalic gene expression when transplanted to more posterior regions of the neural plate (Houart et al. 1998). Furthermore, when such cells are ablated, telencephalic gene expression is reduced or absent supporting the notion that anterior neural plate cells produce signals that induce the telencephalon. Among these signals is a member of the sFRP family of secreted Wnt inhibitors, termed Tlc (Houart et al. 2002). sFRPs have a structure resembling the extracellular domain of proteins of the Wnt coreceptor Frizzled family and it is thought that they bind to Wnt ligands, thereby preventing them from interacting with the Frizzled receptors (Leyns et al. 1997; Wodarz and Nusse 1998). Tlc-expressing cells have the same activity as anterior marginal cells in that they can non-autonomously induce telencephalic gene expression and suppress midbrain-specific gene expression. Furthermore, abrogation of Tlc activity leads to a reduction of telencephalic gene expression confirming an in vivo requirement for this gene. These studies imply that local antagonism of Wnt signals in the anterior neural plate promotes telencephalic identity. They also imply that Wnt ligands are present within the telencephalon-forming region of the anterior neural plate. In theory, such ligands could derive from neighboring non-neural or neural ectoderm or from the underlying mesendoderm or yolk syncytium. Among the candidate genes for being locally acting inhibitors of telencephalic identity are *wnt8b* and *wnt1* (Houart et al. 2002), both of which are expressed in the neural plate caudal to the telencephalon (Kelly and Moon 1995; Kelly et al. 1995b).

masterblind (*mbl*) is one of very few mutations known to directly affect allocation of regional fates within the anterior neural plate. In *mbl* mutant embryos, there is a transformation of telencephalic and eye fates to more posterior diencephalic identity (Heisenberg et al. 1996, 2001; Masai et al. 1997). Recent studies have shown that the *mbl* mutation affects Wnt signaling in the anterior neural plate (Heisenberg et al. 2001; van de Water et al. 2001). The *mbl* mutant phenotype is due to a mutation in the *axin1* gene that is predicted to generate an Axin protein that cannot bind GSK3 (Heisenberg et al. 2001). Axin normally functions in an intracellular complex that includes GSK3 and several other proteins (Ikeda et al. 1998; Kishida et al. 1999). In the absence of Wnt activity, this complex is responsible for degrading β-catenin, thereby preventing activation of Wnt target genes. When Wnts bind their receptor complex, Axin interacts with the Wnt coreceptor LRP5/6 and the activity of the Axin/GSK3 complex is inhibited (Mao et al. 2001). This allows accumulation of β-catenin and activation of Wnt target genes. As the *mbl* mutation affects the ability of Axin1 to bind to GSk3, it is predicted that the Axin1 complex will not function correctly in *mbl* mutant embryos, β-catenin will accumulate and Wnt target gene activation will occur. If correct, then this suggests that telencephalon and eye-forming regions of the neural plate can be converted to diencephalon through overactivation of Wnt signaling. Thus, suppression of Wnt signaling may be important not only for induction of the telencephalon, but also for induction of the eyes. Altogether, these results raise some intriguing questions. It appears that manipulating Wnt activity either within the germ ring/organizer or within the anterior neural plate can both lead to altered specification of anterior neural plate fates. However, it remains unknown how early Wnt signaling in the germ ring is linked to later Wnt signaling in the neural plate. We clearly need to know a lot more about where and when Wnt signals are active if we are to understand how these spatially and temporally complex events are integrated.

5 The Midbrain and Hindbrain

Caudal to the forebrain lies the midbrain or mesencephalon, followed by the isthmus rhombencephali, a region considered the most rostral portion of the hindbrain or rhombencephalon. The midbrain includes dorsally the tectum, which is separated by the tectal ventricle from the ventral tegmentum and the ventrolateral torus semicircularis. The tectum and torus semicircularis derive from the alar plate of the neural tube and predominately serve as major relay centers for sensory information derived from the eyes, ears and lateral line organs, whereas the tegmentum derives from the basal plate and contains several motor nuclei, among them those of the occulomotor (III) and trochlear (IV) nerves that are involved in directing eye movements. In a 1-day-old zebrafish embryo, the isthmus forms a prominent fold, which contains dorsally the cerebellar primordium. As with other brain regions discussed so far, midbrain and isthmus development start during gastrulation and initially occur in close association. During early somitogenesis, the two territories become molecularly distinct and at late somitogenesis the sections of the neural tube giving rise to the midbrain and isthmus fold up as two clearly separable morphogenetic domains, forming the prominent fold referred to as the midbrain-hindbrain boundary (MHB; Fig. 6).

Although the histological features of this area have been well described for the adult zebrafish brain (Wullimann et al. 1996), little is known so far about the connectivity, physiology or underlying function of the various midbrain and isthmic structures. As we argued above for the forebrain, the differences in adult neuroanatomy between fish and mammals make it difficult to determine homologies between brain parts at the adult stage, whereas at embryonic stages the similarities, especially in gene expression patterns, suggest a much more conserved relationship. By studying how this brain area is patterned during embryonic stages, we can therefore hope to understand mechanisms that are common to all vertebrates. For instance, studies of the zebrafish midbrain have long served as an excellent model for development of a simple visual system in vertebrates, because it becomes functional after only about 3 d (Easter and Nicola 1996; Karlstrom et al. 1997).

One major reason for interest in the midbrain-hindbrain region is that it contains a cell population acting as an organizer of cell fate, the MHB organizer or isthmic organizer, that was initially discovered by experimental manipulations in chick. When MHB tissue is transplanted into caudal forebrain, the surrounding host tissue switches fate, adopting isthmic or midbrain character; in the rhombencephalon, cerebellar fate is induced (Martinez and Alvarado-Mallart 1990; for reviews, see Rhinn and Brand 2001; Wurst and Bally-Cuif 2001). These experiments suggest that MHB tissue also acts as an organizing center in its normal location, and that differential competence of the host tissue determines the nature of the induced structures. Work on the MHB organizer has now shed light on the poorly understood mechanisms involved in AP patterning of the brain, and it may serve as a paradigm for organizers acting in other brain regions.

5.1 Midbrain and Hindbrain Development Starts in Gastrulation

As with the other CNS regions, development of the midbrain and hindbrain starts during gastrulation. Fate map studies have indicated that, by the end of gastrulation, the midbrain and hindbrain precursors occupy largely non-overlapping, bilateral, v-shaped domains in the dorsal neuroectoderm (Woo and Fraser 1995); the isthmus is not yet separate from the midbrain domain at this stage (Müller et al. 1996b). Among the earliest genes to be activated are *pax2.1*, *her5* and *wnt1* in the midbrain domain (Krauss et al. 1991, 1992; Müller et al. 1996b), and *fgf8* in the immediately abutting anterior hindbrain domain, at around 70–80% of epiboly, a time when these territories are not yet morphologically distinct (Fig. 6; Reifers et al. 1998). Genetic studies of mutants with discrete lesions in the midbrain-hindbrain domain show that, at this stage of development, the two domains do not yet influence each other.

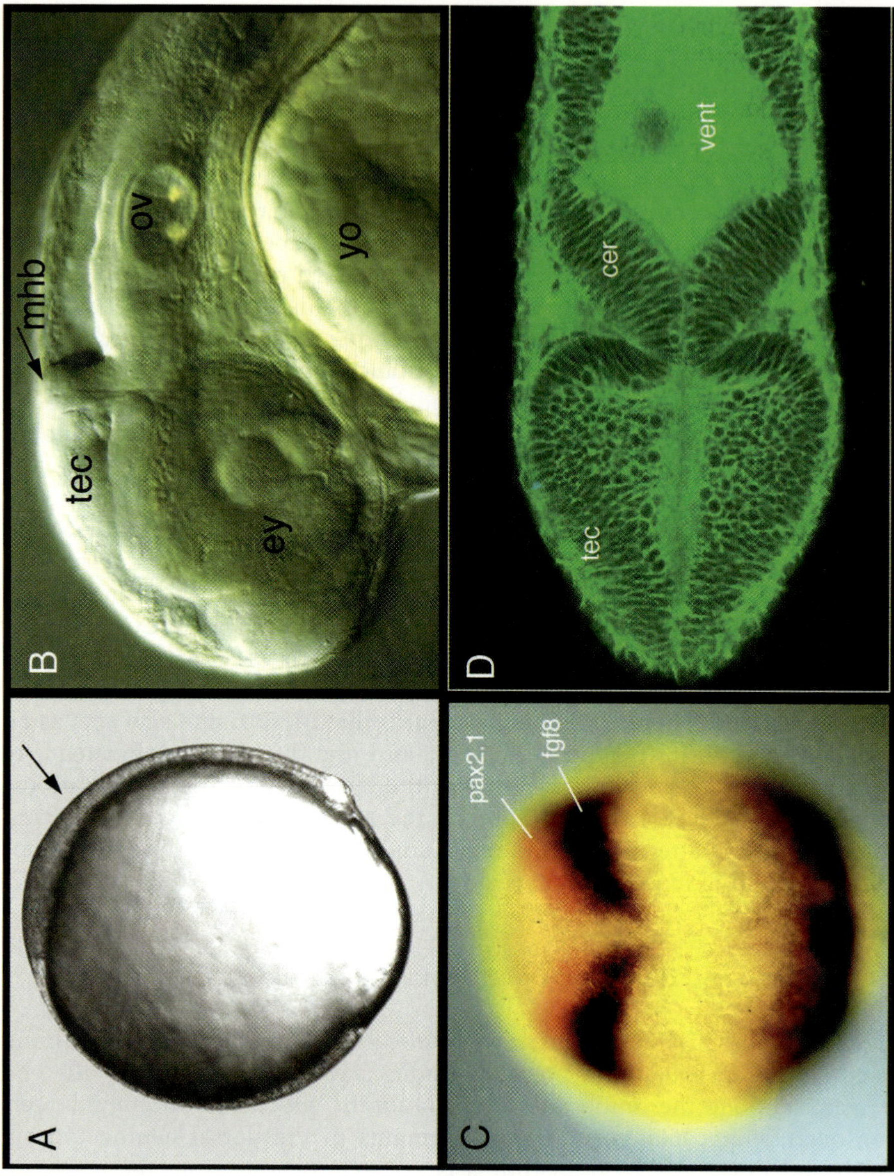
A
B
mhb
tec
ov
ey
yo
C
pax2.1
fgf8
D
tec
cer
vent

no isthmus (*noi/pax2.1*) mutant embryos have no MHB, tectum and cerebellum at 1 d. *noi* mutant embryos fail in the earliest stages of midbrain and isthmus development to activate the *pax2.1* target genes *engrailed2* (*eng2*) and *engrailed3* (*eng3*), but have normal activation of *fgf8* in the adjacent anterior hindbrain, showing that the early hindbrain primordium develops in the absence of functional *pax2.1* (Brand et al. 1996b; Lun and Brand 1998). Conversely, the *acerebellar* (*ace/fgf8*) mutation, or *fgf*-8MO, causes absence of the MHB and cerebellum, but these embryos have normal activation of *pax2.1*, showing that midbrain development is initially normal in the absence of functional Fgf8 (Brand et al. 1996b; Reifers et al. 1998; Araki and Brand 2001; Draper et al. 2001). Furthermore, knockout studies of *Wnt1* in mice are known to cause absence of the midbrain and MHB (McMahon et al. 1992). Zebrafish *wnt1* is activated around the same time as *pax2.1* (Kelly and Moon 1995). Expression of *wnt1* is unaffected in either *noi* or *ace* mutants at the end of gastrulation (Lun and Brand 1998; Reifers et al. 1998), and, conversely, both *pax2.1* and *fgf8* are expressed normally at the end of gastrulation in a *wnt1* deletion mutant (A. Lekven, pers. comm.). These studies have suggested that at least three separate signaling pathways are activated independently of each other around the forming MHB. At later stages of midbrain development, however, these genes come to depend on each other's expression (see below).

5.1.1 Initial AP Subdivision of the Neural Plate

The spatially discrete and independent activation of *pax2.1*, *fgf8* and *wnt1* during gastrulation show that AP patterning information is already present in the neuroectoderm, raising the question of how this information is generated. Based on knockout studies in mice and expression analysis in zebrafish, the homeodomain transcription factor Otx2, which is expressed in the fore- and midbrain primordia during gastrulation, is required for correct spatial activation of MHB genes (Simeone 1998). Murine Gbx2, another homeodomain transcription factor expressed in a complementary fashion in the posterior neuroectoderm, is required for later stages of hindbrain and MHB development (Wassarman et al. 1997), though apparently not for early patterning

◀

Fig. 6A–D. AP patterning of the midbrain and hindbrain. **A** Lateral view of a gastrulating embryo at 80% epiboly; the *arrow* points to the site of MHB formation. **B** View into the living brain of an embryo at 36 h. The various brain parts are clearly distinct from each other. **C** Similar stage embryo as in **A**, dorsal view onto the forming neural primordium. The embryo was hybridized with *pax2.1* probe (*red*), recognizing the midbrain and isthmus primordia and *fgf8* probe (*blue*), recognizing the anterior hindbrain primordium and the germ ring. **D** Confocal section of a live embryo stained with Bodipy-Ceramide to reveal the morphological folding of the brain. Dorsal view of a 30-h embryo, sectioned through the dorsal midbrain and cerebellum. Cells near the ventricular surface of the neuroepithelium are rounded up during mitosis. (Photograph kindly provided by Tobias Langenberg) during gastrulation (A. Simeone, pers. comm.). Local misexpression of Otx2 or Gbx2 in mice causes downregulation of the other gene (Broccoli et al. 1999; Millet et al. 1999), suggesting that mutually repressive interactions contribute to establishing or maintaining the position of the MHB. For technical reasons, a possible function during gastrulation could not be tested in these studies. As in mice, zebrafish have two *gbx* genes, but, in contrast to mice, both appear to be expressed during gastrulation. *gbx1* is expressed earlier than *pax2.1* and *fgf8* in the hindbrain neuroectoderm and *gbx1* misexpression can shift *pax2.1* and *fgf8* expression to ectopic locations and convert anterior neural plate to hindbrain; *gbx2* acts in the same subdomain but later, downstream of Fgf8 (K. Lun, M. Rhinn, M. Brand, unpubl.). An additional regulator acting upstream of *pax2.1* has recently been identified. Bts-1 encodes a zinc-finger transcription factor related to the *Drosophila buttonhead* gene, and is both necessary and sufficient for *pax2.1* regulation (Tallafuss et al. 2001).

5.1.2 Wnt8 Signaling Positions the Midbrain and Hindbrain

The data in mice and zebrafish show that mutual repression of *otx2* and *gbx* is important in positioning the MHB, raising the question of how in turn the correct spatial domains of *otx2* and *gbx1* (*Gbx2* in mice) are generated. As is discussed above for the forebrain and in more detail elsewhere (see Hibi et al., this Vol.), signals derived from the germ ring, including Wnt8, may perform this function (Erter el al. 2001; Lekven et al. 2001). In this model, midbrain and hindbrain derive from neurectoderm cells that are exposed to higher concentrations of Wnt8, because they reside closer to the signal source in the germ ring than forebrain precursors, consistent with the fate map. One attractive possibility is that *gbx1* is a direct response gene to a particular concentration of Wnt8, acting to transduce the posteriorizing Wnt8 signal to specify hindbrain identity. Indeed, clones of Wnt8 misexpressing cells in the gastrula neuroectoderm can cause ectopic activation of *gbx1* in surrounding host cells, and reducing Wnt signaling in the posterior neural plate reduces *gbx1* expression, suggesting that Wnt8 might directly or indirectly control the domain of *gbx1* expression (M. Rhinn, K. Lun, M. Brand, unpubl.). It remains to be tested whether and how additional secreted factors, such as Nodals and retinoic acid or the more recently described midkine-related growth factor2 (Winkler and Moon 2001), contribute to posteriorization of the MHB domain.

5.2 Wnts and Fgfs Maintain and Pattern the Midbrain and Hindbrain

Once posteriorization and initial subdivision of the neurectoderm is achieved, the MHB organizer forms at the newly established molecular interface. At present, it is not known what, if any, additional signals at this interface allow ordered activation or maintenance of gene expression around it. Initially, however, the midbrain and hindbrain domains appear to function independently of each other. In the midbrain, *pax2.1* is a likely direct regulator of *eng2* and *eng3* (Lun and Brand 1998), and engrailed gene-MO therefore phenocopies the midbrain/isthmus phenotype of *noi* mutants (Scholpp and Brand 2001). Fgf8 also functions at the end of gastrulation in the early hindbrain primordium (Reifers et al. 1998; Fürthauer et al. 2001; Raible and Brand 2001), but does not initially affect gene expression in the midbrain. During early somitogenesis, these genes become successively restricted in their expression toward the isthmus, where they then overlap and become mutually dependent in their expression. Around this time, isthmus-specific expression of *pax5*, *pax8* and *fgf17* is first activated, requiring both *pax2.1* and *fgf8* for their activity (Lun and Brand 1998; Pfeffer et al. 1998; Reifers et al. 1998, 2000a; M. Brand, unpubl. results).

Once established, secreted Fgf8 and Wnt1 proteins from the organizer are thought to mediate its organizing influence on surrounding neural tissue. Wnt1 functions as a mitogen and to maintain expression of En genes, but is unable to mimic organizer activity when misexpressed (Dickinson et al. 1994; Danielian and McMahon 1996). *Fgf8*, however, is expressed at the right time and place to mediate organizing activity (Reifers et al. 1998, and references cited therein). In contrast to Wnt1, ectopic application of Fgf8 protein mimics MHB organizer activity and induces isthmic-like structures and MHB-specific gene expression (Crossley et al. 1996; Liu et al. 1999; Martinez et al. 1999; M. Brand, unpubl. results). Because Fgfs can mimic each other's activity in gain-of-function experiments, loss-of-function mutants are important to support a function for Fgf8 in induction and/or patterning of the MHB region. *ace* mutants lack functional Fgf8, the MHB organizer and a cerebellum, and analysis of this mutant showed that Fgf8 is required to maintain marker gene expression in the midbrain and isthmus, but not to induce midbrain (Reifers et al. 1998; Picker et al. 1999; Araki and Brand 2001).

5.2.1 Polarization of the Midbrain

One of the crucial properties of the midbrain tectum is the ability to form a set of retinotopically ordered connections with ingrowing axons of retinal ganglion cells. When an ectopic MHB organizer forms after transplantation of MHB cells, or after bead implantation, the induced tectum is correctly polarized with respect to the ectopic organizer (Marin and Puelles 1994), suggesting that the signal required for polarization also derives from the MHB organizer. Analysis of the midbrain in *ace* mutants showed that the MHB is indeed required for AP polarization of the midbrain, including graded expression of EphrinA2 and EphrinA5 ligands in the midbrain neuroepithelium, and hence for proper retinotectal map formation (Picker et al. 1999). The expression of one likely Fgf8-receptor, Fgfr3, may mediate a possible direct activity of Fgf8 in this process (Sleptsova-Friedrich et al. 2001). Indeed, implantation of a bead coated with Fgf8 protein into the posterior diencephalon at late somitogenesis is sufficient to reorient the highpoint of EphrinA5a graded expression (A. Picker and M. Brand, unpubl.).

5.2.2 Fgf Signaling in the Rostral Hindbrain

Fgf8 secreted from the MHB organizer is also thought to be involved in development of the chick rostral hindbrain (Irving and Mason 2000), and is required for the earliest stages of rostral hindbrain development in zebrafish (Fürthauer et al. 2001; Raible and Brand 2001). Rhombomere 1 (r1) lies closest to the MHB, and is the only rhombomere that does not express any *Hox* genes in chick. However, after transplantation to an ectopic position, r1 tissue can express *Hox* genes, and both MHB tissue and Fgf8 can inhibit this expression (Irving and Mason 2000). Thus, Fgf8 may define, directly or indirectly, the anterior limit of *Hox* gene expression. Both Fgf8 and Fgf3 are expressed initially in a wider domain encompassing the anterior hindbrain, and then become restricted to a segmental expression during early segmentation stages (Reifers et al. 1998; Raible and Brand 2001). Analysis of *erm* and *pea3*, two target genes for Fgf8 and Fgf3 signaling, suggests that Fgfs function during early hindbrain segmentation, likely in a redundant fashion (Raible and Brand 2001; Roehl and Nüsslein-Volhard 2001); however, the phenotypic consequences of this function for later hindbrain development remain to be explored. During neuronal cell type determination, Fgfs may serve yet additional roles: tyrosine hydroxylase expression in the locus coeruleus of the hindbrain is abolished in *ace* mutants. In analogy to the suggested requirement for Fgf8 in development of mammalian midbrain dopaminergic neurons (Hynes et al. 1995), it has been argued that this phenotype may represent a non-autonomous requirement for Fgf8 signaling from the MHB organizer (Guo et al. 1999a).

5.2.3 Feedback Control of Fgf Signaling

Given its potency as a signaling molecule, the activity of Fgf8 must be carefully controlled in the embryo. An emerging theme for several signaling pathways is that extracellular or intracellular inhibitors regulate their activity. *Drosophila sprouty* functions in development of the trachea and eye, as a target gene and feedback-inhibitor for Fgf and EGF signaling (Placzek and Skaer 1999). Several studies reveal a surprisingly good correlation of the expression of vertebrate *sprouty* homologues with regions of ongoing Fgf signaling, including the MHB. As in flies, vertebrate *sprouty* genes can be induced locally with recombinant Fgf8 protein (Minowada et al. 1999; Chambers and Mason 2000; Fürthauer et al. 2001). In *ace/fgf8* mutants, *sprouty4* is never activated at the MHB and anterior hindbrain, suggesting that Fgf8 regulates *sprouty4* expression. In addition, overexpression of *sprouty4* antagonizes the effects of both *fgf8* and *fgf3* injection, suggesting that zebrafish *sprouty4* is a component of an *fgf8,fgf3*-dependent inhibitory feedback loop (Fürthauer et al. 2001).

Additional observations support the existence of such a feedback loop: *fgf8* RNA is upregulated in *ace* mutants (Reifers et al. 2000a; Shanmugalingam et al. 2000) and in zebrafish *aussicht* (*aus*) mutants (Heisenberg et al. 1999); *aus* could therefore encode a component of the feedback loop. Feedback regulation could serve important functions, for instance, to maintain the MHB organizer itself, and indeed this structure is morphologically absent in zebrafish and mouse *Fgf8* mutants (Meyers et al. 1998; Reifers et al. 1998). The feedback loop also involves *Otx2* and *Gbx2*, since local expression of Fgf8 represses *Otx2* and reduction of *Otx* copy number shifts *Fgf8* and *Gbx2* expression anteriorly (Crossley et al. 1996; Simeone 1998; Liu et al. 1999; Martinez et al. 1999). The existence of the feedback loop may explain why *Fgf8* bead implantations are able to re-activate the entire genetic cascade of MHB development.

Considering the potent abilities of Fgf8, it is notable that different Fgf8 isoforms exist (Liu et al. 1999) and that additional *Fgfs* related to *Fgf8* are also expressed in the MHB organizer. *Fgf17* and *Fgf18* are turned on at the MHB after the onset of *Fgf8* (Reifers et al. 2000a; Xu et al. 2000, and references cited therein), suggesting a role in maintaining MHB organizing activity. Indeed, in zebrafish, *fgf17* injections have similar effects to *fgf8* injections; *fgf17* acts downstream of *pax 2.1* and *fgf8*, and both *fgf17* and *fgf18* can be induced ectopically in the forebrain by Fgf8. Mice carrying a null mutation in *Fgf17* have later defects in the cerebellar anlage, a phenotype that is more severe in an *Fgf8* heterozygous background (Xu et al. 2000). Thus, *Fgf8*, *Fgf17* and *Fgf18* may cooperate to maintain the organizing activity and each other's expression at the isthmus. Fgf8 is also a crucial component of the forebrain organizer located in the ANR/row1 (Shimamura and Rubenstein 1997; Shanmugalingam et al. 2000) where it is coexpressed with at least one other *fgf*, *fgf3*, suggesting a similar functional redundancy of Fgf signals (Reifers et al. 2000a; Shinya et al. 2000; Fürthauer et al. 2001; Raible and Brand 2001).

5.2.4 Controlling Competence to Respond to Fgf8 Signaling

Competence of the responding cells is a further level by which the potent abilities of Fgf8 are controlled, as recently uncovered by studying the zebrafish *spiel-ohne-grenzen* (*spg*) mutations. *spg* mutations are known for their effect on early development of the midbrain-hindbrain boundary upstream of *pax2.1* (Schier et al. 1996). Molecularly, *spg* mutations disrupt the gene encoding the POU domain transcription factor Pou2 (Belting et al. 2001; Burgess et al. 2002), which is a likely ortholog of Oct3/4 in mice, a protein thought to control stem cell and germ cell differentiation in mammals (Burgess et al. 2002, and references cited therein). Further studies of the *spg* phenotype have shown that not only is early development of the midbrain-hindbrain boundary disrupted, but also that all of the anterior hindbrain up to the rhombomere 4/5 boundary fails to be specified properly by late gastrulation (Reim and Brand 2002). Key molecules functioning in the formation of the MHB, like *pax2.1*, *spry4*, *her5*, *eng2*, and *eng3*, and in hindbrain development, like *krox20*, *gbx2*, *fkd3* and *pou2*, are all abnormal in *spg* mutant embryos. In contrast, regional definition of the position of the future MHB in the neuroectoderm by complementary expression of *otx2* and *gbx1*, prior to establishment of the complex regulatory cascade at the MHB, is normal in *spg* embryos. In the hindbrain primordium, the earliest defect is a lack of *gbx2* expression in the neuroectoderm in the mutants. This corresponds to the domain of *fgf8* expression, and indeed both *fgf8* and *pou2* are absolutely required in a combinatorial fashion to activate *gbx2* and other markers (*spry4* and *fkd3*) in this neuroectodermal domain (Reim and Brand 2002; K. Lun, M. Rhinn, M. Brand, unpubl.). Importantly, injection experiments and Fgf8 bead implantations have revealed that *spg/pou2* embryos are insensitive to Fgf8 specifically in the early hindbrain neuroectoderm, although the MAP kinase pathway, through which Fgf8 is thought to signal, can function normally in these mutants. Thus, Pou2 is the first example of a factor controlling competence for Fgf8, in this case, in the gastrula neuroectoderm. Much remains to be learned about the mechanism by which Pou2 confers competence, and it will be interesting to explore potential parallels to the actions of Oct3/4 in controlling mammalian stem cell differentiation.

5.3 DV Patterning of the Midbrain and Isthmus

The midbrain and isthmus are clearly subdivided along the DV axis, but surprisingly few studies in zebrafish have addressed the basis for this. Studies of neuronal subtype specification in the chick tegmentum reveal that, as in other sections of the neural tube, Shh provides a ventral positional signal to specify cell-type identity. Likewise, Shh is sufficient to switch dorsally located tectal precursors to a ventral tegmental fate (Agarwala et al. 2001). Misexpression of *hh* genes in zebrafish similarly prevents differentiation into tectum and tegmentum (Ekker et al. 1995), presumably as a result of the overall ventralization of the neural tube. However, although inactivation of Smoothened causes defects throughout the ventral CNS, including the midbrain (Chen et al. 2001a; Varga et al. 2001), it does not affect the subdivision into tectum and tegmentum. This issue has not been addressed directly using tectum-specific markers; however, it is likely that initial subdivision occurs in response to the gradient of BMP activity that regulates DV patterning at all axial levels (Barth et al. 1999). It is currently unknown when and how the information provided by these primary gradients is converted into DV positional information in the midbrain. However, studies of an allelic series of *noi/pax2.1* and of Fgf function in the isthmus provided genetic evidence suggesting that different midbrain and isthmus genes, such as *eng2*, *eng3* and *her5*, require different levels of *noi* and Fgf activity along the DV axis during midsomitogenesis, arguing that by this time regionalization along the DV axis must be underway (Lun and Brand 1998; Carl and Wittbrodt 1999). The isolation of early tectum-specific markers should help to resolve this issue (Seo et al. 1998; Kudoh and Dawid 2001).

5.4 Later Steps of Patterning the Hindbrain

Following its initial induction during gastrulation stages, the hindbrain primordium becomes subdivided into seven segmental units, or rhombomeres, along the AP axis which are apparent at the 18-somite stage (Hanneman et al. 1988). The segmented architecture of the hindbrain is of consequence not only for the neuronal organization in this area, but also for the craniofacial elements derived from the dorsal apex of the neural tube, the neural crest (see Kelsh and Raible, this Vol.; for reviews, see Lumsden and Krumlauf 1996; Schilling and Knight 2001). Similar to the spinal cord, the hindbrain is subdivided along the DV axis into roof plate, alar plate, basal plate and floor plate, which can be further subdivided into medial and lateral regions. There is no morphological distinct border between basal and alar plates, but studies of hindbrain neuronal organization have shown that motor nuclei are predominately found in the basal plate (Trevarrow et al. 1990; Chandrasekhar et al. 1997), whereas sensory cells are predominately located in the dorsal neural tube and interneurons reside in between (Korzh et al. 1993). It is unclear how the embryonic organization relates to the more complex organization of the adult hindbrain (Wullimann et al. 1996). Unique among model vertebrates, just as in the spinal cord, it is possible in zebrafish to identify individual interneurons in the larval hindbrain, based on position, dendritic tree and axonal projection (Kimmel et al. 1985; Metcalfe et al. 1986). The concept of identified neurons has allowed great progress in unraveling the later steps of neuronal specification of identified neurons in *Drosophila* (e.g., Doe and Technau 1993), and, although comparable studies have not yet been carried out in zebrafish, progress is to be expected in this area, particularly with the advent of GFP transgenic lines labeling such identified neurons.

5.4.1 Dorsoventral Patterning

Studies of Shh pathway mutants have shown that, as in other parts of the neural tube, Shh-family members are involved in patterning ventral hindbrain. Mutations in *syu/shh* alone have only a mild effect on ventrally located branchiomotor neurons (Chandrasekhar et al. 1998; Schauerte et al. 1998), but simultaneous MO inactivation of both *shh* and *twhh* results in complete absence of these neurons (Bingham et al. 2001; Etheridge et al. 2001, and references cited therein). Inactivation of Smoothened gives the same phenotype (Chen et al. 2001a; Varga et al. 2001). Moreover, mutations in *yot/gli2* also cause defects in branchiomotor neuron development (Chandrasekhar et al. 1999; Karlstrom et al. 1999). It is not yet clear to what extent medial interneurons are affected by absence of Shh activity. However, dorsal neural tube derivatives appear to form normally, which is not surprising in light of the evidence that they are probably regulated by the BMP pathway (Barth et al. 1999; Nguyen et al. 2000).

5.4.2 Forming and Maintaining Rhombomeres

As in other vertebrates, the identity of hindbrain rhombomeres is thought to be conferred by the combination of *hox* genes they express, but how and when cells become committed to express a particular *hox* gene is still unclear. Zebrafish *hox* gene clusters are studied extensively as a paradigm for genome evolution, although we will not discuss this further here (see Amores et al. 1998; McClintock et al. 2001; Taylor et al. 2001, and references cited therein). Consistent with a role for *hox* genes in conferring identity, misexpression of *hox* genes can cause transformation of rhombomere identity (McClintock et al. 2001), and transplantation experiments between rhombomeres with differing *hox* gene profiles have shown that plasticity of *hox* gene expression correlates with plasticity in rhombomere fate (Schilling et al. 2001). In *Drosophila*, Hox proteins cooperate in a sequence-specific manner with two other homeodomain proteins, Extradenticle and Homothorax. The vertebrate homologues of the *Drosophila* genes are *pbx* and *Meis*, and recent work in zebrafish has shown that these genes have a similar role in vertebrates. The zebrafish gene *lazarus* (*lzr*) is required for proper segmentation of the hindbrain and proper expression and function of *hox* genes. *lazarus* was found to encode a novel *pbx* gene, *pbx4*, which appears to mediate the function of multiple *hox* genes (Pöpperl et al. 2000). Injecting mRNA encoding a dominant negative version of Meis results in a phenotype very similar to *lzr* mutants, arguing that it indeed functions as a transcription co-factor, as suggested by the fly work. Surprisingly, *meis* mRNA injection can in addition rescue the *lzr* mutant phenotype, suggesting that Meis acts not only at the transcriptional level, but also in some way is able to stabilize maternally derived Lzr protein (Waskiewicz et al. 2001).

5.4.3 Extrinsic Signals Controlling Segmentation

Retinoic Acid. Work in other organisms, along with transplantation experiments by Schilling et al. (2001) in zebrafish, suggested that extrinsic signals normally maintain expression of the correct combination of *hox* genes by each rhombomere. Fgfs are likely to be involved in this process, as discussed above. Studies in several vertebrates provide evidence that retinoic acid (RA) acts as a signal in hindbrain identity and *hox* gene regulation (Papalopulu et al. 1991; Gavalas and Krumlauf 2000; Schilling and Knight 2001). *raldh2* catalyzes RA production in the embryo, and knockout studies in mice or receptor blockade in chick, among other experiments, have shown that RA is indeed needed to ensure proper rhombomere formation and identity in the hindbrain (Dupé et al. 1999; Niederreither et al. 1999, 2000; Dupé and Lumsden 2001; for review, see Gavalas and Krumlauf 2000). Two independently isolated zebrafish mutations, called *neckless* and *no fin*, affect the zebrafish *raldh2* gene, and the phenotypes of these mutants support an involvement of RA in posterior hindbrain development, *hox* gene regulation and fin bud formation (Begemann et al.

2001; Grandel et al. 2002). Moreover, Begemann et al. (2001) demonstrated using mosaic analysis that *raldh2* expression is required in the paraxial mesoderm for proper *hox* gene expression in the neural plate. Notably, pharmacological inhibition of RA signaling during early stages of hindbrain development causes absence of rhombomeres posterior to rhombomere 4 (r4), while the more anterior rhombomeres expand in size (Dupé and Lumsden 2001), a phenotype also seen when mouse *raldh2* is knocked out (Niederreither et al. 1999). With later onset of inhibitor treatment, successively more posterior rhombomeres are able to form, which is similar to weaker inactivation of RA receptors, and more akin to the milder phenotype of the zebrafish mutants. It is not yet clear how these phenotypes arise, but they may reflect a further link between RA signaling and the more general problem of posteriorization of the nervous system (see above). It will be interesting to explore how and to what extent RA signaling cooperates with other posteriorizing molecules in hindbrain induction and patterning, and with molecules such as Fgf8 that signal from the anterior border of the hindbrain (Irving and Mason 2000).

valentino *and Ephrin-Dependent Cell Segregation.* The hindbrain posterior to the r3/4 boundary is also affected in *valentino* (*val*) mutants, which lack visible segmentation in this area, and have reduced expression of *krox20* in r5 (Moens et al. 1996). *val* is the zebrafish ortholog of the mouse b-zip transcription factor *Kreisler* (Cordes and Barsh 1994; Moens et al. 1998). During hindbrain formation, a double-segment morphological periodicity is apparent transiently. *val* is required for subdivision of the r5/6 territory into rhombomeres (Moens et al. 1996, 1998), thus providing the first evidence that the double-segment periodicity has a genetic basis. Moreover, *val* mutant cells are excluded from the r5/6 domain if transplanted into wild-type hindbrain, suggesting that, as rhombomere transplantation in chick had suggested previously, an adhesion mechanism controlled by *val* is involved in establishing physical segregation of cells during segmentation. Based on their expression, members of the Eph family of receptor tyrosine kinases and their ligands, the ephrins, were candidates for functioning in rhombomere boundary formation, and previous evidence indicated that an Eph receptor functions in segmentation of the hindbrain (Xu et al. 1995). In the so-called fish-ball assay, bidirectional signaling between receptor and ligand is necessary to achieve segregation of cell populations injected with mRNAs encoding different Ephs or ephrins (Mellitzer et al. 1999; Xu et al. 1999b). Moreover, recent work on Ephs and ephrins in *val* mutants has elucidated a mechanism for how boundary formation of the r5/6 domain might work (Cooke et al. 2001): expression of ephB4a coincides with *val* expression in r5/6, whereas ephrin-B2a is expressed in neighboring r4 and r7. Normal *val* function is needed to establish the mutually exclusive expression domains of these two genes. When ephB4a-expressing cells and ephrin-B2a-expressing cells are juxtaposed in the hindbrain of *val* mutants, boundary formation is restored, suggesting that interactions between ephB4a and ephrin-B2a, possibly under the direct transcriptional control of Val, mediate cell sorting and boundary formation in this part of the hindbrain (Cooke et al. 2001). Further processes are likely to participate to ensure proper hindbrain segmentation, but their function and relative contribution is unclear. For instance, *lunatic fringe* encodes an N-acetylglucosaminyltransferase involved in modulating Notch signaling, for instance at the DV compartment boundary of the *Drosophila* wing, and participates in lineage restriction to a diencephalic segment (Zeltser et al. 2001). *lunatic fringe* is also expressed in zebrafish rhombomeres (Leve et al. 2001; Prince et al. 2001), as is a potential target gene of the Notch pathway, *her6* (Pasini et al. 2001). It will be illuminating to learn the precise functions of these molecules in hindbrain patterning.

5.5 Secondary Modification of the Ground Plan by Neuronal Migration

Although neurons are generated in particular AP and DV locations, their final locations can be dramatically different because they often migrate to other brain regions. An example of this phenomenon is provided by the GABAergic neurons of the murine neocortex, the majority of which migrate from the *Dlx1*, *Dlx2*-positive domain in the subcortical telencephalon (Anderson 1997). Migration of granular cell precursors from the upper rhombic lip has long been known to contribute extensively to formation of the cerebellum in other vertebrates (Altman and Bayer 1997). Studies addressing the mechanistic basis of such migrations have only just begun in other species (Alcantara et al. 2000). A recent time-lapse investigation of the migration path taken by cells from the upper rhombic lip through the cerebellar primordium has revealed a new, ventrally directed migration path along the MHB into the ventral brainstem (Köster and Fraser 2001b), illustrating the potential for this type of work in zebrafish.

6 Summary

We have described the formation of the zebrafish central nervous system. The spinal cord has the simplest organization and was considered first, followed by the forebrain, midbrain and hindbrain. We have discussed many studies that have revealed the molecular mechanisms, including extrinsic signals and intrinsic responses to them, underlying the establishment of nervous system regions and the wide diversity of neuronal cell types of which they are comprised. Wherever possible, we have tried to compare what has been learned from zebrafish with what is known in other vertebrate species. The simplicity of the developing nervous system makes zebrafish embryos particularly amenable to studies of nervous system development. Thus, many aspects of nervous system patterning that were unknown from other vertebrates have been revealed by studies in zebrafish. However, the relationship between embryonic and adult nervous system morphology is still not entirely clear and remains an important avenue for further studies.

Acknowledgements. We thank members of our groups for discussions and comments on this manuscript. Supported by grants from: Wellcome Trust, BBSRC and EC (S.W.W.), NIH (J.S.E.) and Max-Planck Society, Deutsche Forschungsgemeinschaft and European Union (M.B.).

Specification of Zebrafish Neural Crest

Robert N. Kelsh[1] and David W. Raible[2]

1 Introduction

The neural crest is a tissue unique to vertebrates. It consists characteristically of a set of cells that sit transiently in a position dorsal or lateral to the forming or formed neural tube. The cells are ectodermally derived, appearing at the boundary between neural plate and non-neural ectoderm. Soon after their formation, most neural crest cells migrate away from the neural tube to generate a variety of cell types, including peripheral neurons and glia, pigment cells, and the craniofacial mesenchyme that forms cartilage and bone in the head. In this chapter, we will discuss how neural crest cells become specified, highlighting contributions that research in zebrafish has made to our understanding of these processes.

The neural crest of zebrafish is in most regards highly reminiscent of that of other vertebrates. Compared with other model vertebrates, zebrafish neural crest consists of relatively few, rather large cells, and these are easily seen in living embryos (Fig. 1). Zebrafish neural crest forms the same range of derivatives as neural crest in other vertebrates, but the small number of cells in these structures (e.g., embryonic dorsal root ganglia consist of only two to four neurons) may well facilitate dissection of the mechanisms of their development. Neural crest cells appear in a rostro-caudal progression, pause by the neural tube, and then, in the trunk, migrate down one of two major pathways (Raible et al. 1992). These pathways lie between the somites and the epidermis (lateral pathway) and between the somites and the notochord (medial pathway). Migration on the medial pathway is restricted to a specific rostro-caudal position relative to each somite in zebrafish just like in avian embryos, although there are species-specific differences (Morin-Kensicki and Eisen 1997). Migration onto the lateral pathway seems to be regulated by inhibitory signals from somites, just as in avian embryos (Oakley et al. 1994; Jesuthasan 1996). Iontophoretic labeling of single zebrafish neural crest cells is relatively

[1] Developmental Biology Programme, Department of Biology and Biochemistry, University of Bath, Claverton Down, Bath BA2 7AY, UK
[2] Department of Biological Structure, University of Washington, Seattle, Washington 98195-7420, USA

Results and Problems in Cell Differentiation, Vol. 40
L. Solnica-Krezel (Ed.): Pattern Formation in Zebrafish

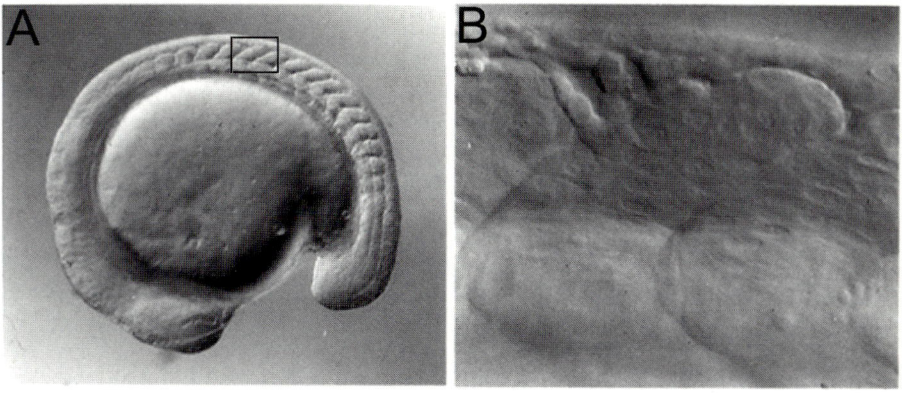

Fig. 1A,B. Zebrafish neural crest cells in the live embryo. **A** Lateral view of 16-somite stage embryo to show position of region shown in close-up in **B**. Dorsal is up, anterior to the left. **B** Medial focal plane of dorsal region of somite 6 (and parts of somites 5 and 7) of trunk showing neural crest cells beginning migration on a medial pathway between the somites and the neural tube

straightforward and has revealed that, as in other vertebrates, all neural derivatives form from cells that migrate on the medial pathway (Raible et al. 1992; Raible and Eisen 1994). Cells that migrate on the lateral pathway form pigment cells, but unlike in mammals and birds, precursors for many zebrafish pigment cells migrate on the medial pathway as well.

1.1 Markers and Their Specificity

One major advantage of zebrafish for developmental biological studies is the single cell resolution of gene expression studies, resulting from the large cell size and optical clarity of the embryo. Table 1 summarizes the markers of neural crest cells and their derivatives that have been characterized from zebrafish to date. It is critical to note that none have been tested exhaustively in combination with each other and hence all should be used cautiously as specific markers of distinct cell fates. Even ignoring expression in non-neural crest-derived cell types, many seemingly "good" markers for specific fates show overlap in a minority of cells (e.g., the transcription factor *mitfa*, melanophore lineage; *xanthine dehydrogenase* (*xdh*), xanthophore lineage; Parichy et al. 2000b; Table 1). It is tempting to speculate that coexpression in a neural crest cell of markers specific for multiple fates may reflect potential of an uncommitted progenitor; as a cell becomes committed to a particular fate, marker expression is refined to only those appropriate for the chosen fate. Thus, expression of the receptor tyrosine kinase *fms* (xanthophore lineage) overlaps in some early migrating cells with *endothelin receptor b1* (*ednrb1*; pigment cell lineages) and *mitfa* (very early melanophore lineage), but more rarely overlaps

Table 1. Marker gene specificity in neural crest and its derivatives. For simplicity, we omit data for adult tissues

Marker	Neural crest cells labeled	Known overlaps of expression[a]	NC fates affected in mutants[b]	Other cell types	Notes	References
crestin	(Almost?) all premigratory trunk NCCs; many (all?) migrating cells; all melanoblasts/early pigmenting melanophores; all xanthoblasts and iridoblasts; enteric nervous system precursors; postmigratory cranial ganglionic neurons, sympathetic neurons, sensory neurons and presumptive satellite glia. Downregulated in most lineages by 72 hpf	Co-expressed with premigratory trunk *fkd6*, premigratory/early migrating *msxb*, *ap2*, migrating/early postmigratory *ednrb1*, *mitf*, *c-kit*, *dct*, *fms*, *c-ret*, postmigratory *dlx2*, anti-Hu, anti-TH	–		Expression absent in premigratory cranial NC	Rubinstein et al. (2000); Luo et al. (2001); A. Pauliny and R.N. Kelsh (unpubL. observ.)
ap2	Premigratory NCCs; many migrating cells	–				Fürthauer et al. (1997)
sna-2	(Almost?) all premigratory NCCs	Co-expressed with *fkd6*	–	Cephalic mesendoderm		Thisse et al. (1995); Kelsh et al. (2000a)
cls/sox10	Most premigratory; (neural?) subset of medial pathway migrating cells; then in (all?) satellite glia and Schwann cells. Late appearance in postmigratory craniofacial cartilage	Some with *dct*, occasional in early pigmenting melanophores, *fkd6* in glia. No overlap with Hu	All non-ectomes-enchymal	Some cells in CNS, otic vesicle, pectoral fin mesenchyme	Turns on after *fkd6*	Dutton et al. (2001); J. Shepley and R.N. Kelsh (unpub. observ.)
kd6 (=*foxd3*)	(Almost?) all premigratory NCCs; off during migration; then in (all?) satellite glia and Schwann cells	No overlap with Hu	–	Somite, ventral CNS, tailbud		Odenthal and Nüsslein-Volhard (1998); Kelsh et al. (2000a)

sna-2	(Almost?) all premigratory NCCs	Coexpressed with *fkd6*, at least in cranial premigratory NCCs	–			
ros/ednrb1	Pigment cell progenitors of all three types at early stages; later strong expression restricted to iridophore lineage	Many cells express *nac/mitfa* and *spa/kit*; many express *fms* and *xdh*. Most or all *dct*-positive cells express *ednrb1* prior to and early in melanization. Many *xdh*-positive cells express *ednrb1* during migration	No embryonic phenotype			Parichy et al. (2000a)
nac/mitfa	Melanophore lineage, from premigratory to pigmenting stages	Coexpressed with *dct*	Melanophore	Retina	Down-regulated in fully pigmented cells	Lister et al. (1999)
spa/kit	Melanophore lineage, from premigratory to fully pigmented stages; postmigratory craniofacial cartilage	*nac/mitfa*	Melanophore	Notochord, haematopoietic precursors, epiphysis, retina, sensory cells in neuromasts	Appears after *nac/mitfa*; maintained in fully pigmented melanophores	Parichy et al. (1999)
dct	Melanophore lineage, from migratory to fully pigmented stages	*nac/mitfa*; none with *fkd6*	–	Pigmented retinal epithelium		Lister et al. (1999); Kelsh et al. (2000b)
fms	Subset of premigratory NCCs. Xanthophore lineage from premigratory to postmigratory stages; possibly uncommitted precursors of other lineages as well	Extensive with *ednrb1*. Some overlap with *nac/mitfa*, occasionally with *spa/kit*, but not *dct*	Xanthophore	Macrophages, osteoclasts	Not clear if maintained in pigmenting xanthophores	Parichy et al. (2000b)
xdh	Xanthophore lineage, from premigratory to pigmented stages	Coexpressed with *fms*. Some overlap with *nac/mitfa*, but not in melanized cells	–			Parichy et al. (2000b); R.N. Kelsh (unpub. observ.)

Table 1. (*Continued*)

Marker	Neural crest cells labeled	Known overlaps of expression[a]	NC fates affected in mutants[b]	Other cell types	Notes	References
gch	Xanthophore lineage, from premigratory to pigmented stages. Expressed early but downregulated in melanophore lineage. Possibly uncommitted precursors of both lineages	Coexpressed with *fms*, *xdh* and *nac/mitfa*. Weak expression in melanophores	–			Parichy et al. (2000b)
Melanosomes	Melanophores	*nac/mitfa, kit, dct*	n/a		Ectopic melanin seen in various non-NCCs after ectopic expression of *nac/mitf*	Kelsh et al. (1996); Lister et al. (1999)
Pterinosomes	Xanthophores	*xdh, gch*	n/a			Kelsh et al (1996); Parichy et al. (2000b)
Reflective platelets	Iridophores	*ednrb1*	n/a			Kelsh et al. (1996); Parichy et al. (2000a)
Anti-Hu Ab	Postmigratory, pan-neuronal	No overlap with *fkd6*	–	CNS neurons		Henion et al. (1996); Kelsh and Eisen (2000); Kelsh et al. (2000a)
Tyrosine hydroxylase Ab	Postmigratory	Anti-Hu sympathetic neurons	–			Raible et al. (1992)
ret	Enteric neuron precursors and early differentiating neurons on gut (?postmigratory); postmigratory branchial arches	Anti-Hu	–	Some cranial ganglia, CNS, retina, pronephric duct, primary motor and sensory neurons		Bisgrove et al. (1997); Marcos-Gutierrez et al. (1997); Shepherd et al. (2001)
gfra1a,b	Enteric neuron precursors on gut (?postmigratory)	–	–	Pronephric duct, primary motor neurons	*GFRa1a* expression much weaker than *GFRa1b*	Shepherd et al. (2001)

Zn-5 mAb	Postmigratory DRG sensory neurons		n/a	Motor neurons		Raible et al. (1992)
msxB,C	Fin mesenchyme	–	–	*MsxB* in median fin epidermis too; pectoral fin	Not clear if all fin mesenchyme are crest derived	Akimenko et al. (1995)
dlx2	Subset of premigratory cranial NCCs; migrating NCCs forming branchial arches and in postmigratory branchial arches	–		CNS, otic vesicle		Akimenko et al. (1994); Miller et al. (2000)
sox9a/9b	Postmigratory craniofacial cartilage (*sox9a/9b*) and progenitors (*sox9a* only). Downregulated as chondrogenesis finishes	Coexpressed with *col2a1*		CNS, otic vesicle, retina, pectoral fin mesenchyme	*sox9a* expression begins earlier and maintained at much higher levels than *sox9b*	(Chiang et al. 2001)
collagen2a1 (*col2a1*)	Postmigratory craniofacial cartilage	Sox9a				Yan et al. (1995)
MsxE	Postmigratory ventral craniofacial cartilage	–	–			Miller et al. (2000)
dHAND	Postmigratory ventral craniofacial cartilage	dlx2	–	Heart, pectoral fin		Miller et al. (2000)
Alcian blue staining	Postmigratory craniofacial cartilage (ECM)	–	n/a	Other cartilage		Schilling and Kimmel (1994)

[a] Apparently complete overlap indicated by "co-expressed with"; extent of partial overlaps indicated semi-quantitatively where noted.

[b] Fates affected in particular mutants show derivatives in which gene is required; absence of effect does not eliminate possible role, since partial redundancy with another gene product may "rescue" absence of function. –, no known mutant.

with the receptor tyrosine kinase *kit* (early melanophore lineage) and does not overlap at all with *dopachrome tautomerase* (*dct*; later melanophore lineage); likewise, *xanthine dehydrogenase* (*xdh*) is coexpressed with *fms*, but overlaps only somewhat with *mitfa* and is absent in melanizing cells. These data suggest that *fms* and *xdh* are rapidly downregulated in the melanophore lineage as cells become committed to this fate, whereas both these markers (but not the melanophore lineage markers) are maintained in cells of the xanthophore lineage.

1.2 Zebrafish Neural Crest Mutants

As for other aspects of embryogenesis, neural crest development has been studied in zebrafish through genetic analysis. The isolation of large collections of neural crest mutants in zebrafish has allowed the identification of key genetic regulators of crest cell fate (Kelsh et al. 1996). Mutant phenotypes of several genes are consistent with the identification of genes critical for the specification of single neural crest derivatives. These include craniofacial cartilage (*chinless*, Schilling et al. 1996b), pigment cell fates (melanophores, *nacre*, Lister et al. 1999) and iridophores (*shady*, Kelsh et al. 1996) and dorsal root ganglial sensory neurons (*nosedive*, Henion et al. 1996; *sensory deprived*, J. Ungos and D.W. Raible, unpubl.). Other mutations affect several different cell types. In the case of embryos homozygous for *colourless*, specification of pigment cell and neural fates is defective (Dutton et al. 2001).

2 Neural Crest Induction

Neural crest is induced from ectoderm as a distinct population of cells lateral to the neuroectoderm. A role for the bone morphogenetic proteins (BMPs) in zebrafish neural crest induction has been established through analysis of *swirl*, *snailhouse* and *somitabun* mutations (Mullins et al. 1996). *swirl* and *snailhouse* encode zebrafish *bmp2b* and *bmp7*, respectively, while *somitabun* encodes the downstream effector *smad5* (Kishimoto et al. 1997; Nguyen et al. 1998b; Hild et al. 1999; Dick et al. 2000; Schmid et al. 2000). These mutations form a phenotypic series, with *swirl* showing the most severe dorsalized phenotypes and *somitabun* and *snailhouse* less so. The *swirl* mutation virtually abolishes BMP signaling, while in the *snailhouse* and *somitabun* alleles examined, some BMP signaling persists. Expression of neural crest markers corresponds to the relative level of BMP signaling (Barth et al. 1999; Nguyen et al. 1998b, 2000). In *swirl* mutant embryos the neural crest markers *foxd3*, *ap2* and *snail2* are greatly reduced or absent. In contrast, in *somitabun* and *snailhouse* mutant embryos, these pan-neural crest markers are expanded toward the ventral side of the embryo. These observations have led to the hypothesis that a gradient of BMP signaling patterns the ectoderm, with non-neuroectoderm formed at high concentrations, neural plate formed at low concentrations and neural crest formed at intermediate concentrations.

Studies in other organisms support a role for BMP signaling in neural crest induction. Treating avian neural tube explants in vitro with BMPs induces neural crest markers and cell migration (Liem et al. 1995). Manipulating BMP levels in *Xenopus* animal caps by injecting dominant-negative BMP receptors also results in neural crest induction (Marchant et al. 1998). Neural crest markers are only induced in animal caps at intermediate levels of BMP signaling, supporting the gradient model of neural crest induction. However, other studies demonstrate that Wnt signaling is required in conjunction with BMPs for neural crest induction in *Xenopus*, suggesting that combinatorial signals are involved (Saint-Jeannet et al. 1997; Chang and Hemmati-Brivanlou 1998; LaBonne and Bronner-Fraser 1998). Whether Wnts are also needed for zebrafish neural crest induction remains to be determined.

Cellular interactions are also involved in the establishment of zebrafish neural crest. Neural crest cells arise from the same ectodermal domain as Rohon-Beard cells, embryonic primary sensory neurons whose cell bodies remain in the dorsal neural tube (Cornell and Eisen 2000). Both neural crest cells and Rohon-Beard cells are specifically affected in *narrowminded* mutant embryos, suggesting that both of these cell types are regulated by the same genetic pathways (Artinger et al. 1999). Several lines of evidence suggest that Notch-Delta signaling is involved in distinguishing between neural crest and Rohon-Beard cells (Jiang et al. 1996; Cornell and Eisen 2000). In zebrafish *deltaA* mutants and in *mindbomb* mutants, Rohon-Beard cells are dramatically expanded while neural crest cells are lost. Blocking Notch-Delta signaling with dominant-negative Delta constructs or dominant-negative SuH also results in decreased neural crest and a concomitant expansion in Rohon-Beard neurons (Cornell and Eisen 2000).

A model for neural crest formation incorporating these data is shown in Fig. 2. A gradient of BMP signaling establishes fields within the ectoderm. At intermediate positions within the gradient, cells are competent to form both neural crest cells and Rohon-Beard cells. Different cell fates are then selected through cellular interactions via Notch-Delta signaling: some cells differentiate as Rohon-Beard neurons, and, by preventing their neighbors from also forming Rohon-Beard cells (by activating Notch), subsequently promote or permit differentiation as neural crest.

3 Cell Fate Specification

After induction from ectoderm, neural crest cells generate derivatives as diverse as neurons, pigment cells and cartilage. How neural crest cells are specified to form these distinct cell types remains an active area of study. It is widely assumed that all neural crest derivatives are formed from multipotent neural crest stem cells and, indeed, neural crest cells with stem cell properties

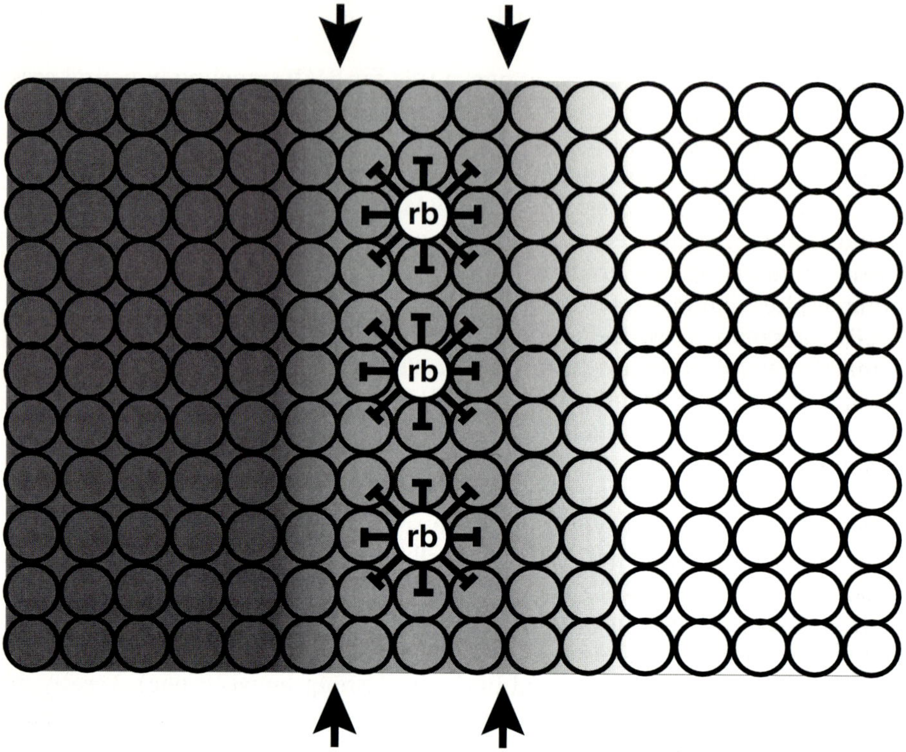

Fig. 2. Cells are competent to form Rohon-Beard cells or neural crest at intermediate concentrations of BMP signaling (*between arrows*). Some cells within this field begin to differentiate as Rohon-Beard cells (*rb*). Then, through Notch-Delta signaling, they promote or permit their neighbors to become neural crest

have been isolated from mammalian embryos (Stemple and Anderson 1992). It remains unclear, however, whether these cells can generate *all* neural crest derivatives, since at present culture conditions that permit generation of only autonomic neuron, glia and smooth muscle derivatives have been described. Likewise in zebrafish it remains to be seen whether a truly multipotent neural crest stem cell exists. However, pluripotent cells have been demonstrated in all vertebrates and hence the mechanism whereby neural crest cells adopt specific fates is a fundamental issue. Adoption of a specific fate is described in terms of fate specification and fate commitment (Kimmel et al. 1991). A cell is defined as being fate-specified as soon as it is distinguishable – by gene expression pattern, cell migration or overt differentiation – from other neural crest cells. In contrast, a cell becomes committed to that fate choice when it cell-autonomously maintains that fate even in a new environment. This can only be tested by transplantation studies.

3.1 When Does Specification Occur?

A first step in understanding fate specification is to determine precisely where and when it occurs. It is unclear from work in chick and mouse embryos whether neural crest cells become specified only after migration or whether at least some fates are specified early (Ciment and Weston 1982; Bronner-Fraser and Fraser 1989; Weston 1991; Anderson et al. 1997; Anderson 2000). In zebrafish, any cranial and most trunk neural crest cells, when iontophoretically labeled before the onset of their migration, give rise to single derivatives, even in cases where sibling cells migrate to different locations, showing that they are fate-restricted even at this early stage (Raible and Eisen 1994; Schilling and Kimmel 1994). Evidence that this fate-restriction might in fact reflect their becoming specified prior to or during migration is strongest for pigment cell fates. Thus, for the melanophore lineage, expression of a key gene in melanophore fate specification, *mitfa* (see below), begins in premigratory neural crest and expression of other molecular markers (*dct* and *kit*) and even melanin is prominent in migrating cells prior to terminal divisions (Raible and Eisen 1994; Lister et al. 1999; Parichy et al. 1999; Kelsh et al. 2000b). Similarly, differentiating xanthophores and iridophores are frequently seen on migration pathways (Parichy et al. 1999, 2000b; Kelsh et al. 2000b; R.N. Kelsh, unpubl. obs.). In addition, *ednrb1*, a gene expressed in most if not all pigment cell progenitors, is prominent in migrating cells (Parichy et al. 2000a). Finally, in *colourless/sox10* mutants, where cells fail to be specified, the precursors remain in a premigratory position, implying that specification must occur prior to, and be necessary for, migration (Kelsh et al. 2000b; Dutton et al. 2001).

Some neural crest derivatives may be specified at different times. It has been suggested that in avian embryos ectomesenchymal fates (forming cartilage, bone or connective tissue) are specified very early (Weston 1983, 1991). Several lines of evidence suggest that in zebrafish, also, ectomesenchymal fates are specified prior to migration. All premigratory cells that generate clones of craniofacial cartilage or fin mesenchyme show fate-restriction and do not generate pigment or neural cells (Schilling and Kimmel 1994; Dutton et al. 2001). Consistent with this, *cls/sox10* mutants are defective in specification of non-ectomesenchymal derivatives, but have no defects in ectomesenchymal fates (Dutton et al. 2001). Likewise, *dlx2*, a marker for craniofacial cartilage precursors in postmigratory stages, is expressed in a subset of premigratory cranial crest and conceivably marks craniofacial cartilage precursors at this earlier stage (Akimenko et al. 1994). More directly, transplantation studies suggest that the *chinless* gene, which is necessary for specification of branchial arch cartilage, primarily acts cell-autonomously in neural crest cells prior to migration (Schilling et al. 1996b).

The timing of neural fate specification is less clear than for ectomesenchymal fates. In avian embryos, very early specification of a subset of sensory neurons has been proposed (Weston 1983). More recently, extensive work on the mechanisms of fate specification of autonomic neurons and glia in mammalian and avian crest suggests that specification to these final fates might occur in postmigratory locations (Shah et al. 1996; Shah and Anderson 1997; Morrison et al. 2000), but early specification of migrating cells to a *neural* fate is consistent with much in vitro and in vivo data (e.g., Bronner-Fraser and Fraser 1989; Dupin et al. 1990; Henion and Weston 1997). In zebrafish, there is currently little evidence to distinguish among these possibilities, but certainly presumptive precursors of the dorsal root ganglia display a characteristic migratory behavior (Raible and Eisen 1994), consistent with their specification to a sensory neuron fate at least during migration. Additionally, *colourless/sox10* expression is rapidly downregulated in pigment cell and ectomesenchymal precursors and hence expression in migrating neural crest cells appears to correlate with neural-fated cells (Dutton et al. 2001). Finally, we note that clonal analyses show that mixed-fate clones, whilst uncommon, are often neural-restricted, although it is not clear whether fate-restriction to a particular ganglionic type (sensory versus autonomic) might occur (Raible and Eisen 1994; Dutton et al. 2001).

In trunk neural crest, cells migrating earlier on the medial migration pathway generate sensory neurons, whereas those migrating later on this or the lateral pathway do not. Transplantation studies demonstrate that these cell populations are intrinsically different, but not committed to, these different fate choices (Raible and Eisen 1996). Consistent with the suggestion above that fate specification begins very early, these transplantation studies support the idea that neural crest cells become biased in their fate choices prior to migration, but are not committed to these choices.

3.2 Progressive Fate Restriction

Two major models of neural crest fate specification have been proposed: (1) direct fate specification (postmigratory pluripotent cells are instructed to adopt a specific fate by environmental signals), and (2) progressive fate restriction (pluripotent cells undergo a process of gradual restriction in the fates open to them (Fig. 3; Ciment and Weston 1982; Weston 1982, 1991; Girdlestone and Weston 1985; Le Douarin et al. 1991). Clonal analysis in primary neural crest cultures has shown an early segregation between neuronal and pigment cell fates in avian neural crest (Henion and Weston 1997). Clone sizes in zebrafish are small, so that full dissection of these processes in clonal studies is unrealistic. However, in the zebrafish trunk, mixed-fate clones are only generated by the earliest migrating cells and precursors for different fates segregate during initial cell divisions (Raible and Eisen 1994).

Analysis of the *colourless* locus, which encodes the HMG-box transcription factor Sox10 (Dutton et al. 2001), provides support for the progressive fate restriction model. Mutants in this locus show defects in all non-ectomesenchymal derivatives (Kelsh and Eisen 2000; Kelsh et al. 2000a). This suggests a progressive fate restriction model in which cells become restricted to either ectomesenchymal or non-ectomesenchymal fates prior to migration. The observed fate restriction of all ectomesenchymal progenitors even at a stage when non-ectomesenchymal progenitors do not show restriction to single fates is consistent with this model, but a direct test involving transplantation of ectomesenchymal progenitors has yet to be performed. The mouse *sox10*^{Dom} mutant phenotype and evidence for direct regulation of *mitf* by Sox10 are consistent with this model applying to mouse as well (Herbarth et al. 1998; Southard-Smith et al. 1998; Kapur 1999; Bondurand et al. 2000; Lee et al. 2000; Potterf et al. 2000; Verastegui et al. 2000).

3.3 Pigment Cell Specification as a Model for Cell Fate Choice

Perhaps the best understanding of neural crest specification in zebrafish comes from work examining melanophore differentiation. Zebrafish genetic screens have produced a wealth of material for understanding the genetic basis of pigment cell development, since the optical properties of zebrafish embryos, coupled with the large size of individual pigment cells and their natural coloration, enable the mutant phenotypes of pigment cells to be perceived in great detail. Some of the very first zebrafish mutants had phenotypes involving pigment cells (Streisinger et al. 1986), and a substantial number of mutations were uncovered in the large-scale screens (Kelsh et al. 1996; Odenthal et al. 1996b). Study of mutant phenotypes with the growing list of markers (see above) and repertoire of labeling and transgenic techniques promises to permit a detailed understanding of how the genome contributes to zebrafish pigment cell fates.

An important gene in regulating zebrafish pigment cell specification is the already mentioned *colourless/sox10*, originally identified by the almost total lack of all pigment cells in mutant embryos. Since *colourless/sox10* function is necessary for the specification of neuron and glial fates as well as pigment cell fates (see above), other factors must act in conjunction with *colourless/sox10* to specify particular fates (Dutton et al. 2001). A strong candidate for such a factor is Wnt signaling, since activation of this pathway in zebrafish premigratory cranial neural crest cells promotes the formation of pigment cells at the expense of neurons and glia (Dorsky et al. 1998). Recent results in mouse and chick are consistent with this mechanism being general to vertebrates (Dunn et al. 2000; Jin et al. 2001). Manipulating the Wnt pathway did not substantially alter ectomesenchymal fates, consistent with the idea that this lineage is segregated early (Dorsky et al. 1998). These data suggest that Wnts and *colourless/sox10* together are involved in the specification of non-ectomesenchymal progenitors. How neural crest cells choose between alternative pigment cell fates (Wnt signaling induces both xanthophore and melanophore fates) and between neuronal and glial fates presents a challenge for future work, although stochastic models of fate choice proposed in the context of mammalian studies may well apply here too (Anderson 2000). These

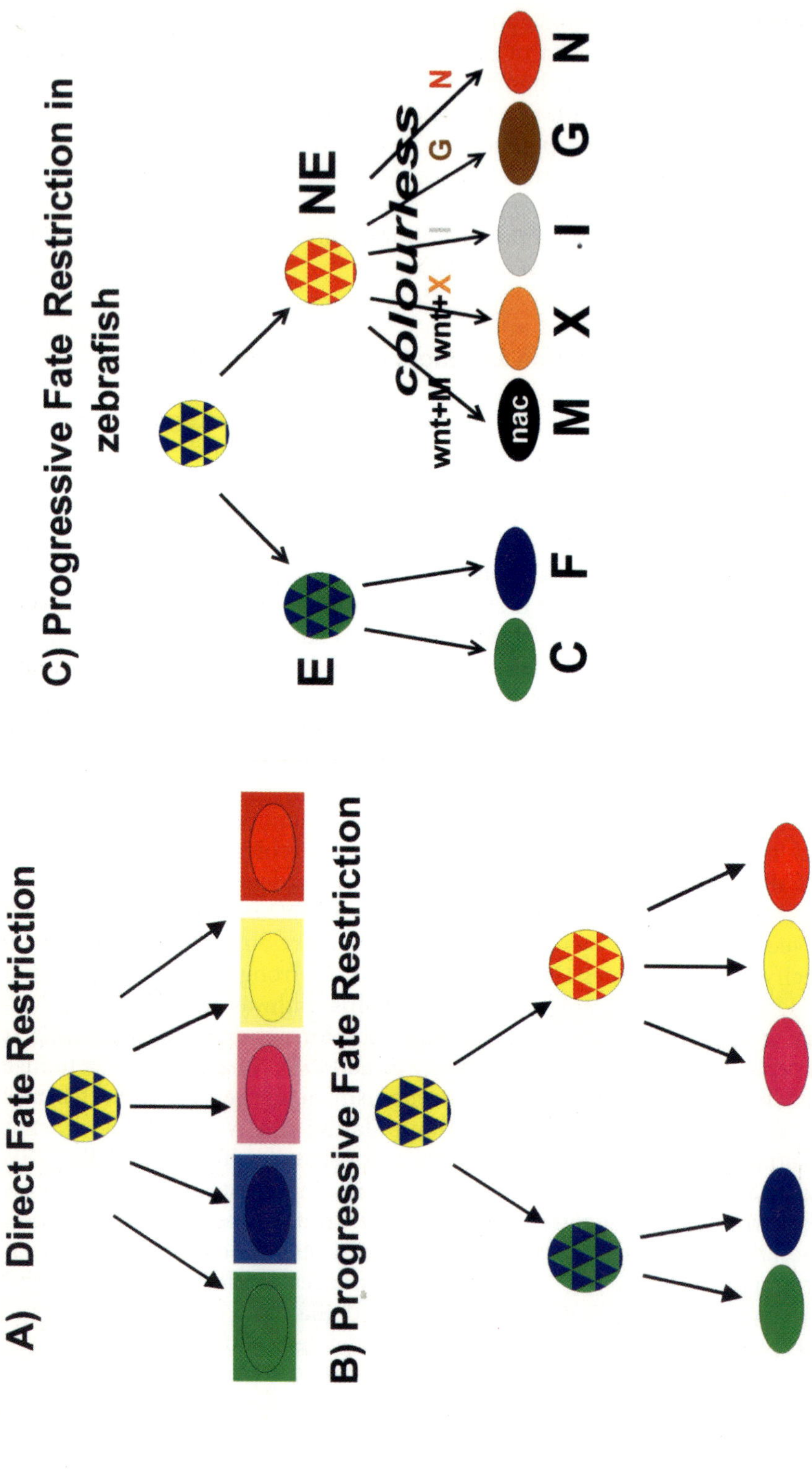
A) Direct Fate Restriction
B) Progressive Fate Restriction
C) Progressive Fate Restriction in zebrafish
E
NE
colourless
wnt+M
wnt+X
I
G
N
nac
C
F
M
X
I
G
N

experiments demonstrate that non-ectomesenchymal precursors in premigratory cranial neural crest are pluripotent, and although in many cases they are also *specified*, they remain responsive to environmental signals that can affect their fate choice, such as Wnts (Fig. 4).

One of the hallmarks of specification is the onset of expression of distinct gene products. One of the earliest genes expressed in melanoblasts is Mitf, a bHLH leucine zipper transcription factor that is necessary for melanoblast differentiation (Hodgkinson et al. 1993; Opdecamp et al. 1997). There are two Mitf genes in zebrafish, and one gene, *mitfa*, is required for melanophore development (Lister et al. 2001). *mitfa* is disrupted in the *nacre* mutation and in mutant embryos melanoblasts cannot be detected by expression of markers such as *dct* and *kit* (Lister et al. 1999). Ectopic expression of *mitfa* causes cells that would not normally differentiate as melanophores to express melanin. Thus a likely key step in the specification of melanophores is the induction of *mitfa* expression. However, it is unlikely that *mitfa* expression signals commitment, since it is downregulated in pigmented melanophores (Lister et al. 1999) and is also expressed in another pigment cell precursor, the xanthoblast (Parichy et al. 1999, 2000b). Work on birds, demonstrating in vitro the dedifferentiation of pigmented melanocytes and their subsequent differentiation into glia, throws into question whether a melanocyte ever becomes truly committed to this fate (Dupin et al. 2000).

Understanding how Mitf expression is induced in neural crest cells should shed light on how melanophores are specified. The zebrafish *mitfa* promoter has consensus Tcf/Lef sites that bind to recombinant Lef in vitro and are

Fig. 3A–C. Neural crest cell fate specification. Two extreme models emphasize the different number of steps between multipotent neural crest stem cells and the final specified cell types. **A** Direct fate restriction: neural crest cells migrate as multipotent stem cells (*blue and yellow checked circle*) to all appropriate locations; local signals (signified by *shaded boxes*) specify particular cell-fate choices. **B** Progressive fate restriction: neural crest cells undertake a series of decisions, each of which restricts a cell's potency to a subset of the previous options. Usually implicit in such models is the idea that at least one of these restrictive decisions is undergone before or during migration. The mechanism whereby fate restrictions are made is not usually explicit, but is likely to include cell-autonomous and stochastic processes, as well as local signals. **C** Model of progressive fate restriction in zebrafish neural crest as suggested by genetic studies. An initial decision between ectomesenchymal (*E*) and non-ectomesenchymal (*NE*) fates generates ectomesenchymal and non-ectomesenchymal progenitors. The former cells generate craniofacial cartilage (*C*) and fin mesenchyme (*F*), amongst other fates. Non-ectomesenchymal precursors are initially pluripotent, but fate restriction to particular fates requires *colourless/sox10* gene function and is influenced by Wnt signaling. The latter biases cells towards pigment fates; the mechanism choosing between these three choices is unclear, but is here proposed to be a series of unknown cofactors (*M, X, I*). The combined signals activate specific factors (e.g., transcription factor encoded by *nacre/mitfa*) and result in fate selection (e.g., melanophore, *M*). Other pigment (*X, I*), glial (*G*) and neuronal (*N*) fates are suggested to result from analogous mechanisms, although the relevant specification factors and target genes have yet to be determined

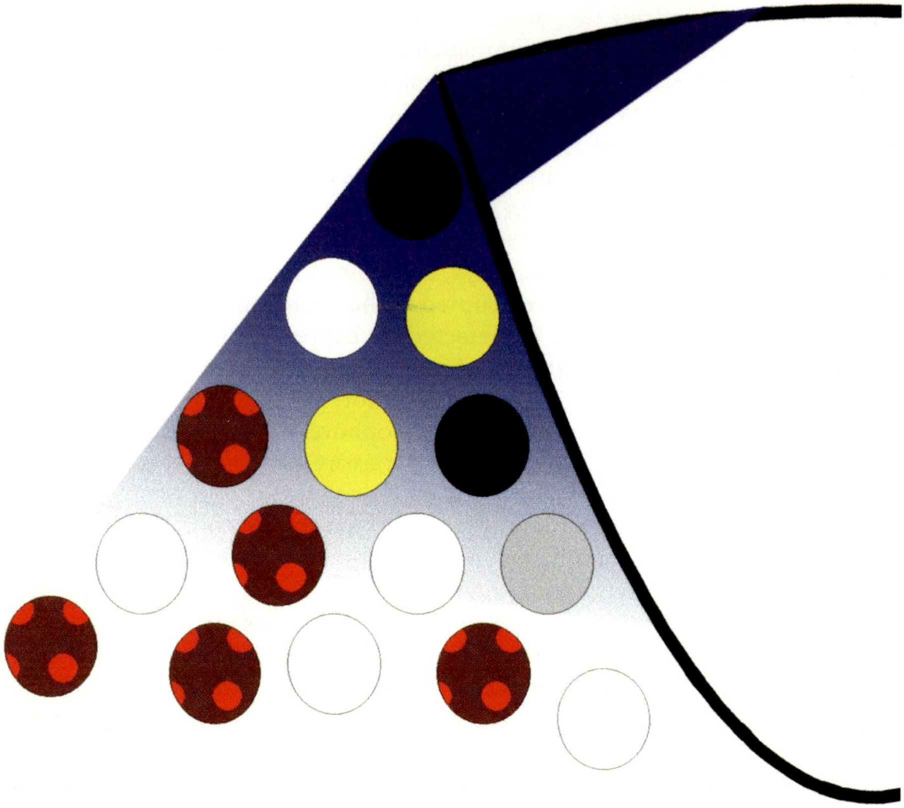

Fig. 4. Wnt-signaling specifies non-ectomesenchymal precursors to pigment cell fates. Dorsal neural keel cells secrete Wnts (*dark blue*) which diffuse through mesenchymal neural crest cells (*circles*) to form gradient. Some cells at this stage are committed ectomesenchymal precursors and are unresponsive to Wnt signaling (*white circles*). Pluripotent non-ectomesenchymal precursors (*colored circles*) are responsive (correlates with colourless/sox10 function) and at high Wnt-signaling doses become specified to pigment cell fates (*solid color* – color code as in Fig. 3). Other non-ectomesenchymal precursors receive lower doses and remain uninfluenced or become specified to a neural (neuron and glia) fate (*mixed brown and red*). It is likely that specification of neuronal and glial fates requires other signals (BMPs?), but this remains to be elucidated

necessary for promoter function in vivo (Dorsky et al. 2000b). The mammalian *Mitf* promoter is also regulated by the Wnt-signaling pathway (Takeda et al. 2000), and, in addition, is responsive to Sox10 (Bondurand et al. 2000; Lee et al. 2000; Potterf et al. 2000; Verastegui et al. 2000). These results suggest that, as part of their role in the combinatorial specification of pigment cells, Sox10 and Wnt signals directly regulate the Mitf promoter.

While some genes such as *colourless/sox10* and *mitfa* are necessary for the specification of pigment cells, other genes are needed for their subsequent migration and survival. *sparse* and *panther* encode the zebrafish receptor tyrosine kinases *kit* and *fms* homologues, respectively (Parichy et al. 1999, 2000b). *kit* has major roles in melanophore and melanoblast migration and survival (Parichy et al. 1999; Kelsh et al. 2000b), whilst *fms* is critical for xanthoblast migration (Parichy et al. 2000b). *kit* function in melanophore development is conserved between zebrafish and mammalian orthologs (Wehrle-Haller and Weston 1995; Parichy et al. 1999). In contrast, *fms* function in xanthophores appears to be novel to fish, although it is perhaps more likely to be an ancient function that is now obscured in mammals, since they have lost the ability to generate the xanthophore lineage.

4 Regional Specification

Up to this point we have considered neural crest cell specification from the perspective of cells choosing individual fates. However, neural crest cells migrating from different axial levels also exhibit regional specification. Although neural crest cells from all levels generate neurons, glia and pigment cells, cells that differentiate as cartilage and connective tissue have region-specific identities. In zebrafish, neural crest cells from cranial regions populate seven pharyngeal arches and form reiterated cartilage elements that have both shared features and unique characteristics (for review, see Kimmel et al. 2001). Neural crest cells that migrate into the first pharyngeal arch, the mandibular, form the lower and upper jaw. Neural crest from the second arch, the hyoid, forms cartilages that support the jaw and oral cavity. Neural crest within arches 3–6 form reiterated cartilages of similar shape that act as gill supports. Neural crest in the posteriormost seventh arch forms a cartilage that sports a set of pharyngeal teeth. In other vertebrates, neural crest cells associated with the posterior arches also contribute to thymus and heart; whether these structures also have contributions from zebrafish neural crest has yet to be determined.

The distinct segmental organization of the hindbrain plays an important role in establishing region-specific identities within the cranial neural crest. Hindbrain segments, or rhombomeres, each express unique combinations of genes, including the *Hox* transcription factors. Neural crest cells that migrate from the hindbrain retain the *Hox* expression pattern of their segment of origin, although regulation of expression in hindbrain and neural crest may be independently controlled (e.g., Nonchev et al. 1996; Maconochie et al. 1999). Targeted mutagenesis of *Hox* genes in mouse has confirmed roles for these genes in neural crest patterning. Inactivation of *Hoxa2* results in a homeotic transformation of crest-derived second arch structures into mirror duplications of first arch structures (Gendron-Maguire et al. 1993; Rijli et al. 1993). Ectopic expression of *Hoxa2* results in a converse phenotype (Grammatopoulos et al. 2000; Pasqualetti et al. 2000). Other *Hox* mutants do not change the apparent segment identity of migrating neural crest but rather have specific reductions in crest migration (e.g., Gavalas et al. 1998) or loss of specific crest-derived cell types (e.g., Manley and Capecchi 1998).

Hox genes were originally identified in *Drosophila* from mutations that resulted in homeotic transformation. As reviewed above, alterations of *Hoxa2* in other vertebrates result in transformations of first and second arch structures. Whether alterations in zebrafish *hoxa2* expression result in similar homeosis awaits further study. However, the phenotype of the zebrafish *valentino* mutant is consistent with the idea that altered *hox* expression in neural crest may in some cases result in homeotic transformation. *valentino* encodes a zinc-finger transcription factor necessary for the early establishment of posterior rhombomeres (Moens et al. 1998), and loss of *valentino* function results in altered *hox* gene expression (Prince et al. 1998b). In *valentino* mutants, an ectopic cartilage forms within the third arch that resembles the interhyal cartilage of the second arch, suggesting an anterior transformation (Moens et al. 1998).

Although mutations in *hox* genes have not been described in zebrafish, analysis of zebrafish *hox* gene function by overexpression studies and by analysis of mutations affecting interacting proteins has given some insight into how they regulate zebrafish neural crest patterning. As in other vertebrates, zebrafish *hox* genes are differentially expressed in hindbrain segments and migrating neural crest cells (Prince et al. 1998a). Overexpression of *hoxb1b* (originally identified as *hoxa1*) resulted in abnormal cartilage growth and fusions between elements of the first and second arches (Alexandre et al. 1996); overexpression of *hoxb2* resulted in almost identical phenotypes (Yan et al. 1998). Neural crest cells anterior to the ear were fused into a single migratory mass instead of remaining as two distinct streams. A similar phenomenon is seen in *lazarus* mutants (Pöpperl et al. 2000). *lazarus* encodes zebrafish Pbx4, a member of the Pbx family of homeodomain proteins that directly interact with Hox proteins, and thus aspects of the *lazarus* phenotype correspond to predicted loss of function in several *hox* genes. In *lazarus* mutant embryos, neural crest cells migrate as an uninterrupted sheet, and first and second arch cartilages are fused together. Together these studies suggest that a major role for zebrafish *hox* genes in neural crest development is to segregate cells into "chondrogenic islands" with distinct characteristics (Kimmel et al. 2001). Surprisingly, although first and second arch cartilages are fused in *lazarus* mutants, they show at least some characters of mandibular and hyoid arches, respectively, and thus appear to be partially specified. This may be a result of maternal *Pbx* expression leading to partial rescue of the phenotype; examination of maternal-zygotic *lazarus* mutants will clarify the precise role of *lazarus* in arch specification.

It may be somewhat surprising that both overexpression of *hox* genes and presumed loss of *hox* function (through the loss of *pbx* expression) have essentially similar phenotypes with respect to cartilage fusions and neural crest migration. One interpretation of these findings is that it is the differences in *hox* expression between the different arches that are necessary for function. Homogenization of the "Hox code" either by widespread *hox* misexpression or by overall *hox* loss of function (through loss of a critical binding partner) results in neural crest cells acquiring more uniform characteristics, so that normally separate migratory streams intermingle and cartilage elements fuse.

The phenotypes described above for altered zebrafish *hox* gene expression may in part be explained by altered eph/ephrin function. Neural crest migration is controlled by Eph receptor tyrosine kinases and corresponding Ephrin ligands (Robinson et al. 1997), and some *eph* receptor genes are directly transactivated by Hox proteins in other tissues (Chen and Ruley 1998). Overexpression of Ephs in *Xenopus* results in fusion of cranial crest migratory streams (Smith et al. 1997). Zebrafish *eph* genes are regulated by *valentino* and play a role in maintaining hindbrain compartments (Cooke et al. 2001), and may also play an important role in zebrafish crest migration.

While it should be noted that *hox* genes may affect neural crest cells indirectly by controlling the development of other tissues that subsequently alter neural crest fates, the fact that neural crest cells themselves express *hox* genes makes it plausible that *hox* genes control neural crest development cell-autonomously. Whether neural crest cells retain positional identity that directs them to differentiate with region-specific characteristics has been analyzed by heterotopic transplantation. When avian neural crest that would normally populate the first arch was transplanted in place of neural crest that would populate the second arch, ectopic first arch structures developed, suggesting that the neural crest was already committed to form region-specific structures (Noden 1983). Transplants of small numbers of cells in mouse have suggested, however, that migrating neural crest cells are plastic in their expression of *Hox* genes (Trainor and Krumlauf 2000).

Whether zebrafish neural crest cells are committed to specific regional identities has been examined by single cell transplant (Schilling et al. 2001). When single neural crest cells were transplanted anteriorly or posteriorly into regions that expressed different *hox* genes, they lost expression of *hox* genes from their segment of origin and acquired expression of *hox* genes appropriate to their new locations. Transplanted cells were also able to contribute normally to cartilage elements appropriate for their new location. However, if groups of cells were transplanted, then some cells retained *hox* expression appropriate for their segment of origin. These results suggest that there may be a "community effect", where interactions among neural crest cells reinforce gene expression. These results allow reconciliation between the avian and mouse experiments described above, since neural crest retained characteristics of its segment of origin when transplanted en masse after crest was removed in the host embryo. These studies also reflect the importance of examining issues of developmental commitment at the level of the single cell. The zebrafish experiments show that establishment of regional identity relies on a variety of signals, including those among neural crest cells to establish and maintain *hox* gene expression (and presumably region-specific fates).

Retinoic acid (RA) has been implicated as an environmental cue involved in regional specification. A zebrafish transgenic line with a RA reporter has demonstrated that cells within the caudal hindbrain respond to exogenous RA

(Perz-Edwards et al. 2001). Treatment of zebrafish embryos with RA results in abnormalities similar to those described for other vertebrates (for reviews, see Durston et al. 1998; Gavalas and Krumlauf 2000). After RA exposure, development of anterior neural-crest-derived structures is perturbed (Holder and Hill 1991; Alexandre et al. 1996; Ellies et al. 1997; Yan et al. 1998). RA directly regulates *hox* gene transcription, and defects in patterning may be the result of altered *hox* expression (Alexandre et al. 1996; Yan et al. 1998). The zebrafish mutant *neckless* is caused by a disruption in retinaldehyde dehydrogenase 2 (Begemann et al. 2001), a key enzyme in RA synthesis, and the associated phenotypes are similar to those described for the RALDH knockout mouse (Niederreither et al. 2000). *neckless* mutant embryos have specific defects in the neural crest that migrates into posterior arches, and posterior crest-derived cartilages do not form (Begemann et al. 2001). The *neckless* gene product is required within the paraxial mesoderm, suggesting this tissue is the source of RA that patterns the neural crest. Taken together, these studies suggest that RA is an important cue in specifying zebrafish posterior arch neural crest through regulating *hox* gene expression.

4.1 Pharyngeal Arch Specification

Development of the pharyngeal arches results from a complex interplay of tissues derived from each of the germ layers (Hall 1980; Noden 1983; Schilling et al. 1996b; Golding et al. 2000; Trainor and Krumlauf 2000). Early in their development, the arches consist of a cranial mesoderm-derived core surrounded by a tube of neural crest that is wrapped in an epithelium derived from endoderm (pharyngeal pouches) on the medial, anterior and posterior faces and derived from ectoderm (epidermis) on the lateral face (Trainor and Tam 1995; Miller et al. 2000). Mutagenesis screens in zebrafish have supplied a wealth of mutants with defects consistent with failures in specification of arch identity (Neuhauss et al. 1996; Piotrowski et al. 1996; Schilling et al. 1996a). Two major classes have been described – those primarily affecting (1) the anterior (mandibular and hyoid; anterior arch group, e.g., *sucker, schmerle, hoover* and *sturgeon*), and (2) the posterior (branchial) arches (flathead group, e.g., *flathead*). It is likely that at least some of these mutants identify genes acting in each of the tissues involved in arch specification; transplantation experiments to test cell autonomy of gene action will permit the ready examination of this point. In time, these collections of mutants promise to elucidate key players in this process.

As described in the previous section, the specification of arch segment *identity* is in large part likely due to information conveyed by the neural crest streams through the actions of the the *Hox* genes, their cofactors and targets. However, careful study of the *van gogh* phenotype has led to the suggestion that the reiterated segmental *pattern* of the arches is imposed primarily by the endoderm (Piotrowski and Nüsslein-Volhard 2000). This is seen most clearly in the posteriormost arches where the migrating neural crest stream from the regions of hindbrain rhombomeres 5–8 (stream 3) is initially not subdivided, as is evident from morphological and gene expression studies (Piotrowski and Nüsslein-Volhard 2000) and from the observed lack of segment restriction at early stages (Schilling and Kimmel 1994). In wild-type embryos, the posteriormost neural crest mass becomes segmented, but this process requires correct segmentation of the endoderm (Piotrowski and Nüsslein-Volhard 2000).

Other genes appear to play a role in the specification of different elements within each arch along the dorsoventral axis. At least one of the "anterior arch" class of mutants appears to be primarily a defect in ventral specification. Thus, Miller et al. (2000) have focused on the differences between dorsal and ventral cartilage elements in the arches. They have shown that *sucker/endothelin-1* is required non-autonomously for specification of ventral cartilage fates, e.g., lower jaw. The *sucker* mutant phenotype can be rescued by injection of human Endothelin-1 into the arches after neural crest migration, showing that specification to a ventral cartilage fate is likely to occur in postmigratory neural crest. The phenotypes of the other three members of the "anterior arch" class mutants are very similar to, although weaker than, *sucker* mutants and thus are likely to be other components of the Endothelin-1-signaling pathway. These observations are consistent with the conclusions of studies of mouse knockouts for genes in this pathway (Kurihara et al. 1994; Clouthier et al. 1998; Yanagisawa et al. 1998). Thus, whether the "anterior arch" class genes actually function in anterior arch specification remains to be seen. It is conceivable that the almost complete specificity of the defects in this class of mutants to the anteriormost two arches reflects such a role; alternatively, perhaps Endothelin-1 signaling mediated ventral arch specification in the posterior arches (*endothelin-1* is expressed in all the arches; Miller et al. 2000) is functionally redundant with another signaling system. Regardless, the phenotypes indicate that arch specification and dorsoventral cartilage specification may be closely coordinated.

5 Summary

Analysis of neural crest cells in wild-type zebrafish and in isolated mutants has complemented studies in other organisms and has led to a greater understanding of the processes of cell fate and regional specification. Analysis of pigment cell development suggests that neural crest cell fate specification results from a combination of intrinsic cues such as differential expression of transcription factors and external factors such as the Wnts. Whether this model can be generalized to accommodate other neural crest cell fates remains an active area of research. Regional specification of neural crest appears to be coordinated by similar combinations of internal factors such as the *hox* genes, and extrinsic cues such as retinoic acid and endothelins. There is still a wealth of zebrafish mutations whose molecular bases remain unknown, and it is likely that the revelation of the underlying genes will both broaden and alter our understanding of specification within the neural crest.

Acknowledgements. We thank Cecilia Moens and David Parichy for comments on the manuscript and Table 1, respectively. Work in the authors' laboratories is funded by the Wellcome Trust and Medical Research Council (R.N.K.) and by NIH and the March of Dimes (D.W.R.).

Neurogenesis and Specification of Neuronal Identity

Bruce Appel[1] and Ajay Chitnis[2]

1 Introduction

Neural development unfolds with remarkable speed in zebrafish embryos where early neurons extend axons in a cell-specific pattern by the first day of development and a functional, though still growing, nervous system permits free-living larvae to hatch from their chorion shells by the second day of development. The relative simplicity of the early nervous system in zebrafish makes it attractive to explore mechanisms that determine how specific types of neurons and glia are made in the correct number and location. In this chapter, we will first briefly describe the organization of different cell types in the early zebrafish nervous system and then discuss mechanisms responsible for establishing their organization. Since an exhaustive description of the anatomy and neurogenesis in the entire neural tube would be beyond the scope of this book, we will emphasize studies related to the development of the spinal cord.

2 Zebrafish Spinal Cord Anatomy

At 24 h postfertilization (hpf), the zebrafish spinal cord has features similar to neural tubes of rodent and chick embryos (Fig. 1). A pseudostratified neuroepithelium forms the lateral walls, bordering a central lumen, or neurocoel. The neuroepithelium is separated ventrally and dorsally by non-neuronal floor plate and roof plate cells, respectively. Neuroepithelial cells are proliferative neural precursors, giving rise to neurons and glia. The earliest-born, differentiated neurons of the zebrafish spinal cord are present at 24 hpf and they have a relatively simple dorsoventral organization: sensory neurons occupy the most dorsal positions, motor neurons occupy relatively ventral positions and

[1] Department of Biological Sciences, Vanderbilt University, VU Station B 351634, Nashville, Tennessee 37235-1634, USA

[2] Unit of Vertebrate Neural Development, NIH/NICHD Laboratory of Molecular Genetics, Bldg. 6B, Rm. 3B-315, 9000 Rockville Pike, Bethesda, Maryland 20892, USA

Results and Problems in Cell Differentiation, Vol. 40
L. Solnica-Krezel (Ed.): Pattern Formation in Zebrafish

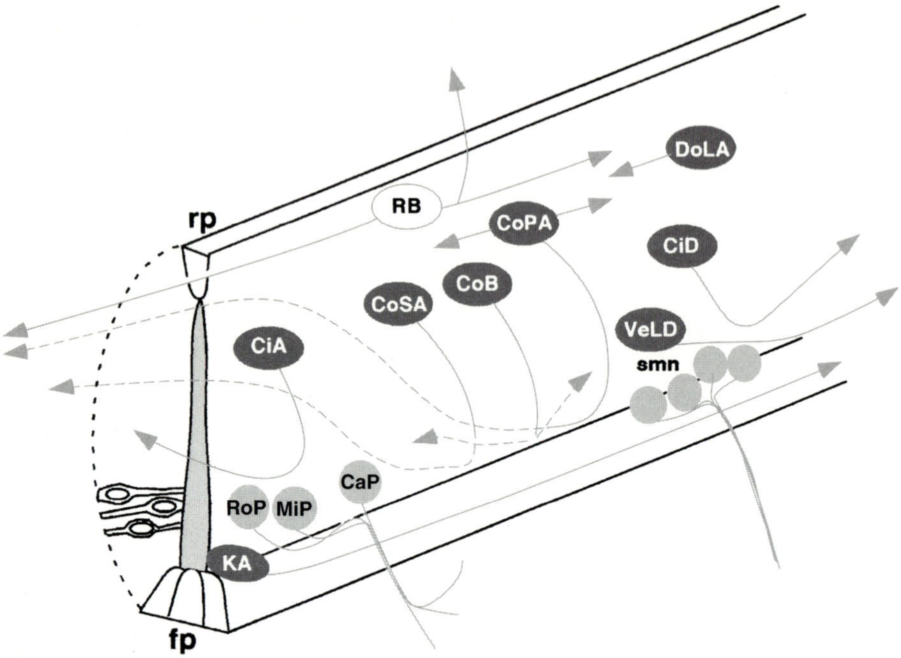

Fig. 1. Cell types in the zebrafish spinal cord. Schematic depicts spinal cord cells at 1 day of development. Interneurons are *dark gray* and motor neurons are *light gray*. *Solid* and *hatched lines* represent ipsilateral and contralateral axon projections, respectively. Three neuroepithelial neural precursor cells are depicted on the contralateral side. Apical surface of the neuroepithelium contacts the lumen (*shaded*). Basal surface of the neuroepithelium is depicted as a *curved, hatched line*. *RB* Rohon-Beard sensory neuron; *smn* secondary motor neuron; *KA* Kolmer-Agduhr neuron; *rp* roof plate; *fp* floor plate

interneurons are typically located at intermediate positions. Along the antero-posterior axis the portion of the spinal cord encompassed by the boundaries of an adjacent, segmentally iterated somite is referred to as a hemisegment. Some types of neurons are distributed in a hemisegmental pattern in that their number and location correlates with somites. The number and distribution of other neurons are not segmentally patterned.

Zebrafish, like other anamniotes, are distinguished by a set of rapidly developing early neurons, which play a critical role in coordinating the activities of larvae soon after they hatch. These early neurons, which are also known as primary neurons, are born during or soon after gastrulation and can be recognized by their large cell bodies, stereotyped positions and cell-specific axonal projections. The following section summarizes the differentiated cell types known to occupy the zebrafish spinal cord by 24 hpf, proceeding from dorsal to ventral. Most of these cells are depicted, schematically, in Fig. 1.

2.1 Roof Plate

Roof plate cells are specialized non-neuronal cells that occupy the dorsal midline of the spinal cord and contact the lumen of the central canal (Altman and Bayer 1984). Though roof plate cells have not been well studied in zebrafish, cell lineage analysis (Pappan and Campos-Ortega 1997) and tenascin-C expression (Tongiorgi 1999) have revealed the presence of these cells in zebrafish embryos. In other vertebrate embryos, roof plate cells produce bone morphogenetic proteins (Bmps) and Slit proteins, which can influence cell fate and axonal trajectories of dorsal neurons (Liem et al. 1997; Holmes et al. 1998; Augsburger et al. 1999).

2.2 Rohon-Beard Sensory Neurons

Rohon-Beard (RB) cells are large sensory neurons that occupy two nearly continuous rows on either side of the dorsal spinal cord midline (Bernhardt et al. 1990). Each RB neuron extends two axons longitudinally within the dorsal spinal cord, one rostrally and one caudally, as well as a peripheral axon that innervates the skin. RB neurons, revealed by gene expression, appear to be specified at neural plate stage, making them among the earliest specified neurons in the zebrafish CNS (Korzh et al. 1993). There are on average two to three RB cells in each hemisegment by 24 hpf; however, they are not distributed in a segmental pattern. RB neurons undergo apoptosis by 3–4 days postfertilization, which may result from reduced NT-3 signaling via the neurotrophin receptor TrkC1 (Williams et al. 2000). It is important to note that this early population of large sensory neurons within the spinal cord is unique to anamniotes like zebrafish and frogs. Later, like chick and rodents, zebrafish develop sensory neurons derived from the neural crest that are part of the segmentally arranged dorsal root ganglia (see Kelsh and Raible, this Vol.).

2.3 Interneurons

At 24 hpf, the zebrafish spinal cord contains at least eight types of interneurons (Fig. 1). Most of them have been identified by dye-filling strategies to reveal cell morphology and immunohistochemistry using an anti-acetylated tubulin antibody, which labels axons and, often, cell bodies. Little is known about the function of individual interneurons; however, several are GABA-immunoreactive indicating they may have inhibitory roles (Bernhardt et al. 1992b). Interneuron cell bodies lie at characteristic dorsoventral positions: DoLA and CoPA are most dorsal, CoSA and CoB are slightly more ventral and CiA and CiD occupy positions ventral to CoSA and CoB. VeLD is located in the ventrolateral spinal cord, just dorsal to primary motor neurons. The acronyms indicate different identifying features, e.g., DoLA and VeLD are Dorsal Longitudinal Ascending and Ventral Longitudinal Descending interneurons, respectively, whereas CoPA and CoSA refer respectively to Commissural Primary and Secondary neurons with Ascending projections. The number of each type of interneuron per hemisegment length is variable, with the possible exception of CoPA, which appears to develop once per hemisegment. VeLDs have a quasi-segmental distribution with cell bodies typically located midway between CaPs (see below). Kolmer-Agduhr (KA) cells are GABA-reactive neurons that contact the lumen of the central canal, ventral to motor neurons and adjacent or slightly lateral to the floor plate. These are an unusual class of neurosecretory neurons: they do not directly innervate other neurons like interneurons or muscles like motor neurons.

2.4 Motor Neurons

Motor neurons are born in two distinct waves, giving rise to primary and secondary populations. Individual primary motor neurons undergo a final S phase between 9 and 16 hpf, whereas secondary motor neurons have their last S phases beginning after 14 hpf and continuing until after 25 hpf (Myers et al. 1986). Primary motor neurons have large cell bodies (~9 µm) that lie above the ventral third of the spinal cord whereas secondary motor neurons have smaller cell bodies (~5 µm) that lie more ventrally, dorsolateral to the floor plate (Myers 1985). The three primary motor neurons have a segmental organization that correlates with the pattern of adjacent axial muscles that they innervate. The Rostal Primary motor neuron (RoP) is adjacent to the anterior edge of each myotome, the Middle Primary motor neuron (MiP) is caudal to RoP, and the Caudal Primary motor neuron (CaP) is located at the midpoint of the adjacent myotome. An additional primary motor neuron, called the Variable Primary motor neuron (VaP), sometimes develops next to a CaP. Most VaPs die, apparently from interaction with the adjacent CaP (Eisen 1992). Each primary motor neuron extends an axon along a cell-specific trajectory to innervate specific parts of the adjacent myotome: CaP innervates the ventral myotome, MiP projects to the dorsal myotome and RoP innervates the transverse myoseptum (Eisen et al. 1990). Secondary motor neurons are much more numerous with possibly as many as 20 occupying each hemisegment (Myers 1985). Most secondary motor neurons innervate adjacent segmented axial muscle. A subset of secondary motor neurons that occupy ventrolateral spinal segments three to six innervates the pectoral fins (Myers 1985).

2.5 Floor Plate

Floor plate cells, glial-like cells that contact the lumen of the central canal, occupy ventral-most spinal cord. When viewed from the side, floor plate cells have a distinctive cuboidal morphology. Floor plate in zebrafish embryos consists of two cell types: a single row of medial floor plate cells bordered on both sides by lateral floor plate cells (Kuwada et al. 1990a; Odenthal et al. 1998, 2000; Schauerte et al. 1998). Like roof plate cells, floor plate cells are the source of secreted proteins that influence the fate and axonal trajectories of adjacent neurons (Colamarino and Tessier-Lavigne 1995; Tanabe and Jessell 1996).

2.6 Glia

Glial cells in the zebrafish embryonic spinal cord have not been investigated extensively. Lineage analysis revealed radial glia, oligodendrocytes and a novel glial cell type called fringed glia (Papan and Campos-Ortega 1999). Additionally, several antibodies that label radial glia have been produced (Trevarrow et al. 1990). Radial glia, which reach from the central canal to the pial surface, are fairly numerous and distributed throughout the dorsoventral extent of the spinal cord (Trevarrow et al. 1990; Appel et al. 2001). Radial glia may be neural stem cells, as studies using other vertebrates show that radial glia can give rise to astrocytes and neurons (reviewed by Alvarez-Buylla et al. 2001). Oligodendrocytes arise from ventral spinal cord and migrate to new dorsoventral positions, similar to oligodendrocytes of chick and rodent embryos (Park et al. 2002).

3 Neurulation and the Early Pattern of Neurons

By the end of gastrulation, dorsal ectoderm is specified as neuroectoderm (see below and Hammerschmidt and Mullins; Wilson et al., this Vol.). At this stage the neuroectoderm is often referred to as the neural plate. Primary neurons, revealed by expression of a neuron-specific gene, *huC* or *elavrC*, a zebrafish homologue of *D. melanogaster elav* (Good 1995; Kim et al. 1996), are born within the neural plate (Fig. 2). These early neurons are distributed in three bilateral longitudinal columns, medial, intermediate and lateral, which give rise to motor neurons, interneurons and sensory neurons, respectively. More rostrally, *huC* expression at the three-somite stage reveals nascent neurons in the midbrain, hindbrain and the trigeminal ganglia. As cells converge and engage in extension movements during neurulation the neural plate becomes longer and narrower, transforming first into a neural keel and then a solid neural cord (Fig. 2; Schmitz et al. 1993). The neural tube forms by cavitation of the neural cord. Zebrafish neurulation, which is similar to the morphogenetic process known as secondary neurulation that shapes the caudal neural tube in amniote embryos, changes the distribution of sensory neurons, interneurons and motor neurons from their lateromedial distribution at the neural plate stage to their characteristic dorsoventral distribution described earlier (Fig. 2).

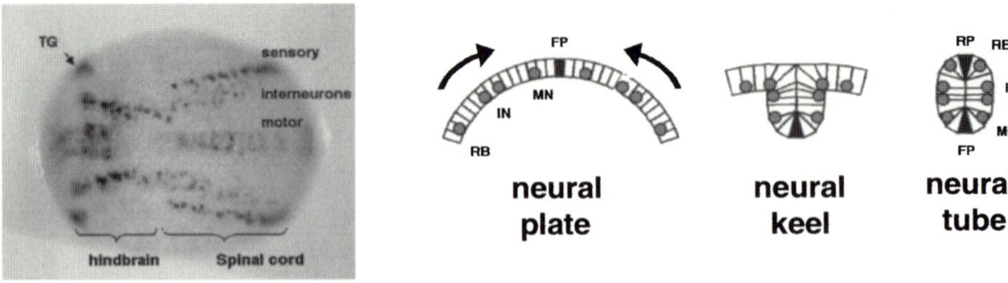

Fig. 2. Photomicrograph showing a dorsal view, anterior to the left, of a neural plate stage zebrafish embryo probed for *huC* expression to reveal the pattern of early neurons. The schematic represents a cross-sectional view of stages in neurulation. Neurulation changes the distribution of Rohon-Beard (*RB*) sensory neurons, interneurons (*IN*) and motor neurons (*MN*) from their lateromedial distribution at the neural plate stage to their characteristic dorsoventral distribution. *TG* Trigeminal ganglion; *FP* floor plate; *RP* roof plate

4 Regulation of Neurogenesis in the Zebrafish Neural Plate

The mechanisms that generate the early patterns of neurons in zebrafish and other vertebrate embryos are remarkably similar to those that regulate neural development in fly embryos. In *D. melanogaster*, cells of the ventrolateral ectoderm are specified as neuroectoderm by suppression of signaling by Decapentaplegic (Dpp), a TGFβ-related ligand. Dorsolateral cells, in which Dpp signaling remains active, are prevented from developing as neural cells and produce ectoderm, instead. Vertebrate homologues of Dpp are Bmps. Molecules that antagonize Bmp signaling induce formation of neuroectoderm from dorsolateral ectoderm during gastrulation. Ventrolateral ectoderm, in which Bmp signaling remains high, forms epidermis (see Hibi et al.; Hammerschmidt and Mullins, this Vol.).

Upon neuroectoderm formation in flies, cell–cell signaling plays an important role in selecting individual neuroectodermal cells for neural fate (reviewed in Campos-Ortega 1995). Early anteroposterior and dorsoventral patterning mechanisms define a reiterated pattern of "proneural" clusters in the neurectoderm where cells express "proneural" genes: bHLH protein-encoding genes of the *acheate-scute* class. Expression of proneural genes gives each cell in a proneural cluster the potential for neural fate. However, as a consequence of dynamic regulation of proneural gene expression by autoregulation, where each cell enhances its own expression of proneural genes (Culi and Modolell 1998), and lateral inhibition, where each cell inhibits proneural expression in neighboring cells (Skeath and Carroll 1992), only one cell develops as a neuroblast within each proneural cluster. Cells in a proneural cluster that are inhibited from becoming neuroblasts become epidermoblasts instead. Lateral inhibition is mediated by a membrane-bound receptor, Delta, whose expression is promoted by proneural genes, and a transmembrane receptor, Notch, whose activation inhibits expression of proneural genes in neighboring cells.

Delta-Notch interaction initiates several specific proteolytic events that produce an intracellular Notch fragment (Notch^{intra}) which, together with Suppressor of Hairless [Su(H)] protein, promotes expression of genes in the *Enhancer of split* [*E(spl)*] complex. E(spl) proteins, in turn, inhibit expression of proneural genes. In this way, cells in which Notch activity is high are prevented from developing as neuroblasts and form epidermoblasts. Conversely, those cells in which Notch signaling is low, e.g., those in which Delta and proneural gene expression is high, develop as neuroblasts. Loss-of-function mutations of *Delta*, *Notch*, or any of the genes in the Notch-signaling pathway, produce "neurogenic" phenotypes where excess neuroblasts are made at the expense of epidermoblasts.

Many zebrafish homologues of fly Notch pathway components have been identified and functional analyses have revealed that similar molecular mechanisms are involved in selecting cells that become neurons within "proneuronal" domains. By the end of the gastrula period, groups of cells that will give rise to the longitudinal distribution of early neurons described above begin to express *neurogenin1* (*ngn1*), which is homologous to the proneural gene *atonal* in *D. melanogaster* (Ma et al. 1996; Blader et al. 1997a; Kim et al. 1997; Korzh et al. 1998). Within the early "proneuronal" domains, and in later stages of neurogenesis, newly specified neurons transiently express three *Delta* homologues, *deltaA*, *deltaB* and *deltaD* (*dlA*, *dlB* and *dlD*, respectively; Dornseifer et al. 1997; Haddon et al. 1998; Appel and Eisen 1998; Appel et al. 2001). A subset of neural precursors also expresses *delta* genes; these could be cells fated to exit the cell cycle and differentiate as neurons (Appel et al. 2001). Additionally, neural plate cells, and neural precursors of 24-hpf embryos, uniformly express several *Notch* homologues (Bierkamp and Campos-Ortega 1993; Westin and Lardelli 1997; Appel et al. 2001). Consistent with the idea that Notch signaling inhibits neural specification, overexpression of *delta* genes inhibited formation of early neurons and activation of Notch1a led to expression of *her4*, a *Hairy Enhancer of split Related* gene, which is related to *Hes* genes of mammals (Takke et al. 1999). Conversely, disruption of Notch signaling, by expression of dominant negative forms of Delta protein or by mutation of *dlA*, caused excess cells within proneuronal domains to express elevated levels of *ngn1* and develop as neurons (Dornseifer et al. 1997; Appel and Eisen 1998; Haddon et al. 1998; Appel et al. 2001). Mutation of another locus, called *mind bomb* (*mib*), produced a neurogenic phenotype similar to that caused by *dlA* mutation (Jiang et al. 1996; Schier et al. 1996). Normally, *ngn1* and *dlA* expression is elevated in a subset of cells within proneuronal domains as cells are selected to become neurons. In *mib* mutants all the cells in a proneuronal domain show elevated *ngn1* and *dlA* expression and most differentiate early as neurons. This change is consistent with a failure of Notch signaling since the neurogenic phenotype in *mib* mutants is accompanied by reduced expression of *her4*, a target of Notch activation. Positional cloning of *mib* showed that it encodes a previously unknown ubiquitin ligase whose function is essential for efficient Notch signaling (Itoh et al. 2002).

One interpretation of these studies is that as a consequence of lateral inhibition some cells in a proneuronal domain acquire a slightly higher level of *ngn1* expression compared to their neighbors. Cells with slightly higher levels of *ngn1* expression become progressively better at inhibiting *ngn1* expression in neighbors and are likely to acquire high levels of *ngn1* expression through autoregulation. Elevated *delta* gene expression in these cells activates Notch receptors in neighboring cells. Consequently, these cells may downregulate *ngn1* expression and maintain a proliferative, neural precursor state. Those cells expressing high levels of *delta* genes and *ngn1* may exit the cell cycle and differentiate as neurons.

Cells selected to differentiate as neurons can be recognized by their expression of a number of genes including *huC*, *deltaB* and *neuroD*, which encodes a basic helix-loop-helix transcription factor like *ngn1* (Lee et al. 1995; Ma et al. 1996; Korzh et al. 1998). As with *ngn1*, ectopic expression of *neuroD* generates ectopic neurons in the ectoderm; however, in contrast to *ngn1*, its ability to induce differentiation of neurons cannot be easily inhibited by Notch signaling (Chitnis and Kintner 1996; Ma et al. 1996). MyT1 is a zinc-finger protein whose expression also becomes prominent in differentiating neurons in *Xenopus* and zebrafish embryos (Bellefroid et al. 1996; Kim et al., unpubl.). Ectopic expression of *MyT1* in *Xenopus* produces a neurogenic phenotype very similar to that seen when Notch signaling is blocked (Bellefroid et al. 1996). These studies suggest that cells selected for a neuronal fate by lateral inhibition may not only initiate expression of neuron specific genes, but may also express genes that facilitate differentiation by making the cells resistant to the effects of Notch activation (Fig. 3).

5 Creating Proneuronal Domains: Regulation of *ngn1* Expression

Mechanisms that regulate *ngn1* expression remain poorly understood; however, screens for zebrafish mutants with an aberrant distribution of early neurons have led to the identification of some genes that regulate *ngn1* expression in the neuroectoderm. A mutation in *narrowminded* (*nrd*) leads to a specific loss of RB sensory neurons and early neural crest cells (Artinger et al. 1999). These mutants are characterized by reduced *ngn1* expression in the lateral proneuronal domain. While the molecular identity of *nrd* remains unknown, ectopic expression of *ngn1* leads to recovery of RB neurons in the mutants. These studies suggest that *nrd* has a critical role in determining *ngn1* expression in the lateral neuroectoderm (Kim et al., in prep.).

In contrast to the loss of sensory neurons observed in *nrd* mutants, *headless* (*hdl*) mutants were identified by their rostral expansion of trigeminal sensory neurons (Kim et al. 2000a). These mutants are characterized by loss of function of T-cell factor 3 (*Tcf3*), a transcription factor that has an essential role in repressing Wnt target genes in the anterior neuroectoderm during early

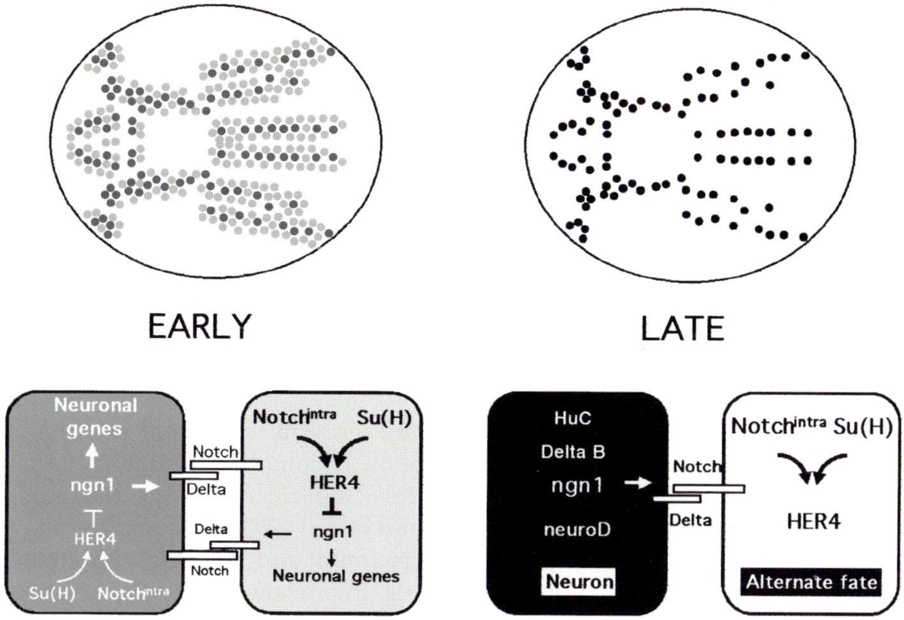

Fig. 3. Lateral inhibition mediated by Notch signaling leads to selection of cells that become neurons. *Top* Cells expressing *ngn1* within proneuronal domains of the neural plate during early and late stages of the selection process. *Bottom* Interactions between adjacent cells that mediate lateral inhibition during early and late stages of selection (see text for details). As a consequence of lateral inhibition, some cells acquire a slightly higher level of *ngn1* expression (*dark gray*) compared to their neighbors (*light gray*). Cells with slightly higher levels of *ngn1* expression become progressively better at inhibiting *ngn1* expression in neighbors and are likely to acquire high levels of *ngn1* expression through autoregulation. These cells eventually become neurons (*black*) while neighbors adopt alternate fates

gastrulation (Kim et al. 2000a). Exaggerated Wnt signaling in *hdl* mutants leads to rostral expansion of the midbrain-hindbrain boundary (MHB) domain and the adjacent neural crest and trigeminal ganglia. Rostral expansion of these tissues correlates with rostral expansion of *iro1* and *iro7*, two *Iroquois* (*Iro*) genes whose early expression defines the territory where the trigeminal ganglia, cranial crest and the MHB domain will form. Previous studies in *D. melanogaster* have shown that *Iro* genes, which encode a class of homeodomain proteins, have an early role in defining the identity of large territories where later they determine the expression proneural genes (reviewed in Cavodeassi et al. 2001). Analysis of *iro1* and *iro7* function revealed that a knockdown of *iro7* inhibits expression of *ngn1* in the trigeminal sensory neurons while a knockdown of both *iro1* and *iro7* results in the additional loss of the cranial neural crest and the MHB domain. These studies suggest that *iro7* is essential for determining expression of *ngn1* in the trigeminal ganglia and, together, *iro1* and *iro7* are essential for development of tissues that define the identity of the compartment where these *Iro* genes are expressed following Wnt signaling (Itoh et al. 2002).

It is important to note that expression of *ngn1* is only observed in restricted subdomains of *Iro* expression. This suggests that while *Iro* genes may be necessary for expression of *ngn1*, additional factors regulate its expression within *Iro* expression domains. A factor with such a role is *zic2*, a *Gli*-related gene that has been suggested as a negative regulator of *ngn1* expression in *X. laevis* (Brewster et al. 1998). Its function has not yet been studied in zebrafish.

6 Dorsal Spinal Cord Development

During neurulation, prospective epidermal ectoderm is adjacent to prospective dorsal neuroectoderm. After neurulation is completed, epidermal ectoderm overlies dorsal neural tube. Non-neural ectoderm and, later, dorsal spinal cord cells express Bmps (reviewed by Lee and Jessell 1999). As described above and elsewhere in this volume (see Hammerschmidt and Mullins; Wilson et al., this Vol.), regulation of Bmp signaling is important for nervous system formation. Modulation of Bmp signaling also might specify different neural cell fates. For example, Bmps can induce neural crest, roof plate and dorsal interneuron formation from neural explants (Liem et al. 1995, 1997), and mouse embryos homozygous for a targeted mutation of *GDF7*, a Bmp-related gene expressed in roof plate, lack a class of dorsal interneurons (Lee et al. 1998). These observations raise the possibility that epidermal expression of Bmps induces development of dorsal-most cell types, neural crest and roof plate, and subsequent dorsal spinal cord Bmp expression induces development of specific dorsal interneurons (Lee and Jessell 1999).

Mutations of zebrafish genes that function in Bmp signaling provide in vivo support for the idea that graded Bmp signaling specifies different dorsal neural fates. Mutations of *swirl* (*swr*) and *snailhouse* (*snh*), which encode Bmp2b and Bmp7, respectively, and *somitabun* (*sbn*), which encodes Smad5, a downstream component of the Bmp-signaling pathway, produce embryos that are dorsalized (Kishimoto et al. 1997; Nguyen et al. 1998b; Hild et al. 1999). The dorsalized phenotype includes expansion of dorsal, neural ectoderm and concomitant reduction of ventral, non-neural ectoderm. Some mutant alleles of these genes differ in the degree to which they affect Bmp signaling and produce strikingly different phenotypes. Embryos homozygous for *swr* mutations or the *sbn*^{sb1aub} allele, in which Bmp signaling appears to be strongly reduced, have greatly enlarged neural plates in which excess *lim*-expressing interneurons form (Barth et al. 1999; Nguyen et al. 2000; Schmid et al. 2000). They also fail to develop anterior neural crest. By contrast, embryos mutant for *sbn*^{tc24} or *snh*^{ty68a}, which apparently reduce Bmp signaling to a lesser degree, have enlarged neural crest territories (Nguyen et al. 1998b; Barth et al. 1999; Schmid et al. 2000). One possible interpretation is that the *sbn*^{tc24} and *snh*^{ty68a} mutations lower Bmp-signaling activity in ventral ectoderm to an intermediate level, converting cells that would normally develop as epidermis to neural crest (Nguyen et al. 1998b; Chitnis 1999). The stronger *swr* and *sbn*^{sb1aub} mutations lower Bmp activity below the threshold necessary for neural crest development to a level that specifies *lim1*-expressing interneurons. Importantly, injection of *din* mRNA, which encodes the Bmp inhibitor Chordin, into *swr* mutant embryos blocks development of *lim1*-expressing interneurons indicating that Bmp activity is required to specify this cell type (Nguyen et al. 2000). Thus, graded Bmp-signaling activity may coordinately establish neural and non-neural ectodermal territories and promote development of different neural cell types from the dorsal region of the neural tube.

7 Ventral Spinal Cord Development

Several lines of evidence indicate that Sonic hedgehog (Shh), produced by notochord and floor plate cells, patterns the ventral neural tube (reviewed by Tanabe and Jessell 1996). Notochord expresses Shh as neuroectoderm is specified; later, floor plate also expresses Shh. Application of Shh to chick neural explants can induce, in concentration-dependent manner, formation of different cell types, including floor plate, motor neurons and distinct ventral interneurons. Blocking antibody experiments and targeted mutation of mouse *Shh* demonstrated that Shh is necessary for specification of ventral cell types, including floor plate and motor neurons. Key effectors of Shh signaling appear to be homeodomain proteins that are expressed in subsets of ventral progenitors by their response to graded Shh activity and cross-regulatory interactions among them (reviewed by Edlund and Jessell 1999).

Zebrafish mutations that disrupt ventral spinal cord development reveal a role for Hedgehog signaling, as well as for TGFβ signaling, in floor plate development. Embryos mutant for *sonic you* (*syu*), which encodes Shh (Schauerte et al. 1998), *smo*, which encodes Smoothened, a Hedgehog co-receptor (Chen et al. 2001a; Varga et al. 2001), or *you-too* (*yot*), which encodes Gli2, a downstream factor in the Hedgehog pathway (Karlstrom et al. 1999), lack lateral floor plate cells (Odenthal et al. 2000; Chen et al. 2001a). However, they do have medial floor plate cells. The same is true for embryos with other mutations thought to disrupt Hedgehog signaling (Odenthal et al. 2000). Thus, in zebrafish, Shh appears to be necessary for lateral but not medial floor plate. One complicating issue is that zebrafish midline cells express three *hedgehog* genes: notochord and medial floor plate express *syu* (Krauss et al. 1993; Schauerrte et al. 1998), notochord expresses *echidna hedgehog* (*ehh*; Currie and Ingham 1996) and medial floor plate expresses *tiggy winkle hedgehog* (*twhh*; Ekker et al. 1995). Thus, medial floor plate might develop in *syu* mutant embryos due to partially overlapping Hedgehog-signaling functions. However, treatment of embryos with cyclopamine, which appears to block all Hedgehog signaling, did not block medial floor plate development (Chen et al. 2001a) and medial floor plate also developed in embryos co-injected with *shh* and *twhh* antisense morpholino oligonucleotides (Etheridge et al. 2001).

These observations raise the possibility that zebrafish medial and lateral floor plate are distinct cell populations, with only the latter having a requirement for Hedgehog signaling. Consistent with this possibility, overexpression of *syu* induced expression of lateral floor plate, but not medial floor plate markers (Odenthal et al. 2000). The idea that zebrafish medial and lateral floor plates are distinct is further supported by analysis of mutations that interfere with Nodal signaling. Embryos lacking zygotic functions of *cyclops* (*cyc*), which encodes a Nodal-like member of the TGFβ superfamily (Rebagliati et al. 1998b; Sampath et al. 1998), *one-eyed pinhead* (*oep*), which encodes an EGF-CFC class protein required for reception of Nodal signals (Zhang et al. 1998b; Gritsman et al. 1999), or *schmalspur* (*sur*), which encodes a FoxH1/Fast1 transcription factor important for Nodal signaling (Pogoda et al. 2000; Sirotkin et al. 2000b) lack medial, but not lateral floor plate (Hatta et al. 1991b; Brand et al. 1996a; Schier et al. 1996, 1997).

In contrast to *Shh* mutant mouse embryos, *syu* mutant zebrafish embryos have motor neurons (Schauerte et al. 1998). Midline expression of *ehh* and *twhh* might be sufficient to induce motor neuron development in the absence of *syu*. Consistent with this possibility, *smo* mutant embryos have a deficit of primary motor neurons and no secondary motor neurons, and cyclopamine-treated embryos lack most primary and secondary motor neurons (Chen et al. 2001a; Lewis and Eisen 2001; Varga et al. 2001). Additionally, mutant embryos lacking both notochord and floor plate and, thus, the sources of Hedgehog signals, develop a few anterior primary motor neurons but no secondary motor neurons (Beattie et al. 1997). These embryos transiently express *shh* during gastrulation; thus, brief exposure to Shh may induce neuroectodermal cells to develop as primary motor neurons whereas extended exposure may be required to induce secondary motor neuron development. In addition, Shh signaling in zebrafish might specify different neuronal fates by signaling through different downstream factors. Embryos mutant for *detour* (*dtr*), which may encode a Hedgehog signaling pathway component (Odenthal et al. 2000), have spinal but not cranial motor neurons (Chandrasekhar et al. 1999). Overexpression of *shh* in zebrafish normally induced formation of excess spinal and cranial motor neurons; however, in *dtr* mutant embryos *shh* overexpression induced only spinal motor neurons. Notably, *dtr* mutant cells in hindbrain could express *netrin1a* in response to *shh* overexpression. Thus, Hedgehog may signal through a bifurcated pathway in hindbrain, with one branch operating through Dtr to specify cranial motor neurons.

8 Elaboration of Cell Fate Specification by Cell–Cell Signaling

Although neural cell fate specification has not yet been thoroughly analyzed in zebrafish embryos mutant for Hedgehog or TGFβ components, the available data are consistent with the idea that signals that act at a distance from dorsal and ventral sources pattern the spinal cord. However, it is unlikely that putative dorsoventral signaling gradients are sufficient to specify even the simple pattern found in the zebrafish spinal cord. One likely source of additional patterning information is cell-cell signaling, mediated by the Delta–Notch system. As described above, Notch signaling regulates the pattern of neuronal birth from neuroectoderm. In doing so, Notch signaling also regulates cell fate specification. At 24 hpf, *dlA* mutant embryos have excess neurons and a deficit of proliferative neural precursors, suggesting that disruption of Notch signaling causes precursors to stop dividing too soon (Appel et al. 2001). Neurons that form in *dlA* or *mib* mutant embryos have identities appropriate for early-born neurons and reduced numbers of later-born neurons, neural crest progenitors and glia (Jiang et al. 1996; Schier et al. 1996; Appel et al. 1999, 2001; Cornell and Eisen 2000). For example, fate mapping showed that neural crest progenitors and RB sensory neurons arise as a mixed population of cells near the lateral boundary of the neural plate (Cornell and Eisen 2000; Kelsh and Raible, this Vol.). Both *dlA* and *mib* mutant embryos have excess RB sensory neurons and deficits of neural crest-derived cells types, indicating that Delta–Notch signaling regulates specification of RB neurons and neural crest progenitors (Jiang et al. 1996; Schier et al. 1996; Appel et al. 1999; Cornell and Eisen 2000). Fate mapping experiments also show that medial neural plate cells are a mixed population, giving rise to primary and secondary motor neurons and oligodendrocyte progenitors (Park and Appel, unpubl. obs.). *dlA* and *mib* mutant embryos have excess primary motor neurons and deficits of secondary motor neurons and oligodendrocyte progenitors (Jiang et al. 1996; Schier et al. 1996; Appel et al. 2001; Park and Appel, unpubl. observ.). Together, these observations suggest that Notch signaling may diversify neural cell fate by regulating the transition of proliferative precursors to specified neurons, glia and neural crest progenitors. One possibility is that cells that escape Notch-mediated inhibition early respond to environmental signals that specify early fates such as primary motor neurons and RB sensory neurons, whereas those that escape late respond to different signals that specify later fates such as secondary motor neurons and neural crest progenitors. In this view, Notch activity performs a permissive function in regulating the time at which a neural cell can respond to other, instructive signals. A non-mutually exclusive alternative is that Notch signaling, in some instances, may be instructive and promote formation of selected cell types, such as oligodendrocyte progenitors, as suggested by recent publications providing evidence that Notch promotes gliogenesis (Furukawa et al. 2000; Gaiano et al. 2000; Morrison et al. 2000; Scheer et al. 2001).

9 Neuronal Specification and Transcriptional Codes

Integration of the intercellular signaling systems described above appears to culminate in finely patterned expression of distinct sets of transcription factors. Although it seems obvious that cell fates are specified by particular transcription factors, the manner in which these factors promote cell differentiation is unclear. Association of cell type with gene expression is best described for expression of genes encoding LIM-class homeodomain proteins in primary motor neurons. Each of the primary motor neurons expresses *lim3* whereas MiP and RoP express *isl1* and CaP and VaP express *isl2* (Appel et al. 1995). Thus, *isl2* expression is specific to primary motor neurons located caudally within hemisegments, which project axons ventrally into the adjacent myotome. Notably, the position that a primary motor neuron occupies determines its *lim* gene expression profile and axon projection: transplantation of prospective MiP neurons to the CaP position caused the transplanted cells to express *isl2* and project axons to ventral myotome (Appel et al. 1995). The role of *isl2* in primary motor neuron development was tested by experiments designed to interfere with Isl2 expression or function (Segawa et al. 2001). CaP and VaP axons in embryos injected with *isl2* antisense morpholino oligonucleotides or mRNA encoding a LIM domain interacting protein failed to extend to the ventral myotome and, instead, usually projected caudally within the CNS. Additionally, CaP and VaP cell bodies were displaced from their normal positions and the cells sometimes expressed GABA. Caudal axon projection and GABA expression are characteristic of VeLD interneurons, raising the possibility that prospective CaP and VaP motor neurons are specified, at least partially, for VeLD fate in the absence of *isl2* function. Interestingly, RB sensory neurons, which also express *isl2*, failed to project peripheral axons but extended central axons normally in injected embryos. These observations support the idea that specific combinations of LIM-homeodomain factors influence cell fate specification and differentiation.

10 Perspectives

Analysis of early neurogenesis in the relatively simple zebrafish nervous system has revealed striking similarities between mechanisms that operate in vertebrates and *D. melanogaster* to specify neural fate in the ectoderm and to select cells for a neural fate within the neuroectoderm. Furthermore, zebrafish mutants have provided new perspectives on mechanisms that operate in vertebrates during neurogenesis and specification of neuronal fate. Comparisons with other model systems, however, reveal a number of important differences and unresolved issues. First, studies in *D. melanogaster* have focused on the role of proneural and neurogenic genes in selection of neuroblasts whereas studies in zebrafish have described the role of these genes in selection of cells that become neurons within the neural plate. Hence, what is the precise role of proneural and neurogenic genes in the formation of zebrafish neural precursors? Second, what factors establish proneural domains and how are their functions integrated with mechanisms responsible for neural induction and patterning on the lateromedial/dorsoventral axis? Third, are zebrafish neural precursors true multi-potential neural stem cells or are they restricted in the types of progeny cells they can produce and, if they are restricted, how does the restriction occur? Fourth, how is the cell cycle controlled to regulate the transition of proliferative neural precursors to specified neurons and glia? Finally, how are many different cell types specified over time? Do early-born neurons produce factors that influence the fate of nearby later-born cells? Answers to these questions will deepen our understanding of the genetic and cellular mechanisms that form functioning nervous systems and contribute to the development of therapeutic strategies to treat neural disease and injury.

Cellular, Genetic and Molecular Mechanisms of Axonal Guidance in the Zebrafish

Christine E. Beattie[1], Michael Granato[2], and John Y. Kuwada[3]

1 Introduction: Pathfinding is Precise and Cell-Specific

There is organization and pattern in the intricate network of axons that comprise the nervous system. This organization is essential to ensure the establishment of appropriate connections between neurons and their targets. Between the initiation of axonogenesis and the point where an axon reaches its target, is the process of axon guidance. In both invertebrates and vertebrates, axon guidance occurs in a stereotyped manner resulting in the formation of patterned nerves. The growing tip of the axon, the growth cone, responds to cues in the environment. These cues, both attractive and repulsive, dictate where growth cones extend and thus where nerves will form (Tessier-Lavigne and Goodman 1996).

Zebrafish offer many advantages for studying axon guidance. They have a relatively simple nervous system compared to higher vertebrates. Due to the optical clarity of the embryo, individual neurons can be analyzed in living embryos. This type of characterization has enabled a detailed analysis of how axon guidance proceeds during development. Moreover, the combination of gene analysis and genetic screens has begun to uncover the genetic pathways governing axon outgrowth.

The early zebrafish embryo contains a simple scaffold of axonal tracts that are pioneered by the growth cones of specific classes of neurons in the spinal cord (Eisen et al. 1986; Myers et al. 1986; Bernhardt et al. 1990; Kuwada et al. 1990a,b) and small clusters of neurons in the brain (Chitnis and Kuwada 1990; Wilson et al. 1990; Wilson and Easter 1991). These early growth cones follow precise and cell-specific pathways to establish this scaffold of tracts. In some cases, such as some classes of spinal interneurons, pioneer growth cones interact with a specific set of cells that act as intermediate targets (see below). In

[1] Neurobiotechnology Center/Department of Neuroscience, Ohio State University, 115 Rightmire Hall, 1060 Carmack Road, Columbus, Ohio 43210, USA

[2] Department of Cell and Developmental Biology, University of Pennsylvania School of Medicine, University of Pennsylvania, 1210 BRB II/III, 421 Curie Blvd., Philadelphia, Pennsylvania 19104, USA

[3] Department of Molecular, Cellular and Developmental Biology, University of Michigan, 830 North University Avenue, Ann Arbor, Michigan 48109-1048, USA

Results and Problems in Cell Differentiation, Vol. 40
L. Solnica-Krezel (Ed.): Pattern Formation in Zebrafish

other cases, such as the epiphysial neurons in the zebrafish brain and Rohon-Beard neurons in the Medaka fish spinal cord, pioneer growth cones extend upon the end feet of neuroepithelial cells found on the superficial surface of the CNS, and do not appear to encounter any organized set of channels or other cellular features which could potentially guide them (Kuwada 1986; Wilson and Easter 1991).

Growth cones that are projected later also follow stereotyped pathways by tracing out cell-specific routes within the early scaffold of axons in the zebrafish embryo (Chitnis and Kuwada 1990). The stereotyped pattern of axonal outgrowth seen in zebrafish embryos is similar to that found in the embryos of a wide variety of animals, including insects (Goodman et al. 1984) and mammals (O'Leary et al. 1990).

In this review, we highlight three areas of the nervous system where stereotyped axon projections have been analyzed in zebrafish: axons of hindbrain and spinal cord interneurons, spinal cord motor axons, and retinal ganglion axons. Using a wide range of experimental techniques and approaches, we are beginning to understand how these stereotyped nerve patterns are established during development thus ensuring a functional nervous system.

2 Axon Pathfinding in the Hindbrain and Spinal Cord

2.1 Redundant Cues Guide Growth Cones in the Spinal Cord

Zebrafish growth cones select specific pathways at choice points within the embryo similar to growth cones in other species. One such site is the floor plate that is found at the ventral midline in the midbrain, hindbrain, and spinal cord. At the floor plate, commissural growth cones make very precise turns (Roberts et al. 1988; Bovalenta and Dodd 1990; Yaginuma et al. 1991). In zebrafish, the growth cones of several classes of primary neurons, including CoPA and VeLD in the spinal cord and reticulospinal neurons in the hindbrain (Kimmel et al. 1982; Kuwada et al. 1990a,b), exhibit cell-specific turns near the ventral midline in the floor plate region (Fig. 1A). For example, CoPA and VeLD neurons initially project growth cones toward the ventral midline, but once near the midline their growth cones diverge. VeLD growth cones turn posteriorly without crossing the midline to pioneer the ventral longitudinal fasciculus (VLF) whereas CoPA growth cones cross the midline, turn anteriorly to follow an anterior/dorsal diagonal pathway, and eventually ascend in the dorsal longitudinal fasciculus (DLF). In embryos missing the morphologically and molecularly distinct midline floor plate cells (Kuwada et al. 1990b; Odenthal et al. 2000), either via laser ablation or due to the *cyclops* mutation (Hatta et al. 1991b), both VeLD and CoPA growth cones extended to the midline but once at the midline they followed aberrant pathways (Bernhardt et al. 1992a,b; Fig. 1B). In the case of the CoPA growth cones, 57% followed aberrant pathways in *cyclops* embryos as compared to 1% in wild-type embryos. The

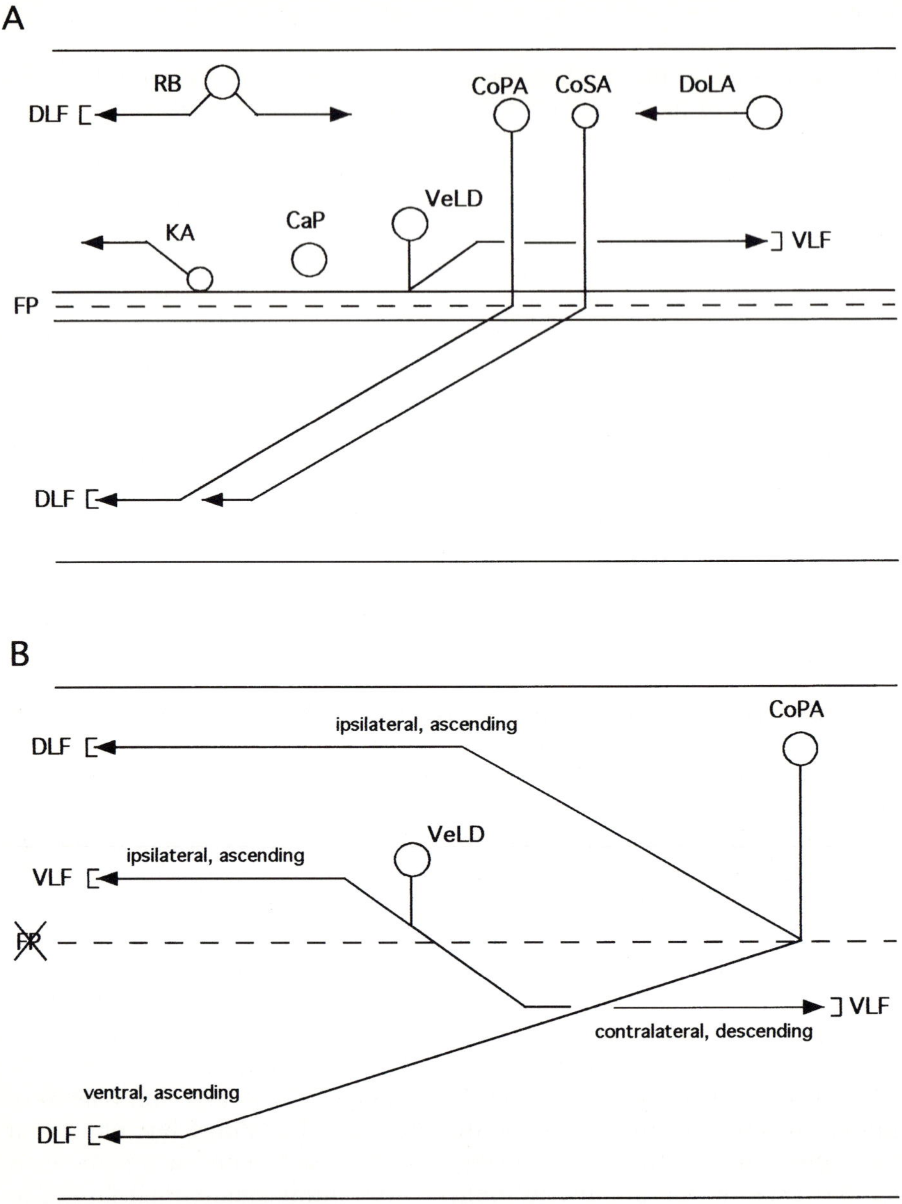

Fig. 1A, B. Schematic diagrams of the spinal cord in a zebrafish embryo shown in an "open-book configuration". In this configuration the spinal cord is shown as if it was cut longitudinally along the dorsal midline, opened up, and flattened. Anterior is to the left. **A** The axons of seven classes of neurons, the commissural primary ascending (*CoPA*), commissural secondary ascending (*CoSA*), ventral longitudinal descending (*VeLD*), Rohon-Beard (*RB*), dorsal longitudinal ascending (*DoLA*), Kolmer-Agduhr (*KA*) and caudal primary motor (*CaP*) neuron; the two early longitudinal axon tracts, the dorsal (*DLF*) and ventral (*VLF*) longitudinal fasciculus; and the midline floor plate (*FP*) cells are illustrated. **B** The aberrant trajectories followed by CoPA and VeLD axons in the absence of the FP. For each neuron the two most common aberrant trajectories are shown. The ventral, ascending CoPAs are abnormal because their axons remain in the ventral cord for a much longer distance compared with normal CoPA axons aberrant pathways taken by CoPA and VeLD growth cones included laterality and directionality errors: CoPA growth cones failed to cross the midline or extended posteriorly and VeLD growth cones crossed the midline or extended anteriorly. Furthermore, neurons such as the CoSA commissural neurons and the Kolmer-Agduhr cells that project growth cones towards or near the floor plate after the primary neurons also followed aberrant pathways in the absence of the floor plate. These effects on growth cones were cell-specific since other classes of spinal neurons such as the motor, RB, and DoLA neurons were not affected. Thus the midline floor plate cells are not necessary for growth cone extension towards the ventral midline, but are necessary for normal error-free pathfinding at the midline. Similarly, in the hindbrain, reticulospinal axons are disarrayed in the absence of the floor plate (Hatta 1992). Therefore, midline floor plate cells function as intermediate targets that provide multiple, cell-specific guidance cues.

The flip side of the experiments in the spinal cord is that a significant number of CoPA and VeLD growth cones followed their normal pathways despite the absence of the midline floor plate cells. The floor plate region in the spinal cord consists of three cells in cross section, the midline cell flanked by two lateral floor plate cells (Kuwada et al. 1990b). Thus, the variable behavior of growth cones in the absence of the midline cell may be due to a quantitative decrease in a cue normally provided by the entire floor plate region. This, however, does not appear to be the case since ablation of one of the lateral floor plate cells induces no errors by CoPA growth cones, and ablation of the midline and lateral floor plate cells in wild-type embryos and cells near the ventral midline in *cyclops* embryos did not prevent CoPA growth cones from extending to the midline along their normal route (Bernhardt et al. 1992a). Furthermore, once at the midline their error rate was not higher than that seen in unmanipulated *cyclops* embryos. Thus, the floor plate does not appear to be the sole source of pathfinding cues at the ventral midline. Redundancy of pathfinding cues is also likely to be found in the zebrafish brain, since elimination of cues increases error rates but a significant number of axons are still normal (Chitnis and Kuwada 1991; Chitnis et al. 1992).

What might be a second source of pathfinding cues near the ventral midline? One likely candidate is the notochord. CoPA growth cones cross the midline by inserting between the ventral surface of the floor plate cells and the basal lamina that surrounds the spinal cord, and the notochord is located just ventral to the spinal cord and contacts the cord during the entire period growth cones are approaching and crossing the midline (Kuwada et al. 1990a). Potentially, the notochord could provide a secreted signal that guides growth cones at the midline. If the notochord is another source of pathfinding cues, then elimination of the notochord would be expected to induce errors in pathfinding and elimination of both the notochord and floor plate would perhaps induce a higher rate of errors than the elimination of the floor plate or notochord alone. This is, in fact, the case. In the absence of the notochord, either due to laser ablation of the notochord in wild-type embryos or to the *no tail* mutation, the error rate for CoPA increased to 26 and 38%, respectively, compared with 1% in control embryos (Greenspoon et al. 1995). In the absence of both the floor plate and notochord due to laser ablation of the notochord in *cyclops* embryos the error rate increased to 73% compared with 57% in unmanipulated *cyclops* embryos. Thus, both the floor plate and notochord provide pathfinding cues at the ventral midline.

2.2 Molecules That Guide Spinal and Hindbrain Growth Cones

Not only is the floor plate the site of turns made by growth cones, it also emits the chemoattractant, Netrin, that attracts commissural growth cones to the floor plate (Kennedy et al. 1994; Serafini et al. 1994, 1996). Netrins also guide commissural axons to the midline of the CNS in *Drosophila* (Harris et al. 1996; Mitchell et al. 1996) and specific axons towards and away from the ventral cord in *C. elegans* (Hedgecock et al. 1990; Ishii et al. 1992). In chicks, there are two Netrins, Netrin1 and Netrin2, expressed in the spinal cord with *netrin1* highly expressed by the floor plate and *netrin2* expressed at lower levels and uniformly by the spinal cord ventral to the commissural neurons with the exception of the floor plate. In mice, only *netrin1* is expressed in the spinal cord and it is limited to the floor plate. Despite this difference, it is likely that Netrins are distributed as a gradient along the dorsal/ventral axis with highest levels at the floor plate in both chicks and mice. In zebrafish, there are two Netrins that are highly and equally homologous to Netrin1 and these have been called Netrin1a and Netrin1b (Lauderdale et al. 1998). Zebrafish Netrin1b (published as Netrin1; Strähle et al. 1997) is 86 and 72% identical with chick Netrin1 and Netrin2, respectively, while Netrin1a (Lauderdale et al. 1997) is 89 and 75% identical to chick Netrin1 and Netrin2, respectively. The expression of *netrins* in the zebrafish spinal cord is similar to *netrins* in the chick. *netrin1b* is expressed by the floor plate while *netrin1a* is expressed by the spinal cord ventral to the DLF including the floor plate. Thus, it is likely that Netrins may be distributed as a gradient in zebrafish as well. At present there is no direct evidence that Netrins attract commissural growth cones to the ventral midline in zebrafish, but when the expression of *netrin1a* is perturbed the behavior of commissural axons is correlatively perturbed. For example, in *floating head* embryos, there are patches of floor plate cells that express *netrin1a* and commissural axons extend towards the nearest patch suggesting that commissural axons are attracted by the Netrin patches (Lauderdale et al. 1997).

If Netrin is required for proper extension of commissural axons to the ventral midline in mice (Serafini et al. 1996) and Netrins can attract commissural axons in zebrafish (Lauderdale et al. 1997), why is the extension of commissural axons to the ventral midline not perturbed in the absence of the floor plate (Bernhardt et al. 1992a,b)? Since *netrin1a* is still expressed in the ventral spinal cord in *cyclops* embryos (Lauderdale et al. 1998), and presumably following laser ablation of the floor plate in wild-type embryos, Netrin1a may be sufficient to guide commissural growth cones to the ventral midline. Since *netrin1a* is expressed uniformly through the ventral spinal cord, Netrin1a is not likely distributed as a gradient in the ventral cord. This suggests that it may be acting more to stimulate axon outgrowth. If this is correct, then other cues must be able to provide directional information in the absence of a Netrin gradient.

What might such cues be? Secreted Bone Morphogenetic Proteins (BMPs) may repulse commissural growth cones away from the dorsal cord (Augsburger et al. 1999). BMPs are expressed by the dorsal spinal cord, and in vitro chick commissural axons are repulsed by explants of the roof plate and COS cells expressing *BMP4* or *BMP7*. There has been no examination yet of the role of BMPs for guidance of commissural axons in zebrafish, but *BMP4* is expressed in the dorsal midline of the tail and *BMP7* by the dorsal neural tube in the tail (Martinez-Barberá et al. 1997; Schmid et al. 2000), leaving open the possibility that they may have a pathfinding role. Thus, it is possible that two gradients, one for the repulsive BMPs with highest levels at the dorsal midline and one for the attractive Netrins with highest levels at the ventral midline, may act coordinately to guide commissural growth cones in a ventral direction in a variety of embryos including the zebrafish.

Another set of molecules that may participate in guiding commissural and other axons in the zebrafish spinal cord are the semaphorins and their neuropilin/plexin receptors. The semaphorins constitute a large, conserved family of molecules of which several members are known to repel specific growth cones and act as guidance molecules for others (Raper 2000). These proteins are either secreted, GPI-linked or transmembrane, and all share a conserved 500 amino acid sema domain. In zebrafish, Sema3A1 and Sema3A2 (formerly SemaZ1a and SemaZ1b), two secreted semaphorins highly homologous to Sema3A in chicks and mammals, repulse spinal motor axons and Sema3A1 repulses the posterior lateral line axons (Shoji et al. 1998; Roos et al. 1999; Yee et al. 1999; Halloran et al. 2000). It is known that two other semaphorins are expressed in the zebrafish spinal cord. Sema3D (formerly SemaZ8), a secreted semaphorin, is expressed by the roof plate cells that occupy the dorsal midline (Halloran et al. 1999) and Sema4E (formerly SemaZ7), a transmembrane semaphorin, in two domains; cells dorsal to the DLF with the exception of the roof plate and a patchy longitudinal stripe of cells just lateral to the midline floor plate cell (Halloran et al. 1998). There is no direct evidence for a guidance role for these semaphorins, but their expression pattern suggests they may regulate pathfinding by a variety of axons in the zebrafish. The roof plate expression of Sema3D suggests that a gradient of Sema3D may act to guide commissural axons away from the dorsal cord, i.e., towards the ventral midline, and so could be working with Netrins and perhaps BMPs to guide commissural axons. Consistent with this hypothesis, changes in expression pattern of *sema3D* in several mutants correlate with changes in pathfinding by several classes of spinal axons. For example, the roof plate expression of *sema3D* is absent in *white tail* mutants and many commissural axons extended longitudinally without first extending ventrally (Kuwada, unpubl. data). Furthermore, there is inappropriate dorsal midline crossing by the peripheral axons of RB neurons in the dorsal epidermis overlying the spinal cord of *white tail* mutants (Halloran et al. 1999). These findings are consistent with a repulsive action of Sema3D on commissural and RB peripheral axons.

Slits and their Robo receptors also play important roles in guiding commissural growth cones near the midline. Slit is a secreted protein that is expressed by cells at the midline of the CNS in both invertebrates and vertebrates and is critical for Robo-bearing commissural growth cones to extend beyond the midline and to prevent them from recrossing the midline in *Drosophila* and *C. elegans* (Brose and Tessier-Lavigne 2000), help direct commissural growth cones to their correct longitudinal pathways in *Drosophila* (Rajagopalan et al. 2000; Simpson et al. 2000), and directly regulate responsiveness of commissural growth cones to Netrin in vertebrates (Stein and Tessier-Lavigne 2001). Slits and Robos have also been described in zebrafish. *slit2* and *slit3* are expressed by the floor plate when growth cones are extending and so may participate in guiding commissural growth cones (Yeo et al. 2001). *robo3* is expressed by cells likely to be commissural and motor neurons in the early spinal cord and *robo1* is expressed by dorsal cells likely to be spinal commissural neurons in later embryos (Challa et al. 2001; Lee et al. 2001). Thus, the expression patterns of *slit* and *robo* genes suggest that these genes act to guide growth cones in the zebrafish spinal cord as well as in other organisms. In fact, overexpression of *slit2* severely affects pathfinding by reticulospinal axons such as the Mauthner axons (H. Okamoto, pers. comm.) demonstrating that Slits regulate zebrafish commissural growth cones. Robos also clearly affect growth cones in zebrafish since mutations in *robo2/astray* severely affect the pathways followed by retinal axons (see below; Fricke et al. 2001). Slit can also affect migratory behaviors since overexpression of *slit2*, which is also expressed during mid-gastrula stages, disrupted convergent extension of the mesoderm (Yeo et al. 2001).

Several other molecules also guide spinal axons in a variety of organisms, but their pathfinding roles in zebrafish are not yet known. Ephrins and their Eph receptors play important roles in a variety of processes including guiding growth cones especially of retinal axons (Wilkinson 2001). Ephrins and Eph receptors are also expressed in the spinal cord and both in vitro and in vivo experiments show that they also guide growth cones in the spinal cord (Yue et al. 1999; Imondi et al. 2000; Kullander et al. 2001; Yokoyama et al. 2001). Several ephrins have been cloned in zebrafish and their role in the formation of rhombomeres and somites has been examined (Mellitzer et al. 1999; Xu et al. 1999b; Durbin et al. 2000; Cooke et al. 2001); however, so far there has been little investigation of any pathfinding functions.

Another family of molecules that is likely to guide spinal growth cones is the mindin/F-spondin family of extracellular matrix proteins. These molecules contain thrombospondin type I repeats, are expressed by the floor plate, can promote neurite outgrowth by commissural axons and repulse motor axons, and are required for correct pathfinding by commissural axons in the spinal cord (Klar et al. 1992; Burstyn-Cohen et al. 1999; Tzarfati-Majar et al. 1999). Several zebrafish members of this family have been cloned and as in other vertebrates are expressed by the floor plate (Higashijima et al. 1997a), but little evidence for any pathfinding role is known in zebrafish. Immunoglobulin superfamily members, besides Robo, also participate in regulating axonal outgrowth in a variety of different organisms (Walsh and Doherty 1997). In the zebrafish spinal cord and hindbrain, Tag1 and DM-GRASP/neurolin, two superfamily members, are expressed by secondary motoneurons and branchiomotor neurons, respectively, when these neurons are extending growth cones and DM-GRASP is expressed by hindbrain commissural neurons that extend axons after the reticulospinal interneurons (Kanki et al. 1994; Fashena and Westerfield 1999; Warren et al. 1999). Antibody block experiments suggest that DM-GRASP/neurolin acts to guide secondary motor axons (Ott et al. 2001; see next section) suggesting that DM-GRASP/neurolin may also guide interneurons. Furthermore, three NCAM-related molecules are also specifically expressed within the nervous system (Mizuno et al. 2001) and are likely to play a role in guiding growth cones.

2.3 Mutations That Affect the Development of Neural Circuits in the Hindbrain and Spinal Cord

Genetic screens to identify genes that disrupt the development of normal behavior have uncovered a number of mutations that affect spontaneous movements, swimming, and the escape response of zebrafish (Granato et al. 1996). Since the neural circuits mediating these behaviors reside within the hindbrain and spinal cord (Eaton et al. 1977; O'Malley et al. 1996; Saint-Amant and Drapeau 1998, 2001), it is likely that a number of these mutants have defects in hindbrain and spinal interneurons. Furthermore, some of the mutants have defects in the outgrowth of motor axons (see next section) suggesting that others may involve defects in outgrowth of axons of hindbrain and spinal interneurons. The analysis of *space cadet* mutants (Lorent et al. 2001) illustrates the potential usefulness of this approach.

space cadet mutants fail to initiate the normal escape response sequence and this is likely due to defects in spiral fiber neurons, a specific population of commissural neurons in the hindbrain (Lorent et al. 2001). In *space cadet* mutants most commissural trajectories in the forebrain, midbrain, hindbrain and spinal cord appear normal, while only a small set of commissural neurons, including retinal ganglion cells and spiral fiber neurons, display what are likely pathfinding defects. Thus, the *space cadet* gene shows remarkable specificity and suggests that other genes may exist that may exhibit similar levels of specificity.

3 Axonal Pathfinding by Spinal Motoneurons

3.1 Zebrafish Motor Axons Follow a Common Pathway and Then Make Divergent Choices

Spinal motor axons display overt patterning, as evidenced by antibody labeling at the end of the first day of development (Fig. 2). Between 18–26 h postfertilization (hpf), three primary motoneurons, CaP, MiP and RoP, innervate each myotome (approximately 30 by 24 hpf). Identified by their cell body positions within the spinal cord, CaP, the Caudal Primary motoneuron, extends its growth cone along the ventromedial myotome. MiP, the Middle Primary motoneuron, extends its growth cone along the dorsomedial myotome and RoP, the Rostral Primary motoneuron, extends it growth cone laterally within the region that will give rise to the horizontal myoseptum (Eisen et al. 1986; Myers et al. 1986). There is a fourth primary motoneuron, VaP for Variable Primary, that is present in approximately 50% of the spinal cord hemisegments (Eisen et al. 1990). This neuron usually dies by 36 hpf. During the second day of development, primary motor axons are joined by secondary motor axons (Myers 1985); thus, both the primary and secondary motoneurons comprise the dorsal, intermediate, and ventral nerves.

Detailed analyses of individually labeled primary motoneurons reveal that their axons extend in a highly stereotyped manner. All four primary motor

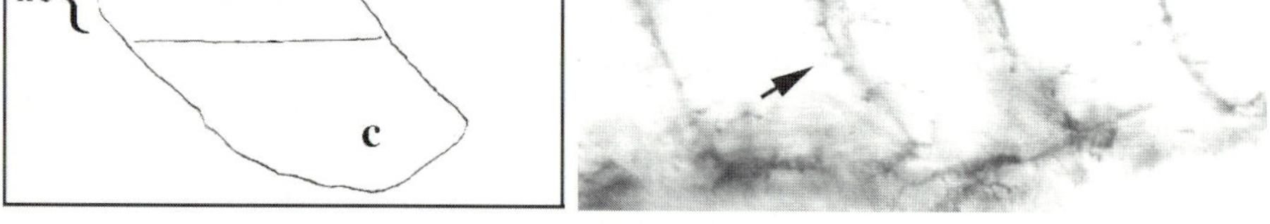

Fig. 2. **A** Schematic diagram of a zebrafish muscle hemisegment at approximately 26 hpf. The cell bodies of CaP (*orange*), MiP (*blue*), and RoP (*green*) reside in the spinal cord (*gray shaded area*). After exiting the spinal cord at the same ventral root, all three axons extend along the common pathway (*a*), a region of the myotome adjacent to the notochord (*nc*). At the nascent horizontal myoseptum (*dashed line*), the motor axons diverge. MiP axons branch and extend along the dorsal myotome (*b*), CaP axons continue to extend ventrally (*c*), and RoP axons branch and extend in the medial myotome territory (*d*). **B** Lateral view of a 48-hpf embryo in whole mount processed for antibody labeling to reveal motor axons. At this stage of development, both primary and secondary motor axons contribute to the dorsal (*white arrow*), ventral (*black arrow*) and medial (not visible in this plane of focus) nerves

axons leave the spinal cord at the same ventral root and extend ventrally along the myotome region, referred to as the common pathway. The CaP axon extends out of spinal cord first, followed by the MiP axon, then finally the RoP axon (Eisen et al. 1986; Myers et al. 1986). At the horizontal myoseptum, their growth cones pause then diverge to extend along their specific pathway regions. For example, CaP continues to extend its axon ventrally whereas the MiP axon branches and extends dorsally. Secondary motor axons follow these same pathways, although less is know about the details of individual secondary motor axon outgrowth. To understand how this pattern of motor axons and nerves is established during development, it is necessary to determine what dictates these stereotyped axon guidance events.

3.2 Molecules That Guide Motor Growth Cones

It has been recognized for some time now that axon guidance is controlled by interactions between the growth cone and the environment. The last decade has seen a dramatic increase in the characterization of molecules functioning in this process (Tessier-Lavigne and Goodman 1996). To determine which molecules also function in zebrafish motor axon guidance, a number of orthologs have been cloned in zebrafish by homology to genes from invertebrates, chick and mouse. Some of these genes are expressed in patterns consistent with a role in motor axon guidance (Bernhardt 1999) and direct testing to determine function has been initiated.

3.2.1 Semaphorins

Two semaphorins, Sema3A1 (formerly SemaZ1a) and Sema3A2 (formerly SemaZ1b), have been directly shown to function in zebrafish motor axon guidance (Roos et al. 1999; Halloran et al. 2000). Both *sema3A1* and *sema3A2* are highly homologous to *sema3A*, which possesses growth cone collapsing activity (Raper and Kapfhammer 1990; Kolodkin et al. 1993; Luo et al. 1993), indicating that they could potentially function as repulsive molecules in zebrafish axon guidance.

sema3A1 is expressed in the dorsal and ventral regions of the developing myotome but not in the nascent horizontal myoseptum, an area between the dorsal and ventral muscle that will eventually form a space separating the two muscle regions (Yee et al. 1999). The primary axons initially extend along the common pathway in the region of the nascent horizontal myoseptum, an area devoid of *sema3A1* expression. When *sema3A1* was specifically misexpressed in a small number of cells in the horizontal myoseptal area, CaP axons stalled upon leaving the spinal cord (Halloran et al. 2000). This data suggests that *sema3A1* is repulsive for CaP motor axons. However, the role of endogenous *sema3A1* is less clear since motor axons leave the horizontal myoseptal region and immediately extend along myotome regions that express *sema3A1*. One intriguing possibility is that interactions with the horizontal myoseptal region may normally change the responsiveness to Sema3A1 expressed by the ventral myotome. In fact, genetic analysis demonstrates that interactions between motor growth cones and cells along the common pathway are necessary for normal pathfinding in this region (see below).

sema3A2 is transiently expressed in the posterior myotome, but is downregulated as CaP axons extend past the nascent horizontal myoseptum (Bernhardt et al. 1998). Since motor axons extend along the anterior half of myotomes, it is possible that *sema3A2* in the posterior half of the myotome is having an inhibitory effect on ventrally extending CaP axons. To test this hypothesis, *sema3A2* was overexpressed throughout zebrafish embryos and consequently throughout the entire myotome. In embryos overexpressing *sema3A2*, CaP axons were missing or severely truncated whereas other axons in the embryo were unaffected (Roos et al. 1999). These results are consistent with the idea that CaP axons respect anteroposterior differences in the myotome and that *sema3A2* may act to keep CaP axons, and presumably later developing secondary motor axons that extend in the ventral nerve, from extending into the posterior myotome. The idea that anterior/posterior myotome patterning is important for the formation of stereotyped zebrafish motor nerves is supported by the finding that mutants in which anterior/posterior patterning is disrupted have severely disorganized primary and secondary motor axons (Gray et al. 2001).

3.2.2 GDNF

Glial-line derived neurotrophic factor (GDNF) is expressed in the ventral myotome whereas one of its receptors, GFR-β1, is expressed in CaP motoneurons, suggesting a role for these molecules in axon guidance (Shepherd et al. 2001). To test this idea, *GDNF* levels in the myotome were manipulated and the effect on CaP axons was observed. Overexpressing *GDNF* RNA throughout the muscle did not influence CaP axonal projections. However, discontinuous or focal patches of misexpression in muscle did cause pathfinding errors in CaP axons. Axons branched aberrantly toward patches of GDNF-expressing muscle cells and, in the more severe cases, failed to extend beyond these cells (Shepherd et al. 2001). Interestingly, decreasing GDNF levels via morpholino injections (Nasevicius and Ekker 2000) had no affect on CaP axons (Shepherd et al. 2001). Thus, the absolute concentration of GDNF may not be important or there may be redundant cues in the ventral myotome. These findings suggest that GDNF may be an attractive cue that directs CaP axons to the ventral myotome.

3.2.3 Neurolin

Zebrafish Neurolin (DM-GRASP), a transmembrane protein with five extracellular immunoglobulin domains (Kanki et al. 1994; Laessing et al. 1994), is expressed on secondary, but not primary, motor axons (Fashena and Westerfield 1999). Proteins containing immunoglobulin domains are known to mediate homophilic or heterophilic interactions and many proteins that function in axon guidance contain these motifs (Tessier-Lavigne and Goodman 1996). To determine whether Neurolin functions in zebrafish motor axon guidance, antibody-blocking experiments were performed (Ott et al. 2001). Secondary motor axons extend along the same pathway regions as primary motor axons, but do so 6–7 h later (Pike et al. 1992). When antibodies against Neurolin (Neurolin Fab) were injected into zebrafish embryos during the second day of development, motor nerves along the common pathway were excessively branched and formation of the ventral nerve was temporarily delayed. Furthermore, when Neurolin Fab injections were performed three or four times between 28–60 h to ensure blocking function at later time points, the rostral and dorsal nerves were reduced or failed to form in approximately 18% of treated embryos. In those embryos where the dorsal and rostral nerves formed, they had pathfinding errors. These studies show that Neurolin is involved in the both the growth and guidance of secondary motor axons.

3.3 Mutations That Disrupt the Formation of Stereotyped Motor Projections

Two genetic screens were designed to uncover genes that control the formation of stereotyped axonal projections in zebrafish (Granato et al. 1996; Beattie et al. 1999). Three mutants identified in these screens have been characterized (Zeller and Granato 1999; Beattie et al. 2000; Zhang and Granato 2000) and have revealed novel aspects of zebrafish axon guidance. As many of the molecules and mechanisms of axon guidance are conserved across species (Chisholm and Tessier-Lavigne 1999), these findings may relate to motor axon guidance in general.

3.3.1 Diwanka Mutants

The proper formation of neuromuscular connections is essential for coordinated movements. Mutations that disrupt the formation of these connections may result in movement defects. *diwanka* mutants were recovered in a screen for mutations that disrupted movement during the first three days of development (Granato et al. 1996). Detailed analysis of individual motor axons in mutant embryos revealed that CaP, MiP, and RoP axons failed to extend along the common pathway. Genetic mosaics were created to elucidate what cell(s) needed *diwanka* function to promote normal axon outgrowth. When genotypically wild-type cells populated the lateral-most region of the myotome, mutant axonal projections were rescued. The lateralmost cells in the myotome originate from adaxial cells located adjacent to the notochord (Devoto et al. 1996). As development proceeds, these cells migrate through the myotome ending up as the lateral-most myotome cell type. Using an adaxial cell marker to analyze genetic mosaics, a strong correlation was found between the presence of wild-type adaxial cells and rescue of *diwanka* axons. Analysis of genetic mosaics revealed that when one to three wild-type adaxial cells were present in the region of the common pathway between the ventral edge of the spinal cord and the nascent horizontal myoseptum, mutant motor axons were rescued. Thus, before they migrate laterally, a subset of adaxial cells function to delineate the common pathway in a *diwanka*-dependent manner. These data stress the importance of the common pathway in establishing stereotyped axonal projections.

3.3.2 Unplugged Mutants

Mutations in *unplugged* also resulted in movement defects (Granato et al. 1996). Antibody labeling of *unplugged* mutants revealed that CaP axons stalled at the horizontal myoseptal choice point and often formed short, aberrant, branches whereas mutant RoP axons did not remain at the choice point as in wild types, but formed ectopic branches or invaded the ventral myotome (Zhang and Granato 2000). These observations suggest that *unplugged* functions to enable CaP and RoP axons to select appropriate cell-specific pathways. Genetic mosaic analysis showed that *unplugged* was not required in CaP and RoP, but was needed in surrounding cells (Zhang and Granato 2000). Similar to *diwanka*, *unplugged* gene function was needed in a population of adaxial cells to rescue the mutant motor axons. When wild-type cells transplanted into *unplugged* mutants contributed to three or more adaxial cells located dorsally just adjacent to the muscle pioneers at the horizontal myoseptum, the *unplugged* phenotype was effectively rescued. Location of transplanted adaxial cells in more dorsal or ventral positions, or the presence of less than three wild-type cells in the correct region, failed to rescue mutant motor axons. Therefore, like Diwanka, Unplugged functions in adaxial cells to establish signals necessary for proper motor axon guidance. However, in the absence of Unplugged function, primary motor axons extend along the common pathway, but, upon reaching the choice point, CaP and RoP growth cones fail to choose their appropriate pathways.

3.3.3 Stumpy Mutants

stumpy mutants were identified in a screen that used immunohistochemistry to reveal defects in the formation of stereotyped primary motor axon projections (Beattie et al. 1999). During the first day of development, *stumpy* mutant axons remained stopped at the horizontal myoseptum and failed to extend along the ventral myotome (Beattie et al. 2000). During the next two days of development, most CaP axons in *stumpy* mutants had progressed beyond the horizontal myoseptum, but stopped at a place along the myotome adjacent to the ventral edge of the notochord; a region where wild-type CaP axons often formed a transient branch (Eisen et al. 1986; Myers et al. 1986). A detailed analysis of MiP and RoP axons also revealed truncated axons and a failure of axons to branch and turn along their pathways. Later developing secondary motor axons also exhibited a similar phenotype with axons being delayed at regions along their pathway where they normally, paused, branched or turned. Taken together, these data suggest that motor axons in *stumpy* mutants are stalling or pausing for abnormal periods of time at previously unrecognized intermediate targets. Intermediate targets are regions along a pathway where axons pause, branch or turn and are thought to be places where information is being imparted to the growth cone. To determine where *stumpy* function is needed, genetic mosaics were created by transplanting single CaP motoneurons between wild types and mutants (Beattie et al. 2000). Surprisingly, it was found that *stumpy* was required in both CaP and in surrounding cells suggesting that *stumpy* is needed both in motoneurons and in the environment. Thus, Stumpy function is essential for motor axon growth cones to leave intermediate targets and extend into distal muscle regions. Without this gene function, motor nerve formation is severely compromised.

4 Axonal Pathfinding in the Visual System

Visual perception depends on the initial formation of precise and stereotypic neuronal connections between the eye and the brain. Such connections are established during embryogenesis by retinal ganglion cell (RGC) axons. During their journey from the eye to the brain, RGC growth cones navigate a considerable distance (~500 μM) through varying territories. Once they enter their target area, the contralateral tectal lobe, growth cones organize topographically such that neighboring RGCs synapse on neighboring target sites, thereby forming a topological image from the retina onto the optic tectum. In this section we will briefly summarize the well-documented retinotectal pathway, and then review some recent genetic approaches to understand the processes of RGC axonal pathfinding and topographic mapping.

The first retinal ganglion cells differentiate around 28 hpf (Nawrocki 1985), and, around 31 hpf, ganglion cells in the ventronasal cluster extend the first axons to form the optic tract (Burrill and Easter 1995). These pioneering RGC growth cones are joined by progressively more axons, and at about 2 days postfertilization (dpf), the optic nerve contains about 1800 axons. Within 20 h after they start extending, RGC growth cones reach their main target, the optic tectum. In addition to the optic tectum, RGC axons innervate nine additional, much smaller, targets. Besides RGC axons the optic nerve also contains retinopetal projections, which appear only after 5 dpf and originate in the region of the rostral telencephalon (Burrill and Easter 1994).

The journey from the eye to the optic tectum is long and many pathfinding decisions await navigating RGC growth cones. A first challenge for RGC growth cones is to navigate along the inner surface of the retina towards the optic nerve head. Once they exit the eye at around 34 hpf (Stuermer 1988), RGC growth cones have to reorient themselves towards the midline, along the dorsoventral axis of the diencephalon. They reach the presumptive optic chiasma at about 32 hpf, where they must cross the midline and project up the contralateral diencephalic wall. Interestingly, RGC growth cones do not extend on preexisting axonal tracts, but pioneer a path adjacent to the tract of the postoptic commissure (Burrill and Easter 1995), suggesting that RGC growth cones might utilize their own set of guidance cues. After crossing the midline, RGC axons from the dorsal and ventral areas of the retina separate into a dorsal and a ventral branch. Both branches extend towards the tectum, and, around 46 hpf, RGC growth cones enter the optic tectum (Stuermer 1988; Burrill and Easter 1994). Here, RGC growth cones are confronted with a last, perhaps most challenging, task: to reproduce their spatial relationship in the retina in the tectum. RGCs originating in the ventral-nasal part of the retina innervate the dorsal-posterior part of the optic tectum, and RGCs in the dorsal-temporal part of the retina innervate the ventral-anterior part of the optic tectum (Stuermer 1988; Baier et al. 1996; Karlstrom et al. 1996; Trowe et al. 1996). For the most part, formation of this precise pattern does not depend on visual input or other forms of electrical activity, suggesting a molecular principle between ingrowing RGC axons and their targets ("chemoaffinity" theory, Sperry 1963).

The formation of the retinotectal projection has long fascinated biologists and has been an intense research area. In recent years, a key question has been to identify the genetic factor controlling RGC guidance and topographic mapping. In 1996, three publications described the results of a systematic genetic screen aimed at isolating zebrafish genes essential for the retinotectal projection (Baier et al. 1996; Karlstrom et al. 1996; Trowe et al. 1996). In this study, the methodology employed to visualize retinotectal trajectories was simple as well as elegant: injection of two different fluorescent tracer dyes into opposite poles of the retina labeled two topographically distinct subsets of RGC axons along their entire length, allowing one to score for proper axonal pathfinding to the tectum *and* for retinotopic mapping (Baier et al. 1996). Injection and analysis of ~125,000 F3 larvae derived from ENU-mutagenized males revealed 114 mutants of at least 30 genes (Baier et al. 1996). The mutants can be subdivided into several groups, depending on which step of the retinotectal pathway they affect. In general, mutations that disrupt pathfinding to the tectum did not affect retinotopic mapping, and vice versa. This suggests that these processes are controlled by different genes. Mutations in 15 genes affected pathfinding, mutants of four genes displayed aberrant fiber sorting in the optic nerve, mutations in four genes affected anterior-posterior or dorsal-ventral retinotopic mapping, mutants of six genes showed defective arbor formation, and mutations in at least three genes affected tectum size. Interestingly, of the 30 genes, only nine genes appeared to be dedicated to establish exclusively the retinotectal projection; mutants of all other genes displayed additional phenotypes, underscoring the notion that many genes play a central role for multiple developmental processes. Together, this unique mutant collection provides an exceptional tool to decipher the genetic, molecular and cellular mechanisms controlling the formation of the retinotectal projection.

What do the mutants tell us about the retinotectal projection? For several of these mutants phenotypic analysis and cloning is currently underway. Here, we will focus on two examples illustrating what has been learned about how RGC growth cones interact with the environment as they extend towards the tectum. We will also briefly mention the potential of the mutants in examining less well-understood processes, such as fiber sorting within the optic nerve and establishment of the dorsoventral retinotopic polarity.

One observation from published studies is that specialized cells along the retinotectal path play an important role in guiding RGC axons. Navigation of commissural growth cones towards and across the CNS midline depends on a complex mixture of signals emanating from specialized midline cells (reviewed in Terman and Kolodkin 1999; Van Vactor and Flanagan 1999). In mutants of four genes, RGC axons, instead of innervating exclusively the contralateral tectum, innervate the ipsilateral tectum (class 1 mutants, see Karlstrom et al. 1996), consistent with the idea that in these mutants axonal guidance at the midline might be impaired. Interestingly, in three of these mutants, the specialized midline floor plate cells failed to develop properly (Brand et al. 1996a). Moreover, one of the mutants, *you-too*, is caused by a mutation in the *Gli2* gene (Karlstrom et al. 1999), a component of the Hedgehog-signaling pathway, known to control floor plate development (Odenthal et al. 2000; Chen et al. 2001a; Varga et al. 2001). Thus, in some of the ipsilaterally projecting mutants, lack of proper patterning or specification of midline cells rather than aberrant RGC guidance might be the primary defect.

At the same time that cells along axonal trajectories provide guidance cues, approaching growth cones require specific receptors to read these signs. In *astray* mutant embryos, RGC growth cones, once they exit the eye, exhibit a variety of pathfinding defects, predominantly at the midline, where they cross and recross (Karlstrom et al. 1996; Fricke et al. 2001). RGC axons also extend aberrantly into anterior and posterior brain regions. Elegant eye transplants between wild-type and *astray* mutant embryos revealed that *astray* gene activity is not required in the brain, but provides its function in the eye, most likely in RGC axons (Fricke et al. 2001). Molecular-genetic mapping placed *astray* on LG 15, in the proximity of *robo2*. Zebrafish *robo2* is a homologue of the *Drosophila roundabout* and the *C. elegans sax-3* genes, evolutionarily conserved receptors of the immunoglobulin (Ig) domain superfamily (Kidd et al. 1998; Zallen et al. 1998). Mutant phenotypes of both of these genes share similarities with *astray*: sax-3 mutants show abnormal axon crossing of the ventral nerve cord, anterior and posterior projection defects, while in *roundabout* mutants axons recross the midline (Kidd et al. 1998; Zallen et al. 1998). The similarity in phenotypes, the requirement of *astray* within the eye, and the fact that zebrafish *robo2* maps in the region of *astray* and is expressed in RGCs (Lee et al. 2001) make *robo2* a likely candidate for the gene mutated at the *astray* locus. Sequence analysis of two mutant *astray* alleles demonstrated that the *astray* phenotype is caused by mutations in the *robo2* gene. Thus, *astray/robo2* is essential for multiple guidance decisions during the formation of the vertebrate visual projection.

After crossing the midline, RGC axons in the optic nerve segregate into the ventral and dorsal branch of the optic tract, respectively. Axons originating from the nasodorsal part of the retina select the ventral branch, and upon entering the optic tectum, continue on a ventral trajectory and terminate in the posteroventral tectum. Vice versa, temporoventral axons choose the dorsal branch and terminate on the anterodorsal tectum (Baier et al. 1996; Trowe et al. 1996). There is growing understanding of the molecular mechanisms underlying topographic mapping along the anterior-posterior axis (O'Leary and Wilkinson 1999; Brown et al. 2000a). In contrast, little is known about the molecules and mechanisms that control sorting of RGC axons into a dorsal and ventral branch and the subsequent mapping along the dorsoventral axis. Fourteen mutations in four genes (*boxer*, *dackel*, *pinscher*, *nevermind*) affect the sorting of axons in the optic nerve (Trowe et al. 1996). In mutants of all these genes, dorsal axons, instead of selecting the ventral branch, extend through the ventral and the dorsal branch. In three of these mutants (*boxer*, *dackel*, *pinscher*), misrouted axons innervate their retinotopically correct target area. In mutants of *nevermind*, dorsal axons extend through the dorsal nerve, but innervate an enlarged target area, reaching from their correct posteroventral to ectopic posterodorsal targets (Trowe et al. 1996). A similar distribution of nasodorsal axons terminating aberrantly along the dorsoventral tectal axis is observed in *who-cares* mutants (Trowe et al. 1996). Together these phenotypes provide compelling evidence that, while sorting of axons within the axonal tract and topographic mapping along the dorsoventral tectal axis are two independent processes, they share some common genes and mechanisms. Thus, mutants of these five genes provide an outstanding opportunity to dissect and understand the molecular and cellular events underlying the establishment of dorsoventral topography in the vertebrate visual system.

5 Conclusions

Genetic analysis of axon pathfinding (Baier et al. 1996; Granato et al. 1996; Karlstrom et al. 1996; Trowe et al. 1996; Beattie et al. 1999, 2000; Zeller and Granato 1999; Zhang and Granato 2000); powerful, inducible and cell-specific expression of transgenes in transgenic zebrafish (Higashijima et al. 1997b; Long et al. 1997; Scheer and Campos-Ortega 1999; Halloran et al. 2000; Scheer et al. 2001); and the recent demonstration of the effectiveness of morpholino antisense "knockdowns" (Nasevicius and Ekker 2000) coupled with the accessibility and the highly detailed characterization of the development of zebrafish have been especially crucial for establishing the in vivo function of putative pathfinding molecules. Undoubtedly, we will see genetic screens in the near future that utilize GFP transgenic zebrafish in which subsets of axons can be readily assayed in living embryos (Higashijima et al. 2000) to identify additional pathfinding genes and the genetic pathways regulating pathfinding. With the fast-paced advancement of genetic mapping in zebrafish and the advent of the zebrafish genome sequencing project, the cloning of the genes identified via mutagenesis should be greatly enhanced making the analysis of pathfinding in zebrafish even more useful in the future.

IV. Aspects of Organogenesis

Somitogenesis

Caroline Brennan[1], Sharon L. Amacher[2], and Peter D. Currie[3]

1 Introduction

Segmentation, the subdivision of the body into morphologically similar units along its axis, is a fundamental mechanism used to establish embryonic organization and form. The segmental nature of vertebrate development is evident in the formation and organization of the hindbrain, the branchial arches, mesoderm and peripheral nervous system. Here we review somitogenesis, the formation and subsequent differentiation of the segmental units of the mesoderm, in zebrafish and relate it to similar processes in higher vertebrates. For recent reviews of somitogenesis in amniotes and discussion of general mechanisms of somite formation, see also Gossler and Hrabe de Angelis (1998), Pourquie (2000) and Rawls et al. (2000).

1.1 Generalized Overview of Somitogenesis

Evidence suggests that segmentation and differentiation of the mesoderm follows a similar path in all vertebrates although the timing of specific events may differ. Subdivision of mesoderm occurs during gastrulation, with dorsally located cells forming axial mesoderm (future notochord) and more laterally located cells forming paraxial mesoderm (future somites). Following gastrulation, the paraxial mesoderm is found as rods of mesenchymally arranged cells on either side of the notochord. As the embryo develops, uniform groups of cells separate from anterior paraxial mesoderm to form segmentally arranged epithelial structures called somites. Once the embryo has reached a critical stage of development (8–10-somite stage in zebrafish), myogenesis is initiated and cells at the anterior end of the presomitic mesoderm (PSM) and

[1] School of Biological Sciences, Queen Mary College, University of London, Mile End Road, London E1 4NS, UK
[2] Department of Molecular and Cell Biology, 555 Life Sciences Addition #3200, University of California, Berkeley, Berkeley, California 94720-3200, USA
[3] Developmental Genetics Section, MRC Human Genetics Unit, Western General Hospital, Crewe Road, Edinburgh EH4 2XU, UK

Results and Problems in Cell Differentiation, Vol. 40
L. Solnica-Krezel (Ed.): Pattern Formation in Zebrafish

within formed somites begin to express myogenic basic helix-loop-helix (bHLH) transcription factors such as MyoD and Myf5. (In zebrafish, there is also an earlier period of myogenic gene expression during gastrulation, see Sect. 5.1). Signals, including Sonic Hedgehog, Wnts and members of the bone morphogenetic family, secreted from surrounding tissues lead to the subdivision of somites into dorsal, ventral, medial and lateral portions that follow separate morphological and differentiation pathways (for review, see Currie and Ingham 1998). In amniotes, somites give rise to all trunk and limb skeletal muscle, the axial skeleton, the dermis of the back, and contribute to trunk smooth muscle and endothelial and connective tissues. In zebrafish, somitic cells also give rise to skeletal muscle, axial skeleton and some connective tissue. The contribution of zebrafish somitic cells to dermis has not been established (see Sects. 5.3–5.5). In addition to contributing to the above-mentioned tissues, the segmental organization of somites imposes a segmental organization on the spinal cord, peripheral nervous system, and trunk intersegmental blood vessels. Thus, the mechanisms underlying formation and differentiation of somites are of fundamental importance to vertebrate development.

Here, we separate somitogenesis into four major events: formation and specification of paraxial mesoderm, patterning of paraxial mesoderm to establish segmental units, formation of morphological boundaries between segmental units, and cellular differentiation within somites. In Sections 2–5, we consider each of these events in turn. First, however, we introduce the morphological events of zebrafish somite formation.

1.2 Morphological Aspects of Zebrafish Somitogenesis

In zebrafish, the first morphological evidence of somite formation is visible just after the completion of gastrulation at approximately 10.5 h postfertilization (hpf, at 28 °C). At this time, a furrow delineating the posterior margin of the first formed somite can be seen by light microscopy within the anterior paraxial mesoderm on either side of the notochord. A somite pair forms every 20 min for the first 5–6 somites and then every 30 min, until approximately 30 somite pairs have been formed (Fig. 1). The somitic furrows form as local regions of de-adhesion in the paraxial mesoderm (Fig. 1A; Wood and Thorogood 1994; Cooper and Kimmel 1998). The loss of adhesion extends laterally until the newly formed somite is separated from the rest of the paraxial mesoderm. The loss of adhesion is associated with a change in shape of cells on either side of the newly formed furrow and results in the formation of an epitheloid sheet surrounding a core of mesenchymal cells.

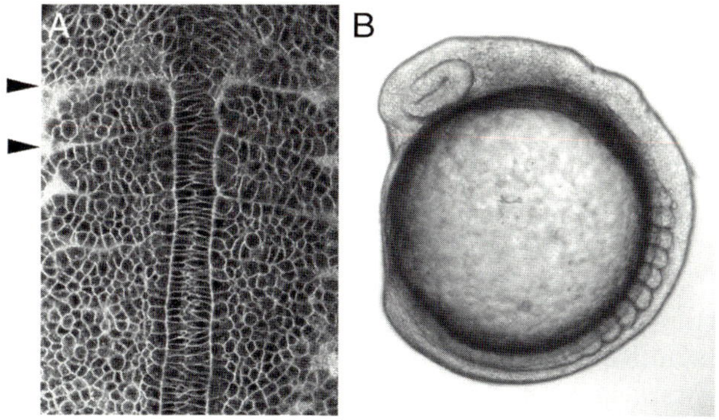

Fig. 1A–C. Morphological aspects of zebrafish somitogenesis. **A** A live embryo (3-somite stage, ~11 hpf) stained with BODIPY-ceramide illustrates two types of morphological boundaries present at this stage: the axial–paraxial boundary separating notochord and adjacent PSM/somitic tissue (*white arrowhead*) and the intersomitic boundaries separating adjacent somites (*white arrow*). See Fig. 4a for a description of the boundary between "adaxial" cells and the rest of the PSM. **B** A live embryo at the 10-somite stage (~14 hpf). **C** A live embryo at the 17-somite stage (~18 hpf)

2 General Anterior/Posterior Pattern and Specification of Paraxial Mesoderm

2.1 *Hox* Gene Expression Patterns and Overall A/P Pattern

A growing number of studies show that *Hox* genes are at least in part responsible for establishing regional identity along the vertebrate embryonic axis. *Hox* gene family members are homeobox genes arranged in genomic clusters. In tetrapod animals, *Hox* genes are arranged in four genomic clusters; in zebrafish, there are at least seven *hox* clusters (Amores et al. 1998; Prince et al. 1998a). Thus, zebrafish have more *hox* gene copies, despite less complexity along the anterior/posterior (A/P) axis (e.g., fewer distinguishable vertebral types). In general, *Hox* genes are expressed in a nested fashion along the A/P axis, with members of each paralog group having a distinctly different anterior expression limit when compared to members of adjacent paralog groups. Comparative studies in mice and avians indicate that anterior expression boundaries of *Hox* genes correlate well with anatomical differences, such as vertebral type and limb bud position (Burke et al. 1995). In contrast to tetrapods, where *Hox* gene anterior expression boundaries are dispersed along the entire A/P axis, the anterior expression boundaries of *hox* genes in zebrafish and stickleback (an acanthorpterygian teleost) are confined largely to the trunk region (Prince et al. 1998a; Ahn and Gibson 1999). In addition, some zebrafish *hox* genes belonging to adjacent paralog groups share the same anterior limit of expression, instead of the typical nested expression pattern (Prince et al. 1998a). Both of these observations suggest that the zebrafish "Hox code" may be simpler, reflecting the simpler axial arrangement (Prince et al. 1998a).

2.2 The Origin of Somitic Cells

As discussed in earlier chapters, embryonic mesoderm forms during gastrulation as epiblast cells involute at the blastoderm margin to form the hypoblast layer. Mesoderm is then further subdivided into dorsal, lateral, and ventral regions. Fate mapping studies indicate that somitic precursors are located in a wide dorsolateral to ventral swath of cells near the blastoderm margin of the gastrula (Kimmel et al. 1990). In general, the dorsoventral position of a somitic precursor (and its progeny) correlates with future position along the A/P axis. Dorsolateral cells tend to become trunk somitic cells and ventrolateral cells tend to become tail somitic cells; however, cell clones spanning both trunk and tail segments are observed (Kimmel and Warga 1987b; Kimmel et al. 1990; Kozlowski et al. 1997). Not surprisingly, fate mapping studies conducted at later developmental times reveal that the future A/P position of somitic precursors is also generally correlated with time of involution. Lateral marginal cells at mid-gastrula stages tend to contribute to trunk (and some tail) somites, but by the late-gastrula period, just 2 h later, marginal cells contribute almost exclusively to tail somites (Müller et al. 1996a). At the end of the gastrula period, trunk somitic cells are located in the paraxial mesoderm, or presomitic mesoderm (PSM), in two blocks of tissue flanking the developing midline notochord. Tail somitic precursors are located within the tail bud, and will form tail PSM through a modified version of gastrulation cell movements (Kanki and Ho 1997). Thus, fate mapping reveals that allocation of presomitic cells to positions along the A/P axis is spacially and temporally organized. As will be discussed later (Sect. 2.3), formation of trunk and tail somites is genetically separable as well.

Fate mapping studies allow one to generally predict where along the A/P axis a somite cell will be located. However, it is important to note that clonally related cells derived from a single marginal somitic precursor can be distributed over several somites (Kimmel and Warga 1987b; Müller et al. 1996a), suggesting that somitic precursors are not fated to a particular somite until after they leave the marginal zone. When lineage tracer dyes are injected into single, non-marginal, *hypoblast* cells, labeled progeny are much more spatially restricted, usually residing within a single somite (Müller et al. 1996a). Thus, as described for chick and mouse PSM cells (Tam and Beddington 1986; Stern et al. 1988; Palmeirim et al. 1997), once cells enter the PSM, they remain together and do not scatter widely. Live time-lapse microscopy has shown that cells near the borders of zebrafish somites may shift position relative to neighboring cells just before somite boundary formation (Henry et al. 2000). This observation might explain why clonally related cells derived from a single, non-marginal hypoblast cell are occasionally found in adjacent somites (Müller et al. 1996a).

2.3 *spadetail*, a Gene Controlling Paraxial Mesoderm Formation

The zebrafish *spadetail* (*spt*) gene, originally identified by mutation, is involved in cell fate and cell movements of the paraxial mesoderm (Kimmel et al. 1989). *spt* mutant embryos lack trunk somites and later are extremely deficient in trunk muscle (Fig. 2; Kimmel et al. 1989). Interestingly, tail somites form relatively normally in *spt* mutant embryos, demonstrating that formation of trunk and tail somites are differentially regulated and can be separated genetically. Genetic mosaic experiments reveal that *spt* mutant trunk somitic precursors migrate abnormally during gastrulation; instead of converging toward the dorsal midline, they migrate posteriorly and contribute to tail fates (Ho and Kane 1990). The *spt* gene encodes a T-box transcription factor similar to *Xenopus* VegT (Griffin et al. 1998; for reviews, see Amacher 1999; Smith 1999). *spt* is expressed before gastrulation and transcripts are rapidly localized to fate map regions correlated with future somitic tissue (Fig. 2D); however, unlike *Xenopus* VegT, *spt* is not expressed maternally (Griffin et al. 1998; Ruvinsky et al. 1998).

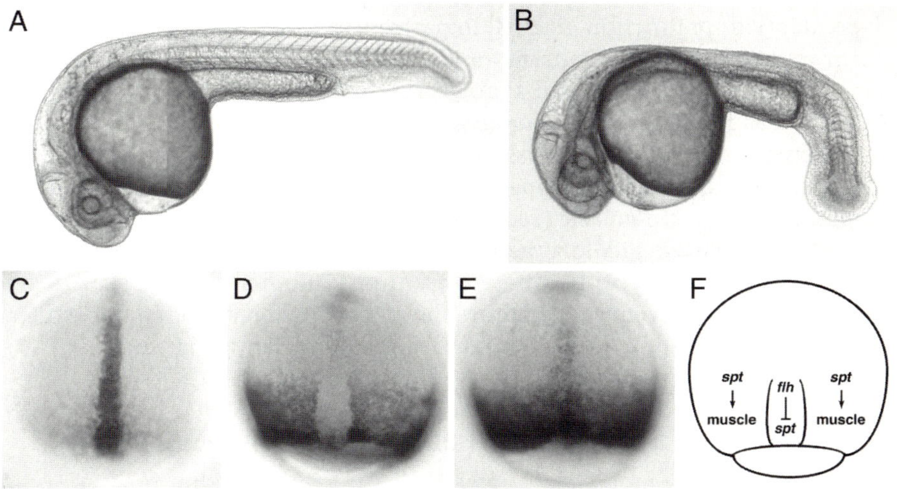

Fig. 2A–F. The *spadetail* gene controls paraxial mesoderm formation. **A** Wild-type and **B** *spadetail* (*spt*) mutant zebrafish embryos at 1 day of development. Note that *spt* mutant embryos lack organized trunk somitic muscle, but do have tail somites. **C–E** *floating head* (*flh*) and *spt* expression in wild-type (WT) and *flh* mutant embryos at mid-gastrula stages. In WT embryos, *flh* is almost exclusively expressed in axial mesodermal cells that will give rise to notochord (**C**), whereas *spt* is expressed in paraxial mesodermal cells (and in axial mesodermal cells that will give rise to the anterior prechordal plate mesoderm) (**D**). In *flh* mutant embryos, *spt* is ectopically expressed in presumptive notochord cells (**E**); this observation, together with the genetic epistasis analysis in *spt;flh* double mutant embryos (see text), are consistent with the model presented in **F**

It is still an open question whether *spt* primarily regulates cell fate, cell movements, or both; thus it is of extreme interest to characterize downstream targets of the Spt transcription factor. Potential downstream genes of VegT have been isolated, but most of these genes appear to be targets of maternally supplied *Xenopus* VegT (reviewed in Smith 1999; Stennard 1998). In zebrafish, *paraxial protocadherin* (*papc*) has been implicated as a *spt* downstream target (for review, see Amacher 1999). *papc* encodes a non-classical cadherin that is expressed during gastrulation in a strikingly similar pattern to that of *spt*, and *papc* expression is nearly abolished in *spt* mutant gastrulae (Kim et al. 1998b; Yamamoto et al. 1998a). Dominant negative experiments demonstrate that loss of *papc* function partially phenocopies the *spt* mutant phenotype (Yamamoto et al. 1998a). Taken together, these data suggest that *papc* acts closely downstream of *spt* and may help explain how the Spt transcription factor affects cell processes such as cell adhesion and cell movements that are regulated at the cell surface.

Is *spt* involved directly in somitic cell fate decisions? Gene expression of putative cell fate regulatory genes, such as the myogenic determination factor *myoD*, is lacking in *spt* mutant gastrula (Weinberg et al. 1996). Double mutant analyses also reveal that *spt* is required for somitic cell fate. In zebrafish *floating head* (*flh*) mutant embryos, axial mesoderm transfates to somitic muscle (Halpern et al. 1995; Melby et al. 1996). The transfating event is absolutely dependent on *spt* function, as axial to paraxial mesoderm transfating does not occur in embryos doubly mutant for *spt* and *flh* (Amacher and Kimmel 1998). The *flh* gene encodes a Not-type homeobox transcription factor that is expressed in notochordal precursors (Fig. 2C; Talbot et al. 1995). *flh* function is required to restrict *spt* expression to paraxial mesoderm (Fig. 2E), suggesting that *flh* is required in axial mesoderm to repress *spt*, thus preventing ectopic formation of somitic muscle (see model, Fig. 2F; Griffin et al. 1998; Yamamoto et al. 1998a). This regulatory pathway demonstrates how axial and paraxial mesoderm initially become different from one another. Once specified, axial mesoderm forms prechordal plate and notochord, whereas paraxial mesoderm becomes segmented in a very temporally and spatially precise manner.

3 Establishing a Segmental Pattern

In vertebrate paraxial mesoderm, each somite forms as groups of cells "pinch off" from the anterior of the PSM at regular intervals. This is in contrast to segmentation of the vertebrate hindbrain or of the A/P axis in long germ band insects where segments form by subdivision of a preexisting field (Lumsden and Krumlauf 1996; Pick 1998). Evidence for models to account for the serial formation of somites has been extensively reviewed elsewhere (e.g., Gossler and Hrabe de Angelis 1998; Pourquie 2000). Therefore, although relevant models are introduced here, they are only discussed in detail in relation to zebrafish somite formation.

3.1 Somitic Periodicity and the Cell Cycle

In birds, frogs, and fish, a single brief heat shock can cause local disruptions in somite boundary formation (Elsdale et al. 1976; Kimmel et al. 1988; Primmett et al. 1988, 1989; Roy et al. 1999). Segmental defects are not observed in the somite forming during the heat shock pulse, but instead are delayed and occur in a somite forming several hours later. Somites anterior and posterior to the defective somite are often normal. These observations suggest that there is a heat-sensitive phase in which cells make a "segmental" decision that is not manifested morphologically until later. Interestingly, in zebrafish and birds, *periodic* somitic disturbances have been observed after a single heat shock application; in avians, the disturbances occur every seventh somite and, in zebrafish, every fifth somite (Primmett et al. 1988, 1989; Roy et al. 1999). Since a somite pair forms every 90 min in chick, a periodicity of 7 somites (~10 h) correlates well with the PSM cell cycle length (Primmett et al. 1988, 1989). In addition, application of drugs that interfere with normal progression of the cell cycle causes disturbances similar to those seen upon heat shock treatment (Stern et al. 1988). A somite pair forms every ~30 min in zebrafish; a periodicity of 5 somites (~2.5 h) correlates less well with general cell cycle length at this stage. However, it is interesting to note that CS131, a zebrafish gene sharing strong similarity to a group of small acidic proteins involved in cell cycle arrest, is expressed strongly in PSM cells in a position to form a somite 1 h later (Durbin et al. 2000). Thus, groups of zebrafish PSM cells may be synchronized at some point during the cell cycle. Since fate mapping experiments fail to reveal a 5-somite periodicity like that observed after heat shock treatment, it seems unlikely that the cells affected by heat shock are clonally related.

3.2 Existence of a Molecular Oscillator

Several models of somite formation propose the existence of a molecular clock or oscillator that functions with a temporal periodicity to translate a gradient of either positional or developmental information into the periodic, or segmental, pattern necessary to form somites. In the positional information model (Meinhardt 1986; Keynes and Stern 1988), cells in the rostral PSM are proposed to oscillate between, and then become stabilized with, anterior and posterior somite identity as a result of action of a gradient of morphogen (giving positional information) and reaction-diffusion mechanisms. Opposition of cells with anterior and posterior identities then results in formation of a somite boundary. This is consistent with grafting experiments which show that somite boundaries form only when cells from anterior-half somites are placed adjacent to cells of posterior-half somites (Stern and Keynes 1987). It is also and is supported by observations of alternating domains of gene expression that correspond to anterior- and posterior-half somites and presumptive somites (Rawls et al. 2000; Stickney et al. 2000; see Sect. 3.4). However, this model requires a gradient of morphogen and thus it cannot easily explain the autonomous formation of somites seen when small pieces of PSM are explanted and cultured in isolation (Menkes and Sandor 1977; Palmeirim et al. 1997). Furthermore, in its simplest form, it is unable to explain why boundaries do not form at the center of each somite. Imposing a pair-rule mechanism, where cells in alternate presumptive segments have different properties or gene expression, on top of the oscillation could resolve this issue (Tautz and Sommer 1995; Müller et al. 1996a).

The "clock and wavefront" model of somitogenesis (Cooke and Zeeman 1976), recently modified by Cooke (1998), proposes that paraxial mesoderm segmentation results from interaction of two separate functions operating within PSM cells. The first component is an autonomous "cellular oscillator" or clock present within each PSM cell. The second component is a "wavefront" of maturation – the result of some critical property that alters in PSM cells as they mature. Since cells in the anterior PSM are older than those of the posterior PSM, the wavefront of maturation progresses in an anterior to posterior direction. Rows of cells at the same A/P level are proposed to oscillate in synchrony and to be competent to form a somite border only at a specific point within each oscillation cycle. As the wavefront of maturation progresses along the embryo, it stabilizes cells with anterior or posterior somite identity, and induces formation of a boundary when it encounters cells at the correct phase of the oscillation. Thus, in this model, the oscillation imposes a stepwise, and therefore segmental, progression to the wavefront by controlling competence to respond to the "wavefront" signal.

Molecular evidence for existence of a cellular oscillator in the PSM has been obtained in chick (Palmeirim et al. 1997; McGrew et al. 1998), mouse (Forsberg et al. 1998; Aulehla and Johnson 1999), and zebrafish (Holley et al. 2000; Jiang et al. 2000; Sawada et al. 2000). Detailed analysis of a chick homologue of the *Drosophila* pair rule gene *hairy*, *c-hairy1*, showed that *c-hairy1* expression oscillates autonomously in PSM cells with a periodicity corresponding to the time required to form one somite. Similar oscillations in expression of *lunatic fringe*, a homologue of the *Drosophila fringe* gene and a component of the Notch-signaling pathway, have subsequently been found in both mouse and chick (Forsberg et al. 1998; McGrew et al. 1998; Aulehla and Johnson 1999). Recently, the expression of a *hairy* homologue, *her1*, and a Notch ligand, *deltaC*, have been shown to oscillate in the zebrafish PSM (Holley et al. 2000; Jiang et al. 2000; Sawada et al. 2000; see Fig. 3). As in chick and mouse, the periodicity of oscillation corresponds to the time needed to form one somite, suggesting a link between Notch signaling and the segmentation clock.

3.2.1 The Notch Pathway and Establishment of Segmental Pattern

Notch is a large transmembrane receptor protein that binds members of the Delta and Serrate transmembrane ligand families. Activation of Notch by one of its ligands causes the receptor to be proteolytically cleaved, resulting in

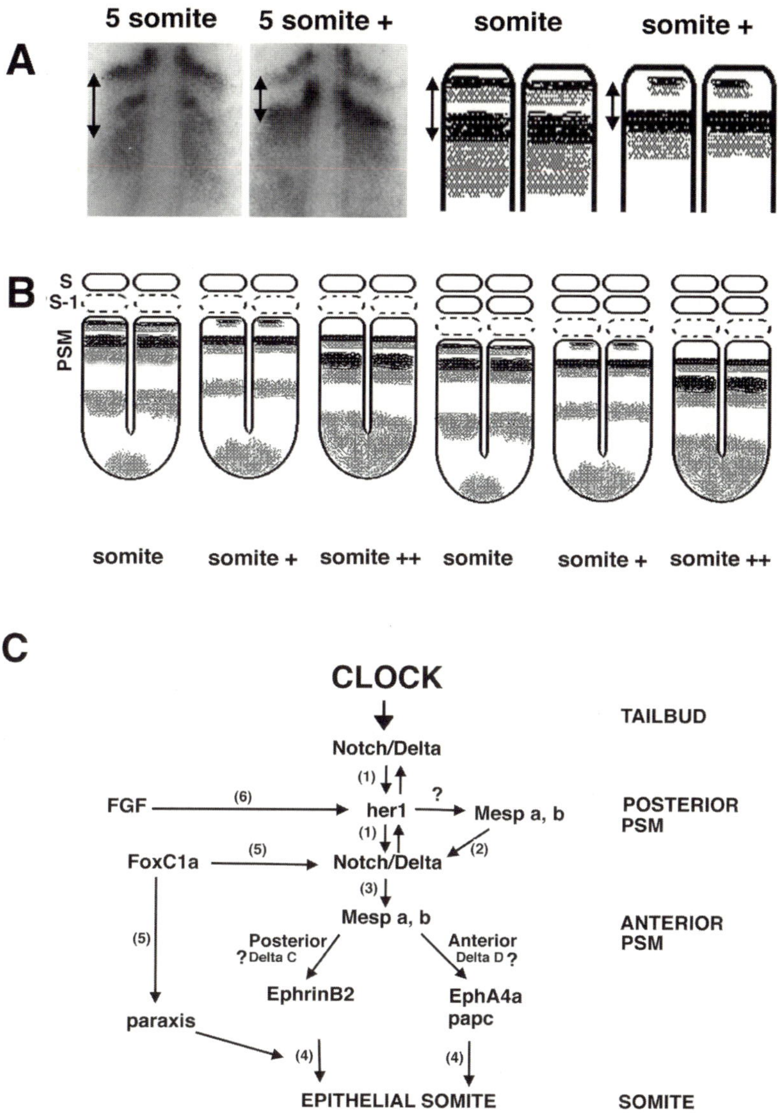

Fig. 3A–C. Cyclical *her1* expression in zebrafish presomitic mesoderm. **A** In situ analysis of *mesp a* (*dark grey*) and *her1* (*light grey*) expression in 5-somite stage embryos. *Panels on the right* are schematic representations of the photographs on the *left*. *mesp b* is expressed in an anterior domain of each anterior presumptive somite (Sawada et al. 2000). Stripes of *her1* alternate with those of *mesp b*, suggesting that *her1* is expressed in each somite primordium. The bands of *her1* expression narrow and get closer together as the somite matures (compare length of arrows). **B** Schematic of expression profiles of *her1* and *mesp b*. *S* Formed somite, *solid line*; *S-1* forming somite, *dotted line*; *PSM* presomitic mesoderm. The *her1* expression domain appears around the tailbud, and moves anteriorly until it finally overlaps with the most anterior stripe of *mesp a*. Both stripes disappear near the point of furrow formation. A stripe of *her1* appears every 30 min (the time taken for one somite to form) and persists for approx. 1.5 h (Holley et al. 2000; Sawada et al. 2000). Modified from Sawada et al. (2000). **C** A working model of hierarchical gene function during somite formation. *Arrows* indicate interactions based on published results. Interactions may be positive or negative, direct or indirect. *Question marks* denote hypothetical interactions. *Numbers in parentheses* refer to references: *1* Takke and Campos-Ortega (1999); *2* Sawada et al. (2000); *3* Holley et al. (2000); *4* Durbin et al. (1998); *5* Topczewska et al. (2001b); *6* Sawada et al. (2001). After Stickney et al. (2000)

translocation of the Notch cytoplasmic domain to the nucleus. Once in the nucleus, cleaved Notch protein acts together with a conserved transcription factor (named Suppressor of Hairless in *Drosophila*, RBPjk/CBP1 in vertebrates) to activate bHLH proteins of the hairy/Enhancer of Split [E(spl)]) family. These bHLH family members in turn complex with a second transcriptional regulator, Groucho, to regulate expression of downstream genes (for review of Notch signaling, see Artavanis-Tsakonas et al. 1999). Recent work has revealed conservation of this pathway and its involvement in somite formation in amniotes (Gossler and Hrabe de Angelis 1998; Barrantes et al. 1999; Pourquie 1999). In zebrafish, injection of RNA encoding a constitutively active Notch receptor protein, Notch1aIntra, leads to ectopic expression of *her1* and of another *hairy/E(spl)* homologue *her4* in paraxial mesoderm, and to formation of irregularly shaped somites (Takke and Campos-Ortega 1999). These results demonstrate a conserved link between Notch and *hairy/E(spl)* gene function during zebrafish somitogenesis.

To date, all genes whose expression oscillates within the vertebrate PSM are components of the Notch pathway. However, the precise nature of the link between the segmentation clock, Notch signaling, and somite formation is not clearly understood.

3.2.2 Insights from Zebrafish

The zebrafish fused-somite class mutants have defects in both segmentation and somite epithelialization (Holley and Nüsslein-Volhard 2000). No somites form in *fused-somites* (*fss*) mutants, while in *beamter* (*bea*), *deadly seven* (*des*), *after-eight* (*aei*), and *mindbomb* [*mib*; also called *whitetail* (*wit*)] mutants, the first 5–8 somites form normally but the more posterior somites do not. The *aei* gene encodes zebrafish *deltaD*, and the *aei* mutants studied represent null mutations (Holley et al. 2000). The *mib* and *des* genes are also thought to encode components of the Notch-signaling pathway (van Eeden et al. 1996, 1998; Holley and Nüsslein-Volhard 2000; A. Chitnis and S.A. Holley, pers. comm.). In *aei* mutant embryos, stripes of *her1* expression are not seen in the PSM. Rather, *her1* is expressed in the tailbud weakly and in a broad diffuse band in the anterior PSM (van Eeden et al. 1998; Holley et al. 2000). If *her1* stripes in a wild-type embryo are the visualization of cellular clock function, *aei* and Notch signaling are required either for (1) operation of the clock, as a link between clock function and *her1* expression, or (2) maintenance of coordination of cyclical expression across the PSM. Two lines of evidence suggest the latter. First, close examination of *her1* and *deltaC* expression in *aei* mutants reveals that, rather than being uniformly expressed in the anterior PSM, *her1* and *deltaC* expression varies in level between individual cells giving a pepper-and-salt appearance (Holley et al. 2000; Jiang et al. 2000; but see Sect. 3.4). The variation in expression is consistent with a model whereby PSM cells continue to oscillate but that synchrony of oscillation is no longer maintained. This role for Notch signaling has been proposed to explain why anterior somites are relatively unaffected compared to posterior somites in mouse and fish Notch pathway mutants – since cells oscillate autonomously, they will gradually fall out of step in the absence of a system to maintain synchrony (Jiang et al. 2000). Anterior somites form first, have less time to fall out of step, and are therefore more resistant to disruption. Second, although errors in somite formation are detected, *her1* stripes continue to be observed in zebrafish embryos that have been injected with RNA encoding a dominant negative Delta construct (DN-Delta) that is predicted to interfere with all Notch-mediated signaling (Takke and Campos-Ortega 1999). Thus, the clock appears to operate in the absence of Notch signaling. The fact that *her1* stripes (and therefore synchronized oscillations) are lost in *aei* mutants, but continue to be observed in DN-Delta-injected embryos (at the ten-somite stage), may reflect the time needed for sufficient levels of DN-Delta protein to accumulate to inhibit Notch signaling, and for PSM cells to fall out of synchrony. Finally, *deltaD* expression is disrupted in a zebrafish deficiency mutant lacking the *her1* gene (M.K. Urban and S.L. Amacher, unpubl. observ.), extending the link between *her* genes and Notch signaling and implicating the existence of a feedback regulatory loop.

3.3 The Fused-Somite Mutant and Operation of a Wavefront

In chick, *c-hairy1* expression oscillates in PSM cells, resulting in a wave of expression that progresses from posterior to anterior. As the "wave" of *c-hairy1* expression progresses anteriorly, the bands of expressing cells narrow and get closer together until *c-hairy1* expression is eventually stabilized in the posterior half of the most rostral presumptive somite and the most recently formed somites. The expression dynamics can be explained by the oscillation gradually slowing and eventually ceasing (Pourquie 1999). Similarly, *her1* and *deltaC* stripes in zebrafish also narrow and get closer together, although rather than being stabilized, *her1* expression fades in anterior PSM and is lost (Fig. 3; Müller et al. 1996a; Holley et al. 2000; Jiang et al. 2000). The gradual slowing and eventual cessation of oscillation are thought to be functions of a "wavefront" of maturation, as proposed by the "clock and wavefront" model. However, the molecular basis of the wavefront is not known. Continued analysis of the fused-somite class of zebrafish mutants, in particular *fss*, will undoubtedly help in answering this question. *fss* mutant embryos lack all morphological evidence of mesodermal segmentation. In *fss* mutant embryos, *her1* stripes emerge from the tailbud, demonstrating the onset of oscillation; however, anterior *her1* stripes are never seen, suggesting that oscillations stop prematurely (Holley et al. 2000). Furthermore, when two *her1* stripes are observed in *fss* mutant embryos, they are significantly closer together than those observed in comparable positions of wild-type embryos. This suggests that the period of oscillation and perhaps clock speed is affected. If in wild-type embryos the cessation of oscillation is due to interaction of a wavefront of maturation with the clock mechanism, the premature cessation of *her1* oscillation in *fss* mutants may represent a disruption in the regulation of this wavefront – an error in the rate of mesodermal maturation. *fss* mutant embryos continue to undergo myogenesis normally, suggesting that it is the link between the clock and wavefront of maturation rather than mesodermal maturation per se that is affected. It is interesting to note that PSM cells in *fss* mutant embryos uniformly express markers of posterior somite identity and lack markers of anterior somite identity (Durbin et al. 2000; Sawada et al. 2000). This suggests that continued oscillation or the *fss* gene product itself is required for specification of anterior somite identity (see Sect. 3.4).

Work by Dubrelle et al. (2001) in chick and Sawada et al. (2001) in zebrafish suggests that FGF/MAPK signaling in the PSM is also crucial for maturation wavefront function and for positioning somite boundaries. In zebrafish, FGFR1 is expressed throughout the posterior PSM and in posterior domains of forming somites. FGF8, a ligand for this receptor, is similarly expressed in the posterior PSM and tailbud and in anterior domains of forming somites. Brief exposure of segmentation-stage embryos to a drug that inhibits FGFR signaling (SU5402, an inhibitor of MAPK) leads to formation of somites that are larger in A/P dimension. In contrast, increasing FGFR signaling within the PSM by implantation of an FGF8-soaked bead leads to formation of smaller somites (Sawada et al. 2001). According to the "clock and wavefront" model, larger somites could be produced either by slowing the rate of oscillation or by increasing the rate of progression of the wavefront. In SU5402-treated embryos, the periodicity of *her1* stripes is not altered, suggesting oscillation of the clock is unaffected. Thus FGF signaling appears to affect the rate of mesodermal maturation – progression of the wavefront. This suggestion is supported by the observation that, although *her1* stripes form in SU5402-treated embryos, they terminate prematurely compared with control embryos. This is similar to the situation seen in *fss* mutant embryos. However, significant differences between the phenotypes observed in *fss* mutant and SU5402-treated embryos suggest that FGF signaling affects mesodermal maturation independently of the *fss* gene product (Sawada et al. 2001). As FGF8 RNA is expressed in a gradient in the caudal PSM that is mirrored by ppMAPK protein levels, and the last formed somite boundary always forms at the same distance from the ppMAPK protein domain, it is possible that FGF signaling provides a positional cue within the PSM. Once FGF signaling falls below a critical level, the mesoderm is capable of responding to maturation signals and of forming a somite boundary.

3.4 Establishment of Anterior/Posterior Somite Polarity

Embryological experiments in chick and analysis of zebrafish somite mutants demonstrate that establishment of A/P polarity is required for formation and maintenance of the somite boundary – presumably as a result of different cell surface properties of anterior and posterior somitic cells (Stern and Keynes

1987; Durbin et al. 2000). Numerous genes are expressed in the anterior PSM of all vertebrates studied in domains that have suitable positions and dimensions to represent presumptive anterior- or posterior-half somites (reviewed in Rawls et al. 2000; Stickney et al. 2000). For example, zebrafish *notch6*, *delta D*, *fgf8* (*ace*), *papc*, *EphA4*, *mesp a*, and *mesp b* are expressed in anterior domains of the most rostral presumptive somites and *myoD*, *ephrinB2*, *deltaC*, *fgfr1* and *notch5* in posterior domains (Westin and Lardelli 1997; Durbin et al. 1998; Haddon et al. 1998; Sawada et al. 2000). Mouse and zebrafish loss-of-function mutations in components of the Notch pathway indicate that Notch signaling is required for organized establishment of A/P somite polarity and for positioning and formation of the somite boundary. For example, in *aei* mutant embryos, markers of A/P somite identity are diffusely expressed in the anterior PSM as early as the 8–10-somite stage, and somite boundaries fail to form or form erratically (Durbin et al. 2000). It is unclear at present whether diffuse expression represents a uniform level of expression in all cells or a weak pepper-and salt pattern (as described in Sect. 3.2.1; Holley et al. 2000; Jiang et al. 2000), and if the precise pattern is dependent on which polarity marker is being examined. Similarly, whether involvement of Notch signaling in organization of A/P somite polarity in *aei* mutant embryos reflects its role in maintaining synchronized cellular oscillations and/or a role linking oscillations to the molecular machinery establishing A/P somite identity (see below) is yet to be determined.

Analysis of *mesp* gene expression in embryos with disrupted Notch signaling strongly suggests a link between Notch signaling and the establishment of anterior somite identity. Zebrafish homologues of the murine *MesP 1* and *2* genes have recently been identified (*mesp a* and *b*, respectively; Durbin et al. 2000; Sawada et al. 2000). In zebrafish, *mesp a* and *b* are expressed in narrow stripes corresponding to the anterior half of rostral presumptive somites. Ectopic expression of *mesp b* leads to expansion of anterior somite compartments at the expense of posterior somite tissue, and no somites form (Sawada et al. 2000). A similar result is obtained when *Mesp* genes are misexpressed in mouse and *Xenopus* (Saga et al. 1996, 1997; Buchberger et al. 1998; Sparrow et al. 1998). In *aei* mutant embryos, PSM *mesp a* expression is lost, and *mesp b* expression, although weak, loses its striped appearance. These findings suggest that *mesp* gene expression and establishment of anterior somite identity are downstream targets of Notch signaling in zebrafish. Again, a similar situation exists in other vertebrates (Kim et al. 2000b; Takahashi et al. 2000b). In mouse, *Mesp2* is thought to establish A/P somite identity by controlling Presenilin1-dependent and independent Notch-signaling pathways in posterior and anterior domains of presumptive somites. Mesp2 activates a Presenilin1-dependent signaling pathway that induces *Delta-like1* expression in the posterior somite and a Presenilin1-independent Notch-signaling cascade that suppresses *Delta-like 1* expression in the anterior somite (Takahashi et al. 2000b). Thus, in mouse, Notch signaling is required for and regulated by *Mesp* gene expression and is thereby thought to control establishment of A/P somite identity. In *Xenopus*, negative feedback control of Notch signaling has also been shown to be necessary to prevent expression of markers of anterior somite fate (like *papc*) in posterior somite compartments. Here *papc* itself appears to cooperate in the maintenance of A/P somite polarity (Kim et al. 2000b).

Members of the forkhead/winged loop helix family of transcription factors have also been implicated in somite formation in both mice and zebrafish. Two forkhead transcription factors, FoxC1 and FoxC2 in mouse, FoxC1a and 1b in zebrafish, are expressed in the PSM of segmentation stage embryos (Topczewska et al. 2001a). Inactivation of FoxC1 and FoxC2 simultaneously in mouse or of FoxC1a alone in fish leads to disruption of A/P somite identity and failure to form an epithelial somite (Topczewska et al. 2001b). In zebrafish embryos depleted of FoxC1a protein, *her1* and *deltaC* stripes are still detected and have the same temporal and spatial characteristics as in wild-type embryos. Thus, neither the operation of the clock nor the rate of progression of the wavefront is affected. However, markers of anterior somite identity, *mesp a*, *Epha4a* and *ephrinb2*, are no longer expressed. Furthermore, expression of *deltaC* and *deltaD* is not maintained in formed somites. Thus, FoxC1a appears to be necessary for stabilization of A/P somite identity and for expression of cell surface proteins required for epithelialization and boundary formation. Expression of *notch5* and *notch6* is also severely reduced in embryos lacking FoxC1a suggesting a pathway by which FoxC1a may regulate A/P somite identity (Fig. 3C). Experiments in mouse have shown that *paraxis*, a gene encoding a bHLH transcription factor, is essential for epithelial somite formation, but is not required for expression of A/P polarity markers such as EphA4 (Johnson et al. 2001). In zebrafish, FoxC1a protein is necessary for *paraxis* expression, suggesting that FoxC1a acts at several levels to regulate somite boundary formation (Fig. 3C).

4 Formation of the Somite Boundary

Somite boundary formation involves the regulation of adhesive properties and a mesenchymal-to-epithelial transition of cells on either side of the boundary (i.e., in cells of adjacent posterior and anterior compartments). Several molecules that show differential A/P expression in presumptive somites of the most anterior PSM have been implicated in the regulation of cell adhesion: FGFR1, Papc and EphR and Ephrins (Kim et al. 1998b, 2000; Yamamoto et al. 1998a; Wilkinson 2001). The role of FGFR1 in regulation of adhesion and somite boundary formation in zebrafish has not been studied. However, expression analyses in wild-type and *fss* mutant embryos have shown a role for Eph receptor signaling in somite boundary formation (Durbin et al. 1998, 2000). Similarly, ectopic expression and dominant negative experiments in *Xenopus* and zebrafish implicate *papc* in somite boundary formation (Yamamoto et al. 1998a; Kim et al. 2000b).

In zebrafish, *EphA4a* and *ephrinB2a* are expressed in anterior and posterior halves, respectively, of each newly formed and forming somite. A somite boundary forms at the site of interaction between posterior *ephrinB2a*-expressing cells of one presumptive somite and anterior *EphA4a*-expressing cells of the next most caudal presumptive somite (Durbin et al. 1998). Given that Eph signaling regulates cytoskeletal organization and adhesion in a number of in vitro systems (Wilkinson 2001), the striped expression of *EphA4a* and *ephrinB2a* in the anterior PSM suggests that these molecules mediate changes in cell adhesion and cell shape associated with boundary formation. Morphological observations of boundary formation in zebrafish are consistent with this hypothesis (Cooper and Kimmel 1998; Henry et al. 2000); zebrafish somite furrow formation first appears as local de-adhesions in the PSM along the line of the nascent boundary accompanied by a mesenchymal to epithelial-like transition that spreads laterally. The site and timing of this de-adhesion correlates with *EphA4a* and *ephrinB2a* expression, and loss of Eph signaling leads to loss of somite boundary formation (Durbin et al. 1998). Analysis of zebrafish embryos doubly mutant for *knypek* and *trilobite* shows that somite boundary formation occurs in the absence of any central mesenchymal cells (Henry et al. 2000). Thus the force that drives boundary formation must come from interaction of boundary cells themselves, not from an increase in compaction of central cells as proposed in chick (Linask et al. 1998). Furthermore, transplantation of *EphA4*-expressing cells into a field of *fss* mutant PSM cells (which uniformly express *ephrinB2a*) is sufficient to induce a furrow between transplanted and host cells (Durbin et al. 2000). The similar expression patterns of *EphA4* and *ephrinB2* in mouse, chick and zebrafish suggest that such a role for Eph signaling in somite formation may be conserved (Durbin et al. 1998; Rawls et al. 2000).

In amniotes, roles for cadherin-mediated adhesion in somite boundary formation and myogenesis have been suggested (Linask et al. 1998; Shin et al. 2000). Cadherins are a central component of major sites of cell-cell adhesion, the adherens junctions. During mesenchymal-to-epithelial transitions, adherens junctions are re-distributed to the apical and lateral cell membranes. Within chick somitic mesoderm, the redistribution of adherens junctions, as determined by localization of β-catenin, occurs just prior to visualization of somite boundaries and has been proposed to be important for boundary formation (Linask et al. 1998). Furthermore, mouse embryos null for both cadherin11 and N-cadherin form somites that tend to separate into anterior- and posterior-half somites along the line of internal *EphA4/ephrinB2* expression (Gossler and Hrabe de Angelis 1998; Linask et al. 1998), suggesting a role for cadherin-mediated adhesion in preventing separation of A/P-half somites. To date, a role for members of the classical cadherin family of adhesion molecules in zebrafish somite morphogenesis has not been established. However, disruption of expression or function of the non-classical cadherin, *papc*, does lead to disruption of somite boundary formation in zebrafish and *Xenopus* (Yamamoto et al. 1998a; Kim et al. 2000b).

The results discussed here strongly suggest the involvement of Eph/ephrin signaling and *papc* function in establishment and maintenance of morphological somite boundaries in zebrafish. However, additional mechanisms must exist to explain why a boundary does not form at the interface between anterior and posterior compartments within an individual somite. Ca^{2+}-mediated cell communication via gap junctions is one possible mechanism. In zebrafish, waves of high intracellular Ca^{2+} propagate through blocks of anterior PSM cells of approximately one somite's width just prior to somite formation (C. Brennan, unpubl. obs.; Creton et al. 1998; Gilland et al. 1999). The role of this Ca^{2+} increase has not been established. However, since both loss of Ca^{2+}-dependent cadherin-mediated cell adhesion and cell communication via gap junctions lead to formation of somites that tend to separate into anterior- and posterior-half somites (Gossler and Hrabe de Angelis 1998; Linask et al. 1998; Becker et al. 1999), it is tempting to speculate a role for Ca^{2+} signaling in this process.

5 Induction and Patterning the Presomitic and Somitic Mesoderm

Once formed, each somite is partitioned into dorsal and ventral compartments containing progenitors for individual embryonic structures. In avians, the ventral portion of the somite undergoes a mesenchymal transition and gives rise to the sclerotome that migrates and differentiates into components of the axial skeleton. The dorsal somite remains epithelial and forms the dermomyotome. The dorsal extent of the dermomyotome gives rise to the dermis, whereas skeletal muscle derives from the remaining portion that contributes to the myotome (reviewed in Currie and Ingham 1998).

Our understanding of the processes that control somite specification in zebrafish embryos has been dominated almost entirely by studies of the myotome. In part this is due to the architecture of the somite itself, which in fish contributes mainly to myotome. While we concentrate mainly on the results of such analyses, we will also consider how other less well-described somite populations are derived and specified.

5.1 Embryonic Myotome Formation and the Initiation of Myogenesis

5.1.1 Formation of the "Adaxial" Cell Compartment and Presomitic Myogenic Induction

The cellular events that coordinate myotome formation are initiated in zebrafish prior to completion of the cell movements of gastrulation. Precocious commitment to the myogenic fate, as suggested by expression of myogenic regulatory factors (MRFs) such as MyoD and Myf5, is exhibited by paraxial mesoderm cells flanking the extending embryonic shield (Weinberg et al. 1996; Chen et al. 2001b; Coutelle et al. 2001). It is believed that cells within this field converge to the midline during axis extension and generate a row of approximately 20 cells per somite (Devoto et al. 1996), termed the "adaxial cells", that flank the forming notochord. Their adaxial position, as well as their large cuboidal epithelial-like morphology, identifies the first cellular compartment within the PSM (Fig. 4a). Towards the end of gastrulation, Myf5 expression also occurs more broadly within the lateral PSM. By contrast, MyoD expression in lateral non-adaxial cells does not begin until segmentation has initiated and is only seen in the posterior two-thirds of forming somites (Weinberg et al. 1996). Myf5 expression also becomes restricted to this posterior domain, but is quickly downregulated during somitogenesis while *myoD* transcripts are maintained within mature somites (Fig. 4).

Although the timing of commitment of paraxial mesodermal cells to the myogenic fate has not been directly assessed in the zebrafish embryo, the early

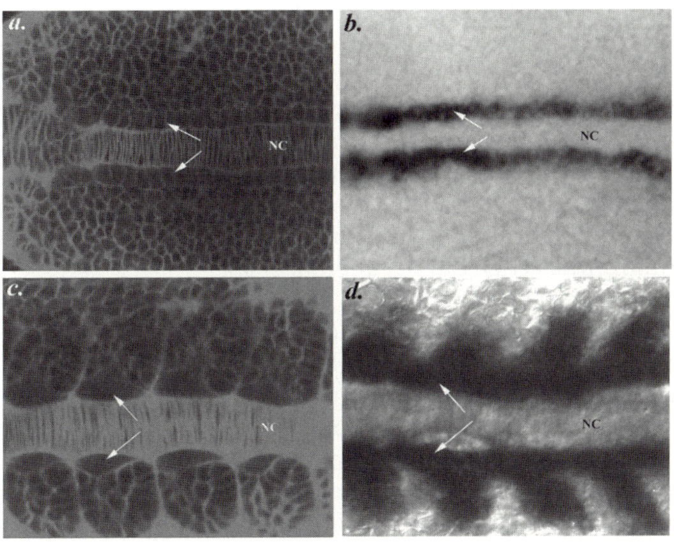

Fig. 4a–d. Somite formation and myogenesis. **a** Live 1-somite stage embryo that has been stained with BODIPY-ceramide. The adaxial cells (*arrows*) are clearly distinct from the remainder of the presomitic mesoderm because of their location adjacent to the notochord and their epithelial-like cubodial morphology. **b** At this stage of development adaxial cells are also distinguished by their precocious expression of myogenic regulatory factors such as MyoD, here revealed by in situ hybridization with an antisense probe against the zebrafish *myoD* gene. **c** 15-somite stage embryo stained with BODIPY-ceramide reveals the epithelial somite. Adaxial cells (*arrows*) which have begun to differentiate now possess a triangular morphology. **d** 15-somite stage embryo hybridized with an antisense *myoD* probe. Myogenesis initiates in the lateral posterior region of each somite shortly after segmentation. All views are dorsal with anterior to the left

induction of MRF gene expression within adaxial cells suggests this is a strikingly early event. By contrast, MRF expression in amniote embryos does not occur until somitogenesis is well established. This remains a puzzling difference between fish and amniote embryos in the way they coordinate muscle formation. Early commitment of PSM cells to the myogenic lineage is shared with the anuran *Xenopus*, which exhibit broad PSM expression of both Myf5 and MyoD, but no obvious adaxial cell compartment (Harvey 1990; Rupp and Weintraub 1991; Hopwood et al. 1991). It is possible that the modality of non-somitic muscle specification is driven by a requirement for similar larval locomotor strategies in fish and anuran embryos. The precocious differentiation of zebrafish adaxial cells into striated muscle may also be required to coordinate the overall architecture of the developing fish myotome (Blagden et al. 1997; see below).

5.1.2 Muscle Pioneer Cells and Myotomal Architecture

The first elongating and striating cells of the teleost myotome (termed "muscle pioneers") arise from adaxial cells (Waterman 1969; Felsenfeld et al. 1991), reinforcing the notion of an adaxially derived myotomal scaffold. Muscle pioneers are clearly distinguishable from other adaxially derived slow muscle cells by expression of the zebrafish *engrailed* (*eng*) *1* and *2* genes (Ekker et al. 1992). En expression is not solely restricted to muscle pioneer cells, as had been previously described (Hatta et al. 1991a), as lower level expression of both En transcript and protein is detected in fast MyHC-expressing myocytes immediately lateral to forming muscle pioneers (Roy et al. 2001; P.D. Currie, unpubl. obs.). However, En transcript expression in fast muscle cells rapidly diminishes during somitogenesis (P.D. Currie, unpubl. obs.).

Muscle pioneer cells differentiate in an A/P wave that mirrors somite differentiation, forming adjacent to the notochord at the dorsal-ventral midline of the zebrafish myotome. At this level, a specialized myotomal structure – the horizontal myoseptum – forms. Horizontal myosepta are present in all gnathostome (jawed), but not cyclostome (jawless), fish and divide the differentiating myotome into dorsal and ventral muscle domains (Bone 1989). The close association between muscle pioneer cell differentiation and horizontal myoseptum formation has led to speculation that muscle pioneers play a role in controlling horizontal myoseptum formation (Halpern et al. 1993). Muscle pioneer cell striation and elongation are also coincident with a myotomal shape change from a block-like form to a characteristic chevron shape. How muscle pioneer cells might control these different events is poorly understood, but it is proposed that the chevron form is critical for relatively short muscle fibers to collectively extend over several skeletal segments and generate powerful alternate lateral muscle undulations (the major propulsive force of the fish; Hardistry 1979). Zebrafish mutants that lack muscle pioneer cells lack chevron-shaped somites and possess somites that remain block-like or u-shaped.

The role of muscle pioneers in other aspects of somite architecture is less clear. The growth cones of primary trunk motor neurons grow along a common myotomal path, touching the muscle pioneer cells before they diverge to contact individual insertion points within the myotome. However, laser ablation of muscle pioneer cells prior to axon outgrowth has provided no evidence for a role in primary axon guidance although arborization of individual neurons is altered (Melançon et al. 1997). Growth cone guidance along both the common and divergent pathways appears to be controlled instead by signaling from dorsal adaxial cells. Mutations within two genes, *diwanka* and *unplugged*, reveal a requirement for these gene products within adaxial cells for correct motor axon guidance along both pathways (Zeller and Granato 1999; Zhang and Granato 2000).

The debate about muscle pioneer cell function has recently come into sharper focus. Studies in chick embryos reveal the existence of early born muscle cells prefiguring the mature myotomal architecture (Kahane et al. 1998). These "pioneer" myoblasts bear a number of superficial hallmarks of zebrafish pioneer cells, and it is worth noting that En expression is found within a subset of myotomally derived muscle cells in all vertebrates that have so far been examined (Davis et al. 1991). While the function of these cells in myotome substructure is still unresolved, and the role of En in these cells remains mysterious (Degenhardt and Sassoon 2001), the evolutionary conservation of En-positive myocytes within a phylogenetically diverse range of vertebrates suggests an important functional role.

5.2 Fiber Type and Myotome Morphogenesis

5.2.1 Fiber Type Formation in Separate Myotomal Compartments

The adaxial and lateral myogenic compartments not only activate myogenesis in temporally distinct phases, but also give rise to specific cell fates. All vertebrate muscle cells are composed of two basic fiber types, those expressing fast twitch or slow twitch varieties of myosin heavy chain (MyHC). All adaxial cell derivatives, including muscle pioneer cells, express a slow MyHC isoform and differentiate into slow twitch muscle. Dye-labeling experiments have revealed that midway through segmentation most adaxial cells migrate from their site of origin next to the notochord, traversing the entire extent of the myotome, to come to lie at its most lateral surface (Devoto et al. 1996). Here, these slow MyHC-expressing myoblasts form a subcutaneous layer of slow twitch muscle (Bone 1978; Devoto et al. 1996; Blagden et al. 1997). By contrast, slow MyHC-expressing muscle pioneer cells remain medial up to 48 h after differentiation, reinforcing their distinct role in somite morphogenesis (Waterman 1969; Devoto et al. 1996; Blagden et al. 1997). The remainder of the lateral myotome differentiates as fast twitch muscle and differentiation begins after slow muscle cell migration (Fig. 5a; Devoto et al. 1996; Blagden et al. 1997). However, fast

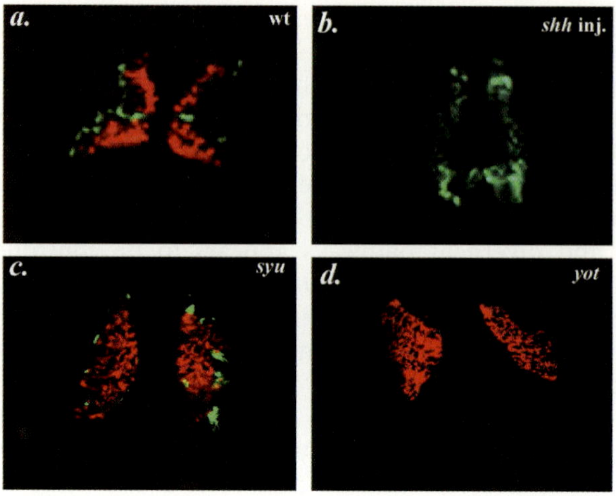

Fig. 5a–d. HH signaling controls muscle cell fates in the embryonic myotome. Cross sections at the level of the yolk extension reacted with antibodies against slow MyHC (*green*) and fast MyHC (*red*). Dorsal is to the top. **a** Wild-type embryo revealing the normal arrangement of fiber types. **b** Embryo in which *shh* mRNA has been injected at the 2-cell stage. Slow MyHC-expressing fibers are expanded at the expense of fast, which are entirely absent. **c** Homozygous *syu* mutant embryos in which slow muscle is reduced but not entirely eliminated. **d** Homozygous *yot* mutant embryo, which completely lacks slow MyHC-expressing fibers embryonically

twitch muscle formation does not require slow muscle differentiation as mutants that completely lack slow muscle still form fast muscle on schedule (Blagden et al. 1997).

The cellular phenotype of these two spatially separate muscle populations also differs. Slow muscle fibers remain mononucleate during their lateral migration (as do the non-migratory muscle pioneers) and, for some yet to be determined period, after migration is completed. In contrast, fast muscle cells start fusing to form multinucleate fibers shortly after differentiation begins (Roy et al. 2001). The reason for this difference is unclear, but slow fibers may need to remain mononucleate during migration, a process that may be impossible for large multinucleate syncytial fibers.

Although amniote fast and slow twitch muscle cells do not lie in spatially separable populations but are peppered throughout the myotome, recent studies reveal the existence of chick primary "pioneer" myoblasts. These cells express both slow and fast MyHC, which differs from later born cells that express only fast MyHC and intersperse between pioneer fibers (Kahane et al. 1998; Kalcheim et al. 1999). These findings support the hypothesis that the evolutionary origin of amniote myotome formation lies within the simplified muscle cell arrangement present in zebrafish. However, it must be acknowledged that amniote myotome formation remains controversial with several different models in evidence (Denetclaw et al. 1997; Denetclaw and Ordahl

2000). Furthermore, the overall architecture of the myotome relative to midline structures differs between amniotes and zebrafish. Whether these structural differences are reflected in the deployment of different signaling pathways to generate different somitic cell types is unknown.

5.2.2 Myotomal Patterning Mutants and the Molecular Mechanisms Controlling Slow Muscle Cell Specification

Mutational screens have identified many genes required for zebrafish embryonic development (Driever et al. 1996; Haffter et al. 1996). Several mutants, in addition to *flh* (described in Sect. 2.3), affect notochord formation (Halpern et al. 1993; Talbot et al. 1995; Odenthal et al. 1996a; Stemple et al. 1996). Invariably, mutant embryos that lack a differentiated notochord also show paraxial mesoderm defects. In particular, such embryos fail to form the horizontal myoseptum, show deficits in muscle pioneer and slow muscle fiber formation, and display u-shaped somite morphology, suggesting that the notochord controls formation of these cell types. For two mutants, *no tail* (*ntl*) and *doc*, this notion has been directly tested by transplanting wild-type notochord precursors into mutant embryos. These cells rescue horizontal myosepta and muscle pioneer cells in mutant hosts, but only within the somite directly adjacent to transplanted notochord cells (Halpern et al. 1993; Odenthal et al. 1996a). These observations strongly suggest that notochord-derived signals act at short range to control teleost myotome specification.

A number of proteins with inductive properties are expressed within the zebrafish notochord. Notable are members of the Hedgehog (HH) family of secreted glycoproteins, known to be required for patterning a variety of vertebrate embryonic tissues (Hammerschmidt et al. 1997). In zebrafish, three different HH proteins, Sonic hedgehog (SHH), Echidna hedgehog (EHH), and Tiggywinkle hedgehog (TwHH), are expressed in the midline (Krauss et al. 1993; Currie and Ingham 1996; Ekker et al. 1995). At some level, all these proteins contribute inductive cues to specific myotomal cells (Currie and Ingham 1996; Blagden et al. 1997; Du et al. 1997; Nasevicius and Ekker 2000; Norris et al. 2000). Embryos homozygous for a mutation in the *shh* gene itself, *sonic you* (*syu*; Schauerte et al. 1998), lack muscle pioneers and have fewer superficial adaxially derived slow muscle fibers (Fig. 5c; Lewis et al. 1999). Conversely, ubiquitous *shh* expression during early embryogenesis results in *myoD* activation throughout the PSM (Weinberg et al. 1996) and subsequent differentiation of the entire myotome into slow twitch muscle fibers (Fig. 5b; Blagden et al. 1997; Du et al. 1997).

Genetic inactivation of intracellular components of the HH signal transduction machinery results in a complete lack of adaxially derived muscle fates. Mutations at the *you-too* (*yot*) locus inactivate the zebrafish GLI-2 protein (Karlstrom et al. 1999), a member of a class of transcription factors required for HH signal transduction. Homozygous *yot* mutants lack all embryonic slow twitch muscle, including muscle pioneers, a phenotype that is shared with another recently isolated mutation *slow muscle omitted* (*smu*; Barresi et al. 2000). The activity of the *smu* gene has been placed upstream of cytosolic components of HH signal transduction, but downstream of the HH signal itself, suggesting a requirement for *smu* activity in HH signal reception. Confirming such a role, the *smo* phenotype has recently been shown to result from mutation within the zebrafish *smoothened* gene, the signaling component of the HH receptor complex (Chen et al. 2001a; Z. Varga, pers. comm.).

Several other mutants phenotypically similar to *syu*, *yot* and *smo* exist and collectively constitute a phenotypic class known as the u-type mutants, so named for their altered "u"-shaped somite morphology. There is good evidence to suggest that the majority of these mutants identify genes required for generation or transduction of Hedgehog signals (van Eeden et al. 1996; Schauerte et al. 1998; Lewis et al. 1999). One exception is the mutation *u-boot* (*ubo*) which acts epistatically downstream of the HH signal transduction pathway. Homozygous *ubo* mutants induce the adaxial compartment normally, as assayed by *myoD* expression (Lewis et al. 1996; Roy et al. 2001). However, the differentiation of most slow muscle fibers is subsequently blocked and these cells display an altered morphology and morphogenesis. In contrast, neither notochord-derived HH signaling nor the notochord itself is required for fast muscle cell formation (Fig. 5; Blagden et al. 1997). A separate, yet to be defined, mechanism controls specification of these cells.

5.3 Migratory or Hypaxial Muscle Formation in Zebrafish Embryos

In amniotes, somitically derived cells also contribute to muscle populations other than axial muscles, such as limb and body wall muscle. Myoblasts that contribute to these "hypaxial" muscles are derived from lateral aspects of the somite and are specified by signals distinct from those that pattern the somite at the midline. Detailed information on the developmental mechanisms that coordinate hypaxial muscle formation has relied heavily on data from avian and mammalian species (reviewed in Currie and Ingham 1998). These studies reveal that future limb myoblasts undergo an epithelial-to-mesenchymal transition in the lateral dermomyotome, before delaminating from the somites. Cells migrate singly or as small groups into the limb bud, where they proliferate and then differentiate to form the future limb dorsal and ventral muscle masses (Williams and Ordahl 1994). Migrating cells specifically express two transcription factors, Lbx and Mox2, mutations in which result in a lack of limb myoblasts (Mennerich et al. 1998; Dietrich et al. 1998; Mankoo et al. 1999; Schäfer and Braun 1999).

Until very recently, existence of homologous somitically derived muscle populations in fish had not been determined. In fact, the most detailed understanding of the origins of fin muscle-forming cells comes from studies performed during the late 1800s on pectoral fins of the chondricthyan shark *Scyliorhinus canicula* (Dohrn 1884; Braus 1899; reviewed in Goodrich 1958). A detailed characterization of embryos of different stages suggests that pectoral fin muscle is derived from direct extensions of myotome cells into the developing fin bud (Fig. 6a,b). Based on these and other studies, the "text-book" view has been that all fish generate fin muscle in this manner, and that the tetrapod migratory mode of limb muscle formation evolved to generate the more complex muscle required for limb-dominant locomotor strategies (Kardong

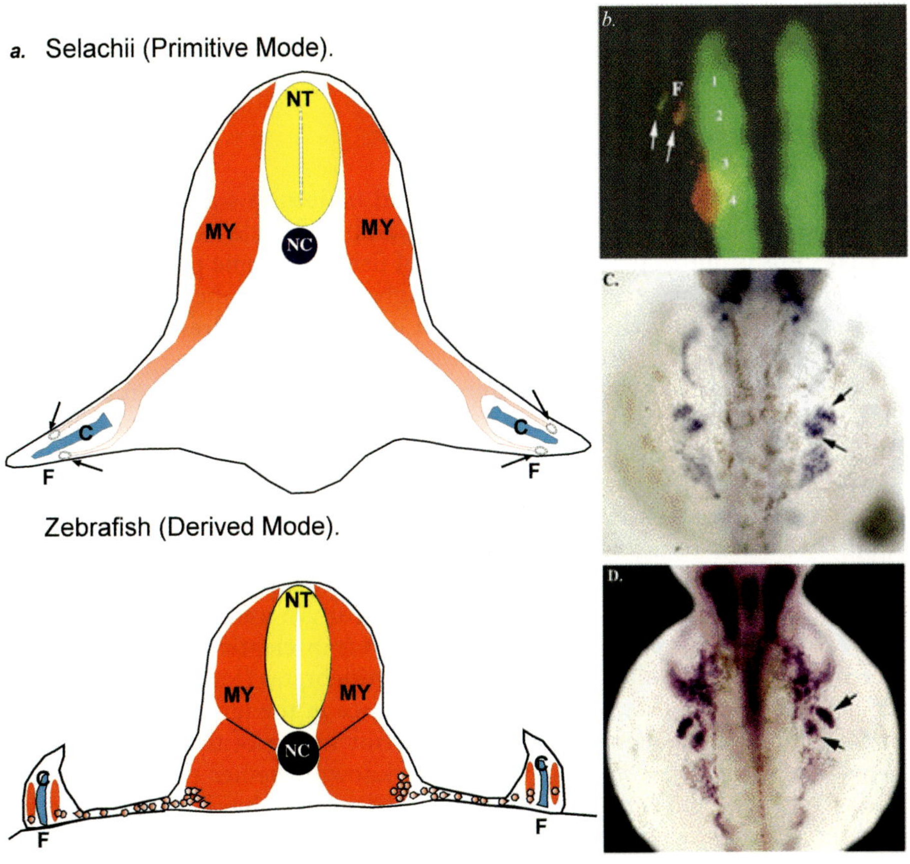

Fig. 6a–d. Hypaxial muscle formation in fish embryos. **a** Primitive fish such as Selachii utilize epithelial extensions (*arrows*) from the myotome (*MY*) to generate the muscles of the fin (*F*). Teleosts utilize an epithelial-to-mesenchymal transition of ventrolateral cells of the myotome to generate the muscles of the fins, a mechanism identical to that employed in amniote embryos to generate the muscles of the limbs. *c* Chondrocytes; *NC* notochord; *NT* neural tube. **b** Zebrafish fin myoblasts derived from fin level somites. Di I labeling (*red*) of the lateral region of somite 4 within a strain of fish driving GFP (*green*) under control of the muscle-specific alpha-actin promoter results in coincident labeling of Di I and GFP positive cells within the fin 24 h after labeling. 48 hpf, dorsal view. **c** MyoD expression is discrete within the forming fin revealing the dorsal and ventral muscle masses (*arrows*). 36 hpf, dorsal view. **d** Migrating fin myoblasts express LBX, a gene which marks migrating myoblasts in amniote embryos. 36 hpf, dorsal view

1995; Amthor et al. 1998). However, zebrafish fate mapping and time-lapse analyses show that teleost fin myoblasts do originate from fin level somites and these cells also express genes such as *Lbx* and *Mox* (Neyt et al. 2000). Collectively, these studies reveal that zebrafish embryos employ an identical morphogenesis to that utilized by amniote embryos to generate fin muscles. However, examination of fin muscle formation in *S. canicula* reveals the existence of direct extending muscle fibers, headed by epithelial extensions within the developing fin field. *S. canicula* somite extensions do not express *Lbx*, reinforcing the primitive nature of the genetic hierarchy that coordinates Selachian fin muscle formation.

5.4 Sclerotome Formation

While much is known about the origins and molecular signals coordinating sclerotome formation in amniotes, comparatively little is known in fish. Perhaps this is due in part to the scarcity of fish sclerotome progenitors – fate mapping reveals the existence of surprisingly few, and to the absence of molecular markers, although this latter deficit is now being redressed. Also, it has been difficult to be definitive about labeled cells contributing to the axial skeleton as chondrogenesis within the embryonic fish axis begins late in larval life. Despite these handicaps, intriguing observations have been made on fish sclerotome formation and are included here for discussion. Utilizing iontophoretic labeling methods in mid-somite stage zebrafish embryos, Morin-Kensicki and Eisen (1997) revealed that sclerotome arises in a ventromedial section of the somite, a similar position to that from which amniote sclerotome is derived. Cells labeled in this position early during somitogenesis migrate dorsally to surround the neural tube and notochord. The origin and migration of these cells mirrors the expression profile of the bHLH transcription factor gene *twist*, which marks sclerotome precursors in other vertebrates (Morin-Kensicki et al. 1997). These authors also note an A/P difference in the temporal sequence of migrating cells. Cells derived from the anterior region of the somite migrate before posteriorly derived cells, but it is unknown whether these cells define distinct sclerotomal fates. Stickney et al. (2000) note an interesting correlation between sclerotome migration and slow muscle and muscle pioneer cell movements, and suggest either population could regulate the morphogenesis of the other, although a causal relationship between these two events has yet to be established.

There is some evidence to support a more direct link between slow muscle and sclerotome formation, as attenuation of HH signaling can lead to sclerotome expansion and a concomitant loss of muscle pioneer cell formation (Hammerschmidt and McMahon 1998). It is worth noting that individually labeled posterior ventromedial cells not only give rise to sclerotome, but also can give rise to muscle, suggesting the existence of an uncommitted precursor pool. Complete blockage of HH signal transduction, however, leads to loss of both cell types, mirroring the requirement for HH signaling in sclerotome induction in amniotes (Fan and Tessier-Lavigne 1994). How a subtlety of HH signaling induces one cell type versus the other from epithelial somites remains unresolved.

Recent observations point to a role for the notochord in axial bone formation. Preliminary studies suggest that, during early vertebral body (centrum) formation, an acellular osteoid matrix is deposited in a periodic, striped pattern around the notochord, that is independent of somite periodicity. Explanted notochords have also been shown to secrete this matrix in a segmental pattern in vitro (A. Fleming et al., unpubl. observ.). Further evidence for the role of notochord in axial skeleton ossification comes from studies in another teleost, the Japanese medaka, where it has been shown that centra also arise from "acellular" bone that is deposited around the notochord (Ekanayake and Hall 1987, 1988). In this case, however, labeling studies suggest that the matrix is deposited by cells surrounding the notochord rather than by the notochord itself. These observations hint at the fact that ossification of the axial skeleton, previously thought to be entirely the domain of sclerotome-derived axial chondrocytes, may involve substantial input from the notochord itself, and even impart the metamerism of axial components independent of the somites. Whether sclerotome cells operate in a substantially different manner in fish compared to the role that these cells perform in amniotes has yet to be established.

5.5 Other Somite-Derived Cell Types

Despite careful characterization of fish myotomal and sclerotomal progenitors by dye-labeling techniques, other somite derivatives, that have been described in amniote species, have not been well characterized. For example, the dermatome, which in chick embryos has been shown to give rise to the dermis of the back and arises from the most dorsal aspect of the dermomyotome, does not seem to have any clear functional equivalent in zebrafish. Somewhat surprisingly, the epithelial structure of the dermomyotome itself, which in amniotes gives rise to both the myotome and dermatome, does not seem to have a direct functional ortholog in zebrafish. Furthermore, given the copious amount of zebrafish somite fate mapping data, one must assume that if a zebrafish somitic dermatome compartment does exist, it must constitute a small and highly proliferative component of the somite that has been so far refractile to existing fate mapping techniques. As Stickney et al. (2000) note, Waterman (1969), in a detailed structural study of the zebrafish somite, does attribute a few cells of the zebrafish somite in an aspect dorsal and external to the myotome, but whether these cells can contribute to the dermis is unknown. In fact, surprisingly little is understood about the origin of fish dermis, despite a wealth of fate mapping analysis revealing the origins of other zebrafish embryonic tissues.

Morin-Kensicki and Eisen (1997), in their fate mapping studies of the sclerotome mentioned above, provide the only other piece of evidence of a different cell type produced from the somite. These studies reveal a somitic origin for some of the cells of the blood vessels, a phenomenon also recently described to occur in chick embryos (Ambler et al. 2001).

6 Questions for the Future

Although there have been significant advances in our understanding of zebrafish somitogenesis, there are many open questions. Although one gene (*spadetail*) has been identified that regulates paraxial mesoderm formation, we await the identification of additional regulatory genes and of genes that act downstream of *spadetail*. Once the paraxial mesoderm forms, it becomes segmented in a temporally and spatially precise manner. Evidence is mounting that supports a "clock and wavefront" model of segmentation with Notch signaling playing an important role in both synchronizing the clock and in establishment of A/P somite identity. However, the molecular nature of the clock, what drives the clock, how the clock is linked to Notch signaling, and the molecular basis of the maturation wavefront that stops the clock, have yet to be determined. Further analysis of the zebrafish fused-somite class of mutants and mutants isolated in additional genetic screens will help answer these questions. For example, identification of the *fss* gene product will greatly aid our understanding of the link between the segmentation clock and operation of the wavefront. This mutant has been placed on the zebrafish genetic map and experiments to identify the gene are underway. Work in zebralish shows FGF signaling regulates *her1* expression and the stopping of the clock. How this is achieved is not understood. It will also be interesting to determine the role of FGF signaling in somite formation in explants of PSM tissue. Finally, it is unclear whether the cell cycle is involved in regulation of segmentation.

Once A/P somite identity is established, somite boundaries form as the result of differential cell surface properties. Some of these cell surface molecules have been identified. However, whether these molecules interact synergistically or redundantly to restrict cell mixing between presumptive somites is not known. Furthermore, it is by no means clear why a boundary fails to form at the center of each somite.

Also little is clearly understood about how different cell lineages and individual cell fates within lineages are determined within the zebrafish somite once it has been formed. We even remain unclear about the nature of cell types that are derived from the epithelial somite and the molecular cues that control the formation and migration of these cells. While progress has been made through the analysis of mutations and molecules that determine slow muscle formation, the adaxial cells remain a small subset of somitically derived cells. A real challenge for the future is to chart the morphogenesis and molecular processes that control the specification of the remainder of the cells of the zebrafish somite.

Acknowledgements. We thank Drs. A. Sawada, S. Holley, A. Chitnis, J. Dubrulle, H. Takeda, and O. Pourquie for access to results prior to publication. Research in the Amacher laboratory is supported in part by a Basil O'Connor Starter Scholar Award (#5-FY00-628) from the March of Dimes Birth Defects Foundation and by the University of California Cancer Research Coordinating Committee funds. Work mentioned here from the Brennan laboratory is supported by the the BBSRC (68/G15805) and the Royal Society, London. Research in the Currie laboratory is supported in part by an MRC Career Development Fellowship, the BBSRC, the Scottish Hospitals Endowment Fund, The Muscular Dystrophy Campaign UK and the EMBO Young Investigators Programme. We thank R.P. Currie for comments and suggestions on the manuscript.

Cardiovascular System

Deborah Yelon[1], Brant M. Weinstein[2], and Mark C. Fishman[3]

1 Introduction

One of the hopes of the large-scale genetic screens in zebrafish was that they would be informative regarding the process of patterning organ systems. That single gene mutations could reveal a logic to developmental patterning was evident from previous genetic screens in *Drosophila*, and these genetic findings have been successfully extrapolated to vertebrates by dint of evolutionary gene conservation. However, the *Drosophila* screens revealed little about organogenesis. How are the size and global form of organs controlled? What controls the onset of physiological function, especially when integrated among organs? At which points do development of form and function intersect? Are the parts of a complex organ generated as individual units that can be ablated by mutations in single genes? Alternatively, are pathways so interwoven and entangled that all mutations lead to an uninterpretable jumble of amorphous tissue?

One reason for initial skepticism about the power of genetic screens vis-a-vis organ systems was that mutations affecting an organ of interest would be too pleiotropic to interpret. This was particularly worrisome for the cardiovascular system. It is the first organ system to form and function and, in mammals, any perturbation of embryonic cardiovascular function causes hypoxic deterioration within minutes. The zebrafish embryo, however, lives by diffusion of oxygen from its environment for several days (Pelster and Burggren 1996). The complementary concern was that systems as important as the heart would be built with a redundancy so great as to obscure any defects from single gene mutations. Fortunately, single gene mutations in zebrafish have proven to be informative about cardiovascular development.

[1] Skirball Institute of Biomolecular Medicine, New York University School of Medicine, 540 First Avenue, New York, New York 10016, USA

[2] Unit on Vertebrate Organogenesis, Laboratory of Molecular Genetics, NICHD/NIH, 6 Center Drive, Bethesda, Maryland 20892, USA

[3] Cardiovascular Research Center, Massachusetts General Hospital, 149 13th Street, Charlestown, Massachusetts 02129, USA

Results and Problems in Cell Differentiation, Vol. 40
L. Solnica-Krezel (Ed.): Pattern Formation in Zebrafish

We focus here on progress emerging from the last few years of work with zebrafish, in particular with regard to the heart and vessels, although the same pattern formation issues can, and are, being addressed for other organs such as kidney and gut (Drummond, this Vol.). Clearly, mutations of genes and pathways responsible for organogenesis could have important ramifications for disease and repair in humans.

1.1 Background in Classical Embryology: Some of the Questions

Recent discoveries in zebrafish must be placed in the context of classical embryology that has established some of the paradigms in search of a molecular explanation. The region of the mesoderm containing precursors for the heart is the heart field (Fishman and Chien 1997). This was assayed experimentally in several organisms by tissue explantation or transplantation as the region specified (or determined, respectively) to become heart. The field has a defined border, within which there is a gradient of propensity to become heart. Importantly, the field is quite plastic, and regulates to perturbation. Even after removal of the entire heart field, embryos of many species have been noted to form a normally sized heart. Of course, these experiments do not define the origin of precursors in vivo or of the reparative cells during regulation. Both of these can be addressed at a single cell level only by lineage analysis, of the type pioneered in the transparent zebrafish embryo. These classical experiments also do not address the genetic control of heart field formation and regulation, but genetic studies in zebrafish have begun to reveal relevant players.

Classical embryology studies have also established that the heart field, located within the lateral plate mesoderm, is closely apposed to the anterior endoderm (Fishman and Chien 1997). Removal of the endoderm in several species prevents heart formation, indicating that the endoderm sends key signals to nascent cardiocytes. The nature of those signals, however, was poorly understood prior to genetic studies in zebrafish. As cells migrate medially, just before merging to form the primitive heart tube, a pattern emerges in the field, with pre-atrial cells separate from pre-ventricular (as assayed by molecular markers, or by characteristic beat rates in culture). It was not clear how cells with these two fates assemble into a long continuous tube, with an immediate functional polarity pumping blood from inflow (atrium) to outflow (ventricle). Again, zebrafish genetics have helped to clarify this step.

In all embryos, the heart grows at a rate proportional to the metabolic needs of the animal (Fishman and Chien 1997). What controls this proportionate growth is not known. In the adult, growth (in the sense of hypertrophy) is controlled by use (blood pressure and systolic volume), but it is not known when this form/function linkage is established and whether activity helps to mold heart growth. In the embryo, growth is a complex phenomenon. In the ventricle, for example, linear accretion of cells is only one dimension of growth. Shortly after the chambers form, new myocytes are added to the ventricle especially in a concentric direction, causing a thickening of the wall needed to generate higher systemic blood pressure. Specific mutations in zebrafish speak to each of the steps. There are clearly non-autonomous signals involved in this patterning. For example, the single-layered endocardial lining to the heart, the endocardium, contributes at least two types of myocardial growth patterning signals. One, exemplified by neuregulin, stimulates a trabecular (corrugated) pattern of thickening, presumably enhancing local oxygenation from the chamber side. The other is the focal stimulation of specialized development of interchamber cushions, the forerunners of the diaphanous valves that help to guarantee unidirectional blood flow.

The first vessels of vertebrates, a major artery and vein in the trunk, form by assembly of free angioblasts. This process, termed vasculogenesis, is poorly understood. Where do the angioblasts originate? What drives their cell fate decision to become artery or vein? What regulates tubulogenesis? Is it autonomous or due to signals from neighboring tissues? Once formed, these vessels branch, both intersegmentally and into what will become organs, such as the kidney, and incorporate as well into a head circulation. Even the order of appearance of these vessels, whether perfectly patterned from the onset or rather rearranged over time, as for axonal patterns in the nervous system, is not well described. What guides the specific branch points and growth patterns? Several factors, notably VEGF, have been suggested in other organisms to cause growth, but are these sufficient, and are they also the patterning agents? Cell tracking, vascular angiography, and mutational analysis in zebrafish have begun to address these issues.

2 Zebrafish: a Propitious Embryo for Cardiovascular Studies

Several features of the zebrafish embryo make it an appealing model organism for the study of cardiac and blood vessel development (Stainier and Fishman 1994; Vogel and Weinstein 2000). First, as mentioned above, embryonic zebrafish can survive without a functional circulatory system (Pelster and Burggren 1996). Also, since the transparent zebrafish embryo is fertilized externally, it is possible to observe cardiac and vascular cells extensively throughout their development. Finally, as stressed throughout this volume, the small size, fecundity, and brief generation time of the zebrafish make it possible to conduct large-scale genetic screens (Westerfield 1995). Altogether, these features allow the identification and study of mutations affecting many aspects of zebrafish cardiovascular development. Taking this approach, several groups have successfully identified key regulators of cardiac and vascular pattern formation.

Here, we review the development of the zebrafish heart and vasculature, describe molecular, genetic and experimental tools available for their analysis, and highlight recent findings from fish that have contributed to a deeper understanding of the patterning of the cardiovascular system during development. We begin by review of the key steps of pattern formation during zebrafish cardiac development, emphasizing molecular markers of developing myocardium and important regulators of myocardial patterning. We then describe genetic studies of patterned elements of cardiac function in the zebrafish. Last, we review the genesis of blood vessels in the zebrafish and some of the genes, mutants and experimental tools used to analyze embryonic blood vessel formation.

3 Patterning the Heart

Fate maps of the zebrafish embryo indicate that myocardial and endocardial progenitors are both located in the ventrolateral marginal zone of the early blastula (512-cell stage), with the endocardial progenitors found more ventrally than most of the myocardial progenitors (Stainier et al. 1993; Lee et al. 1994). In the late blastula (40% epiboly), the myocardial and endocardial progenitors are clustered laterally within the three tiers of blastomeres closest to the embryonic margin (Warga and Nüsslein-Volhard 1999). At this stage, endothelial progenitors are also found near the margin, but are spread over a broader circumference of the embryo and distributed among the first six marginal tiers (Warga and Nüsslein-Volhard 1999). These marginal blastomeres are among the first to involute when gastrulation begins at 50% epiboly (Warga and Kimmel 1990). In this section, we will focus on the differentiation and morphogenesis of the myocardial progenitors following gastrulation. (Major steps of this process are summarized in a timeline in Fig. 1C.)

3.1 Formation of the Myocardium in Zebrafish

During early somitogenesis, the myocardial progenitors, together with the rest of the lateral plate mesoderm, converge toward the embryonic midline and extend toward the anterior of the embryo (Stainier and Fishman 1992; Stainier et al. 1993). This gradual movement is reflected in the changing expression patterns of several genes that are strongly expressed within the anterior lateral plate mesoderm (ALPM), including *gata4*, *gata5*, *gata6*, *hand2*, *oep*, *tbx5*, and *tbx20* (Tables 1 and 2). The precardiac mesoderm can be distinguished within the ALPM by its expression of the transcription factor gene *nkx2.5* (Chen and Fishman 1996; Lee et al. 1996) and the growth factor gene *bmp4* (Chen et al. 1997; Fig. 1A). Additionally, the growth factor gene *fgf8* is expressed in and/or near the precardiac mesoderm soon after somitogenesis begins (Reifers et al. 2000b).

By mid-somitogenesis, the bilateral populations of precardiac mesoderm are located near the embryonic axis, at the anterior-posterior level of the prechordal plate/notochord boundary. Most of the precardiac mesoderm lies anterior to the rostral tip of the notochord, but the posterior portion of the

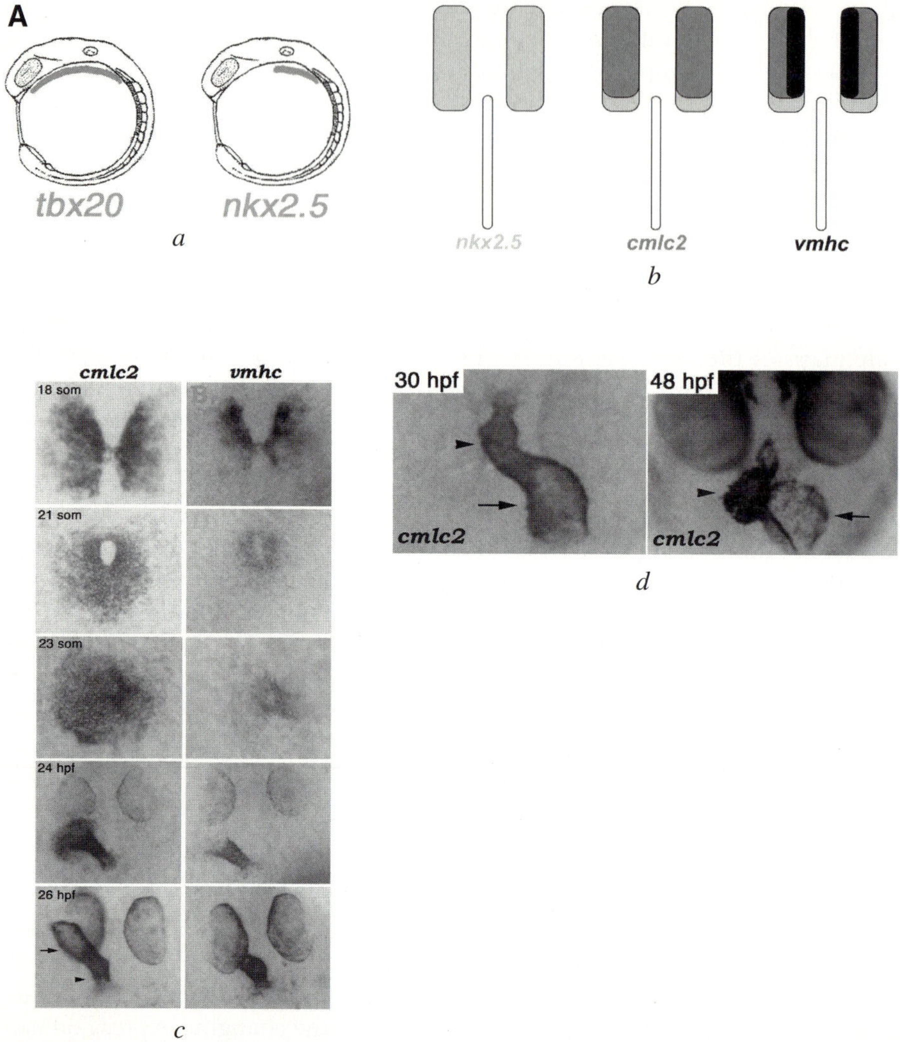

Fig. 1A. *a nkx2.5* is expressed in a portion of the ALPM. Schematics are lateral views at the 10-somite stage, comparing expression patterns of *tbx20* (found in the ALPM) and *nkx2.5* (found in a portion of the ALPM that includes the myocardial precursors). See Ahn et al. (2000); Griffin et al. (2000). *b* Markers of myocardial differentiation subdivide the precardiac mesoderm. Schematics compare expression patterns at the 15-somite stage, showing dorsal views of the ALPM, with anterior at the top. Notochord is depicted as a *white rod*. Overlaps between *nkx2.5*, *cmlc2*, and *vmhc* expression are shown in *different shades of gray* (Yelon et al. 1999). *c* Cardiac fusion and heart tube assembly involves dynamic cell movements. Images are dorsal views, anterior at the top, comparing expression of *cmlc2* and *vmhc* at stages during the process of heart tube assembly. (Adapted from Yelon et al. 1999). *d* Cardiac looping reorients the ventricle and atrium. Images are head-on views, comparing cardiac morphology in a looping heart (30 hpf) and a looped heart (48 hpf). The ventricle is indicated with an *arrowhead*, and the atrium is indicated with an *arrow*. (Adapted from Yelon et al. 1999)

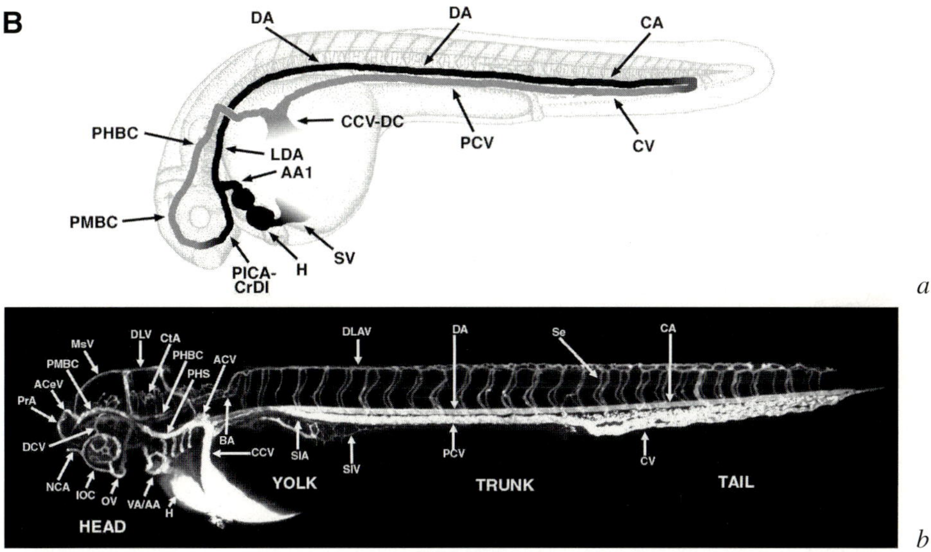

Fig. 1B. Vascular anatomy of the developing zebrafish. *a* Diagram of active blood vessels in the zebrafish embryo at approximately 26 hpf. (Adapted from Isogai et al. 2001). Lateral view with vessels on the right side of the embryo not shown. One circulatory loop goes caudally from the first aortic arch through the trunk and tail via the dorsal aorta/caudal artery, returning via the caudal vein/posterior cardinal vein. A second cranial circulatory loop goes rostrally from the first aortic arch into the internal carotid artery and its cranial division, returning via the primordial mid- and hindbrain channels. *b* Confocal microangiogram of a developing zebrafish at approximately 3 dpf, compiled from four separate reconstructions pasted together. (Adapted from Isogai et al. 2001). Lateral view, with vessels labeled. Many additional blood vessels have formed by this stage and circulatory patterns have become much more complex. All panels are oriented with rostral to the left and dorsal up. A glossary of the names corresponding to all labeled vessels is provided in Tables 1 and 2 of Isogai et al. 2001, and online at *The Interactive Atlas of Zebrafish Vascular Anatomy* (http://eclipse.nichd.nih.gov/nichd/lmg/redirect.html)

precardiac mesoderm lies beside the notochord (Serbedzija et al. 1998). Terminal differentiation of the myocardial precursors begins around the 13-somite stage (Yelon et al. 1999). Differentiating cells initiate expression of genes that encode cardiac sarcomere proteins, such as *cmlc1* and *cmlc2* (Reiter et al. 1999; Yelon et al. 1999; Tables 1 and 2). Most, but not all, of the *nkx2.5*-expressing precardiac mesoderm will undergo myocardial differentiation and contribute to the myocardium of the heart (Fig. 1A). However, the cells that are located adjacent to the notochord do not express terminal differentiation markers (Serbedzija et al. 1998; Yelon et al. 1999), and these cells give rise to head mesenchyme instead of myocardium (Serbedzija et al. 1998).

The differentiating myocardial precursors are divided into two subpopulations by the expression pattern of *vmhc*, which encodes a ventricle-specific myosin heavy chain (Yelon et al. 1999). Following tube assembly, the heart will

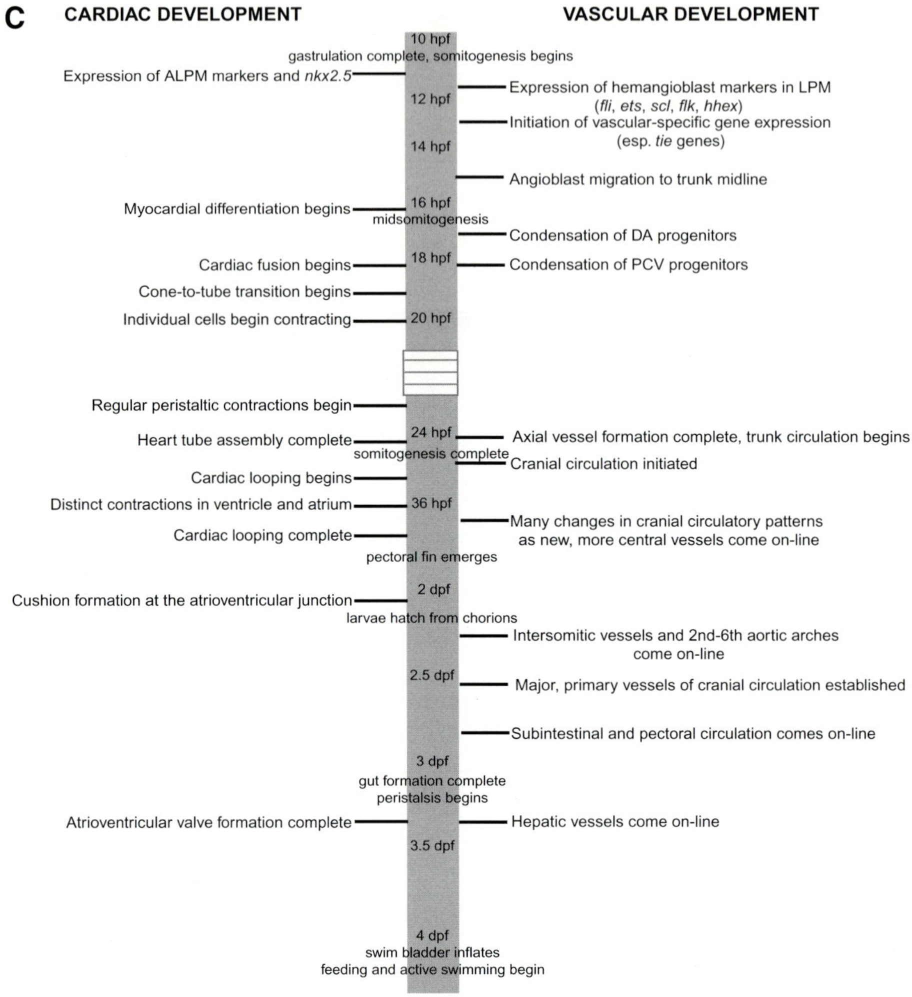

Fig. 1C. Timeline of developmental milestones in zebrafish cardiac and vascular development. The approximate timing of major events in cardiac (*left*) and vascular (*right*) development is shown relative to age and other major developmental milestones (*center*)

be composed of two major chambers, a ventricle and an atrium. In all vertebrates, the ventricular myocardium differs from the atrial myocardium by many molecular, histological, and physiological criteria (DeHaan 1965; Satin et al. 1988; Lyons 1994; Fishman and Chien 1997; Franco et al. 1998). In zebrafish, the expression of *vmhc* in a medial subpopulation of myocardial precursors is the earliest known indication of chamber-specific differentiation (Yelon et al. 1999). The medial *vmhc*$^{+}$-positive cells are thought to be the ventricular precursors, and the lateral *vmhc*$^{-}$-negative cells are thought to be the atrial precursors (Fig. 1A).

Table 1. Zebrafish genes expressed in developing myocardium

Gene name	Type of gene	Earliest site of cardiac expression	Onset of cardiac expression	Function[a]	Reference(s)
gata4	Transcription factor	Anterior lateral plate mesoderm	Early somitogenesis	?	Serbedzija et al. (1998)
gata5	Transcription factor	Anterior lateral plate mesoderm	Early somitogenesis	A, B, C	Reiter et al. (1999)
gata6	Transcription factor	Anterior lateral plate mesoderm	Early somitogenesis	?	Reifers et al. (2000b)
hand2	Transcription factor	Anterior lateral plate mesoderm	Early somitogenesis	B	Angelo et al. (2000); Yelon et al. (2000)
oep	EGF-CFC protein	Anterior lateral plate mesoderm	Early somitogenesis	A, B	Zhang et al. (1998b); Gritsman et al. (1999); Reiter et al. (2001b)
tbx5	Transcription factor	Anterior lateral plate mesoderm	Early somitogenesis	?	Begemann and Ingham (2000); Ruvinsky et al. (2000)
tbx20	Transcription factor	Anterior lateral plate mesoderm	Early somitogenesis	?	Ahn et al. (2000); Griffin et al. (2000)
nkx2.5	Transcription factor	Precardiac mesoderm	Early somitogenesis	C	Chen and Fishman (1996); Lee et al. (1996)
bmp4	Growth factor	Precardiac mesoderm	Early somitogenesis	D	Chen et al. (1997)
fgf8	Growth factor	Near or within precardiac mesoderm	Early somitogenesis	A, B	Reifers et al. (2000b)
cmlc1	Myosin light chain	Myocardial precursors	Mid-somitogenesis	E	Reiter et al. (1999)
cmlc2	Myosin light chain	Myocardial precursors	Mid-somitogenesis	E	Yelon et al. (1999)
vmhc	Myosin heavy chain	Ventricular precursors	Mid-somitogenesis	E	Yelon et al. (1999)

[a] Functions during zebrafish myocardial development (if known) are coded as follows: *A*, regulates *nkx2.5* expression; *B*, regulates myocardial differentiation/patterning; *C*, when overexpressed can induce expression of cardiac differentiation markers; *D*, may influence heart laterality; E component of cardiac sarcomere

Table 2. Zebrafish genes expressed in developing vasculature

Gene name	Type of gene	Expressed in:	Function[a]	Reference
vegfr-2/flk-1	Receptor tyrosine kinase	Endothelium	1, 2, 3	Fouquet et al. (1997)
vegfr-3	Receptor tyrosine kinase	Venous/lymphatic endothelium	2	Thompson et al. (1998)
vegf (vegf-A)	Ligand	Endoderm, somites, hypochord, etc.	1, 2, 3	Liang et al. (1998)
tie-1	Receptor tyrosine kinase	Endothelium	2, 3, 4	Lyons et al. (1998)
tie-2	Receptor tyrosine kinase	Endothelium	2, 3, 4	Lyons et al. (1998)
angiopoietin-1	Ligand	Mesenchyme	2, 3, 4	Pham et al. (2001)
angiopoietin-2	Ligand	Mesenchyme	2, 3, 4	Pham et al. (2001)
Eph-B4	Transmembrane receptor	Venous endothelium	2, 3	Lawson and Weinstein (unpubl.)
ephrin-B2	Ligand	Arterial endothelium	2, 3	Lawson and Weinstein (unpubl.)
notch-3	Transmembrane receptor	Arterial endothelium	?	Itoh and Chitnis (2001)
deltaC	Ligand	Arterial endothelium	?	Smithers et al. (2000)
scl	Transcription factor	Endothelium	1	Thompson et al. (1998)
fli-1	Transcription factor	Endothelium	?	Thompson et al. (1998)
ets-1	Transcription factor	Endothelium	?	Thompson et al. (1998)
hhex	Transcription factor	Endothelium	?	Liao et al. (2000)
gridlock	Transcription factor	Arterial endothelium	1	Zhong et al. (2000)

[a] Functions during vascular development in zebrafish or in other vertebrates are coded as follows: 1, angioblast differentiation, migration, vasculogenic tubule formation; 2, angiogenic remodeling of plexuses; 3, angiogenic sprouting; 4, formation of perivascular sheath

Around the 18-somite stage, the bilateral populations of cells meet each other at the embryonic midline, beginning the process of cardiac fusion. Then, a dynamic series of cell movements reorganize the ventricular and atrial precursors to form a tube with ventricular and atrial cells composing its opposite ends (Yelon et al. 1999). Specifically, fusion begins as the posterior portions of the medial edges of the two sheets of myocardial precursors come together, followed by an anterior closure that creates a central lumen (Fig. 1A). Thus, the myocardial precursors form a shallow cone, the apex of which is raised dorsally around the lumen. Next, the myocardial cone is gradually transformed into a linear tube. This process begins when the apex of the cone tilts posteriorly and to the right, shifting the axis of the cone from a dorsal-ventral plane to an anterior-posterior plane. This rightward tilt is the earliest morphological indication of left-right asymmetry in the zebrafish embryo. (Molecular left-right asymmetries, such as left-sided expression of *pitx2*, *cyclops*, and *lefty*, are apparent earlier; the topic of left-right axis formation is covered elsewhere, see Wright and Halpern, this Vol.). The tilted cone then extends, and its anterior-posterior axis elongates until the base of the cone coalesces into a tube.

After only 1 day of development, the zebrafish heart tube exhibits regular, peristaltic contractions that drive circulation from its atrial inflow region through the ventricular outflow tract (Warren and Fishman 1998). During the next 12h, continuing morphogenesis converts the linear heart tube into a looped organ, with the ventricle positioned to the right of the atrium (Fig. 1A). Additionally, chamber-specific differentiation continues, including a marked thickening of the ventricular wall (Hu et al. 2000). By 36h postfertilization (hpf), the ventricle and the atrium exhibit distinct sequential contractions. By 48hpf, the separation of the ventricle and the atrium is further defined by the formation of cardiac cushions at the atrioventricular boundary; these cushions will later differentiate into the atrioventricular valve.

These stages of myocardial morphogenesis and the expression patterns of myocardial markers in zebrafish bear a strong resemblance to the morphogenesis and expression patterns observed in other vertebrates, including mammals (Fishman and Chien 1997). Later stages are less similar, since the hearts of vertebrates with lungs undergo additional septation events that are not found in fish. The similarity of the initial stages of cardiac patterning across vertebrate taxa suggests that the genetic pathways regulating heart formation may be highly conserved. In fact, recent work in *Drosophila* has suggested that the genetic pathways establishing cardiomyocyte cell fate have been well conserved throughout metazoan evolution (Chen and Fishman 2000).

3.2 Genetic Regulation of Myocardial Development in Zebrafish

The dynamic process of heart formation involves a number of tightly regulated specification, differentiation, and morphogenesis events, ranging from the early induction of *nkx2.5* expression to the later chamber assembly and induction of valve formation. The genetic control of many of these events has been investigated in the zebrafish. Here, we emphasize the studies of zebrafish mutants that have advanced our understanding of the initial pattern formation events that establish the myocardial precursors (Table 3).

3.2.1 Requirements for *nkx2.5* Induction

Studies of zebrafish mutations have indicated four genes – *gata5*, *fgf8*, *bmp2b*, and *oep* – that play roles during the induction of *nkx2.5* expression. The transcription factor gene *gata5* is expressed broadly within the ALPM, including the region that expresses *nkx2.5* (Reiter et al. 1999). Mutation of the *faust/gata5* locus results in a severe reduction of *nkx2.5* expression (Reiter et al. 1999). Few myocardial cells differentiate in *faust* (*fau*) mutants, and differentiation of ventricular precursors is particularly inhibited. Overexpression of *gata5* can cause expanded and/or ectopic expression of *nkx2.5* as well as several myocardial markers (Reiter et al. 1999). Altogether, these data indicate that Gata5 is necessary for the induction of precardiac mesoderm and also sufficient for this induction when overexpressed. However, *gata5* is normally expressed in many cells that do not become myocardium, such as other regions of the ALPM and also the endodermal progenitors, so it is likely that *gata5* cooperates with other essential regulators during the specification of cardiac precursors.

Like Gata5, the growth factor Fgf8 is also important for the initial induction of *nkx2.5* expression in precardiac mesoderm. *fgf8* expression can be detected in the precardiac mesoderm and neighboring cells during early somitogenesis, and mutation of the *acerebellar/fgf8* locus causes a reduction of *nkx2.5* expression (Reifers et al. 2000b). Also, *acerebellar* (*ace*) mutants exhibit significant reduction of differentiated myocardial tissue, with the ventricle being more strongly affected than the atrium. Implantation of Fgf8-soaked beads near the ALPM can slightly expand the expression of *nkx2.5* (Reifers et al. 2000b). These data suggest that Fgf8 is necessary, but not sufficient, for *nkx2.5* induction. Perhaps regionally restricted Fgf8 signaling collaborates with the more widely distributed Gata5 activity to select the precardiac portion of the ALPM.

The growth factor Bmp2b is important for the assignment of ventrolateral fates in the zebrafish embryo (Kishimoto et al. 1997; Nguyen et al. 1998b). Thus, *swirl/bmp2b* mutants have a severe reduction of ventral mesoderm, including *nkx2.5*-expressing precardiac mesoderm (Kishimoto et al. 1997; Reiter et al. 2001b). Accordingly, *swirl* (*swr*) mutants have very little differentiated myocardium. *swr* mutants exhibit reduced *gata5* expression during and after gastrulation, and overexpression of *gata5* has been shown to rescue *nkx2.5* expression and myocardial differentiation in *swr* mutants (Reiter et al. 2001b). Together, these observations suggest that Bmp2b signaling can act through Gata5 to regulate early steps of myocardial development.

Table 3. Zebrafish mutations affecting cardiovascular development. For information on other relevant mutations, see Chen et al. (1996) and Stainier et al. (1996)

Mutation name	Type of gene	Cardiovascular phenotype	Additional relevant phenotypes	References
faust (*fau*)	Transcription factor	Reduced *nkx2.5*; reduced myocardial differentiation (especially ventricular); cardia bifida	Reduction of gut endoderm	Reiter et al. (1999)
acerebellar (*ace*)	Growth factor	Reduced *nkx2.5*; reduced myocardial differentiation (especially ventricular)		Reifers et al. (2000b)
one-eyed pinhead (*oep*)	EGF-CFC protein	Reduced *nkx2.5*; reduced myocardial differentiation (especially ventricular); cardia bifida	Absence of gut endoderm	Zhang et al. (1998b); Gritsman et al. (1999); Reiter et al. (2001b)
swirl (*swr*)	Growth factor	Reduced *nkx2.5*; reduced myocardial differentiation		Kishimoto et al. (1997); Reiter et al. (2001b)
hands off (*han*)	Transcription factor	Reduced myocardial differentiation (especially ventricular); cardia bifida		Yelon et al. (2000)
pandora (*pan*)	?	Reduced myocardial differentiation (especially ventricular)		Yelon et al. (1999); Griffin et al. (2000)
miles apart (*mil*)	G-protein-coupled receptor	Cardia bifida		Kupperman et al. (2000)
casanova (*cas*)	Transcription factor	Cardia bifida	Absence of gut endoderm	Alexander et al. (1999); Kikuchi et al. (2001)
bonnie and clyde (*bon*)	Transcription factor	Cardia bifida	Reduction of gut endoderm	Kikuchi et al. (2000)
heart and soul (*has*)	Atypical PKC	Aberrant heart tube assembly		Yelon et al. (1999); Horne-Badovinac et al. (2001); Peterson et al. (2001)
jekyll (*jek*)	UDP-glucose dehydrogenase	No valve formation		Walsh and Stainier (2001)
slow mo (*smo*)	?	Chronic bradycardia (slow heart beat)		Baker et al. (1997); Warren et al. (2001)

Table 3. (*Continued*)

Mutation name	Type of gene	Cardiovascular phenotype	Additional relevant phenotypes	References
island beat (*isl*)	L-type calcium channel	Silent ventricle; atrial fibrillation		Rottbauer et al. (2001)
pickwick (*pik*)	Titin	Weak heart beat; dilated cardiomyopathy		Xu and Fishman (unpubl.)
cloche (*clo*)	?	Absence of nearly all blood and vascular endothelial cells		Stainier et al. (1995); Liao (1997, 1998); Thompson et al. (1998); Parker and Stainier (1999)
gridlock (*grl*)	Transcription factor	Blockage in anterior dosal aorta; loss of caudal circulation at 1 dpf		Weinstein et al. (1995); Zhong et al. (2000)
kurzschluss (*kus*)	?	Arterial-venous shunt from posterior aortic arches; loss of caudal circulation at 3 dpf		Chen et al. (1996)
violet beauregarde (*vbg*)	?	Malformation of central cranial vasculature; loss of caudal circulation at 2 dpf		(unpubl.)
sonic you (*syu*)	Ligand	Fail to form a dorsal aorta	Somite patterning defects	Van Eeden et al. (1996); Brown et al. (2000b)
you-too (*yot*)	Transcription factor	Fail to form a dorsal aorta	Somite patterning defects	Van Eeden et al. (1996); Brown et al. (2000b)
chameleon (*con*)	?	Fail to form a dorsal aorta	Somite patterning defects	Van Eeden et al. (1996); Brown et al. (2000b)
slow muscle omitted (*smu*)	Receptor	Fail to form a dorsal aorta	somite patterning defects	Barresi et al. (2000)
floating head (*flh*)	Transcription factor	Fail to form a dorsal aorta	No notochord	Fouquet et al. (1997); Sumoy et al. (1997)
no tail (*ntl*)	Transcription factor	Fail to form a dorsal aorta	No notochord	Sumoy et al. (1997)

Like Bmp2b, the EGF-CFC protein Oep is required for *gata5* expression in the ALPM. Oep is an essential cofactor for Nodal signaling, and *oep* is expressed broadly both maternally and zygotically (Zhang et al. 1998b; Gritsman et al. 1999). The defects in mutant embryos lacking zygotic Oep (*Zoep*) include reduction of *gata5* expression, reduction of *nkx2.5* expression, and reduction of myocardial differentiation, especially ventricular differentiation (Reiter et al. 2001b). *gata5* overexpression can rescue *nkx2.5* expression and myocardial differentiation in *Zoep* mutants (Reiter et al. 2001b), indicating that Nodal signaling, like Bmp2b signaling, can act through Gata5 to regulate early steps of myocardial development.

3.2.2 Requirements for Myocardial Differentiation

Since some *nkx2.5*-expressing cells do not differentiate into myocardial tissue, it is likely that additional factors, besides those required for *nkx2.5* induction, contribute to the selection of the myocardial precursors. *fau* and *ace* mutants exhibit reduced expression of terminal differentiation markers, but these defects may be secondary to the loss of *nkx2.5* expression in these embryos. The effects of *nkx2.5* loss of function have not yet been examined in zebrafish, but *nkx2.5* overexpression can induce the expanded and/or ectopic expression of some myocardial markers (Chen and Fishman 1996).

In contrast to the activities of Gata5 and Fgf8, the transcription factor Hand2 is dispensable for *nkx2.5* induction, but essential for myocardial differentiation (Yelon et al. 2000). *hand2* is expressed in a large portion of the ALPM, and mutation of the *hands off/hand2* locus results in a significant reduction of differentiated myocardial tissue, especially ventricular tissue, without affecting the initial expression of *nkx2.5* in precardiac mesoderm (Fig. 2A). It is not yet clear how the broad expression of *hand2* in the ALPM regulates the selection of myocardial precursors within the *nkx2.5*-expressing populations of precardiac mesoderm; further study of the binding partners and transcriptional targets of Hand2 will be required to clarify its role.

To date, *hand2* is the only zebrafish cardiac gene known to regulate myocardial differentiation without regulating *nkx2.5* expression. Of course, non-autonomous influences are also likely to contribute to the selection of myocardial precursors. In this vein, it is interesting to note that the notochord appears to play a repressive role during myocardial differentiation. Normally, *nkx2.5*-expressing cells that are adjacent to the notochord do not undergo myocardial differentiation, but when notochord cells are removed (via laser-mediated ablation or mutation of the *no tail* locus), this population can contribute to the myocardium (Goldstein and Fishman 1998). So, negative signals from the notochord (and/or positive signals from the prechordal plate) can influence myocardial differentiation, but the molecular nature of these signals has not yet been identified.

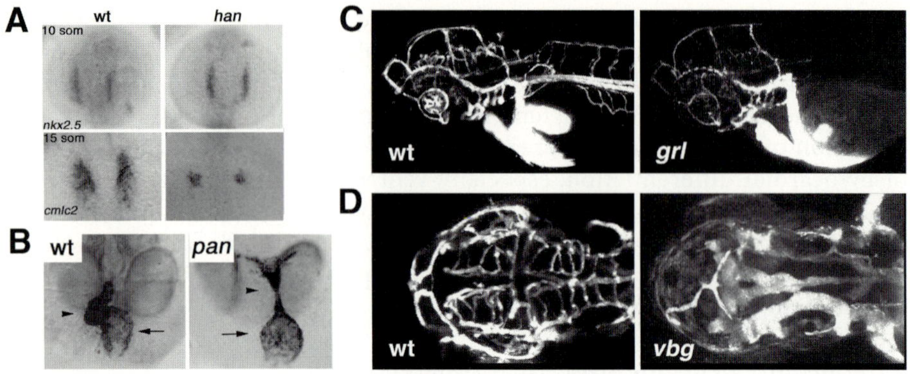

Fig. 2A–D. Defects in cardiac and vascular zebrafish mutants. **A** *hands off/hand2* mutants have normal *nkx2.5* expression, but exhibit severe defects in myocardial differentiation (shown by a deficit in *cmlc2*-expressing cells). Images are dorsal views, anterior to the top. Adapted from Yelon et al. (2000). **B** *pandora* mutants exhibit an abnormally small ventricle (*arrowhead*) and a bulbous atrium (*arrow*). Images are head-on views at 48 hpf. Adapted from Yelon et al. (1999). **C** *gridlock* mutants lack blood circulation to the trunk and tail due to a defect in the dorsal aorta, although cranial circulation is essentially normal. Images are lateral views of confocal microangiograms of the anterior half of 2.5-dpf wild-type and *grl* mutant embryos. Adapted from Weinstein et al. (1995). **D** *violet beauregarde* mutants develop severe abnormalities in cranial blood circulation, including massive enlargement of primary cranial vessels by 2.5 dpf. Images are dorsal views of confocal microangiograms of the head vasculature of 2.5-day-old wild-type and *vbg* mutant embryos. (Courtesy of B.L. Roman)

3.2.3 Requirements for Chamber-Specific Differentiation

Fate mapping studies have suggested that the ventricular and atrial lineages are physically separate in the zebrafish mid-blastula (at the 2000-cell stage; Stainier et al. 1993). To date, the expression of *vmhc* in a subset of *cmlc2*-expressing cells is the earliest known indication of molecular diversification of the ventricular and atrial precursors (Yelon et al. 1999), and several genes appear to be involved in regulating *vmhc* directly or indirectly. *hand2* is one important influence during the induction of *vmhc* expression (Yelon et al. 2000). *gata5*, *fgf8*, and *oep* also influence the formation of ventricular precursors (Reiter et al. 1999, 2001b; Reifers et al. 2000b), although it is not clear whether the ventricular defects in *gata5*, *fgf8*, and *oep* mutants could be secondary to the loss of *nkx2.5* expression. Several different mutations [including *pandora* (Stainier et al. 1996; Yelon et al. 1999) and others (Alexander et al. 1998; D. Yelon, unpubl.)] are known to disrupt *vmhc* expression and ventricle formation (Fig. 2B); cloning and characterization of the mutated genes will likely reveal additional players relevant to ventricular development.

Additional studies suggest that multiple signaling pathways contribute to cardiac chamber formation. For example, retinoic acid (RA) signaling may play a critical role during atrial differentiation in zebrafish. Exogenous RA treatments can transform ventricular tissue into atrial tissue and/or inhibit ventricular development, suggesting that endogenous RA signaling promotes atrial development (Stainier and Fishman 1992). Clarification of the role of endogenous RA during the diversification of the ventricular and atrial lineages will require further investigation. Other relevant signals have been suggested by studies of the zebrafish gene *jekyll*, which is required for valve formation. *jekyll* mutants appear to have defects in endocardial-myocardial communication that is critical for the formation of a distinct cell type at the atrioventricular boundary (Walsh and Stainier 2001). The *jekyll* locus encodes UDP-glucose dehydrogenase (Walsh and Stainier 2001), an enzyme required for glycoprotein synthesis that is necessary for Fgf and Wg signaling in *Drosophila* (Hacker et al. 1997; Lin et al. 1999). Further studies will be required to reveal the precise molecular nature and timing of the signaling events regulated by *jekyll* during zebrafish cardiac patterning.

3.2.4 Requirements for Heart Tube Assembly

After pattern formation within the ALPM is complete, cardiac fusion is required to assemble the patterned myocardial precursors into a patterned heart tube. Interactions between anterior endoderm and ALPM are required for cardiac fusion to proceed: mutations that reduce or eliminate gut endoderm also affect cardiac fusion. For example, *casanova*, *bonnie and clyde*, *oep*, and *faust* mutants exhibit a condition called "cardia bifida" in which cardiac fusion fails to proceed and two populations of myocardial cells are found at lateral positions (Alexander et al. 1999; Reiter et al. 1999, 2001b; Kikuchi et al. 2000). Cardia bifida can also be observed in mutants that form normal amounts of gut endoderm. For example, *hands off* mutants exhibit cardia bifida in the presence of a normal amount of endoderm, suggesting that *hand2* plays a mesoderm-autonomous role in the regulation of cardiac fusion (Yelon et al. 2000). Additionally, mutation of the *miles apart* locus, which encodes a lysosphingolipid G protein-coupled receptor, causes cardia bifida without disrupting gut formation (Kupperman et al. 2000). It is not yet clear which tissue(s) require activity of this receptor in order to facilitate cardiac fusion.

Beyond cardiac fusion, additional morphogenetic steps are required to assemble a properly patterned heart tube. One of these critical steps is disrupted by *the heart and soul* (*has*) mutation. In *has* mutants, abnormal cardiac fusion, followed by an abnormal cone-to-tube transition, ultimately results in the atrium forming outside of the ventricle (Fishman and Chien 1997; Yelon et al. 1999). Recently, positional cloning has revealed that the *has* locus encodes an atypical PKC molecule, similar to PKCλ (Horne-Badovinac et al. 2001; Peterson et al. 2001). In invertebrates, PKCλ activity is required for epithelial polarity (Tabuse et al. 1998; Hung and Kemphues 1999; Wodarz et al. 2000). Because many epithelia are perturbed in *has*, including retina, gut, and the CNS, the timing and source of the cardiac defect are not clear. The small molecule concentramide phenocopies the *has* cardiac defect without perturbing other systems or affecting PKCλ (Peterson et al. 2000, 2001). Like *has*, concentramide blocks fusion of posterior heart field cells at the cone stage, apparently by shifting the field toward the anterior. The nature of the effects of concentramide suggests that proper patterning of the heart tube may rely on midline guidance signals in a manner akin to midline sorting of axons in the nervous system. It will be interesting to determine whether and how proper epithelial polarity is required for the normal assembly of the heart tube, and how concentramide fits into epithelial polarity pathways.

4 Pattern and Orientation to the Onset of Function

Pattern of the cardiovascular system is evident not only in its morphology, but also in its function. Unidirectional blood flow begins in the embryo. This polarity of function depends on patterning of both electrical and contractile features of the heart, such that the beat originates from one site, and propagates through the chambers, pausing at the A–V junction to permit atrial emptying prior to ventricular contraction. Blood leaves the heart at higher pressures than when it returns, reflecting the differences in structure and function of the two chambers. Arteries have different functions than veins. The zebrafish is particularly suited to analysis of these features, and the dependence of the murine embryo on blood flow makes it difficult to assign distinct functional roles to particular genes. Several zebrafish mutations speak to this "functional patterning". These are important models for understanding integrative biology, and prevalent diseases, such as arrythmias and heart failure, for which we lack handles on molecular pathogenesis.

Pacemaking is a property of all cells of the heart, but the fastest pacemaker drives the heart. Normally, this is in the sino-atrial node, and impulses then traverse the atrium, pause at the atrioventricular node, thereby permitting atrial contraction to occur before the propagated impulse excites the ventricle. *slow mo (smo)* mutant embryos have a chronic bradycardia (slow heart beat). *smo* mutant cardiomyocytes have all of the normal complement of channels, except that for the hyperpolarization-activated current termed Ih (Baker et al. 1997). *smo* is not yet cloned, but does not appear to be the Ih channel itself (Warren et al. 2001). *smo* is not a lethal mutation, and *smo -/-* embryos survive to adulthood. Bradycardia and Ih diminution persist in adult *smo -/-* animals showing that innervation does not suffice to overcome the rhythmicity defect.

One gene critical for the function and growth of the earliest embryonic heart tube is that for the α1C L-type calcium channel. *island beat* (*isl*) mutant embryos have a mutation in the α1C L-type calcium channel gene that abolishes entirely the L-type calcium current but not the T-type current in the heart (Rottbauer et al. 2001). The ventricle is silent, but the atrium manifests a chaotic rhythm and an electrocardiogram that resembles atrial fibrillation.

(This is of interest because biopsies of tissue from humans with atrial fibrillation show a deficit in expression of this channel, suggesting that an imbalance of L-type and T-type channels may be responsible.) There is no propagation of the beat through the ventricle, normally or with stimulation, showing the importance of this channel for conduction. Finally, the ventricle remains diminutive in size. Other mutations that cause ventricular silence do not affect cell number. Hence, calcium entry though this channel is a link between function and form. This is one of the few known controllers of size of early organs.

For several dozen mutations the principal notable phenotype is a weak heart beat, seemingly equivalent to dilated cardiomyopathies. Heart failure in humans has a significant genetic predisposition, but only a few responsible genes are known. Some of these may primarily affect myofibrillar assembly. For example, all described *pickwick* (*pik*) alleles have cardiomyopathy, and some also include a skeletal myopathy. *pik* maps to the *titin* gene, the longest gene known, so not readily tractable for mutational analysis. One cardiac-specific *pik* allele encodes a mutation within a cardiac-specific exon (Xu et al. 2002), proving that *pik* is *titin*. Titin provides the organizational matrix for assembly and alignment of myosin with actin, and additionally is the spring for diastolic elasticity of the heart. Hence, transplanted *pik* cells bulge outward when in a wild-type heart, unable to provide the normal recoil force.

5 Vascular Pattern in the Zebrafish

Endothelial-lined blood vessels are believed to have arisen during vertebrate evolution from more primitive chordates. Although a primitive heart and "blood" (hemolymph) are present in *Drosophila*, hemolymph circulates through spaces that do not possess a cellular endothelial lining. Parallels have been noted between the tracheal system of *Drosophila* and vertebrate blood vessels, but there are clearly fundamental structural, functional, and molecular differences between these two systems. As noted above, the genetic tractability of the zebrafish and the physical accessibility and optical clarity of its embryo make it ideal for both genetic and experimental analysis of vascular embryogenesis. A number of powerful vascular-specific imaging tools and methods complement the optical clarity of the fish embryo. These include confocal microangiography (Weinstein et al. 1995; Isogai et al. 2001), digital motion analysis of vascular form and function (Schwerte and Pelster 2000), and vascular cell tracking in vivo using transgenic zebrafish expressing green fluorescent protein (GFP) in blood vessel cells (Motoike et al. 2000, and unpubl. results). The zebrafish embryo and the genetic and experimental tools available in this organism represent an unprecedented opportunity to study the formation of the vasculature.

5.1 Formation of Blood Vessels in the Zebrafish

As in other vertebrates, the earliest progenitors of blood vessels in the zebrafish arise from lateral plate mesoderm. Vascular and hematopoietic cells are thought to have at least a transient common early progenitor in developing vertebrate embryos, the hemangioblast (Ribatti et al. 2000). Expression of hemangioblast markers (see below) is detected in the lateral plate mesoderm beginning at approximately the 3–5-somite stage. Specific endothelial progenitors ("angioblasts") separate from hematopoietic progenitors by the 10-somite stage, as assessed by expression of specific molecular markers. Angioblasts begin a process of active migration to the trunk midline, reaching the midline and beginning to condense into two cords of cells at approximately the 15-somite stage. Dorsal aorta progenitors condense together first, just ventral to the embryonic hypochord and notochord, followed shortly thereafter by the aggregation of progenitors for the posterior cardinal vein, below the dorsal aorta and just dorsal to the trunk endoderm (Fouquet et al. 1997). Both vessels form open (patent) lumens by approximately 24 hpf and circulation commences through the trunk thereafter. These "axial vessels" are the first vessels to function in vertebrate embryos, and are the primary vessels carrying blood to and from the trunk and tail. Blood initially flows only through the trunk and tail, but by approximately 26 hpf a cranial circulatory branch appears, utilizing the internal carotid artery and venous primary head sinus (Fig. 1B).

All of the vessels that have formed up to this stage in the fish have direct homologues in other vertebrates. These and other details about the vascular anatomy of the zebrafish are described in the *Interactive Atlas of Zebrafish Vascular Anatomy* available online at http://eclipse.nichd.nih.gov/nichd/lmg/redirect.html and in Isogai et al. (2001). This detailed description of the vascular anatomy of the embryonic and early larval zebrafish, prepared using confocal microangiography, shows the many attributes of the early development of the vasculature in zebrafish that are similar to other vertebrates.

Over the course of the rest of the first week of development rapid and extensive growth of additional vessels occurs. The larger vessels forming at these later stages generally also have analogs in mammals, as, for example, the parachordal and vertebral vessels of the trunk. Many vessel tracts arise in concert with particular organs or structures. The intestinal vasculature appears at 2.5–3 days postfertilization (dpf), the pectoral artery at late 2.5–3 dpf, and hepatic vessels at 3+ dpf. The overall pattern of the zebrafish vasculature changes relatively little between approximately 2.5 and 7 dpf, by which time the embryo has become a swimming, feeding larva. Preliminary corrosion cast analysis of the adult zebrafish vasculature indicates that the pattern of major blood vessels in the 7-dpf zebrafish is similar to the final adult configuration (S. Isogai, unpubl. results). Thus, essentially the entire sequence of blood vessel development is readily accessible to observation and experimental manipulation.

5.2 Molecular Analysis of Blood Vessel Formation in the Zebrafish

Zebrafish homologues have been identified for genes known to be involved in vascular development in other species (Tables 1 and 2). These genes include early molecular markers of lateral mesoderm and its derivatives such as the ets transcription factors *fli-1* and *ets-1* (Thompson et al. 1998; Brown et al. 2000b; N.D. Lawson and B.M. Weinstein, unpubl. results), among the earliest markers of presumptive "hemangiogenic" lateral mesoderm. Homologues of the *scl* and *hhex* genes are also expressed, at slightly later stages (Liao et al. 1998, 2000; Porcher et al. 1999). Members of the vascular endothelial growth factor (VEGF) ligand family and their receptors, the VEGFR tyrosine kinases, have also been identified in the zebrafish (Fouquet et al. 1997; Sumoy et al. 1997; Liang et al. 1998). In mammals, VEGF-VEGFR signaling has been shown to play a central role in both embryonic vessel formation and adult angiogenesis (Larrivee and Karson 2000). Homologues of the *Tie-1* and *Tie-2* receptor tyrosine kinase genes and their ligands, the angiopoietins, have also been identified in the fish (Lyons et al. 1998; Pham et al. 2001). Tie-angiopoietin signaling plays a critical role in regulating interaction between vascular endothelial cells and the vascular smooth muscle cells that envelop and surround them (Yancopoulos et al. 2000). The temporal and spatial expression patterns in zebrafish embryos of the genes noted above and in Tables 1 and 2 are largely comparable to those in other vertebrate embryos. Relatively few tests of the embryonic function of these genes have been performed in fish so far.

The fish embryo has provided a useful window into some of the earliest events in vessel formation, including the specification of angioblasts from lateral plate mesoderm and formation of the axial vessels. Lineage studies (Serluca and Fishman 2001) and the expression patterns of *scl* and *fli-1* (Fouquet et al. 1997; Gering et al. 1998; Brown et al. 2000b) support the idea that lateral mesodermal cells give rise to pronephric and "hemangioblast" precursors, which subdivide into hematopoietic and angiogenic progenitors. Overexpression experiments have also been used to probe the functional roles of some of these early genes in the fish embryo. Ectopic overexpression of *scl* by injection of in vitro synthesized mRNA into zebrafish embryos results in overproduction of hematopoietic and endothelial precursors, perturbation of vasculogenesis, and loss of pronephric and somitic tissue (Gering et al. 1998). Likewise, overexpression of the *hhex* gene leads to premature and ectopic expression of vascular- and hematopoietic-specific genes such as *fli1*, *flk1*, and *gata1* (Liao et al. 2000). The results of these and other overexpression experiments suggest that there is a great deal of cross-regulation between many of these factors and additional studies will be needed to sort out the particular functional contributions of each factor to angioblast and/or hemangioblast development.

The zebrafish has also provided some novel insights into the role of axial mesoderm in formation of the dorsal aorta. Analysis of zebrafish notochord mutants *floating head* (*flh*) and *no tail* (*ntl*) has demonstrated that axial mesoderm plays an important role in formation of the dorsal aorta, but appears to be dispensable for formation of the posterior cardinal vein (Fouquet et al. 1997; Sumoy et al. 1997). This may reflect a requirement for the notochord to support formation of an adjacent midline structure called the hypochord. The hypochord is a thin strip of cells found immediately ventral to the notochord in fish and amphibian embryos, just above the site of dorsal aorta assembly (Cleaver et al. 2000). The hypochord expresses a number of factors known to be important signaling molecules in vascular development, including angiopoietin-1 and VEGF-121, a soluble diffusible isoform of VEGF (Cleaver and Krieg 1998; Pham et al. 2001). VEGF-121 can act as a long-range signaling factor to attract migratory angioblasts, suggesting that hypochord VEGF-121 may be important for guiding lateral mesodermal angioblasts to the trunk midline. It should be pointed out, however, that a functional role for the hypochord in vascular development has yet been shown experimentally. Axial mesoderm may also be playing a more direct role in dorsal aorta formation via Hedgehog signaling. The notochord is the major axial source of Sonic hedgehog signaling in the zebrafish embryo, and this signal is absent in *flh* and *ntl* mutants. Analysis of a group of zebrafish mutations with defects in different members of the Sonic hedgehog signaling pathway, the "*you*" mutants (van Eeden et al. 1996; Barresi et al. 2000), has revealed that, in addition to their other Hedgehog-dependent defects, these mutants fail to make a dorsal aorta (van Eeden et al. 1996; Brown et al. 2000b; A. Vogel and B. Weinstein, unpubl. results).

5.3 Genetic Analysis of Blood Vessel Formation in the Zebrafish

Genetic screens have yielded an array of mutations that perturb vascular development (Chen et al. 1996; Stainier et al. 1996). The *cloche* mutation causes a profound reduction in both vascular and hematopoietic progenitors (Stainier et al. 1995; Thompson et al. 1998). Positional cloning of *cloche* is underway, but it is still not clear how *cloche* functions relative to pathways of hemangioblast and angioblast formation. *cloche* is required cell-autonomously for both the formation of vascular endothelium (Stainier et al. 1995) and for the differentiation and/or survival of red blood cells (Parker and Stainier 1999), although there is evidence it also plays a cell-non-autonomous role in hematopoietic development, perhaps as a result of a requirement for interaction with endothelium in order to generate blood cells (Parker and Stainier 1999). Experimental studies suggest that *cloche* is upstream of *flk* (Liao et al. 1997) and downstream of *scl* (Liao et al. 1998) and *hhex* (Liao et al. 2000).

Three additional mutations, *gridlock*, *kurzschluss*, and *violet beauregarde*, cause vascular patterning defects. Embryos affected by these mutations all eventually lose caudal (trunk and tail) circulation, while in many cases retaining substantial cranial circulation. However, the nature of the defects that give rise to the deficit in caudal blood flow is quite different. *kurzschluss* mutant embryos form improper arterial-venous connections (shunts) between the fifth and sixth aortic arches and the large adjacent venous primary head sinus (A.M. Vogel and B.M. Weinstein, unpubl. results). This circulatory shunt diverts blood flow from other parts of the embryo and results in progressive loss of circulation to these other regions. The molecular nature of this defect is still uncharacterized.

In *gridlock* mutants, a localized blockage to caudal flow through the dorsal aorta occurs at the point where the paired lateral dorsal aortas normally merge to form a single midline dorsal aorta at the anterior trunk midline (Weinstein et al. 1995). In most cases, this is accompanied by the formation of arterial-venous shunts that link one or both lateral dorsal aortas just anterior to the blockage point to the large adjacent cardinal veins. The block to caudal circulation occurs in a vascular location analogous to that affected in the common human congenital cardiovascular defect coarctation of the aorta, making this mutant of great potential clinical interest. The *gridlock* gene encodes a hairy-related bHLH factor (Zhong et al. 2000) homologous to the mammalian *HRT2* gene (Nakagawa et al. 1999). In mammalian cell culture, HRT2 functions downstream of Notch signaling (Nakagawa et al. 2000). In the zebrafish trunk, *gridlock* mRNA is strongly expressed in the dorsal aorta but not in the posterior cardinal vein, supporting the idea that this gene is specifically involved in dorsal aorta formation (Zhong et al. 2000).

In *violet beauregarde* mutants the early, primary vessels appear to form normally, and wild-type and mutant embryos are phenotypically indistinguishable until at least 1.5 dpf. However, mutants develop severe deficiencies in the secondary cranial central vessels that form after this stage and primary cranial vessels become abnormally distended and form improper arterial-venous anastomoses. The *violet beauregarde* gene was also recently cloned and shown to be a zebrafish homologue of the type I TGF-beta superfamily receptor Alk1 (B.L. Roman and B.M. Weinstein, unpubl. results). Alk1 is one of two known loci in humans mutated in the autosomal dominant inherited vascular disorder "hereditary hemorrhagic telangiectasia" (HHT; Johnson et al. 1996). Targeted disruption of mouse *Alk1* causes vascular abnormalities with evidence of arterial-venous anastomoses and other malformations of the anterior vasculature (Oh et al. 2000).

As noted above for the notochord and "*you*" mutants, primary genetic defects in non-vascular tissues have been extremely useful for dissecting the mechanisms of embryonic blood vessel formation, where the affected tissues provide cues or signals necessary for the specification, patterning, or differentiation of adjacent blood vessels. The relatively small number of mutants described above was obtained from mutant screens that did not specifically target blood vessel mutants. Additional screens directed specifically toward uncovering vascular mutants are currently underway, and preliminary evidence suggests that these screens may yield additional vascular mutants, including mutants with few or no blood vessels, mutants with too many blood vessels, and mutants with disorganized blood vessels.

5.4 Experimental Analysis of Vascular Form and Function: Imaging Blood Vessels in Situ

As noted briefly above, one of the major advantages of the zebrafish as a model system for studying vascular development is the accessibility and extraordinary optical clarity of its embryo. Imaging methods permit real-time, non-invasive visualization of the dynamics of vessel formation and function in intact, living embryos, and allow detailed observation of every portion of the vasculature, even vessels deep within the brain of the developing zebrafish. The anatomy of actively functioning blood vessels can be visualized using confocal microangiography (Weinstein et al. 1995), performed by injecting fluorescent microspheres into the circulation of a developing zebrafish and then collecting "optical slices" of the fluorescently labeled vessels using a confocal microscope. This technique has been used to derive a complete atlas of the vascular anatomy of the zebrafish throughout embryonic and early larval development (Isogai et al. 2001), to study the nature of the vascular defects in zebrafish mutants (Weinstein et al. 1995; Reifers et al. 2000b; Zhong et al. 2000), and to assess the effects of toxicological treatments on vascular development (Cheng et al. 2001). Another useful set of tools are transgenic zebrafish lines in which the expression of GFP or other fluorescent proteins is driven by vascular-specific promoters such as those for *tie-2* (Motoike et al. 2000). Transgenic zebrafish provide a number of advantages over angiographic visualization of blood vessels. While angiography delineates the patent lumens of functioning blood vessels, in transgenics all endothelial cells, and even the angioblast progenitors, may be made fluorescent well before they have actually formed vascular structures. This permits study of the origins and assembly of blood vessels during development. In addition to angiography and transgenesis, vessels can also be visualized in the zebrafish using digital motion analysis, a method using successive frame subtraction of digitally collected light microscopic images to generate "shifting vector" casts of the functioning vasculature (Schwerte and Pelster 2000). In addition to providing another means for imaging and tracing the vascular anatomy of developing zebrafish, this method allows characterization of the physiology of the embryonic and larval circulatory system (Fritsche et al. 2000).

6 Prospects for Future Zebrafish Cardiovascular Research

As a principal focus of the first large-scale genetic screens in zebrafish, the cardiovascular system has proved gratifying in two ways. First, the work on this system demonstrated that a vertebrate organ of great complexity can be tractable to genetic dissection, with individual mutations being informative and overall logic of assembly being apparent. Second, the ease of extrapolation to other vertebrates, both in terms of logic and specific genes, meant that there are powerful evolutionary and clinical implications.

The rapid pace of zebrafish cardiovascular research suggests that ongoing efforts in this arena will continue to be fruitful. Because the screens still fall well short of saturation, and many facets of cardiovascular formation remain elusive, gene discovery by new screens and cloning will be rewarding for many years to come. Improved resolution by transgenic cell marking will improve the resolution of screening and analysis. Each new gene discovered provides an entrance point to a pathway and other components to be discovered. In this search, genetic screens could be complemented by small molecule screens and by the subsequent identification of cellular targets of small molecules.

As future studies reveal the molecular underpinnings of cardiac and vascular patterning, it will be especially exciting to apply these findings to an untapped arena, the genetic basis of the onset of cardiovascular function. There is a rhythmic pattern to function, and vectorial flow and pressure transmission. How are these cardiovascular forces first instituted? How might they determine growth and patterning of vessels and of the heart, and perhaps of other tissues? Organ growth and its dependence on function are important issues for many organ systems, and we are now poised to make progress on these topics in the context of the zebrafish cardiovascular system. Such findings are likely to have important clinical implications, not only with regard to congenital heart disease, but also in reparative medicine, in which principles of developmental biology are used to make new tissues from progenitor cells.

Acknowledgements. We thank the members of the Yelon, Weinstein, Stainier, and Fishman laboratories for permission to cite unpublished work.

The Pronephros

Ian Drummond

1 Introduction

1.1 Variation and Evolution of the Kidney

In the course of vertebrate evolution, the kidney, as the primary osmoregulatory organ, has allowed for the exploration and colonization of new environments ranging from seawater to dry land (Smith 1953). The universal demand for tissue fluid homeostasis and osmoregulation in varied habitats has led to the development of a variety of kidney forms. With the exception of aglomerular marine kidneys, all vertebrate kidneys are composed of three functional units: the glomerulus, the tubules (proximal and distal), and the collecting system derived from the nephric duct (Fig. 1A). The glomerulus is the blood filter; capillary tufts covered in specialized basket- or sieve-like epithelial cells (podocytes) are suspended in an epithelial capsule (the nephrocoele or Bowman's capsule), which receives the cell- and protein-free filtrate of the blood. Kidney tubules connect to the capsule and allow passage of filtered fluids to the exterior and, at the same time, perform the delicate operation of recovering needed metabolites and ions so as to maintain blood ion concentrations within the narrow limits required for the proper functioning of all other body cells. Collecting ducts continue the job of metabolite recovery begun in the tubules and provide the physical link to the exterior. Each of these three structures appears to develop as a result of a distinct series of morphogenetic events controlled by tissue patterning mechanisms during organogenesis.

1.2 A Brief History of the Kidney

The pronephros, mesonephros and metanephros represent three distinct kidneys of increasing complexity that form in an anterior to posterior pro-

Unit/CNY 8000, Massachusetts General Hospital, 149 13th St., Charlestown, Massachusetts 02129, USA

Results and Problems in Cell Differentiation, Vol. 40
L. Solnica-Krezel (Ed.): Pattern Formation in Zebrafish

gression during avian and mammalian embryogenesis (Saxén 1987). The pronephros is the first kidney to form during embryogenesis. In vertebrates with free-swimming larvae, including amphibians and teleost fish, the pronephros is the functional kidney of early life (Howland 1921; Tytler 1988; Tytler et al. 1996; Vize et al. 1997) and is required for proper osmoregulation (Howland 1921).

The pronephros was first described by Müller in 1829 and the glomus, a variant of the glomerulus where capillary tufts span several body segments and hang freely in the coelom, was identified by Bidder as its vascular component in 1846 (reviewed in Adelmann 1966). The formation of the pronephros in several amphibian species was described in detail more than a century ago by Field, and the structure of the trout pronephros was described around the same time by Balfour (Balfour 1880; Field 1891). Studies since that time have focused primarily on the amphibian and teleost pronephros. The existence of a pronephros beyond the status of a vestigial organ in higher vertebrates remains debated (Torrey 1965; Saxén 1987).

In freshwater teleosts (guppy, trout, and zebrafish) as well as marine teleosts (killifish, salmon, turbot, and herring), the pronephros consists of only two nephrons with glomeruli fused at the embryo midline just ventral to the dorsal aorta (Fig. 1B; Balfour 1880; Goodrich 1930; Marshall and Smith 1930; Armstrong 1932; Newstead and Ford 1960; Agarwal and John 1988; Tytler 1988; Hentschel and Elger 1996; Tytler et al. 1996; Drummond et al. 1998; Drummond 2000). Two pronephric tubules connect directly to the closed nephrocoele (Bowman's space) and paired bilateral pronephric ducts convey the altered blood filtrate outside the animal. The pronephric nephrons are typically located under somite three. A connection between the nephrocoele and the coelom is lacking (Tytler 1988; Drummond et al. 1998) and therefore the teleost pronephros does not possess the "nephrostomes" of amphibian and avian pronephroi (see below). Presumably, the nephrostomes in teleosts have become the neck of the pronephric tubule connecting to the nephrocoele (Bowman's space) and the nephron forms a closed system of blood filtration, tubular resorption, and fluid excretion.

The pronephros is thought to have evolved from a more primitive protovertebrate kidney, often referred to as the "archinephron" (Saxén 1987). The archinephron is believed to have been a segmentally reiterated structure consisting of a hollow nephrocoele surrounding a capillary tuft, connecting ventrally to the coelom and laterally, via segmental openings in the coelom wall (nephrostomes) and tubules, to a common duct (Fig. 1C; Goodrich 1930; Saxén 1987). Each segmental tubule was thought to have fused into the pronephric duct, which opened caudally into the cloaca (Goodrich 1930). This general tissue organization of glomeruli suspended in the coelom (external glomera) and reiterated nephrostomes and tubules persists in the amphioxus, amphibian, and chicken pronephroi (Fig. 1D; Field 1891; Davies 1950; Romanoff 1960; Fox 1963; Nakao 1965; Vize et al. 1995). It has been argued that the vertebrates inherited the essential components of the archinephron, as well as the overall

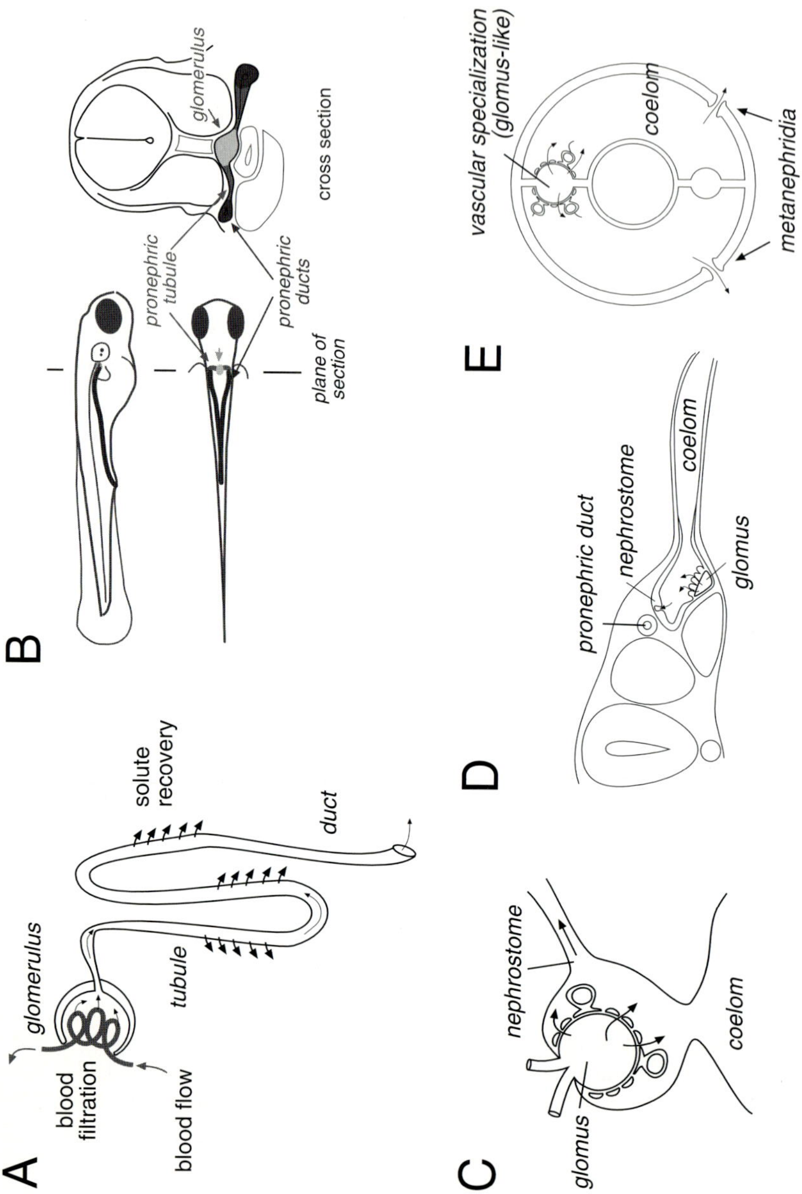
A
blood filtration
glomerulus
blood flow
tubule
solute recovery
duct
B
glomerulus
pronephric tubule
pronephric ducts
cross section
plane of section
C
nephrostome
glomus
coelom
D
pronephric duct
nephrostome
coelom
glomus
E
vascular specialization (glomus-like)
coelom
metanephridia

functional organization of the excretory system, from invertebrate ancestors (Ruppert 1994). Invertebrate metanephridia (annelids) consist of a tubule leading from the coelom to the exterior, functionally associated with vascular outpouchings in the coelom covered by podocyte-like cells (Fig. 1E; Ruppert and Smith 1988). These vascular specializations are structurally and functionally similar to the vertebrate glomus or "external" glomeruli observed in the frog and chick pronephroi. At the cellular level, invertebrate metanephridial podocytes have extensive foot processes and form filtration slits on basement membranes adjacent to the blood capillaries. Evidence that annelid metanephridia are functional "kidneys" has been derived from studies demonstrating ultrafiltration of injected iron dextran (MW 4000–6000) by glomerular podocytes in the acorn worm, *Saccoglossus kowalevskii* (Balser 1985; Ruppert and Smith 1988; Hansen 1996).

In the course of evolution, the number of segments contributing to the pronephros has been reduced and the archinephron is thought to have lost its connection to the coelom, allowing the nephrocoele to specialize as Bowman's capsule (Goodrich 1930; Fox 1963). With the integration of the previously separate glomerular capillary tuft, the pronephric tubule, and duct into a closed filtration/recovery system comes the concept of the nephron as the functional unit of the kidney. The discrete origins of the nephron components would suggest that the mechanisms of cell differentiation in different nephron cell types, for instance, podocytes vs pronephric tubule cells, might be quite distinct. On the other hand, the derivation of nephron components from adjacent undifferentiated tissue in the mesoderm would indicate that developmental patterning mechanisms are required to set and maintain the boundaries between the components of the forming nephron.

Fig. 1A–E. Variation and evolution of the kidney. **A** The nephron is the functional unit of the kidney and is composed of three major subunits: the glomerulus, tubules and collecting ducts. **B** Diagram of the three elements of the larval zebrafish pronephros at 72 hpf showing the position of glomerulus under the notochord, the pronephric tubules extending laterally to connect with the pronephric ducts which serve the function of the collecting system. **C** Proposed structure for a segmentally repeated, ancestral "archinephros" composed of a vascular tuft within a nephrocoel which remained open to the coelom. The nephrocoel is drained by the ciliated nephrostome and associated archinephric tubules and ducts. **D** Cross section of the chick pronephros shows multiple "external" glomera positioned in a nephrocoel space that is open to the coelom. Blood filtrate passes through the nephrocoel (coelom) space and exits via multiple nephrostome openings on the dorsal surface of the nephrocoel to collect in the pronephric ducts. **E** An invertebrate metanephridium is a ciliated pore or funnel that communicates between the coelom and the exterior. Metanephridia are functionally paired with segmentally repeated outpouchings of the vasculature which are covered on the coelomic surface with podocyte-like cells, creating a vascular filtration site similar to the vertebrate glomus

2 Morphogenesis and Patterning of the Zebrafish Pronephros

Kidney organogenesis can be divided into four general stages: (1) the commitment of undifferentiated mesodermal cells to a nephrogenic fate and the formation of an organ primordium, (2) the growth and epithelialization of the Wolffian or pronephric duct which will eventually serve as the collecting system, (3) the induction and formation of the nephron and the cell patterning events associated with the differentiation of glomerular and tubular cell types, and (4) the formation of the glomerular capillary tuft by ingrowing endothelial cells and the onset of blood filtration and nephron function (Saxén 1987). The stages in the development of the zebrafish pronephros are illustrated schematically in Fig. 2.

The morphogenesis of the pronephros is directed by sequential patterning processes that control its final form and function. Early stages of kidney cell specification are clearly influenced by the broadly acting morphogen gradients that pattern the mesoderm. Later stages of pronephric development, for instance, boundary formation between different components of the nephron and formation of the glomerulus, appear to be more local events involving interactions between the forming nephron domains and cell types. Although current evidence is incomplete, key features of pronephric development and patterning have emerged from studies using simple histology, cell lineage tracing, gene expression patterns, and analysis of zebrafish mutants affecting this process.

2.1 Patterning of the Mesoderm and Formation of the Pronephric Primordium

The region of the early embryo that is fated to form the pronephros has been determined for a number of vertebrates using cell lineage tracing. Cell labeling in zebrafish gastrula-stage embryos has demonstrated that cells destined to form the pronephros lie just dorsal to the heart progenitors in the shield-stage mesodermal germ ring, and appear to overlap somewhat with cells fated to form blood (Fig. 3A; Kimmel et al. 1990). This position in the prospective mesoderm is very similar to the region defined in several different amphibian embryos as the source of future kidney cells (Delarue et al. 1997). These cells emerge during and shortly after gastrulation as a band of tissue, the intermediate mesoderm, at the ventrolateral edge of the paraxial mesoderm (Fig. 3B,C). Although currently there is no data on when pronephric cells are specified in zebrafish embryos, by extension of what is known from amphibian pronephric development (Brennan et al. 1998, 1999), it is likely that signaling events occurring at this early stage of development have a major impact on all subsequent differentiation of the pronephros.

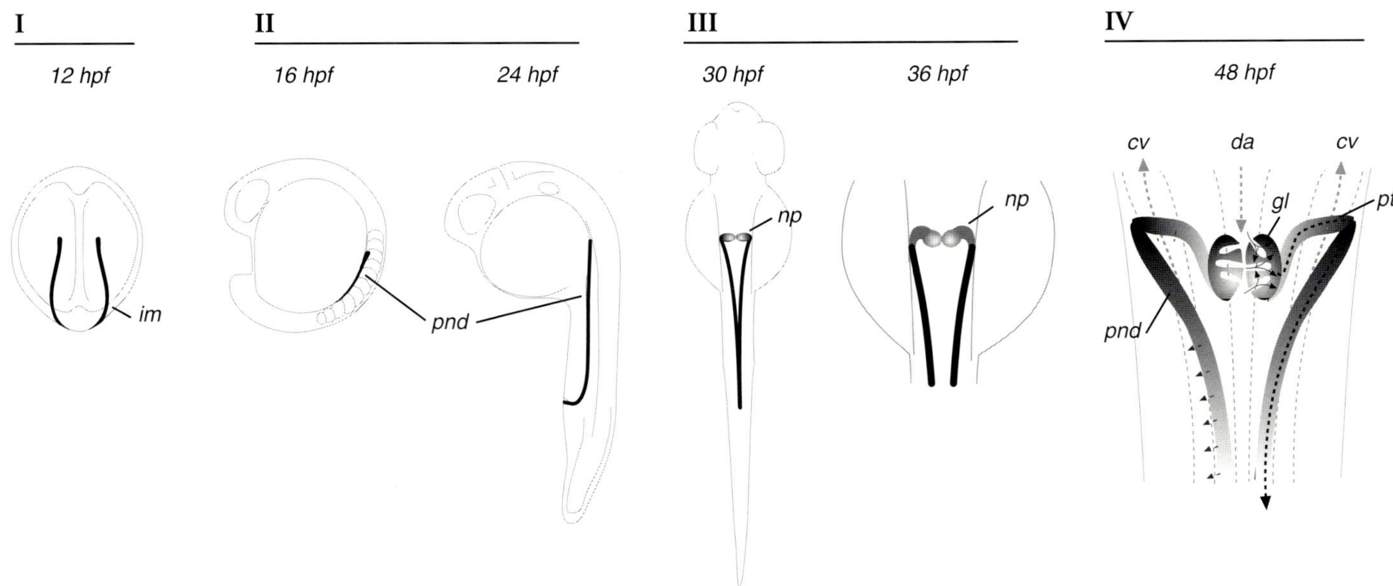

Fig. 2. Developmental stages in the formation of the zebrafish pronephros. **I** 1–8-somite stage (10–13 hpf): commitment of the intermediate mesoderm. Expression patterns of *pax2.1* and *lim-1* define the posterior regions of the intermediate mesoderm (*im*) and suggest that a nephrogenic field is established early in development. **II** Early somitogenesis to Prim-5 stage (12–24 hpf): formation of the pronephric duct and nephron primordia. Epithelialization of the pronephric duct (*pnd*) follows behind somitogenesis and is complete by 24 hpf. Nephron primordia (*np*) form at the anterior tips of the pronephric ducts by 24 hpf. **III** Prim-5 to High-Pec stage (24–42 hpf): nephron primordia (*np*) ventral to somite 3 separate from the coelom and undergo integrated morphogenesis, giving rise to pronephric glomeruli (*gl*) and tubules (*pt*). **IV** High-Pec stage to Hatching (42–48 hpf): the onset of pronephric function. Vascularization of the pronephros by ingrowth of capillaries from the dorsal aorta (*da*) starts at 40–42 hpf and continues over the next 24 h correlating with the onset of glomerular filtration. Blood from the dorsal aorta (*da*) is filtered at the glomerulus (*gl*; *arrows*). The filtrate passes to the pronephric tubules (*pt*) and ducts (*pnd*) where it is modified by active transport processes (*arrows*) and the recovery of small metabolites to the cardinal veins (*cv*). *Arrows* in the vasculature indicate the direction of blood flow. For stages of zebrafish development, refer to Kimmel et al. (1995)

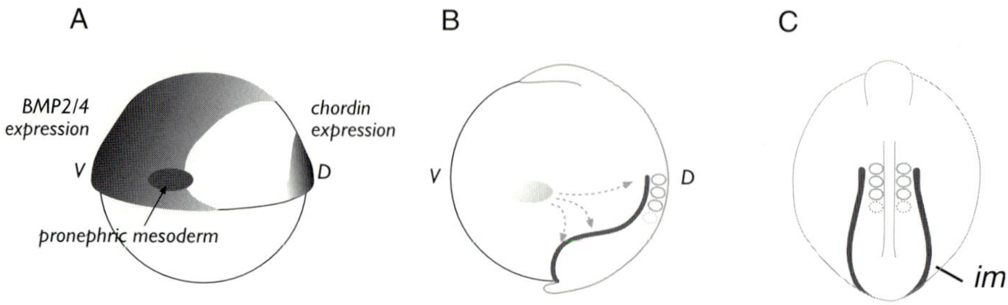

Fig. 3A–C. Origin of the pronephric mesoderm. **A** Fate mapping studies reveal that cells destined to become the pronephros lie just dorsal to the heart progenitors and appear to overlap somewhat with cells destined to form blood in the germ ring (shield-stage embryos). Expression of bone morphogenetic protein 2b (BMP2b) and BMP4 on the ventral side of the embryo is required for expression of *pax2.1* in the intermediate mesoderm at later stages. **B** Cell migration during gastrulation results in the kidney progenitor cells coming to reside adjacent to the paraxial mesoderm. **C** Position of the intermediate mesoderm (*im*) during early stages of somitogenesis

Pronephric mesoderm at the blastula and gastrula stages is patterned by well-established morphogen gradients in the early embryo which include BMP expression on the ventral side of the embryo and β-catenin /BMP inhibitor function in the dorsal embryonic shield (Fig. 3A; see Kimelman and Schier; Hibi et al.; Hammerschmidt and Mullins, this Vol.; Hammerschmidt et al. 1996b,c; Neave et al. 1997; Nikaido et al. 1997). The so-called dorsalized mutants, *swirl* (*bmp2b*), *snailhouse* (bmp7), *somitabun* (*smad5*), and *lost-a-fin* (*alk8*), all lack signals necessary for ventral mesodermal development and show a reduction or elimination of *pax2.1* positive, presumptive kidney intermediate mesoderm (Mullins et al. 1996; Kishimoto et al. 1997; Nguyen et al. 1998b; Hild et al. 1999; Bauer et al. 2001; Mintzer et al. 2001; Payne et al. 2001). Conversely, the ventralized mutant *chordino* lacks the dorsal shield determinant Chordin and shows an expansion of the *pax2.1* expression domain at later stages (Hammerschmidt et al. 1996c). Taken together, the data from these mutants suggests that BMP signaling, most likely in concert with additional signals from the shield region, is required to specify kidney mesoderm.

Continued development of somitic and intermediate mesoderm requires the activity of the trunk organizer. As part of, or in response to, signals from the trunk organizer, T-box transcription factor genes, *notail* (*ntl*), *spadetail* (*spt*), and *tbx6*, function in concert to pattern the trunk and tail mesoderm (Ho and Kane 1990; Hammerschmidt et al. 1996a; Griffin et al. 1998; see Warga and Stainier, this Vol.). In shield-stage embryos, all three of these genes are expressed in the ventral and lateral germ ring (Griffin et al. 1998). The function of *spadetail* and *tbx6* are probably most relevant to pronephric development as these two genes are later expressed broadly in the hypoblast (4–6-somite stage) out to the lateral edge of the forming mesoderm and are more specifically involved in trunk development (Griffin et al. 1998). Mutations in *spadetail* cause loss of trunk muscle (Ho and Kane 1990; Hammerschmidt et al. 1996a; Griffin et al. 1998), and some reduction (but not complete loss) of pronephric mesoderm (I. Drummond, unpubl.). A more complete analysis of the combined loss of *spt* and *tbx6* will be required to fully assess their role in trunk and, specifically, pronephric development.

The dramatic loss of all ventral-posterior embryonic structures in embryos expressing a dominant negative FGF receptor is most likely due to the requirement for FGF signaling in the initiation or maintenance of *notail*, *spadetail*, and *tbx6* gene expression (Griffin et al. 1995, 1998). Overexpression of FGF ligands (eFGF, FGF8) has somewhat paradoxical effects, resulting in dorsalized embryos and inhibition of ventral gene expression including *pax2.1* in the pronephric mesoderm (Griffin et al. 1995; Fürthauer et al. 1997). This effect is most likely due to the capacity of early FGF overexpression to inhibit ventral BMP expression at gastrulation (Fürthauer et al. 1997). It is likely that many different FGF genes are expressed in distinct spatial and temporal patterns and their specific roles in mesoderm patterning cannot be deduced from ectopic overexpression (Reifers et al. 1998). The identity of the zebrafish FGF gene responsible for regulating trunk organizer gene expression is currently unclear. Another secreted factor found to influence the trunk organizer and T-box gene expression is Wnt8 (Lekven et al. 2001). The *wnt8* gene is expressed in the germ ring at gastrulation, transiently in the region of the intermediate mesoderm just prior to somitogenesis, and in a more prolonged fashion in the tail bud (Kelly et al. 1995b). Deletion mutants of the bicistronic *wnt8* locus show significant reduction in *tbx6* expression and loss of ventral-posterior mesoderm (Lekven et al. 2001). Inhibition of *wnt8* mRNA translation in *ntl* and *spt* mutant embryos using injected morpholino antisense oligonucleotides has synergistic effects resulting in more profound loss of ventral-posterior mesoderm (Lekven et al. 2001). Thus, the trunk organizer signals by way of at least two secreted factors, an FGF and Wnt8, and control the expression of T-box genes that are required for proper development of trunk mesoderm.

2.2 Mediolateral Patterning of the Intermediate Mesoderm and Induction of the Pronephros

In both amphibians and teleost fish, the first morphological sign of pronephric cell differentiation is the formation of a pronephric primordium as a mass of intermediate mesoderm lying ventral and lateral to the anterior somites (Armstrong 1932; Nieuwkoop and Faber 1994; Kimmel et al. 1995; Vize et al. 1995, 1997; Drummond et al. 1998). In most vertebrate embryos, this primordium is the source of cells destined to populate all components of the pronephric nephron (Saxén 1987). In zebrafish, the area of the intermediate mesoderm fated to become the kidney appears to be relatively large, extending from somite 1 to at least somite 10 (Fig. 4; Serluca and Fishman 2001). This suggests that, although the position of the pronephric primordium is roughly the same as in other vertebrates, the mechanisms of morphogenesis,

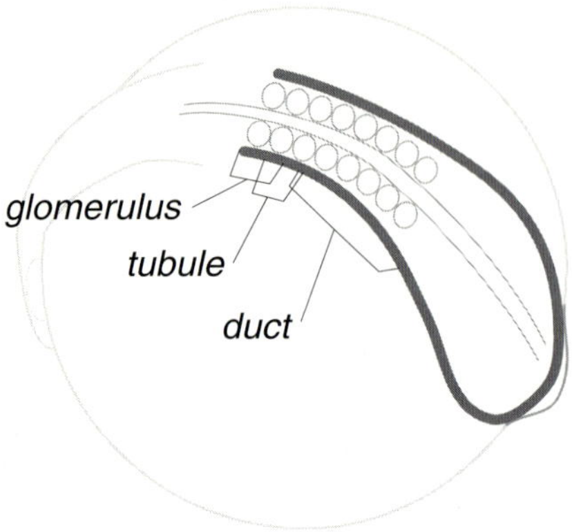

Fig. 4. The pronephric field. Lineage tracing using caged fluorescein activation at various axial levels of the intermediate mesoderm shows that glomerulus progenitors are present adjacent to somites 1 and 2, pronephric tubule progenitors are present adjacent to somites 2 and 3, and pronephric duct progenitors are distributed more caudally from somites 3 to 10. No evidence of caudal cell migration has been observed. (Serluca and Fishman 2001)

particularly with regard to the growth of the pronephric duct, may differ somewhat (see below).

In addition to the pronephric primordium, the intermediate mesoderm of the trunk also gives rise to the blood and endothelial cells of the major trunk blood vessels (Weinstein et al. 1996). At the tailbud stage, markers of blood cell development such as *gata2* are expressed in a cell population that overlaps with *pax2.1*-expressing, presumptive kidney intermediate mesoderm cells (Detrich et al. 1995; Gering et al. 1998). Later in development, the blood islands where erythropoesis takes place lie just ventral to the formed pronephric duct (Detrich et al. 1995; Ransom et al. 1996). The *scl* gene, a basic helix-loop-helix transcription factor essential for blood cell development in both mouse and fish, is expressed in a pattern that partially overlaps with *pax2.1* (Gering et al. 1998; Liao et al. 1998). Close inspection of the *pax2.1* and *scl* expression domains suggests that mediolateral patterning signals (i.e., acting at a 90° angle to both the dorsoventral axis and the anterioposterior axis) direct the fate of intermediate mesoderm cells to the blood/endothelium or kidney lineages (Fig. 5A). *scl* expression is strongest in the medial half of the intermediate mesoderm, while the most lateral intermediate mesoderm only expresses *pax2.1* (Fig. 5B; Gering et al. 1998). Overexpression of *scl* causes excess blood cell development at the expense of somite and pronephric duct cell development (Gering et al. 1998). These studies suggest that, prior to overt differentiation, intermediate mesoderm cells are pluripotent and that transfating may occur if the balance of mediolateral patterning signals is perturbed.

Expression of *pax2.1* in the intermediate mesoderm and later development of kidney structures occur independently of the notochord and Sonic hedgehog signaling, and does not require the proper differentiation of the endoderm. These conclusions are based on studies of *floating head*, *sonic-you*, *you-too*, and *one-eyed pinhead* mutant embryos where *pax2.1* expression and/or pronephric duct morphology, if not perfectly normal, is still observed (Majumdar and Drummond 2000; I. Drummond, unpubl. observ.). The question then becomes: what tissue serves to regulate expression of the pronephric-specific genes and

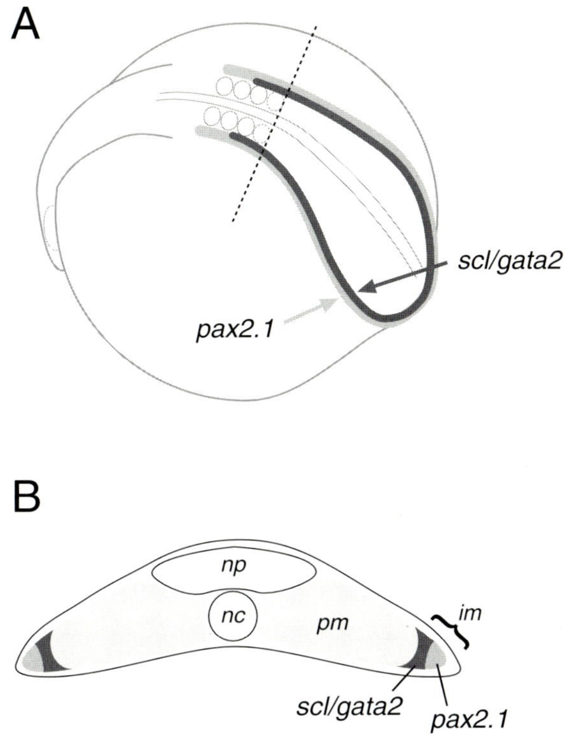

Fig. 5A, B. Mediolateral differentiation of kidney and blood lineages. **A** Diagram of a whole-mount view depicting the overlapping expression of *pax2.1* (pronephros) and *scl/gata2* (blood) markers. *pax2.1* is observed to be expressed more laterally than *scl.* **B** In cross section a lateral domain of the intermediate mesoderm primarily expresses *pax2.1* while more medial tissue expresses blood lineage markers *scl* and *gata2*

induce the formation of the pronephros from the intermediate mesoderm? Microsurgical implantation of barriers in the chick embryo between the intermediate mesoderm and somite tissue results in loss of chick *pax2* expression in the intermediate mesoderm, while physically blocking notochord signals or signals from the lateral plate has no effect (Mauch et al. 2000). In chick embryos with double axes, excess anterior somite tissue appears to have the opposite effect, inducing an expansion of the *pax2* expression domain (Mauch et al. 2000). Similar conclusions can be drawn from experiments on the frog: anterior somite tissue can induce pronephric tubule formation in explanted unspecified mesoderm (Seufert et al. 1999). These embryological experiments from other systems present a starting point for further examination of the source and identity of signals that induce pronephric development using zebrafish genetics.

Gene expression patterns in the pronephric intermediate mesoderm during early somitogenesis suggest potential candidate signals for pronephric induction and/or patterning. DeltaC, a ligand for Notch, is expressed in the intermediate mesoderm adjacent to somites 1–4 at the 7-somite stage, suggesting that Notch signaling may play a role in patterning the pronephric primordium (Fig. 6; Smithers et al. 2000). This expression domain corresponds to the pre-

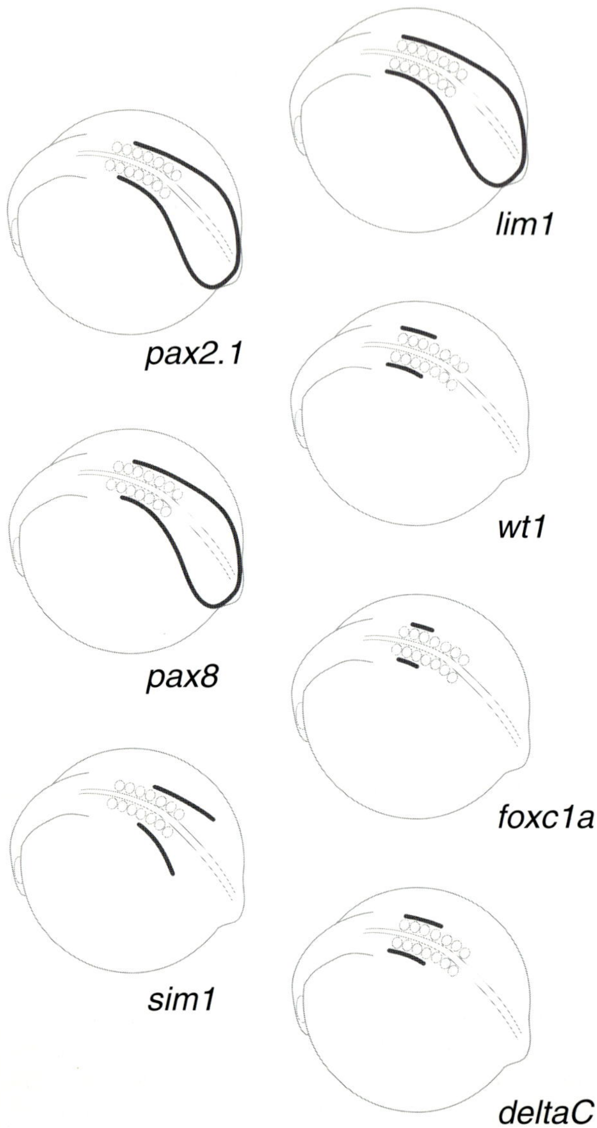

Fig. 6. Gene expression patterns in the pronephric intermediate mesoderm. See text for details

cursors of the glomerulus and tubules, and extends anterior of the *pax2.1* expression domain (Fig. 4; Smithers et al. 2000; Serluca and Fishman 2001). Cloning of the *gridlock* gene has suggested the involvement of Notch signaling in vascular cell fate decisions in intermediate mesoderm cells (Zhong et al. 2000); it may be that lateral inhibition mechanisms in the trunk also pattern the pronephros. Evidence for Notch regulation of cell fate in the *Xenopus* pronephric primordium suggests that Notch signaling affects the tubule vs duct cell fate decision (McLaughlin et al. 2000). The zebrafish frizzled gene, *fz8a*, presumably encoding a receptor for a locally acting Wnt signal, is expressed throughout the intermediate mesoderm at the 5–6-somite stage (Kim et al. 1998a). The timing of *fr8a* expression is significantly later than *wnt8* expression (see above) suggesting that a different Wnt ligand may signal via *fr8a*. Whereas *wnt4* is reported to be weakly expressed at about this time in bilateral anterior stripes of intermediate mesoderm (Ungar et al. 1995), we have not been able to confirm this (Liu et al. 2000; I. Drummond, unpubl. observ.).

A variety of transcription factors are expressed in the presumptive pronephric mesoderm. *pax2.1* is the most extensively studied marker of pronephric mesoderm and is expressed in a continuous band of intermediate mesoderm from the cloaca up to the posterior boundary of somite 2 (Fig. 6; Krauss et al. 1991; Püschel et al. 1992; Heller and Brandli 1997, 1999; Pfeffer et al. 1998; Carroll et al. 1999; Drummond 2000; Majumdar et al. 2000; Mauch et al. 2000). *pax8* is expressed in a similar pattern (Fig. 6), with *pax8* expression initiated in the intermediate mesoderm during gastrulation slightly earlier than *pax2.1* expression (Pfeffer et al. 1998). Mutations in *pax2.1* (*no isthmus*) do not prevent initial development of the pronephric duct epithelium (Majumdar et al. 2000) but do have extensive effects on later tubule development and nephron patterning (see below). The *lim1* gene is also expressed during early somitogenesis at all A-P levels of the pronephric mesoderm extending from somite 1 to the future cloaca (Fig. 6; Toyama and Dawid 1997). Posterior *lim1* expression is downregulated by the 12-somite stage while anterior expression (adjacent to somites 1 and 2) just rostral the *pax2.1* expression domain persists until 24 h postfertilization (hpf; Toyama and Dawid 1997). The early and extensive expression of *lim1* in the intermediate mesoderm suggests an important role for this gene in pronephric development. In the small number of *lim-1*-deficient mouse embryos that have been examined, the genitourinary tract is completely missing (Shawlot and Behringer 1995). In mouse embryo chimeras, *lim-1* is required for full expression of lateral mesodermal fates (Tsang et al. 2000). The Wilms tumor suppressor gene, *wt1*, is expressed beginning at the 2–3-somite stage, and later at the 8-somite stage, in the intermediate mesoderm extending from somites 1–4 (Fig. 6; Serluca and Fishman 2001). In the mature pronephros, *wt1* is expressed exclusively in podocytes (Drummond et al. 1998; Majumdar and Drummond 1999, 2000; Majumdar et al. 2000; Serluca and Fishman 2001; see below); its early expression in the intermediate mesoderm at this A-P level corresponds well with fate mapping experiments demonstrating a glomerular fate for cells adjacent to somites 1 and 2 (Serluca and Fishman 2001). *Sim1*, a basic helix-loop-helix factor, is a zebrafish homologue of the Drosophila *simple minded* gene. Zebrafish *sim1* is expressed in a more limited and posterior domain of the intermediate mesoderm, from somite 4 extending caudally approximately 6–8 somites toward the cloaca (Fig. 6; Serluca and Fishman 2001). Finally, the *foxc1a* gene, encoding a member of the forkhead/winged helix transcription factor family, is expressed in the intermediate mesoderm adjacent to somites 2 and 3 at the 7-somite stage

(Fig. 6; Topczewska et al. 2001a). Although the functions of most of these genes in zebrafish pronephric development can only be inferred, their expression patterns provide some indication that the intermediate mesoderm is patterned into glomerulus, tubule and duct domains in an anterior-to-posterior direction. As noted above, lineage analysis confirms this (Serluca and Fishman 2001). New genetic screens and application of antisense technology may soon add significantly to the understanding of gene function at this stage of pronephric development.

2.3 Development of the Pronephric Duct

Development of the pronephric duct is initiated at the boundary of somites 2 and 3 on both sides of the embryo and temporally follows the anterior to posterior progression of somitogenesis (Fig. 2; Kimmel et al. 1995). By 24 hpf, the duct is complete, existing as an epithelial tube just dorsal to the yolk extension, from somite 3 to the cloaca (Kimmel et al. 1995; Drummond et al. 1998). In most vertebrates, the duct forms primarily by a caudal migration of cells from the pronephric primordium followed by epithelialization and connection of the duct lumen with the exterior at the cloaca (Saxén 1987). In some species, a contribution of posterior intermediate mesoderm cells to the duct has been observed; intercalation with migrating cells and in situ differentiation forms a duct with a more mosaic character (Cornish and Etkin 1993). In zebrafish, it appears that the duct forms primarily through recruitment of intermediate mesoderm at all axial levels followed by in situ epithelial differentiation; lineage tracing experiments have not detected any contribution from caudally migrating cells (Serluca and Fishman 2001). In all species, duct formation represents a mesenchyme-to-epithelial transformation; a process central to kidney formation in all vertebrates (Saxén 1987). By the end of this transition the epithelial cells are polarized with an apical brush border and a basolateral membrane domain containing ion transport proteins that are essential for the osmoregulatory function of the duct (Fig. 7A; Drummond et al. 1998). Interestingly, duct cells are not morphologically homogeneous. Single, multi-flagellated cells are dispersed along the length of the duct, appearing side by side with the more typical duct cuboidal epithelial cells. The function of these cells appears to be to drive fluid down the duct by beating their flagella in a corkscrew motion parallel to the axis of the lumen (I. Drummond, pers. observ.). The pronephric duct forms the basis of the collecting system to which all embryonic and future adult nephrons will attach (Saxén 1987).

In addition to the continued expression of *pax2.1*, *pax8*, *sim1*, and to some extent *lim1*, new expression of the *ret1* gene is observed in the forming pronephric duct as it develops in an anterior-to-posterior progression (Bisgrove et al. 1997; Marcos-Gutierrez et al. 1997). *ret1* encodes a tyrosine kinase cell surface receptor for glial cell line derived neurotrophic factor, GDNF, a TGF-β superfamily member (Schuchardt et al. 1994, 1995; Moore et

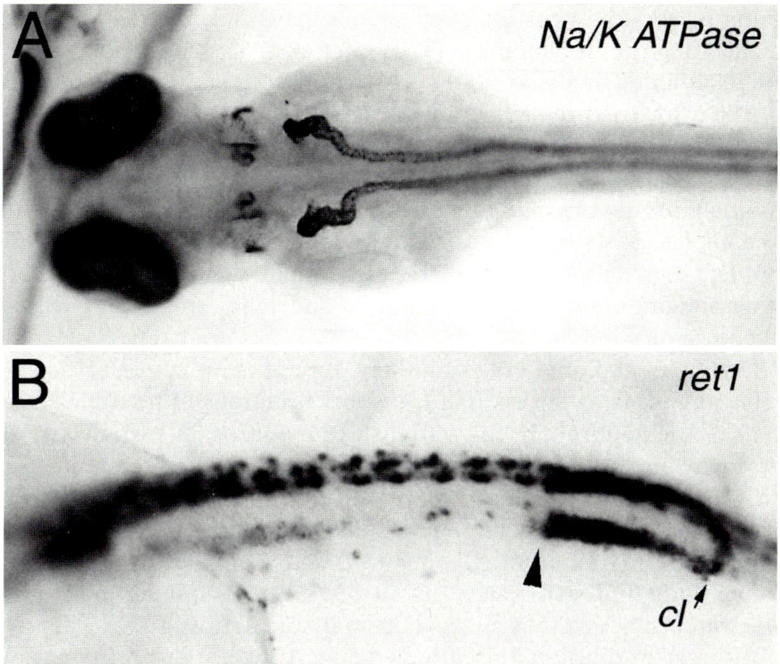

Fig. 7A,B. The pronephric ducts. **A** Expression of the Na/K ATPase alpha1 subunit in the pronephric ducts and tubules is detected with the monoclonal antibody alpha6F. **B** In situ hybridization using a *ret1* probe reveals expression in the caudal region of the pronephric ducts connecting to the cloaca (*cl*). The boundary of expressing and non-expressing cells is very sharp (*arrowhead*). Spinal cord neurons also express *ret1*

al. 1996; Pichel et al. 1996; Sanchez et al. 1996; Marcos-Gutierrez et al. 1997; Shepherd et al. 2001). At 24 hpf, when duct epithelialization is complete, *ret1* expression marks the most posterior quarter of the pronephric duct (Fig. 7B; Marcos-Gutierrez et al. 1997). The border between expressing and non-expressing cells is remarkably abrupt. It is currently unknown if these cells have unique origins different from the rest of the duct or whether A-P patterning mechanisms generate the sharp expression boundary. The function of the ligand GDNF has been studied using antisense oligonucleotides and found to be required for enteric neuron development but dispensable for pronephric development at least at the level of pronephric duct morphology (Shepherd et al. 2001). It has been shown that other genes implicated in developmental signaling pathways are expressed in the pronephric ducts. A retinoic acid (RA) response element/green fluorescent protein reporter transgene shows activity in the pronephric duct as early as the 18-somite stage (Perz-Edwards et al. 2001). Whether retinoids are involved in cell fate decisions in the pronephros or are only required to support terminal differentiation of pronephric duct epithelial cells is currently unknown. Dietary deficiency of vitamin A and mutations in retinoic acid receptors are both sufficient to severely perturb mouse metanephric development (Wilson et al. 1953; Mendelsohn et al. 1999; Burrow 2000; Batourina et al. 2001). RA-treated frog embryos show an expansion of the *xlim-1* expression domain and, in some cases, larger pronephric tubules (Taira et al. 1994). Currently it is known that loss of function of the major RA-synthesizing enzyme RALDH2 in zebrafish *neckless* mutants does not eliminate pronephric *pax2.1* expression (P. Ingham, pers. comm.). If a specific role for RA exists in pronephric development it may be at a later stage in development or, alternatively, RALDH2 function may be redundant in zebrafish and other enzymes may compensate for its deficiency specifically with regard to pronephric development.

A relatively unexplored area of kidney development concerns how epithelial patterning leads to the distinct transport functions of proximal and distal segments of the nephron. In the zebrafish pronephros, this would translate to an anterior-to-posterior pattern in the tubule and duct epithelium. To appreciate this question it is necessary to consider that the function of the kidney in freshwater vertebrates, including zebrafish, is to facilitate the elimination of water from the body fluids (Hickman and Trump 1969). Freshwater vertebrates maintain a body fluid osmolarity of 250–330 mOs, containing much more salt than the water in which they swim (Hickman and Trump 1969). This creates a constant threat of dilution and lethal edema. A watery death is prevented by voiding large amounts of water in very dilute urine. In general, osmoregulation by the fresh water teleost kidney is achieved by copious filtration of the blood by the glomerulus and active transport of essential salts and other blood solutes by the kidney tubules and ducts back into the venous circulation (Hickman and Trump 1969). The most distal tubules and ducts are active in salt uptake but impermeable to water, such that water remains in the duct lumen and is excreted. Consistent with its role in active ion resorption, the pronephric duct, and later the tubule epithelia, abundantly express the Na^+/K^+ ATPase, a driving force for many transport processes, in their basolateral cell membranes (Fig. 7A; Drummond et al. 1998). The linear A-P orientation of the zebrafish pronephric duct may serve as a simplified model system for studies of proximal/distal nephron patterning.

2.4 Nephron Formation

The morphogenesis of the pronephric glomerulus and tubule is closely linked to the formation of the coelom. Although not easily observed in fish embryos, coelom formation has been well studied in amphibians and chicken: the coelom is formed by a splitting of the lateral plate mesoderm. A mesenchyme-to-epithelial transition of cells in the center of the lateral plate forms the coelomic cavity (Meier 1980; Gilbert 1991; Funayama et al. 1999). The emergence of a primary pronephric fold between the ventral splanchnic and dorsal somatic mesodermal layers is observed ventral to somite 2 at the pharyngula stage (24 hpf) in zebrafish (Fig. 8A) and at the 14-somite stage in another teleost, *Fundulus heteroclitus* (Armstrong 1932; Drummond et al. 1998). In teleosts including zebrafish, the primary pronephric fold pinches off completely from the coelom at the anterior limit of the formed pronephric ducts and ventral to the second somite, forming a nephron primordium (Fig. 8B; Armstrong 1932; Tytler 1988; Tytler et al. 1996; Drummond et al. 1998). At this stage, the nephron primordium appears in cross section to be a flattened sac of cells ventral to the border between somites 2 and 3. Between 28 and 32 hpf, the pronephric tubules begin to form from the lateral aspect of the nephron primordium, between the anterior tips of the pronephric ducts and the most medial cells destined to become the glomerulus (Fig. 8C,D; Drummond et al. 1998). By 40 hpf, the previously uniform appearing nephron primordia are partitioned into medial, encapsulated groups of cuboidal cells closely apposed to the dorsal aorta (podocyte progenitors) and short lateral tubules continuous with the duct (Fig. 8E,F; Drummond et al. 1998). Cell labeling at the 8–10-somite stage shows that intermediate mesoderm cells that will contribute to the pronephric glomerulus lie adjacent to somite 1 while tubule progenitors are found adjacent to somite 2 (Serluca and Fishman 2001).

As the pronephric nephron primordia form and pinch off from the coelom (24 hpf), expression of the Wilms tumor suppressor gene, *wt1*, becomes restricted to the primordia ventral to somite 3 and later to the glomerular epithelium as the primordia coalesce at the embryo midline (Fig. 9D,F; Drummond et al. 1998; Majumdar and Drummond 1999, 2000; Majumdar et al. 2000; Serluca and Fishman 2001). At about the same time, new areas of intense expression of *pax2.1* appear in the lateral aspects of the nephron primordium where the future tubules will be (Fig. 9A–C; Drummond et al. 1998; Majumdar et al. 2000).

The zebrafish mutant *no isthmus* (*noi*) lacks a functional *pax2.1* gene (Brand et al. 1996b) and *noi* embryos fail to form pronephric tubules (Majumdar et al. 2000). However, in addition to the tubules deficiency, *noi* embryos show an abnormal, posteriorly expanded expression of the podocyte markers *wt1* and *vegf* in cells of the pronephric duct (Majumdar et al. 2000). The implication from these studies is that a normal function for *pax2.1* is to repress podocyte gene expression in the pronephric tubules and ducts and restrict podocyte development to medial nephron primordia cells. The *noi* phenotype suggests that *wt1* and *pax2.1* play an important role in demarcating the boundary between glomerular and tubular epithelium in the forming nephron.

2.5 Cell Interactions in the Vascularization of the Glomerulus

Vascularization of the glomerulus in teleosts occurs after the completion of pronephric duct and tubule development (Marshall and Smith 1930; Armstrong 1932; Drummond et al. 1998; Tytler 1988; Tytler et al. 1996). In zebrafish, the bilateral glomerular primordia coalesce at 36–40 hpf ventral to

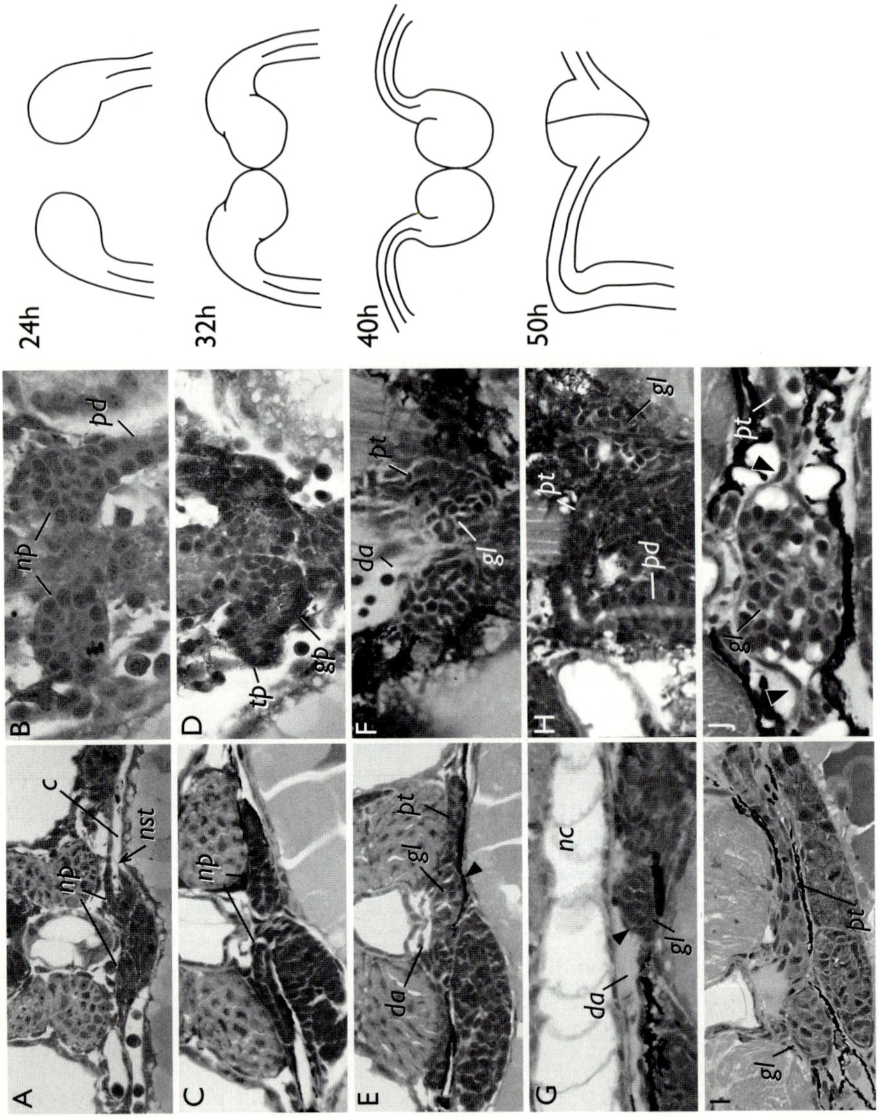
A
c
np
nst
B
np
pd
C
np
D
tp
gp
E
da
gl
pt
F
da
pt
gl
G
nc
da
gl
H
pt
pd
gl
I
gl
pt
J
gl
pt
24h
32h
40h
50h

the notochord, bringing the presumptive podocytes into intimate contact with capillary-forming endothelial cells of the dorsal aorta (Fig. 10; Drummond et al. 1998; Majumdar and Drummond 1999; Drummond 2000). Vessel sprouts from the aorta are the source of the glomerular capillary tuft in all functional pronephroi (Balfour 1880; Tytler 1988; Tytler et al. 1996; Drummond et al. 1998; Majumdar and Drummond 1999; Drummond 2000). Between 40 and 48 hpf, podocytes elaborate foot processes and show extensive interdigitation with the basement membrane supporting the overlying aortic endothelial cells (Fig. 10D; Drummond et al. 1998). Further capillary growth and invasion of endothelial cells result in extensive convolution of the glomerular basement membrane, capillary formation, and the onset of glomerular filtration (Drummond et al. 1998).

In contrast to earlier stages of pronephric patterning, the morphogenesis of the glomerulus appears to be controlled by more local cell-cell interactions. Evidence that podocytes act as the organizer of vessel ingrowth can be seen in (1) the expression patterns of genes known to play an important role in angiogenesis, and (2) the recruitment of endothelial cells to ectopic clusters of podocytes in mutant embryos that lack the dorsal aorta, the normal blood supply for the pronephric glomerulus.

Zebrafish pronephric podocytes express vascular endothelial growth factor (VEGF; Fig. 11; Majumdar and Drummond 1999, 2000), which is known to be

Fig. 8A–J. Pronephric nephron development. **A** Cross section of wild-type pronephroi at 24 hpf shows the forming nephron primordium (*np*) as an invagination of the coelomic lining. The lumen of the nephron precursor, or nephrocoel, is still connected to the coelom (*c*) at this stage by a transient nephrostome (*nst*; *arrow*). **B** Sagital section at 24 hpf shows the anterior pronephric duct (*pd*) abutting on the lateral aspect of the nephron primordium (*np*). **C** At 32 hpf the nephron primordium is closed off from the coelom and appears as a double-layered sac of cells with a central lumen. **D** Sagittal view at 32 hpf reveals early stages of nephron morphogenesis with lateral-most cells in the position of future tubules (tubule primordium; *tp*) becoming distinct from medial cells in the position of the future glomerulus (glomerular primordium; *gp*). **E** By 40 hpf the nephron primordium has become partitioned into a glomerular domain medially directly under the dorsal aorta (*da*) and tubule domains laterally with a distinct basement membrane surrounding the developing glomerulus (*arrow*). Pigment cells envelope the pronephros at this stage. **F** Sagittal view of the 40-hpf pronephric nephron shows the medial paired glomeruli (*gl*) with the pronephric tubules (*pt*) fused to their posterior side. *da* Dorsal aorta. **G** Longitudinal sections through the forming glomerulus (*gl*) at 40 hpf show its position below the notochord (*nc*) and its intimate association with the dorsal aorta (*da*) and cells (*arrow*) that surround the forming glomerular basement membrane. **H** Sagittal view of the pronephros at 50 hpf with the glomeruli (*gl*) fused at the midline and the pronephric tubule (*pt*) arcing over to connect to the pronephric duct (*pd*). **I** The pronephros at 2.5 days postfertilization (dpf) shows paired glomeruli (*gl*) ventral to the aorta and pronephric tubules (*pt*) connecting laterally to the pronephric ducts. **J** By 3 dpf nephron formation is essentially complete and the direct connection between Bowman's space and the lumen of the pronephric tubules is evident (*arrows*). On the *right* are schematized versions of the micrographs in **B, D, F,** and **H,** derived from serial sections, illustrating the morphological transitions that occur during pronephric nephron morphogenesis. In all cross sections the top of the figure is dorsal and in all sagittal sections the top of the figure is anterior

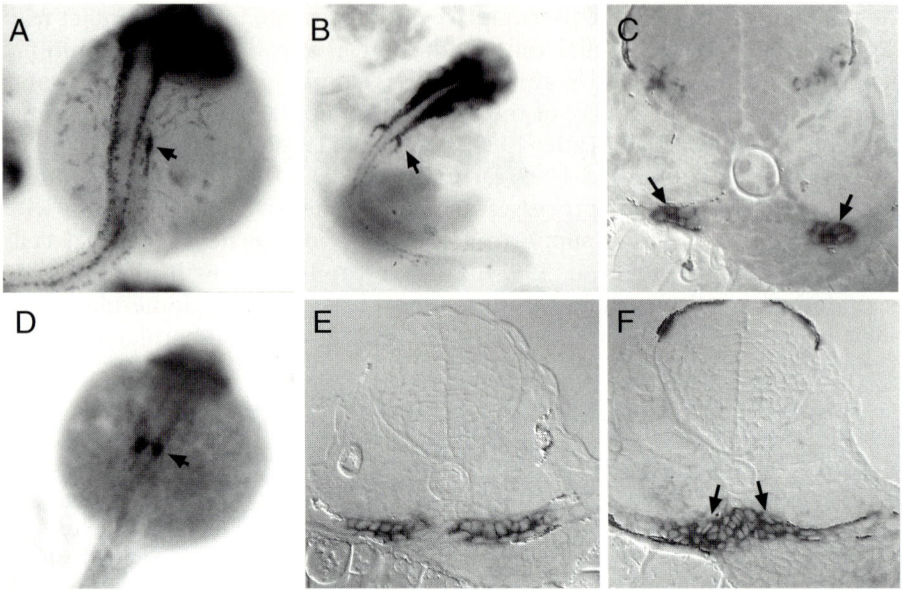

Fig. 9A–F. Expression of *pax2.1* and *wt1* during nephron formation. **A** Whole-mount in situ hybridization shows *pax2.1* is expressed in the anterior pronephric ducts (*arrow*) at 24 hpf as well as in the spinal cord. **B** At 30–32 hpf, a new domain of *pax2.1* expression is observed anterior and medial to the pronephric ducts (*arrow*), in the position of the future pronephric tubules. **C** Histological sections of *pax2.1* expression in the presumptive tubule primordium cells show that these cells constitute the lateral half of the nephron primordium and that the medial halves of the nephron primordium are negative. **D** Whole-mount in situ hybridization shows *wt1* expression in the paired nephron primordia at 24 hpf. **E** At 30–32 hpf, *wt1* is uniformly expressed throughout the nephron primordia. **F** By 36 hpf, *wt1* expression becomes restricted to a mass of cells at the midline in the position of the future glomerulus (*arrows*). Lateral cells within the forming nephron in the position of the future tubule do not express *wt1*

important for blood vessel formation in a variety of vertebrate systems (Shalaby et al. 1995; Carmeliet et al. 1996; Ferrara et al. 1996). In a complementary manner, capillary-forming endothelial cells express *flk1* (Fig. 11), a VEGF receptor, and an early marker of the endothelial differentiation program (Majumdar and Drummond 1999). In zebrafish embryos at 40 hpf, *flk-1*-positive endothelial cells can be observed invading the glomerular epithelium (Fig. 11). Endothelial cells also express alkaline phosphatase which can be used in zebrafish to visualize the formation of blood vessels (Fig. 11; Drummond 2000). In zebrafish *floating head* mutant embryos, the normal source of glomerular blood vessels, the dorsal aorta, is absent (Fouquet et al. 1997) and the pronephric nephron primordia do not fuse at the midline. Instead, they remain at ectopic lateral positions in the embryo (Fig. 11C; Majumdar and Drummond 2000). These podocytes continue to express *wt1* and *vegf* and appear to recruit *flk-1*-positive endothelial cells (Fig. 11D) and form a functional glomerulus

Fig. 10A–D. Vascularization of the zebrafish pronephros. **A, B** Schematic view of the process of pronephric vascularization. **A** 24–30 hpf: unvascularized nephron primordia exist as flattened sacs of cells with a central lumen. The medial cells of the primordium abut on the dorsal aorta and express *wt1* while lateral cells express *pax2.1*. **B** 30–40 hpf: endothelial cells appear between the paired primordia which have developed distinct midline glomerular and lateral tubular domains. **C** 40–48 hpf: ingrowth of capillaries from the dorsal aorta (*da*) invaginates the capsule surrounding the forming podocytes, giving rise to the GBM of the functioning glomerulus (*gl*). **D** Injection of rhodamine dextran into the circulatory system of 48-hpf embryos reveals infoldings of the glomerular basement membrane (*gbm*) where endothelial cells from the dorsal aorta (*da*) are invading the nephron primordium

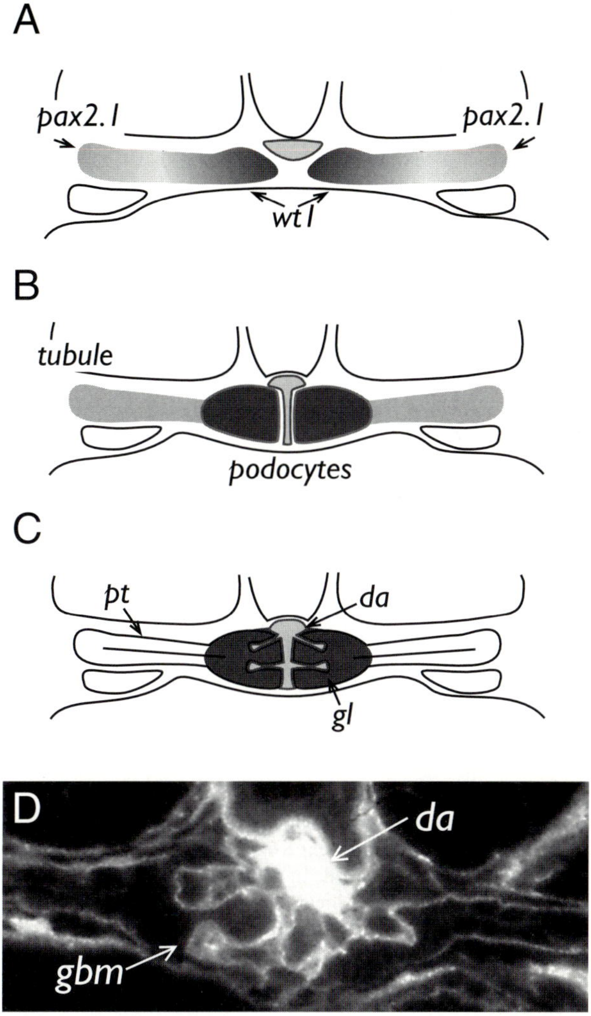

(Majumdar and Drummond 2000). These results support the idea that podocytes, by expressing *vegf*, play a primary role in attracting and assembling the glomerular capillary tuft.

The question of whether complementary signals are emitted from endothelial cells that would stimulate podocyte development has been addressed in studies of the mutant *cloche* (*clo*), which is almost completely lacking in endothelial cells (Stainier et al. 1995). When assayed by the expression of molecular markers and by an ultrastructural analysis of podocyte morphology, the differentiation of podocytes in *cloche* proceeds normally despite the complete absence of glomerular endothelial cells (Fig. 11; Majumdar and Drummond 1999). *cloche* podocytes express *wt1* and *vegf* and form extensive foot processes

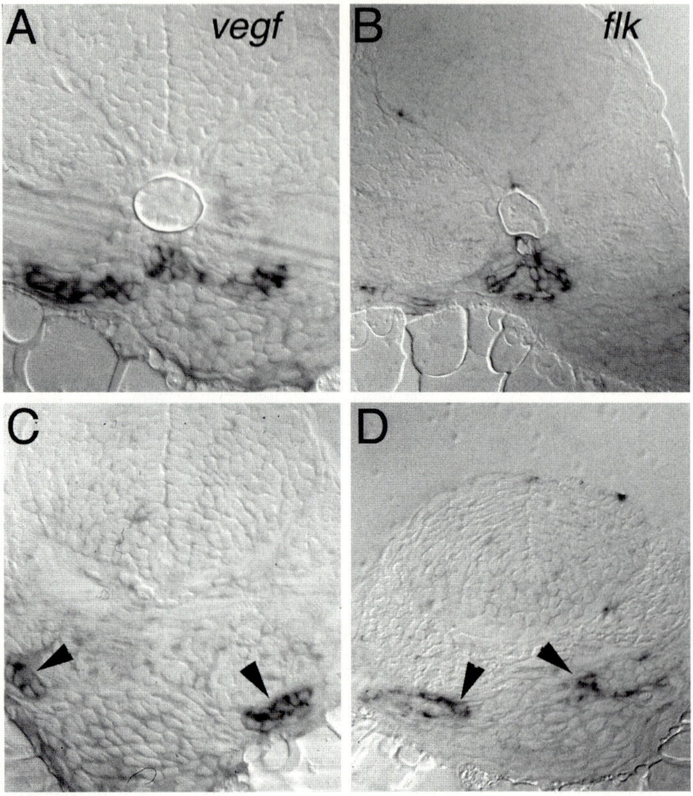

Fig. 11A–D. Expression of angiogenic genes during pronephric glomerular vascularization. **A** Differentiated podocytes express VEGF in wild-type embryos. **B** *flk-1* labels glomerular capillary-forming endothelial cells in wild-type embryos. **C** VEGF-positive presumptive podocytes are found in ectopic lateral locations in *flh* embryos (*arrows*). **D** *flk-1*-positive cells surround laterally positioned nephric primordia in *flh* embryos (*arrows*)

arranged as pedicels along a glomerular basement membrane (Majumdar and Drummond 1999). These findings suggest that, once initiated, podocyte differentiation can proceed independently of endothelial cells or endothelial cell derived signals. Many unanswered questions remain, however, concerning the significance of signaling between podocytes and endothelial cells for later glomerular maturation, for instance, in the structure and composition of the glomerular basement membrane (Miner and Sanes 1994).

In contrast to their lack of involvement in podocyte foot process development, endothelial cells do appear to play an essential role in mediating the requirement for vascular flow in the remodeling of glomerular form and the regulation of vessel ingrowth. Studies examining glomerular formation in mutants specifically lacking cardiac function, e.g., *silent heart* and *island beat* (*isl*), demonstrate a key role for vascular shear force in driving glomerulogenesis. In mutants that lack circulation, endothelial cells fail to invade the encapsulated podocytes and instead are found surrounding the glomerular primordium (Serluca et al. 2002). The cells themselves appear normal in several ways; for instance, expression patterns of *wt1* and *vegf* in podocytes and *flk-1* in endothelial cells remain unaltered and individual cell ultrastructure appears normal (Fig. 12; Serluca et al. 2002). The glomerular morphogenesis defect

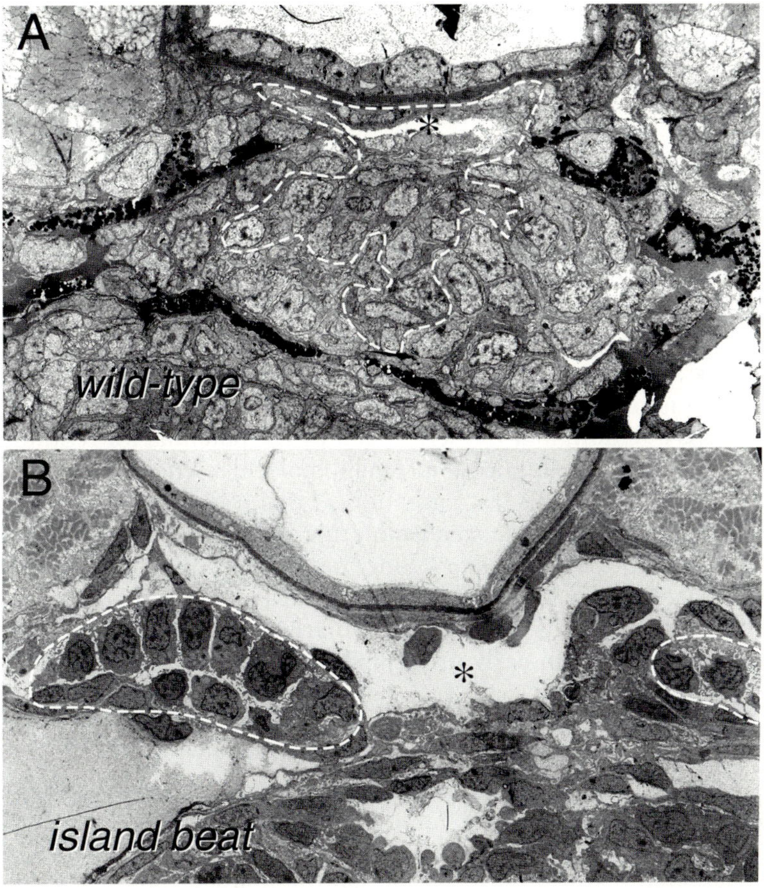

Fig. 12A, B. Vascular flow is required for angiogenic growth of vessels in the glomerulus. **A** Wild-type glomerulus at 48 hpf showing ingrowth of endothelial cells from the aorta (*outlined with white dashed line*), invading the podocyte clusters. **B** *Island beat* embryos are defective in cardiac contractility and lack blood flow. Podocyte clusters (*outlined*) remain in lateral clusters while endothelial cells surround but do not invade the podocytes. To demarcate endothelial cells and podocytes, the glomerular basement membranes were traced and podocytes and endothelial cells were digitally painted using Photoshop can be phenocopied by pharmacological and surgical manipulations that disrupt flow through the aorta. This failure in glomerular morphogenesis is likely related to the expression of matrix metalloproteinase-2 (MMP-2) in endothelial cells since (1) expression of MMP-2 in the trunk vasculature is modulated by vascular flow, and (2) when MMP-2 activity is inhibited by TIMP1 (tissue inhibitor of metalloproteinase-1) injected into the vasculature, glomerulus formation is blocked (Serluca et al. 2002). Thus, the final phase of pronephric development, just prior to the onset of glomerular filtration and organ function, is regulated by the force required for filtration itself.

3 Summary and Perspectives

The zebrafish pronephros, although distinct in overall organization from the meso- and metanephric kidney, is constructed of cell types with very similar form and function to those found in kidneys of higher vertebrates. The genetic analysis of zebrafish pronephric development that is beginning to unfold thus offers a new perspective and a new set of tools for the analysis of kidney cell differentiation and organ patterning. The completion of the zebrafish genomic sequence will greatly facilitate the cloning of genes identified in past and future mutagenesis screens. At that point the discovery of new gene functions in pronephric development will be limited only by the inventiveness and design of the screen and by the potential for redundancy in function due to gene duplication. Small molecular chemical screens may provide a route around this last limitation, as multiple copies of a gene may be similarly affected by a single small molecule inhibitor (Peterson et al. 2000). Certainly, much work remains to define the function of genes for which only expression patterns are known. The use of nuclease-resistant antisense oligonucleotides ("morpholinos") will be particularly useful for dissecting the function of genes involved in early pronephric patterning (Nasevicius and Ekker 2000).

In addition to the discovery of genes essential for early patterning events, genes required for generating proper pronephric epithelial cell polarity and epithelial cell function have also been found in zebrafish mutagenesis screens (Drummond et al. 1998). Parallels to human disease (polycystic kidney disease) can be seen in the phenotype of pronephric cyst mutants where the normal cuboidal epithelium of the pronephric tubules and ducts is flattened and epithelial lumens are dramatically expanded (Drummond et al. 1998; Drummond 2000). Some mutants manifest pleiotropic defects affecting both the kidney and the eye, referred to as "renal-retinal dysplasia" when observed in human patients (Godel et al. 1980; Drummond et al. 1998). As the genes underlying this human syndrome have yet to be characterized, the zebrafish may well provide an expeditious route to the annotation of human gene function.

Acknowledgements. I would like to thank past and present members of my laboratory, Arindam Majumdar, Aiping Liu, and Tomoko Obara-Ishihara, for their contributions to understanding pronephric development. I would also like to thank my collaborators Fabrizio Serluca and Mark Fishman for their contribution to this work and for providing an open and stimulating intellectual environment. I.A.D is supported by grants from the National Institutes of Health RO1 DK53093 and PO1 DK54711.

The Zebrafish Eye: Developmental and Genetic Analysis

Stephen S. Easter, Jr.[1] and Jarema J. Malicki[2]

1 Introduction

The eye is a complex organ in which different tissues cooperate to form an optical image, to transduce the image into an array of neural messages, to analyze the messages, and to transmit them to the brain. All but the first of these functions occur in the retina, which is the main subject of this review.

The mature zebrafish retina is essentially similar to that of other vertebrates, containing the same classes of neurons and glia and sharing some astonishing powers. Under low levels of illumination, the photoreceptors detect single photons of light, the absolute physical limit for light-sensors; however, when the light level is high, the retina changes its sensitivity and continues to function usefully. The cones parse the electromagnetic spectrum into several bands to provide the basis for color vision, and the interneurons process these channels to produce color contrast. The interneurons also compare light received by photoreceptors at different locations to create spatial contrast. All of this information is transmitted to the ganglion cells, among the least numerous neurons in the retina, and whose axons project to the brain.

The rate of zebrafish retinal development is noteworthy, as it changes from a proliferative neuroepithelium into a functional retina over 2.5 days. At about 12 h postfertilization (hpf), the retinal primordium is first evident as a planar pseudostratified epithelium, and by 3 days postfertilization (dpf), the eye – including photoreceptors and the associated retinal synaptic network – is sufficiently developed that the hatchling can see and respond to patterns. We focus on the developmental events in that brief period. Earlier development is events are covered elsewhere (see Wilson et al., this Vol.).

The acceptance of zebrafish as an animal suitable for developmental analysis is largely due to the efforts of the late George Streisinger, who recognized

[1] Dept. of Molecular, Cellular, and Developmental Biology, University of Michigan, 830N. University Ave., Ann Arbor, Michigan 48109-1048, USA

[2] Dept. of Ophthalmology, Harvard Medical School, 243 Charles St., Boston, Massachusetts 02114, USA

Results and Problems in Cell Differentiation, Vol. 40
L. Solnica-Krezel (Ed.): Pattern Formation in Zebrafish

early the utility of ocular mutants (e.g., Streisinger et al. 1989); we are pleased to continue along the path that he pioneered.

2 Morphogenesis

2.1 Optic Vesicle

In the anterior end of the neural keel the dorsal and ventral cells remain near the midline to become the dorsal forebrain (including the telencephalon) and the ventral diencephalon (hypothalamus), respectively. The cells in between move laterally to form the optic primordia (Varga et al. 1999; Chuang and Raymond 2001), which are first evident about 12 hpf, when they resemble airplane wings, a pair of broad flattened lateral extensions of the anterior neural tube (Fig. 1A,B). Over the next few hours, the "wing" detaches from the neural tube (neural keel), starting posteriorly, along most of its length, and remains

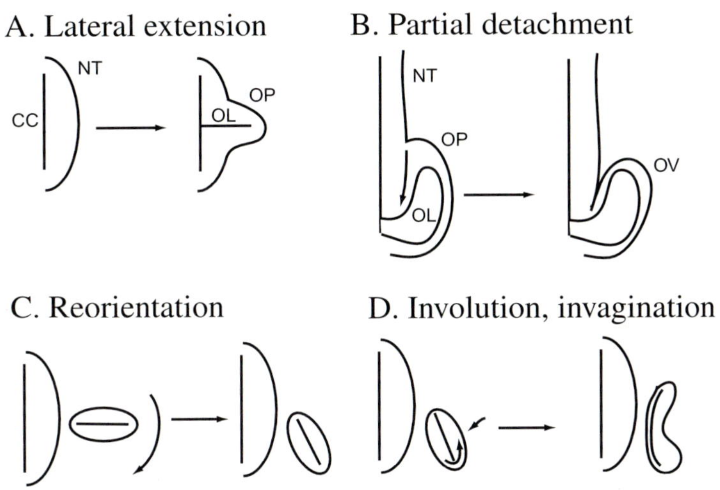

Fig. 1A–D. Summary diagram of the main events of eye morphogenesis. In all panels, the *vertical line on the left* represents the central canal, a thin dorsoventrally aligned space. **A** Transverse view of half of the anterior neural tube showing the lateral movement of cells to form the winglike optic primordium, its two epithelial layers separated by the optic lumen (*solid horizontal line*). **B** Dorsal view of the anterior neural tube, showing the wing-like optic primordium partially detaching from the neural tube along the *arrow* to form the optic vesicle attached to the neural tube by the optic stalk. **C** Transverse view of the neural tube and optic vesicle showing the sag of the vesicle relative to the neural tube. **D** Same field as in **C**, but a few hours later. The *curved arrow* in the vesicle indicates cellular involution as cells from the medial layer move around the ventral edge of the vesicle to enter the lateral layer. Simultaneously, the flat vesicle invaginates (*external arrow*) to become the concave eye cup with two disparate layers, thick retina outside and thin pigmented epithelium inside. *CC* Central canal; *NT* neural tube (neural keel); *OL* optic lumen; *OP* optic primordium; *OV* optic vesicle connected through the optic stalk at the front of the neural tube (Fig. 1B). By 16 hpf, the neural tube has a vertically oriented central canal with bilateral extensions through the optic stalks into the optic lumina of the two vesicles (Fig. 2A). The narrow optic lumen separates the flat optic vesicle into two columnar epithelial layers, one dorsal, one ventral, and both about 20–30 μm thick. Note that the zebrafish optic vesicle is structurally quite different from the thin-walled hollow spherical structure that usually comes to mind when the term "vesicle" is used. The optic lumen is barely visible in micrographs (Fig. 2A) and the outer boundary is flat rather than spherical.

2.2 Eye Cup

The flat optic vesicle is transformed into the hemispherical eye cup over 16–24 hpf (Schmitt and Dowling 1994; Easter and Nicola 1996; Li et al. 2000a). The vesicle tilts relative to the neural tube, and the surfaces that originally faced dorsally and ventrally face more laterally and medially, respectively (Figs. 1C, 2B). The dorsolateral surface becomes concave and progressively deepens into a cup that tightly surrounds the inwardly bulging lens placode (Fig. 2C). By 24 hpf, the lens has detached from the overlying ectoderm and is surrounded by the eye cup everywhere except at the presumptive pupil and choroid fissure (Fig. 2D). The hyaloid blood vessels have invaded the choroid fissure and ramified on the cup's inner surface, the presumptive inner limiting membrane of the retina (S.S. Easter, pers. observ.). The original dorsolateral and ventromedial epithelial layers have changed character. The former is now recognizable as the retina, a thick pseudostratified columnar epithelium, and the latter is thin, squamous, and packed with melanin pigment granules, which identify it as the retinal-pigmented epithelium. In the adult, the remnant of the choroid fissure lies along the vertical meridian in the ventral retina, thus defining the 6 o'clock position that divides the ventral retina into nasal (anterior) and temporal (posterior) halves. The choroid fissure in the embryonic eye serves as the first marker of adult coordinates. The optic stalk extends eccentrically from the ventronasal quadrant of the retina to the presumptive forebrain. The conversion of optic vesicle into an eye cup is achieved with essentially constant volume (Li et al. 2000a) and cell number (Li et al. 2000b) in the two layers.

The morphogenesis of the eye cup involves the movement of cells from the ventromedial to the dorsolateral layer. The two layers in the optic vesicle were generally assumed to be the precursors of the two layers of the mature eye; that is, dorsolateral layer/retina and ventromedial layer/pigmented epithelium. A fate map of the 14-hpf optic vesicle onto the 24-hpf eye (made by marking small clusters of cells with a fluorescent dye) showed that central retina originates from the dorsolateral layer of the vesicle, but surprisingly some of peripheral retina originates from the ventromedial layer. Although the pathway of this cellular movement has not been followed in real time, the images of fixed tissue suggest that the most likely route is via involution around

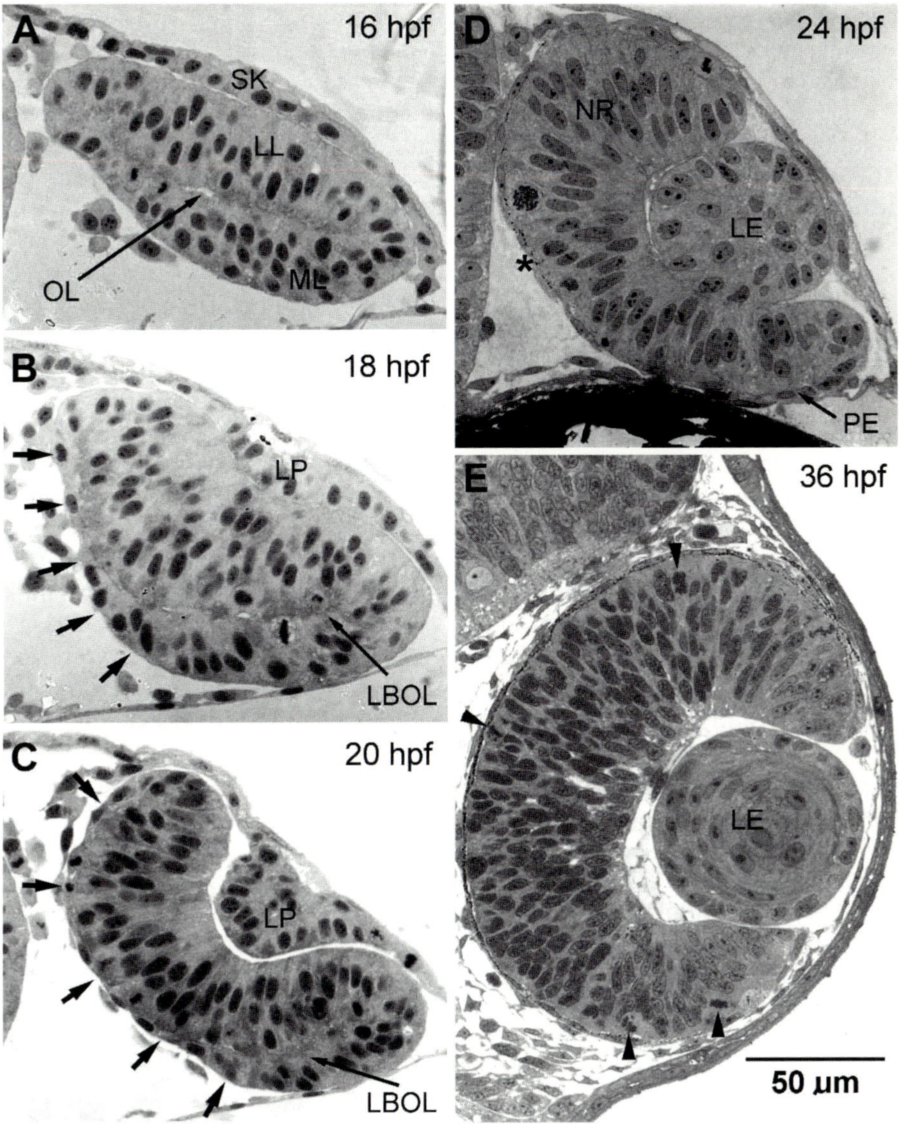

Fig. 2A–E. Semi-thin sections illustrating the morphogenesis of the flat optic vesicle into the nearly spherical eye cup. Dorsal up, medial left in all sections. The flat vesicle at 16 hpf (**A**) changes as some of the columnar cells on the ventromedial surface become flattened (*arrows* in **B**). The flattened cells cover more area at 20 hpf (**C**), and by 24 hpf (**D**) the squamous epithelium is pigmented (*). At 36 hpf, the lens has adopted the onion-like layered morphology, the number of cells in the retina is increased, and mitotic figures are still evident (*arrowheads*). *LBOL* Lateral border of the optic lumen; *LE* lens; *LP* lens placode; *OL* optic lumen; *PE* pigmented epithelium. (Reproduced from Li et al. 2000a)

the hairpin turn where the two layers are continuous (Fig. 1D). The first sign of movement appears at about 18 hpf, when a portion of the ventromedial layer begins to thin, producing the first squamous cells in a population that had been exclusively columnar (Fig. 2B). This squamous zone gradually spreads, as columnar cells segregate laterally toward the border with the neural retina (Fig. 2C,D; Li et al. 2000a). The columnar cells in both layers express the gene *rx1*, which is specific to the retina, but the squamous ones do not (Chuang et al. 1999), supporting the idea that the columnar cells in the ventromedial layer have already assumed a retinal identity and are en route to that layer. The motive force behind this cell rearrangement – whether the columnar cells pulled the squamous ones or the squamous ones pushed the columnar ones – is unknown. Changes in these cellular rearrangements are probably at the root of the abnormalities associated with the perturbation of retinoic acid levels – retinal "halving" and "duplication" (Hyatt et al. 1992, 1996; Marsh-Armstrong et al. 1994).

Several transcription factor genes and gene families that are, all transcription factors that are normally expressed in the eye, have been shown experimentally to play important roles in eye development. (1) *Pax6* is necessary for the transformation of the optic vesicle into the eye in both humans and mice, and homologous genes in flies, *eyeless* (*ey*) and *twin of eyeless* (*toy*), are essential for eye formation (Gehring and Ikeo 1999). Zebrafish have two *Pax6*-related genes: *Pax6.1*, originally called *pax [zf-b]*, and *6.2*, that are expressed in complementary and partially overlapping regions of the optic vesicle. Both fish genes can induce ectopic eyes in flies (Nornes et al. 1998); reducing the field of expression of these genes by manipulating *sonic hedgehog* (*shh*) function is associated with a reduction of eye size (Macdonald et al. 1995); and ectopic optic cups express at least one of the genes (Loosli et al. 1999). (2) The fly *sine oculis* (*so*), important to the development of the visual system, has many – perhaps 9 - homologues in zebrafish (Seo et al. 1999). Two of them, *six3* and *six6*, are expressed in and around the optic vesicle (Seo et al. 1998). Ectopic expression of murine *Six3* induces an ectopic lens in zebrafish (Oliver et al. 1996) and enlarged optic cups and supernumerary optic cups in medaka (Loosli et al. 1998). (3) *rx1*, *2*, and *3* are all expressed in the eye primordium (Chuang et al. 1999), and the knockout of a homologous gene in mice prevents the formation of an optic vesicle (Mathers et al. 1997). Overexpression of zebrafish *rx1* and *rx2* leads to enlarged ocular tissue at the expense of axial forebrain (Chuang and Raymond 2001). (4) *vsx-2/alx*, a homologue of the murine *Chx10*, is expressed early in the retina, and antisense oligonucleotides produce ocular malformations (Barbarino et al. 1997; Passini et al. 1997).

Loss-of-function and gain-of-function phenotypes of several other genes, not all of which are expressed in the eye at this early stage, have been associated with altered or missing eyes. (1) The homeodomain transcription factor, *islet-3*, is normally expressed in the forebrain at the time of optic vesicle formation, and overexpression of its LIM domains at this time prevents the formation of the optic vesicles presumably in a dominant-negative fashion. The phenotype most likely results from a loss of cell movement from the neural tube (Kikuchi et al. 1997). (2) The basic HLH transcription factor *neurogenin1* (*ngn1*) is expressed in early neurogenesis at a variety of sites and times, but not in the eye primordium. When it is overexpressed, optic vesicles fail to develop. This phenotype is associated with an expanded zone of expression of *islet-1*, a marker of the trigeminal ganglion, suggesting that the eye field identity has been changed to trigeminal (Blader et al. 1997a). (3) The paired-like homeodomain gene *Pax2.1* (initially called *pax [zf-a]*) is normally expressed in the optic stalk and choroid fissure. When *Pax2.1* is inactivated in the *no isthmus* (*noi*) mutant, the choroid fissure remains open (Macdonald et al. 1997). (4) The eyes in the mutant *out of sight* (out^{m233} allele) are recognizably smaller than normal on the second dpf, and show no obvious sign of excess cell death, suggesting that the proximate cause of its small size is subnormal proliferation (Malicki et al. 1996). (5) In *pandora* (pan^{m313} allele), the optic nerve exits at the ventral margin of the eye, suggesting that part of the ventral retina has failed to develop (Malicki et al. 1996 and unpubl. results). This phenotype resembles the "halved" eye that Marsh-Armstrong et al. (1994) produced by interfering with retinoic acid synthesis, and raises the possibility that the gene product of *pandora* may be involved in the retinoic acid signaling pathway.

2.3 Lens

The progression from embryonic eye cup to functional eye is traced below largely through the changes in the retina, but of course there are other structures essential to the mature eye. Here, we briefly describe the development of one of the most important of these, the lens.

The lens develops as a solid cell mass that acquires a spherical shape and buds off from the surface ectoderm by around 24 hpf (Fig. 2D). The cells are initially radially oriented around the lens center, but reorganize into a layered arrangement resembling an onion (Fig. 2E) after which most of the cells in the center lose their nuclei and leave behind a proteinaceous core that mostly contains specialized polypeptides, the cristallins (Posner et al. 1999). It is initially a weak lens; it forms an image far behind the eye, and therefore the eye could be said to be far-sighted. Because the photoreceptors do not yet exist, the eye is blind at this stage anyway. As the lens ages, its focal length shortens, and, by about 72 hpf, the beginning of the first day outside of the chorion, the image is formed at the plane of the outer segments, just where it has to be for good vision. The fish can first see at that time, as evidenced by the startle and optokinetic reflexes. These same changes occur in total darkness, and are therefore independent of visual experience (Easter and Nicola 1996).

Numerous mutants with lens phenotypes (most degenerative) have been described (Heisenberg et al. 1996), but few have been investigated in detail. An interesting observation about lens development comes from the studies of two zebrafish midline mutants, *you-too* (*yot*) and *iguana* (*igu*). They develop ectopic lenses on the midline where the adenohypophysis (anterior pituitary lobe) would normally appear, presumably because of the suppression of Shh signaling in this area (Kondoh et al. 2000). This observation suggests that the developmental genetic circuitry of adenohypophysis and lens may be similar.

The lens may also serve a developmental role as well as an optical one, as an interesting comparative study suggests. The Mexican tetra, *Astyanax mexicanus*, comprises two geographically isolated morphotypes, one that lives in brightly lit surface waters and the other in dark caves. Both form lens and eye cup early, but in the cave fish they subsequently degenerate. Transplantation of the lens rudiment from the surface variety rescues the eyeless phenotype in the cave fish, and transplantation from the cave fish to the surface fish produces eyelessness, thus establishing a neuralizing function for the lens (Yamomoto and Jeffery 2000).

3 Neurogenesis

3.1 The Fan Gradient

Retinal neurogenesis begins in a tiny cluster of ventronasal cells and spreads to cover the entire retina along a front resembling the rotating hand of a clock or an opening hand-fan (Fig. 3). A variety of neurogenic events, such as the expression of a proneural gene, the withdrawal from the cell cycle, and

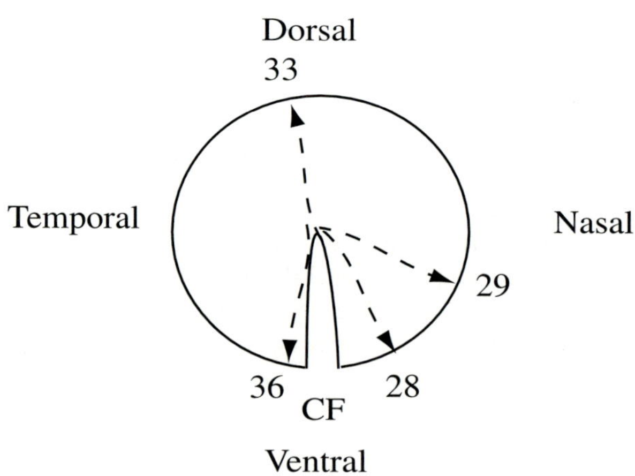

Fig. 3. Schematic of the advance of the neurogenic wave. The cartoon represents a lateral view of a right eye, with the anatomical directions and the choroid fissure (*CF*) indicated. The shafts of the *dashed arrows* indicate the locations of the front of terminal mitoses at the various ages (in hpf) indicated by the *numbers*; note the similarity to the opening of a fan. Similar series could be drawn for other aspects of retinal maturation; e.g., *ath5* expression (initiated earlier) and axogenesis (initiated later)

differentiation, are all initiated at different times, but always in the same location: ventronasal retina adjacent to the junction with the optic stalk. Soon after the onset at that site, adjacent cells on the dorsal side of this precocious region behave similarly, and, slightly later, their dorsal neighbors do likewise. This wave continues to expand dorsally and then pivots around the central end of the choroid fissure into temporal retina, and ultimately reaches the ventrotemporal border of the choroid fissure. The various events are phased differently: proneural gene expression precedes cell cycle withdrawal, which precedes differentiation. Thus, a gradient of maturity exists, in which the cells near the ventronasal site of initiation are the most advanced, and those away from it progressively less advanced. The front sweeps through 360°; in terms of clock-face coordinates, the hand has advanced from 6 to 6; in terms of the fan, the fan has fully opened. We refer to this developmental gradient as the "fan gradient".

It has been suggested that this neurogenic wave is the fish homologue of the events associated with the advancing morphogenetic furrow in the eye imaginal disk of *Drosophila* (for review see Wolff et al. 1997). The advance of the furrow involves terminal mitoses, expression of the proneural gene *atonal*, and the secretion of Hedgehog, all of which are shared with the advancing front in fish retina (see below). These parallels between the genetic networks underlying eye development in fish and *Drosophila* have stimulated considerable discussion about homologies across the phyla. Much more remains to be done, and the case may well prove to be a strong one, but it is still tentative because several of the main players in *Drosophila* have not been shown to have comparable roles in fish (see Wawersik and Maas 2000). Mutations of *eyes absent* (*eya*) and *dachshund* (*dachs*) produce an eyeless phenotype in flies, but, in fish, *eya* is not expressed in the eyes (Sahly et al. 1999) and no zebrafish homologue to *dachs* has yet been described. In addition, *decapentaplegic* (*dpp*), a member of the TGFβ family, is essential for initiating retinal differentiation in the fly, but no comparable role for a *dpp* ortholog has been found in fish retina, although abundant homologues are known to be expressed elsewhere. Finally, the *ey/toy* pair in flies does not seem to have a counterpart in fish. The *Pax6.1/2* pair seems superficially similar to the *ey/toy* pair, because both represent gene duplications, but certainly not the same duplication. The *Pax6.1/2* duplication probably occurred in the jawed fish radiation (Nornes et al. 1998), whereas the *ey/toy* duplication occurred along a separate clade, from primitive to advanced insects (Czerny et al. 1999).

3.2 Ganglion Cell Layer

One of the earliest signs of retinal maturation is at 25 hpf, when the proneural transcription factor, *atonal homologue 5* (*ath5*), is first expressed in the ventronasal area, adjacent to the optic stalk. Expression of *ath5* spreads along the fan gradient to cover the whole retina by about 36 hpf, after which it gradually disappears in central retina to linger just inside the ciliary germinal zone by 72 hpf (Masai et al. 2000). The neural markers *neuroD* and *islet-1* (Korzh et al. 1998) are expressed a few hours later, also along the fan gradient. Overexpression of *ath5* produced abnormally high levels of *neuroD*, confirming a proneural role for *ath5* (Masai et al. 2000). The *ath5* null mutant *lakritz* (*lak*; Kay et al. 2001) lacks retinal ganglion cells and has abnormally high numbers of cells in the inner nuclear layer of the mature retina. BrdU labeling has shown that the absence of ganglion cells in mutant homozygotes results from the loss of the first wave of terminal mitoses, when ganglion cells are normally produced (see Sects. 3.4.1 and 3.4.2), but subsequent waves producing neurons in other layers are not affected. One attractive interpretation of these results suggests that *ath5/lak* drives cells to make the terminal division that normally produces ganglion cells, but is not directly involved in the production of other retinal neurons. This is consistent with earlier work on orthologs of *ath5/lak* in both *Xenopus* (Kanekar et al. 1997) and mice (Brown et al. 2001; Wang et al. 2001).

The induction and maintenance of *ath5/lak* expression have also been examined (Masai et al. 2000). Results from mutants, coupled with ablations and transplantations, suggest that the proximate control of the initiation of *ath5/lak* expression is provided by the p*ax2.1*-expressing cells in the optic stalk adjacent to the eye cup. The inductive activity of these cells is dependent on signaling from the prechordal plate, perhaps through Shh. The advance of the *ath5/lak* wave of expression was also investigated experimentally; when isolated eyes were cut into temporal and nasal halves and cultured, *ath5/lak* expression initiated and spread normally in the nasal half but was absent in the temporal half, suggesting that the spread is initiated normally in the ventronasal region and requires a cell-non-autonomous mechanism to spread.

The first retinal neurons, predominantly and perhaps exclusively ganglion cells, are produced at about 29 hpf. Neuroepithelial cells at all retinal locations divide mitotically, but in the ventronasal retina a few (less than 10) finish their terminal S-phase between 27–28 hpf, presumably divide soon thereafter, and one or both of their progeny migrate to the inner retina to settle in the future ganglion cell layer (Fig. 4A). Adjacent neuroepithelial cells follow suit along the fan gradient, with the result that presumptive ganglion cells occupy the inner layer of the entire central retina by about 36 hpf (Fig. 4B; Hu and Easter 1999), prior to overt stratification (Fig. 2E).

Two genes, *ath5/lak* and a member of the *Notch* family, have been shown to influence ganglion cell numbers. *ath5/lak* is expressed for a few hours before the terminal mitoses, and, in the *ath5/lak*-null mutant, which lacks ganglion cells, this wave of terminal mitoses is absent. Thus, the first neurons are ganglion cells only if they withdraw from the cell cycle during the appropriate time interval, consistent with either or both of two roles for *ath5/lak*. First, it may promote terminal mitosis at a time when external signals favor ganglion cell fate selection. Alternatively, its expression may bias the cell to take a ganglion cell fate (Kay et al. 2001). The role of *Notch* is less clear. The receptors *Notch-1* and *Notch-3* and their ligand *Delta-C* are all expressed in the retina at about this time (Bierkamp and Campos-Ortega 1993; Sullivan et al. 1997; Haddon et al. 1998). Misexpression of a constitutively active Notch receptor causes a decrease in the number of ganglion cells and an increase in glial cells (Scheer et al. 2001), evidence that Notch plays a role in directing cells from a ganglion cell fate to glial. This is consistent with work in other vertebrates (Vetter and Moore 2001).

shh contributes to the spread of neurogenesis. It is initially expressed in the retinal ganglion cells at 28–30 hpf and spreads along the fan gradient to cover the retina by 52 hpf (Neumann and Nüsslein-Volhard 2000). In the amorphic *shh* mutant, *sonic you* (syu^{t4}), *shh* expression is initiated but does not spread, indicating that *shh* regulates its own expression. Mosaic analysis shows that *shh* expression is cell-autonomous in the ventral retina but cell-non-autonomous in more dorsal regions. This may reflect a similar cell-cell signaling to the one previously documented for *ath*. Surprisingly, both the production of neurons (indicated by Zn5 immunoreactivity) and the activation of the mitogen-activated protein kinase (dp-ERK), a downstream signaling target of Shh, *do* spread in the *syu* mutant, albeit more slowly and less completely than normal, indicating that Shh is involved in, but not essential for, their advance. A second member of the *hedgehog* gene family, *tiggywinkle hedgehog* (*twhh*), is also expressed in the retina, and probably substitutes for the defective *shh*. Exposure of wild-type embryos to cyclopamine, which blocks all Hh signaling, blocked the spread of all three indicators (expression of *shh*, Zn5 immunoreactivity, and dp-ERK). Thus, *shh* and *twhh* apparently control together the spread of the neurogenic wave that produces ganglion cells. Their relation to *ath5/lak* expression has not been examined beyond noting that the time of onset of *shh* expression, 3–5 h after that of *ath5/lak*, indicates that retinal *shh* cannot be upstream of the initiation of *ath5/lak* expression.

3.3 Inner Nuclear Layer

At about 38 hpf, the first inner nuclear layer neurons are born in the same ventronasal region that produced the first ganglion cells (Fig. 4B), and spread along the fan gradient until about 48 hpf (Hu and Easter 1999). The progress of the front in this second wave of terminal mitoses is less orderly than the first one. By the time the first inner nuclear layer neurons are born, the ganglion cells to which they will be connected have axons that extend most of the way to their termination sites (see Sect. 5.1).

The genetic control of the production of inner nuclear layer neurons has not been examined in any depth. Genes known to affect ganglion cell development frequently influence the inner nuclear layer as well. The $ath5/lak^{th241}$ mutant, for example, has abnormally large numbers of amacrine and bipolar neurons and Muller glia (Kay et al. 2001). Constitutive overexpression of activated *Notch* reduces the number of neurons in the inner nuclear layer as it

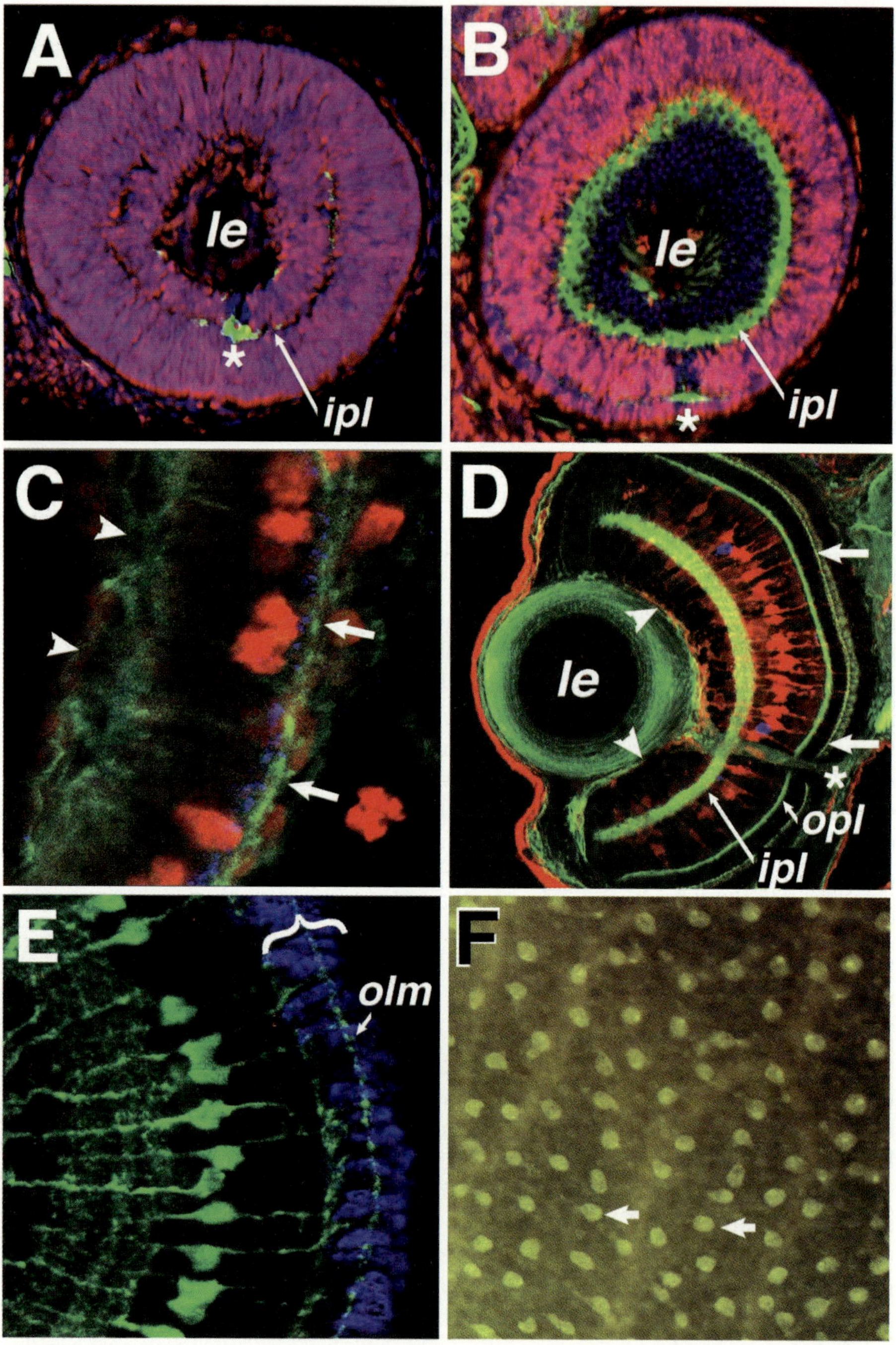
A
le
*
ipl
B
le
ipl
*
C
D
le
*
opl
ipl
E
olm
F

does in the ganglion cell layer, and favors the production of glia (Scheer et al. 2001).

The inner nuclear layer is affected in the mutants *motionless* (*mot*) and *foggy* (*fog*) that were initially reported to display abnormally low numbers of interplexiform neurons (Guo et al. 1999b). Further work revealed that *fog* specifically inhibits cell differentiation in the inner and outer nuclear layers while the ganglion cell layer is normal or maybe even hypertrophic. Positional cloning showed that *fog* encodes an *Spt5*-related transcription elongation factor (Guo et al. 2000). Despite the progress in molecular characterization of this gene, it is still not clear how the *fog* mutation affects cell fate decisions in the retina. One would like to know, for example, whether cell proliferation continues in the *fog* retina once ganglion cells are differentiated. This and other questions will have to await further analysis of this interesting gene.

3.4 Outer Nuclear Layer

The first postmitotic cells to settle in the outer nuclear layer are born by 48 hpf (Nawrocki 1985; Kljavin 1987), again in the ventronasal site, and production of these cells spreads along a disorderly fan gradient to fill the central retina by about 60 hpf (Hu and Easter 1999). By about 50 hpf, a small number of these

Fig. 4A–F. Neurogenesis and pattern in the zebrafish eye. **A, B** Parasagittal sections through zebrafish eyes labeled with a nuclear stain (*blue*), an antibody against BrdU (*red*), and a marker of neurons (Zn12 antibody, *green*). Dorsal is up, anterior right. The embryos were injected with BrdU at 28 hpf (**A**) or 38 hpf (**B**) and survived for 14 h. The first cells to exit the cell cycle are found in a small ventronasal patch (*dark blue, asterisk* in **A**). When BrdU is injected 10 h later (**B**), the entire ganglion cell layer is postmitotic (*dark blue*, central annulus in **B**), but the cells destined to make up the inner nuclear layer are still proliferative (*red*), with the exception of a small ventronasal patch (*dark blue area, asterisk* in **B**). Neurogenesis takes place in the context of a neuroepithelial sheet (**C**). The polarity of this tissue is evident in the distribution of M-phase nuclei (*red*, visualized with anti-phosphohistone H3 antibody), centrosomes (*blue*, visualized with anti-gamma tubulin antibody), and actin bundles (*green*, visualized with Alexa 488-conjugated phalloidin). Ca. 2 days after the onset of neurogenesis, the pseudostratified neuroepithelium of the optic cup is differentiated into a functional retina that contains seven major cell classes organized in precise layered architecture. Two of these cell classes are indicated in **D**: the interplexiform neurons (*blue*, visualized with an anti-tyrosine hydroxylase antibody) and Muller glia (*red*, visualized with an anti-carbonic anhydrase antibody). The three major cellular laminae of differentiated retina are separated from each other by layers of neuronal processes: *ipl* and *opl* (stained *green* with phalloidin in **D**). In the outer retina, the photoreceptor cells (*blue*, visualized with the Fret43 antibody and indicated by a *bracket*) and Muller glia (*green*, carbonic anhydrase) form junctions at the outer retinal membrane (*olm*) as shown in **E**. The precision of retinal architecture is also evident within individual cell layers and can be appreciated on sections tangential to the eye-cup hemisphere. Synaptic termini of the bipolar cells (stained with anti-PKC antibody, *arrows*) are shown in **F**. *ipl* Inner plexiform layer; *le* lens; *opl* outer plexiform layer; *olm* outer limiting membrane. **A** and **B** are from Hu and Easter (1999); **C** is from Pujic and Malicki (2001); **F** is courtesy of David Cameron cells express rod-opsin and, by 52 hpf, others express cone-opsin, therefore the initial cohort of postmitotic cells is likely to include both classes of photoreceptors (Raymond et al. 1995). The spread of cone-opsin expression expands along the fan gradient, consistent with the idea developed from other indicants that cell differentiation is phased relative to time of birth. In contrast, rod-opsin expression spreads in a disorderly ventral to dorsal direction, eschewing the fan pattern. Immunostaining with antibodies specific to cones, and rods revealed similar patterns as the in situs (Schmitt and Dowling 1996).

Both *twhh* and *shh* are expressed in the pigmented epithelium, beginning between 39–45 hpf in the ventronasal quadrant and spreading to all quadrants by 54–81 hpf (Stenkamp et al. 2000). The Hh receptor, *patched2* (*ptc2*), is expressed in the retina at this time, and the fact that the photoreceptors are born at this time suggests a role for the Hhs in photoreceptor development. In support of this idea, antisense oligos to *shh* and *twhh* slow down the advance of photoreceptor differentiation.

3.5 Prolonged Neurogenesis and Regeneration

Retinogenesis is very prolonged in zebrafish. The first phase, from 28–60 hpf, produces the tiny hatchling retina, about 100 μm in diameter (Fig. 5A). The second phase goes on for years and produces most of the adult retina, a much larger structure with the hatchling retina at its center. During the second phase, DNA synthesis and associated neurogenesis are for the most part restricted to the circumferential germinal zone, a thin annulus of proliferative cells at the retinal margin (Marcus et al. 1999). Some of their progeny withdraw from the cell cycle and differentiate in situ, thus enlarging the retina by the addition of an annulus of new neurons of all types. The proliferative daughters remain on the outermost edge of the retina where they continue to proliferate past 2 years of age (Cameron 2000), producing new generations of retinal tissue, each of which lies peripheral to the generation that just preceded it.

The proliferative cells in central and peripheral retina are both similar and different. They have in common that both produce all types of retinal neurons. They differ with respect to self-renewal, which is dependent on the fraction of proliferative daughters that they produce. In the central retina, three waves of terminal mitoses, each lasting about 10 h, produce thousands of neurons and leave the central retina depleted of proliferative cells, suggesting that either the proliferative progeny died prior to hatching, or the progeny did not include proliferative cells. Cell death is negligible during this time (Li et al. 2000b), excluding the cell death explanation; therefore, nearly all the progeny of the last wave of mitosis must withdraw from the cell cycle and differentiate. If the progeny of the cells in the circumferential germinal zone were to favor withdrawal from the cell cycle at the same rate as those of central retina, then retinal growth would stop within a few days owing to a depletion of proliferative cells. But growth in the periphery continues; the cells there proliferate for years and

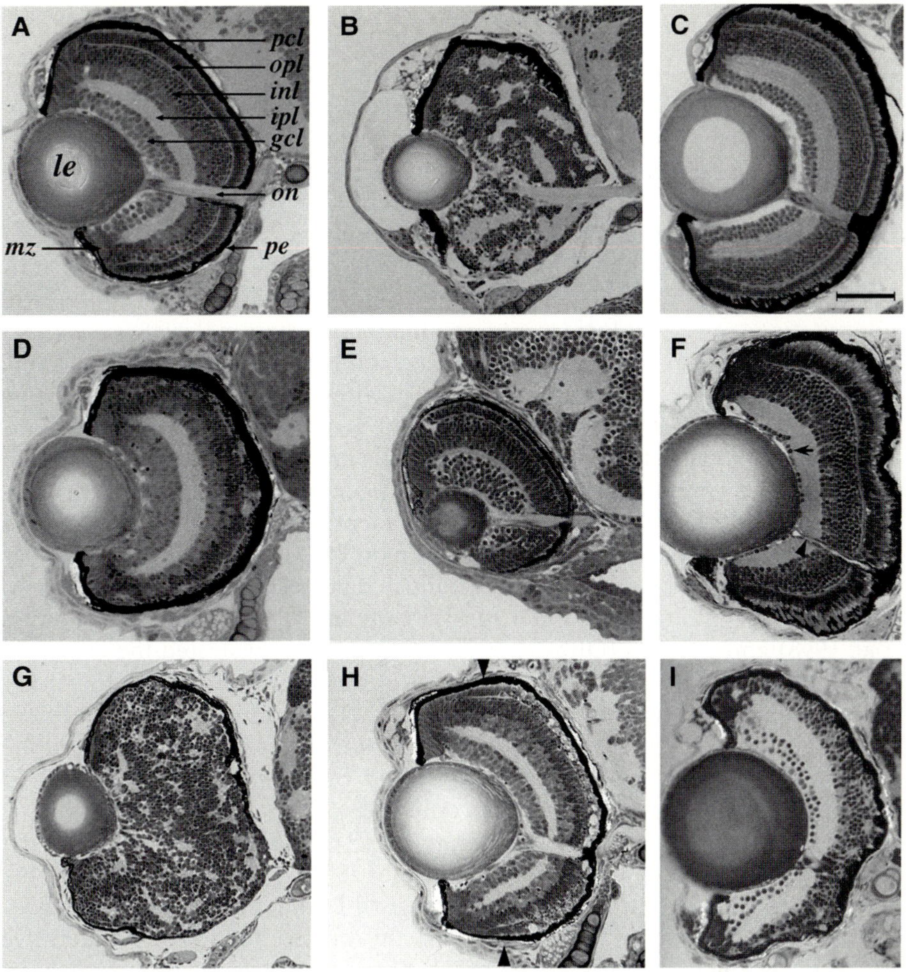

Fig. 5A–H. Examples of mutations affecting development of the zebrafish retina. **A, G** In the wild-type retina, several classes of neurons can be distinguished in simple histological preparations at 3 dpf (**A**) as well as at 5 dpf (**G**). This is certainly true for photoreceptor cells (*pcl*) and ganglion cells (*gcl*). Additional cell categories can be distinguished upon closer inspection. **B** In *mikre oko*, many photoreceptor cells do not maintain elongated morphology and eventually die. This phenotype is usually more severe in the retinal periphery. **C** In *nagie oko*, the laminar pattern of neurons is almost entirely disorganized. **D** The phenotype of *oko meduzy* also involves a severe neuronal disorganization. **E** Although all neuronal laminae are present, the *out of sight* retina is markedly smaller as compared to the wild type in (**A**). **F** In the *brudas* eye, photoreceptor cell loss is first obvious in the central retina and only later spreads into the periphery. At 3 dpf, some intact photoreceptors survive at the retinal margins (*arrowheads*). **H** *lakritz* is characterized by a specific loss of ganglion cells (*arrow*) and a severe reduction of the optic nerve (*arrowhead*) compared to wild-type retina at the same stage (**G**). **I** In contrast to *brudas*, the loss of photoreceptor cells in *not really finished* is most severe in the periphery. Some relatively intact photoreceptor cells persist in the central retina at least until 5 dpf. Sections were prepared from embryos collected at ca. 3 dpf (**A–F**) or ca. 5 dpf (**G–I**). Dorsal is up and midline is to the right in all panels (indicated by *arrows* in **C**). *gcl* Ganglion cell layer; *inl* inner nuclear layer; *ipl* inner plexiform layer; *le* lens; *mz* marginal zone; *on* optic nerve; *opl* outer plexiform layer; *pcl* photoreceptor cell layer; *rpe* retinal pigmented epithelium. *Scale bars* 100 μm. (Reprinted with permission from Malicki 2000)

ultimately produce a number of retinal neurons that vastly exceeds the number in the hatchling retina that existed at the time of first sight. This long-term production of neurons and the maintenance of a proliferative population (self-renewal) are possible only if at least half of the progeny in the circumferential germinal zone are proliferative, quite a different bias from central retina. Thus, the proliferative programs of central and peripheral retina differ, the former biased in favor of withdrawal from the cell cycle, and the latter in favor of continuation of proliferation. The underlying cause of this difference is unknown, but mutations in several loci (*piegus*, *round eye* and *marginal eye*) cause selective degeneration in the circumferential germinal zone (Malicki et al. 1996; Fadool et al. 1997), suggesting that the proliferative programs in the two regions may be under the control of different genetic mechanisms.

In contrast to all other retinal neurons, rods are produced in the functional retina that lies central to the circumferential germinal zone. The rod precursors, a special class of progenitor cells first identified in cichlids and goldfish (Johns and Fernald 1981; Raymond and Rivlin 1987), lie in the outer nuclear layer, synthesize DNA, and divide to produce daughters that differentiate into rods and enter into the neural network already in place. Work in trout (Julian et al. 1998) and goldfish (Otteson et al. 2001) has demonstrated the existence of another population of proliferative cells in central retina, in the inner nuclear layer. They proliferate more slowly than the rod precursors, and have been suggested to be their progenitors. Thus, the vigorous proliferation in the circumferential germinal zone is augmented by a smaller proliferation in central retina that normally produces only rods.

Fish, like amphibians, can regenerate their retinas completely following injury (Cameron 2000; Cameron and Carney 2000), but the source(s) of the new cells remain obscure. Following a surgical lesion, the first step in regeneration is the formation of a blastema on the wound boundary. Proliferative cells similar in appearance to those in the circumferential germinal zone appear all round the wound and, over the next days to weeks, they gradually fill in the hole created by the surgery. The rapid appearance of proliferative cells on the wound is incompatible with their having come from the margin, which is frequently a long way away, probably too far for such a quick migration. The only proliferative cells in central retina are the rod precursors and the proliferative cells in the inner nuclear layer (both normally proliferative) and the Muller glia (which proliferate only in response to injury; Braisted et al. 1994). One or more of these three must be the source of the cells in the wound blastema. Normally, only rods are produced in central retina, never cones or inner retinal neurons; however, following injury, all types of cells are produced, therefore the fates of some progeny of these proliferative cells must change. One interpretation is that the injury causes the production of a molecule that either instructs the progeny to adopt another fate (e.g., ganglion cell rather than rod or glia) or antagonizes whatever influence normally restricts their fates to rods or glia. A more complete understanding of how the sluggish precursors are stimulated to proliferate more, and how the fates of their progeny are determined, is obviously desirable, and may contribute to the solution of the problem of clinical repair of the nervous system.

3.6 Modulation of the Rate of Proliferation

The rate of cell production depends on two factors: the fraction of the cells that are proliferating and the length of the cell cycle. Initially, the proliferative fraction is 1 and the cell number doubles over a time that depends on the length of the cell cycle. Although most studies in the vertebrate central nervous system have concluded that the length of the cell cycle increases monotonically (for review, see Jacobson 1991), which slows the rate of proliferation, these studies have not generally included the period prior to the onset of terminal mitoses, which begin at 28–29 hpf in the zebrafish retina (Nawrocki 1985).

The number of cells in the zebrafish optic vesicle/eye was followed from 16–36 hpf, and did not show the expected monotonic deceleration. Over 16–24 hpf, cell number increases only about 12% and over 24–36 hpf by nearly 300% (Fig. 2D,E). The initially sluggish increase could be explained by cells either dying or emigrating from the eye; however, these possibilities have been excluded. A third alternative, that significant numbers of cells temporarily withdraw from the mitotic cycle as in larval fly imaginal disks, was excluded by showing that most cells were labeled with BrdU. The only explanation remaining for the changing rate of proliferation is a transient lengthening of the cell cycle during 16–24 hpf followed by an abrupt shortening thereafter. This alternative was supported by the observation that the mitotic index rose abruptly after 24 hpf (Li et al. 2000b). The acceleration of proliferation coincides roughly with the completion of the morphogenesis of the eye cup. This schedule of low proliferation during morphogenesis followed by extensive proliferation thereafter has also been noted in the development of zebrafish pharyngeal cartilages (Kimmel et al. 1998).

3.7 Pulsatile Production of Neurons

It was shown in Sections 3.1–3.4 that the central retina is produced by three sequential traveling waves of terminal mitoses. Each wave begins 10 h after its predecessor and takes about 10 h to complete the full 360° circuit from the ventronasal to the ventrotemporal side of the choroid fissure. Each wave contributes cells to one of the three nuclear layers, in the order: ganglion cell, inner nuclear, and outer nuclear. The apparent exclusivity of each wave (i.e., the observation that cells for only a single nuclear layer are produced in any wave) is surprising in the light of earlier work in avian and mammalian retina that showed that cells of various layers were produced simultaneously (Young 1985; LaVail et al. 1991; Belecky-Adams et al. 1996). The evidence in favor of this exclusivity in early zebrafish retina is clearest in the precocious ventronasal column of neuroepithelial cells, the site of initiation of all three nuclear layers. Figure 4A shows that some cells destined to become ganglion cells made their terminal S-phase before 28 hpf. Keep in mind that the embryo survived until 42 hpf, an additional 14 h, during which time BrdU was available for DNA synthesis, so if any of the cells in the precocious ventronasal column made a terminal division during that time and settled in the ganglion cell layer, they would be recognized as red nuclei. But red nuclei are absent there. Thus, the ganglion cells born before 28 hpf constitute all of the ganglion cells in that column. When BrdU was injected at 38 hpf, all of the cells destined for the ganglion cell layer were postmitotic (Fig. 4B) and almost none of the cells destined for the inner nuclear layer were postmitotic, except in the ventronasal column. The clear distinction – blue ganglion cell layer, red inner nuclear layer – indicates that all the terminal divisions between 28–38 hpf produced ganglion cells, and inner nuclear layer cells were not born until about 38 hpf, when a new wave of terminal mitoses began. This wave mimicked its predecessor, beginning in ventronasal retina and ending about 10 h later in ventrotemporal retina, and it produced inner nuclear layer cells. Finally, at about 48 hpf, the first outer nuclear layer cells were produced. Thus, the neurons in this column were produced in three pulses separated by two intervals of about 10 h. Neurogenesis in other columns follows a similar program, with pulses occurring at different times. This pulsatile neurogenesis is unexpected, as most models of neurogenesis involve a more-or-less continuous production of postmitotic cells. The underlying mechanism behind this pulsatile behavior is unknown.

4 Pattern and Patterning of Cellular Architecture in the Retina

4.1 Pattern of Differentiated Retina

Cells of the vertebrate retina are organized in a precise laminar pattern. The major features of this pattern, easy to distinguish on transverse sections, are three cellular laminae separated by layers of synaptic processes. As in other vertebrates, the photoreceptors occupy a distinct cell layer adjacent to the pigmented epithelium while the ganglion cells form a layer adjacent to the lens. In between these two cell classes, retinal interneurons form the inner nuclear layer, delimited on both sides by layers of neuronal processes – the inner and the outer plexiform layers (Fig. 4D).

The major classes of retinal neurons can be subdivided into cell types. Similarly, the three major cellular and two major plexiform laminae can be subdivided further into more narrowly defined layers associated with particular cellular classes or types. The photoreceptors are subdivided into five different types in zebrafish: short single cones, long single cones, short and long member of the double-cone pair, and rods (Branchek and Bremiller 1984), each characterized by its unique morphology and opsin expression. Starting at 3–4 dpf, the various parts of photoreceptor cells, e.g., outer segments and nuclei, segregate into different sublayers. This stratification is most obvious in the adult (Raymond et al. 1993). The details of stratification vary with the state of light adaptation, but, in the light-adapted retina, rod nuclei, for example, are proximal to long single-cone nuclei. The rod outer segments, however, are distal to long single-cone outer segments and these, in turn, are distal to short single-cone outer segments. Stratification is also present within the plexiform layers. Analysis of bipolar cell termini within the inner plexiform layer revealed at least five strata that can be distinguished based on the type of innervation they receive (Connaughton and Nelson 2000).

The intricacy and precision of the retinal architecture is also obvious in the arrangement of individual cell types within cell layers: in adult zebrafish retinae, rows of cones are separated by rods (Larison and Bremiller 1990). Within each of the cone photoreceptor rows, cells are arranged in a precise sequence: short single cone, double cone, long single cone, double cone . . . , etc. The neighboring rows are shifted with respect to each other in such a way that a short single cone of one row is flanked on both sides by long single cones of the neighboring rows. Amacrine cells of the zebrafish retina also display a non-random distribution although not as precise as that seen in the photoreceptor cell layer. Statistical analysis of amacrine cell patterns suggests that their spacing results from repulsive interactions between cells of the same type. Similar repulsive interactions may be present among bipolar cells, as evidenced by uniform distribution of their synaptic termini (Fig. 4F). Studies conducted so far indicate that amacrine cells of different types are randomly distributed with respect to each other (Cameron and Carney 2000).

4.2 Formation of Retinal Architecture

Neurogenesis involves a dramatic reorganization of retinal structure. Prior to the appearance of the first postmitotic neurons, the retina consists of a single epithelial sheet of cells. The stratified architecture of the differentiated retina forms between the basal (vitreal) and the apical (ventricular) surfaces of this sheet, concurrent with cell fate decisions of neurogenesis. The first postmitotic neurons, the ganglion cells, undergo their final mitotic division at the apical surface of the neuroepithelium and immediately afterwards migrate towards their ultimate destinations near the basal surface (Hinds and Hinds 1974; Bodick and Levinthal 1980). Following their initial migration, the position of ganglion cells does not change in any substantial way. Thus, even at the onset of neurogenesis, positional cues are present within the retinal neuroepithelium and ganglion cells interpret them appropriately to find their destinations. The nature of these positional cues and the mechanisms that interpret them are not understood.

The inner nuclear layer forms in a similar fashion to the ganglion cell layer. The inner nuclear layer neurons migrate away from the apical surface of neuroepithelium to destinations adjacent to the ganglion cell layer (Hinds and Hinds 1983). In contrast to other cell classes, photoreceptor cells and the Muller glia are unique in that they do not migrate in the course of neurogenesis but differentiate in contact with the apical surface of the retina (Hinds and Hinds 1979). The junctional complexes that maintain this contact persist in the differentiated retina as the outer limiting membrane and are presumed to play an important structural role.

A rudimentary pattern is already present in the retina at the neuroepithelial stage. The wild-type retinal neuroepithelial sheet is characterized by a distinct polarity evident in the apical localization of centrosomes, adherens junctions, and M-phase nuclei (Hinds and Ruffett 1971; Chenn et al. 1998), all of which can be specifically visualized with the help of appropriate markers (Fig. 4C). As the stratified appearance of the mature retina forms within this epithelial sheet, it is tempting to speculate that the neuroepithelial polarity plays a role in the distribution of the guidance cues that cells utilize to form the mature stratified pattern. The phenotypes of several zebrafish mutants are consistent with this idea.

A striking disarrangement of retinal neurons is caused by mutations in several zebrafish loci. The severity of their phenotypes indicates that these mutations affect key patterning mechanisms. In mutants of *glass onion* (*glo*), *heart and soul* (*has*), *mosaic eyes* (*moe*), *nagie oko* (*nok*), and *oko meduzy* (*ome*), individual nuclear or plexiform layers are no longer distinguishable (Malicki et al. 1996). By staining for cell-class specific markers in four of these mutants, it has been shown that the photoreceptors, ganglion cells, amacrine cells, and Muller glia are specified; however, instead of forming layers, they are scattered throughout the entire thickness of the retina (Malicki et al. 1996; Malicki and Driever 1999; Jensen et al. 2001; Pujic and Malicki 2001; Wei and Malicki 2002). Defects of several other loci, e.g., *belladonna* (*bel*) and *chameleon* (*con*), cause milder patterning defects that appear confined to ganglion cells. The retinal architecture of the *syu*^{t4} mutant allele, affecting the *shh* locus, also appears disorganized although this patterning phenotype has not been characterized in detail (Neumann and Nüsslein-Volhard 2000).

Interestingly, mutants of *glo*, *has*, *moe*, *nok* and *ome* all produce patterning defects in the neuroepithelial sheet prior to neurogenesis. This is evidenced by the ectopic location of structures that in the wild-type neuroepithelium localize to the apical surface: M-phase nuclei, centrosomes and adherens junctions (Malicki et al. 1996; Malicki and Driever 1999; Jensen et al. 2001; Pujic and Malicki 2001; Horne-Badovinac et al. 2001; Peterson et al. 2001; Wei and Malicki 2002). These observations indicate that the polarity of the retinal neuroepithelial sheet has been compromised in these mutants. What are the roles of *glo*, *has*, *moe*, *nok* and *ome* in the patterning of the retina? One possibility is that the sole function of these loci is the maintenance of neuroepithelial polarity. If this is the case, then the patterning phenotype in the mature retina is the result of an earlier neuroepithelial defect, possibly because compromised neuroepithelial polarity leads to a mislocalization of positional cues that neurons utilize to form proper patterns. As the nature of these cues in the retina is currently unknown, this idea is difficult to test directly at this time. Alternatively, in addition to their role in neuroepithelium, the genes in question may function in the formation of structural elements of the retina, and the outer limiting membrane is the most plausible candidate. It is the set of junctional complexes in the photoreceptor cell layer that appears to originate from the apical cell junctions of the retinal neuroepithelium. Significantly, the outer limiting membrane appears to be completely absent in *ome* and the related neuronal patterning mutants because mutant photoreceptor cells are completely dissociated from each other. This is noteworthy because, in other studies that involve disorganized retinae, photoreceptors form "rosettes", concentric clusters of cells held together by cell junctions similar to these seen in the outer limiting membrane (Sheffield and Moscona 1970; Nakayama et al. 1996; Tomita et al. 1996). Ectopic photoreceptor cells of *ome*, *moe* and *nok* do not form "rosettes", suggesting that junctions of the outer limiting membrane are absent in these mutants. [Rosettes do form in *glo* (Pujic and Malicki 2001), and have not been investigated in *has*.] The absence of rosettes suggests that *moe*, *nok* and *ome* may play a role in the integrity of the outer limiting membrane and argues that their function extends beyond the neuroepithelial stage.

Another interesting aspect of *glo*, *moe*, *nok* and *ome* is their involvement in cell-cell interactions that maintain the polarity of the retinal neuroepithelial sheet. This involvement is evidenced by mosaic analyses in which mutant (or wild-type) cells are transplanted into a wild-type (or mutant) host. For all four genes, mutant neuroepithelial cells display a normal phenotype when surrounded by wild-type tissue. In mosaic animals, the centrosomes of *glo*, *ome* or *nok* mutant cells localize to the correct apical positions indicating that they are phenotypically wild type (Malicki and Driever 1999; Pujic and Malicki 2001; Wei and Malicki 2002). Similarly, in genetically mosaic retinae containing *moe* neuroepithelial cells surrounded by wild-type retina, mutant cells mostly divide in apical locations characteristic of the wild type (Jensen et al. 2001). These experiments provide evidence for the role of cell-cell interactions in the maintenance of neuroepithelial polarity in the eye primordium. Which cells participate in these interactions? This question has been addressed with most success in the case of the *moe* locus. In genetically mosaic animals, wild-type pigment epithelial cells rescue the phenotype, again scored as the position of dividing nuclei, of the neighboring *moe* mutant retinal neuroepithelial cells (Jensen et al. 2001). This result demonstrates that retinal pigment epithelium plays a role in the apical-basal polarity of the neighboring neural retina.

Mosaic analysis of the neuroepithelial phenotypes leaves several questions unanswered. First, although the interactions between the pigmented epithelium and the retina are sufficient to rescue the *moe* mutant phenotype of the latter, it is not clear whether they are necessary for the maintenance of its wild-type phenotype. This issue could be addressed by the analysis of mosaic animals in which mutant pigmented epithelial cells abut wild-type tissue. Second, one would like to know whether wild-type pigmented epithelial cells are able to rescue the neuronal (as opposed to neuroepithelial) retinal phenotype at later stages of development. Finally, one has to note that wild-type pigmented epithelium does not appear to rescue the *ome* retinal phenotype. The basis of this difference is not known.

5 Terminal Differentiation of Cellular Morphology

The initial cell fate decisions in the retina are followed by an elaborate differentiation process. In zebrafish, this final period of neurogenesis has been described in depth only for ganglion and photoreceptor cells. Below we describe the normal development of these two cell types and its aberrations in mutant animals.

5.1 Ganglion Cell Axogenesis

The first axons are produced soon after terminal mitosis by the small cluster of early ganglion cells in the ventronasal cluster, and are joined by progressively more as new ganglion cells are added along the fan gradient (Burrill and Easter 1995). They express neurolin, a member of the immunoglobulin superfamily of cell adhesion molecules that is important for both axon guidance and fasciculation; indeed, the RNA transcript for this gene first appears at 28 hpf and therefore provides the earliest sign of ganglion cell differentiation identified so far (Laessing and Stuermer 1996; Leppert et al. 1999). Several hours later they express *robo2*, a member of the *roundabout* family of axon guidance receptors, which is involved in multiple pathway decisions (Fricke et al. 2001).

The first optic axons reach the presumptive chiasma at about 32 hpf, and arrive at their final targets at varied times thereafter. The first ganglion cells lie adjacent to the optic stalk and, like most other central neurons, they send their axonal growth cones along a superficial trajectory deep to the surface, separated from the surrounding basement membrane by the thin end feet of neuroepithelial cells (Bodick and Levinthal 1980; Burrill and Easter 1995). They maintain this superficial position as they enter the stalk and advance to the chiasma (Wilson et al. 1990). Later axons fasciculate with their predecessors and continue up the contralateral diencephalic wall alongside the tract of the postoptic commissure, the first longitudinal tract in the forebrain. The axon terminals arborize in at least 10 different fields, the largest and most distant of which is the optic tectum (Stuermer 1988; Burrill and Easter 1994). The tectal field is covered with retinal arbors by about 72 hpf. Numerous genetic defects affecting the retinofugal projection have been isolated (Baier et al. 1996; Karlstrom et al. 1996; Trowe et al. 1996), but a description of these is beyond the scope of this review. More information about the guidance and terminations of optic axons can be found elsewhere in this volume (see Beattie et al., this Vol.).

5.2 Photoreceptor Differentiation

Photoreceptor cells display several specialized structures that are present in few, if any, other retinal cell classes (Fig. 6). The ventricular (apical) terminus of a photoreceptor cell consists of an elongated stack of membrane folds called the outer segment. It is this structure that harbors the components of the phototransduction apparatus, including opsins. The outer segment connects to the cell body via a narrow constriction tightly surrounding a cilium (so-called connecting cilium). Basal to the connecting cilium, the photoreceptor expands again, approximately matching the width of the outer segment. At least four regions can be distinguished basal to the connecting cilium: the mitochondria-rich ellipsoid, the contractile myoid, the perinuclear area, and the synaptic terminus (Rodieck 1973; Dowling 1987). Postsynaptic processes of bipolar and horizontal interneurons form bundles that invaginate photoreceptor synaptic terminals. Inside these invaginations, bipolar and horizontal processes tend to associate with each other in a one-to-two ratio, forming so-called triads. Within a triad, a bipolar cell process is sandwiched between two horizontal cell extensions. On the photoreceptor side of the synaptic membranes, triads are associated with specialized structures termed synaptic ribbons. It is thought

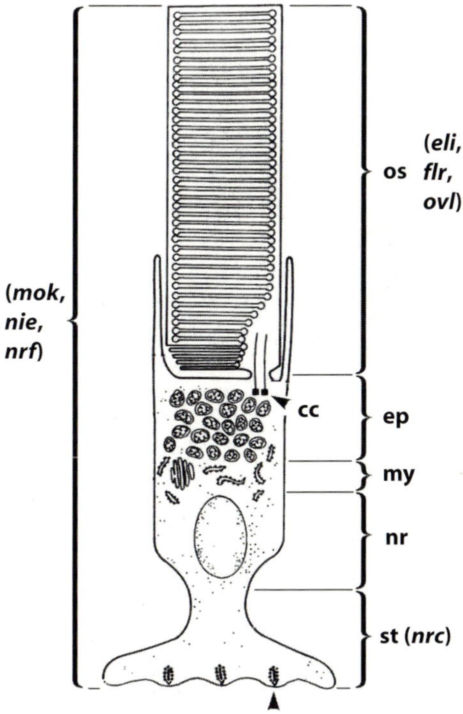

Fig. 6. Structure of the photoreceptor cell. The most prominent features of vertebrate photoreceptor cells are the outer segment (*os*); the ellipsoid (*ep*); the myoid (*my*); the nuclear region (*nr*); and the synaptic terminus (*st*). The latter is characterized by the presence of synaptic ribbons (*sr*). The photoreceptor outer segment connects to the inner segment via a narrow constriction containing a cilium (*cc*). Numerous zebrafish mutations affect photoreceptor cells. Some of them produce defects largely specific to a single photoreceptor cell feature (*eli, flr, nrc, ovl*), others involve more generalized defects (*mok, nie, nrf*). This figure lists only mutants characterized by electron microscopy. Drawing based on Basinger et al. (1976)

that the function of these is to assure a graded and continuous neurotransmitter release (Juusola et al. 1996).

The differentiation of zebrafish photoreceptor cells, starting from the cell cycle exit and culminating in the formation of outer segments and synaptic ribbons, occurs in a narrow window of less than 20 h. The photoreceptors first exit the cell cycle between 43 and 48 hpf, and by 48 hpf develop elongated morphology clearly distinct from other retinal cell classes (Nawrocki 1985; Larison and Bremiller 1990; Hu and Easter 1999). Expression of rod opsin begins around 50 hpf, and blue and red opsins follow shortly thereafter at 52 hpf (Raymond et al. 1995). In total, six distinct opsin genes are expressed in the zebrafish photoreceptor cells: blue, red, rod, ultraviolet and two types of green (Vihtelic et al. 1999). An additional opsin molecule, detectable in horizontal cells, is hypothesized to play a role in regulating the phase of a retinal circadian clock (Kojima et al. 2000), but it has not been followed developmentally. The first membrane folds of the photoreceptor outer segments appear by 54 hpf, and the synaptic ribbons by 62 hpf (Raymond et al. 1995; Schmitt and Dowling 1999). The elaboration of cell-type-specific morphological features within the photoreceptor cell class begins with the appearance of morphologically distinct UV cones between 3 and 4 dpf (Branchek and Bremiller 1984). In the mature retina, the different types of photoreceptor cells can be distinguished both by morphological criteria and by the location of their nuclei and the outer segments (Branchek and Bremiller 1984; Raymond et al. 1993).

The genetic screens in zebrafish have identified more mutations affecting photoreceptor cells than all other retinal cell types taken together. This may be so because photoreceptor cells are particularly numerous and bulky, making their loss particularly evident as a decrease of eye size. As other neuronal cell classes occupy much less of the retinal volume, their loss does not result in an obvious eye-size change. The *ath5/lak*^{th241} mutant, for example, which lacks ganglion cells, does not have a noticeably smaller eye, and was originally discovered on the basis of abnormal pigmentation (Kelsh et al. 1996). Over 30 mutant alleles affecting photoreceptor cells are currently available (Malicki et al. 1996; Fadool et al. 1997; Li and Dowling 1997; Becker et al. 1998; Brockerhoff et al. 1998; Drummond et al. 1998; Neuhauss et al. 1999), and fall into at least three patterns of cell loss. In some mutants, e.g., *mikre oko* (*mok*), *niezerka* (*nie*), and *not really finished* (*nrf*), photoreceptors are initially lost in the retinal periphery. In others, centrally located cells are affected first. This is the case in *brudas* (*bru*), *elipsa* (*eli*), *oval* (*ovl*), *photoreceptors absent* (*pca*), and *fleer* (*flr*), for example. In yet another group of mutants, represented by *krenty* (*krt*) and *sinusoida* (*sid*), patches of photoreceptors are missing. The significance and the underlying causes of these distinct patterns of cell loss are not clear. The initial loss of photoreceptor cells in the central retina may be explained by the spatiotemporal pattern of cell differentiation within the photoreceptor cell layer: the centrally located cells are the first to differentiate and the first to accrue pathological changes. Peripheral cell loss, on the other hand, could be explained by a reduced supply of a hypothetical survival factor. If the central cells deplete such a factor on a "first come, first served" basis, the latecomers do not have access to it at all and therefore develop abnormalities first.

The zebrafish photoreceptor defects display variable specificities and times of onset. The earliest phenotype described so far is observed in *mok* (Malicki et al. 1996), in which the photoreceptors exhibit severe morphological abnormalities by 72 hpf and only infrequently develop rudimentary outer segments (Doerre and Malicki 2001). At the opposite extreme is the phenotype of *night blindness a* (*nba*), a semidominant mutant in which photoreceptors begin to disappear in the heterozygotes by 4 months of age (Li and Dowling 1997). *nba* homozygotes suffer from an early nonspecific degeneration of the entire CNS. The vast majority of zebrafish mutants investigated in detail affect different photoreceptor types with equal strength (Doerre and Malicki 2002). So far, the only exception to this rule is the *partial optokinetic response b* (*pob*) mutant that specifically affects red cone survival (Brockerhoff et al. 1997). *pob* red opsin cones, while present at 3 dpf, are no longer detectable at 5 dpf. Further studies of this mutant may offer insights into the molecular basis of photoreceptor type diversity. This interesting but poorly understood subject is much in need of investigation.

While some of the zebrafish photoreceptor mutants produce general defects of cell morphology, others affect very specific features. Mutations of the *eli* locus, for example, specifically affect development of the outer segments. While the inner segments of *eli* photoreceptor cells are well developed, as evidenced by the abundance of mitochondria, the outer segments appear to be completely missing (Doerre and Malicki 2002) Although the defects of *flr* and *ovl* photoreceptors are somewhat less specific, they also appear to predominantly affect the outer segment formation (Doerre and Malicki 2002). While *eli*, *flr* and *ovl* affect the outer segment, the *no optokinetic response c* (*nrc*) mutation produces a specific defect in the synaptic termini of photoreceptor cells (Allwardt et al. 2001). The processes of *nrc* bipolar and horizontal cells do not form invaginations into the photoreceptor synaptic termini and photoreceptor synaptic ribbons "float" in the cytoplasm instead of forming attachments to the presynaptic membrane.

Cell–cell interactions have previously been shown to play a role in photoreceptor development and survival (Mullen and LaVail 1976; Watanabe and Raff 1990; Stiemke et al. 1994). Mosaic analysis of zebrafish photoreceptor mutants also demonstrated the presence of cell-cell interactions. While the majority of zebrafish photoreceptor phenotypes are cell-autonomous, some exhibit cell-non-autonomous components (Doerre and Malicki 2002). The *mok* and *nie* mutant photoreceptor cells survive better when surrounded by wild-type cells. Remarkably, while outer segments are absent in *mok* retinae or their development is severely stunted, in mosaic animals *mok* cells surrounded by wild-type tissue produce robust outer segments (Doerre and Malicki 2001). This phenotypic rescue is confined to the outer segment as the more proximal photoreceptor cell features, such as the inner segment, the nuclear region and the synaptic terminus retain mutant morphology. Mosaic studies demonstrate that the outer segment rescue is most likely not mediated by the retinal pigment epithelium, suggesting that cell-cell interactions independent of the pigmented epithelium play a role in photoreceptor outer segment development.

Both outer segments and synaptic termini are complex features of the photoreceptor cell, absolutely essential for its function. One would like to know what mechanisms initiate the evagination of cell membrane that leads to the formation of outer segments, or what molecular factors assure that membranes of outer segments fold in such an orderly fashion. Similarly, it would be valuable to understand the genetic pathways underlying the assembly of ribbon synapses. Some of the first insights into these processes may come from the analysis of zebrafish mutants.

6 Summary

In this review, we have attempted to cover all the major points of zebrafish eye development, and have found that, for the most part, it has much in common with other eyes, in both vertebrates and the fly. In addition to the confirmation and extension of earlier studies, however, the work on zebrafish has provided some new insights that should be assessed for their applicability to the development of other vertebrates. Among these are the modulated cellular proliferation in the optic vesicle, the complex spatiotemporal pattern of central retinal neurogenesis, the emergence of spatial order among the photoreceptors, the genetic controls of cell fates, and the genetic mechanisms underlying retinal stratification. Substantial though it is, this contribution will grow rapidly in the next few years as the advances of zebrafish genetics are accelerated by progress of genomics, especially the zebrafish genome project.

Acknowledgements. We thank David Cameron for the unpublished pictures, and Bill Harris, Francesca Pignoni, Herwig Baier, and Zac Pujic for useful comments on an earlier draft of the paper.

References

Abdelilah S, Driever W (1997) Pattern formation in janus-mutant zebrafish embryos. Dev Biol 184:70–84

Abdelilah S, Solnica-Krezel L, Stainier DY, Driever W (1994) Implications for dorsoventral axis determination from the zebrafish mutation *janus*. Nature 370:468–471

Aberle H, Bauer A, Stappert J, Kispert A, Kemler R (1997) Beta-catenin is a target for the ubiquitin-proteasome pathway. EMBO J 16:3797–3804

Acampora D, Mazan S, Lallemand Y, Avantaggiato V, Maury M, Simeone A, Brulet P (1995) Forebrain and midbrain regions are deleted in *Otx2-/*-mutants due to a defective anterior neuroectoderm specification during gastrulation. Development 121:3279–3290

Adams RJ, Faruque D, Concha ML (2002) Dynamics of cell rearrangement in early zebrafish gastrula. (submitted)

Adelmann H (1966) Marcello Malpighi and the evolution of embryology. Cornell University Press, Ithaca

Agarwal S, John PA (1988) Studies on the development of the kidney of the guppy, Lebistes reticulatus. 1. The development of the pronephros. J Anim Morphol Physiol 35:17–24

Agarwala S, Sanders TA, Ragsdale CW (2001) Sonic hedgehog control of size and shape in midbrain pattern formation. Science 291:2147–2150

Agathon A, Thisse B, Thisse C (2001) Morpholino knock-down of *antivin1* and *antivin2* upregulates nodal signaling. Genesis 30:178–182

Agius E, Oelgeschlager M, Wessely O, Kemp C, De Robertis EM (2000) Endodermal Nodal-related signals and mesoderm induction in *Xenopus*. Development 127:1173–1183

Ahn DG, Gibson G (1999) Expression patterns of three spine stickleback *Hox* genes and insights into the evolution of the vertebrate body axis. Dev Genes Evol 209:482–494

Ahn DG, Ruvinsky I, Oates AC, Silver LM, Ho RK (2000) *tbx20*, a new vertebrate T-box gene expressed in the cranial motor neurons and developing cardiovascular structures in zebrafish. Mech Dev 95:253–258

Akimenko MA, Ekker M, Wegner J, Lin W, Westerfield M (1994) Combinatorial expression of three zebrafish genes related to *distal-less*: part of a homeobox gene code for the head. J Neurosci 14:3475–3486

Alcantara S, Ruiz M, DeCastro F, Soriano E, Sotelo C (2000) Netrin 1 acts as an attractive or as a repulsive cue for distinct migrating neurons during the development of the cerebellar system. Development 127:1359–1372

Alexander J, Stainier DYR (1999) A molecular pathway leading to endoderm formation in zebrafish. Curr Biol 9:1147–1157

Alexander J, Stainier DY, Yelon D (1998) Screening mosaic F1 females for mutations affecting zebrafish heart induction and patterning. Dev Genet 22:288–299

Alexander J, Rothenberg M, Henry GL, Stainier DY (1999) *casanova* plays an early and essential role in endoderm formation in zebrafish. Dev Biol 215:343–357

Alexandre D, Clarke JD, Oxtoby E, Yan YL, Jowett T, Holder N (1996) Ectopic expression of Hoxa-1 in the zebrafish alters the fate of the mandibular arch neural crest and phenocopies a retinoic acid-induced phenotype. Development 122:735–746

Allwardt BA, Lall AB, Brockerhoff SE, Dowling JE (2001) Synapse formation is arrested in retinal photoreceptors of the zebrafish *nrc* mutant. J Neurosci 21:2330–2342

Altman J, Bayer SA (1984) The development of the rat spinal cord. Adv Anat Embryol Cell Biol 85:1–164

Altman J, Bayer S (1997) Development of the cerebellar system in relation to its evolution, structure and functions. CRC Press, Boca Raton

Altmann CR, Brivanlou AH (2001) Neural patterning in the vertebrate embryo. Int Rev Cytol 203:447–482

Alvarez-Buylla A, Garcia-Verdugo JM, Tramontin AD (2001) A unified hypothesis on the lineage of neural stem cells. Nat Rev Neurosci 2:287–293

Amacher SL (1999) Transcriptional regulation during zebrafish embryogenesis. Curr Opin Genet Dev 9:548–552

Amacher SL, Draper BD, Summers B, Kimmel CB (2002) The zebrafish T-box genes *no tail* and *spadetail* are required for development of trunk and tail mesoderm and medial floor plate. Development (in press)

Amacher S, Kimmel C (1998) Promoting notochord fate and repressing muscle development in zebrafish axial mesoderm. Development 125:1397–1406

Amaya E, Musci TJ, Kirschner MW (1991) Expression of a dominant negative mutant of the FGF receptor disrupts mesoderm formation in *Xenopus* embryos. Cell 66:257–270

Ambler CA, Nowicki JL, Burke AC, Bautch VL (2001) Assembly of trunk and limb blood vessels involves extensive migration and vasculogenesis of somite-derived angioblasts. Dev Biol 234:352–64

Amores A, Force A, Yan Y-L, Joly L, Amemiya C, Fritz A, Ho RK, Langeland J, Prince V, Wang Y-L, Westerfield M, Ekker M, Postlethwait JH (1998) Zebrafish *hox* clusters and vertebrate genome evolution. Science 282:1711–1714

Amthor H, Christ B, Weil M, Patel K (1998) The importance of timing differentiation during limb muscle development. Curr Biol 21:642–52

Anderson DJ (1997) Cellular and molecular biology of neural crest cell lineage determination. Trends Genet 3:276–280

Anderson DJ (2000) Genes, lineages and the neural crest: a speculative review. Philos Trans R Soc Lond B Biol Sci 355:953–964

Anderson DJ, Groves A, Lo L, Rao M, Shah NM, Sommer L (1997) Cell lineage determination and the control of neuronal identity in the crest. Cold Spring Harb Symp Quant Biol LXII:493–504

Anderson R, Copeland TK, Scholer H, Heasman J, Wylie C (2000) The onset of germ cell migration in the mouse embryo. Mech Dev 91:61–68

Ando H, Furuta T, Tsien RY, Okamoto H (2001) Photo-mediated gene activation using caged RNA/DNA in zebrafish embryos. Nat Genet 28:317–325

Ang SL, Jin O, Rhinn M, Daigle N, Stevenson L, Rossant J (1996) A targeted mouse *Otx2* mutation leads to severe defects in gastrulation and formation of axial mesoderm and to deletion of rostral brain. Development 122:243–252

Angelo S, Lohr J, Lee KH, Ticho BS, Breitbart RE, Hill S, Yost HJ, Srivastava D (2000) Conservation of sequence and expression of Xenopus and zebrafish dHAND during cardiac, branchial arch and lateral mesoderm development. Mech Dev 95:231–237

Appel B, Eisen JS (1998) Regulation of neuronal specification in the zebrafish spinal cord by Delta function. Development 125:371–380

Appel B, Korzh V, Glasgow E, Thor S, Edlund T, Dawid IB, Eisen JS (1995) Motoneuron fate specification revealed by patterned LIM homeobox gene expression in embryonic zebrafish. Development 121:4117–4125

Appel B, Fritz A, Westerfield M, Grunwald DJ, Eisen JS, Riley BB (1999) Delta-mediated specification of midline cell fates in zebrafish embryos. Curr Biol 9:247–256

Appel B, Givan LA, Eisen JS (2001) Delta-Notch signaling and lateral inhibition in zebrafish spinal cord development. BMC Dev Biol 1:13

Araki L, Brand M (2001) Antisense-morpholino-induced knockdown of fgf8 phenocopies of the *acerebellar* (*ace*) phenotype. Genesis 30:157–159

Armstrong PB (1932) The embryonic origin of function in the pronephros through differentiation and parenchyma-vascular association. Am J Anat 51:157–188

Arora K, Levine MS, O'Connor MB (1994) The screw gene encodes a ubiquitously expressed member of the TGF-β family required for the specification of dorsal cell fates in the *Drosophila* embryo. Genes Dev 8:2588–2601

Artavanis-Tsakonas S, Rand MD, Lake RJ (1999) Notch signaling: cell fate control and signal integration in development. Science 284:770–776

Artinger KB, Chitnis AB, Mercola M, Driever W (1999) Zebrafish narrowminded suggests a genetic link between formation of neural crest and primary sensory neurons. Development 126:3969–3979

Asaoka M, Sano H, Obara Y, Kobayashi S (1998) Maternal Nanos regulates zygotic gene expression in germline progenitors of *Drosophila melanogaster*. Mech Dev 78:153–158

Ashe HL, Levine M (1999) Local inhibition and long-range enhancement of Dpp signal transduction by Sog. Nature 398:427–431

Augsburger A, Schuchardt A, Hoskins S, Dodd J, Butler S (1999) BMPs as mediators of roof plate repulsion of commissural neurons. Neuron 24:127–141

Aulehla A, Johnson RL (1999) Dynamic expression of lunatic fringe suggests a link between notch signaling and an autonomous cellular oscillator driving somite segmentation. Dev Biol 207:49–61

Bachiller D, Klingensmith J, Kemp C, Belo JA, Anderson RM, May SR, McMahon JA, McMahon AP, Harland RM, Rossant J, De Robertis EM (2000) The organizer factors Chordin and Noggin are required for mouse forebrain development. Nature 403:658–661

Backstrom S (1954) Morphogenetic effects of lithium on the embryonic development of *Xenopus*. Arkiv Zool 6:526–527

Bafico A, Liu G, Yaniv A, Gazit A, Aaronson SA (2001) Novel mechanism of Wnt signalling inhibition mediated by Dickkopf-1 interaction with LRP6/Arrow. Nat Cell Biol 3:683–686

Baier H, Klostermann S, Trowe T, Karlstrom RO, Nüsslein-Volhard C, Bonhoeffer F (1996) Genetic dissection of the retinotectal projection. Development 123:415–425

Baker K, Warren KS, Yellen G, Fishman MC (1997) Defective "pacemaker" current (Ih) in a zebrafish mutant with a slow heart rate. Proc Natl Acad Sci USA 94:4554–4559

Balfour FM (1880) A treatise on comparative embryology. Macmillan, London

Balser EJ (1985) Ultrastructure and function of the proboscis complex of *Saccoglossus* (*Enteropneusta*). Am Zool 25:41

Ballard WW (1986) Stages and rates of normal development in the Holostean Fish, *Amia calva*. J Exp Zool 238:337–354

Bally-Cuif L, Goutel C, Wassef M, Wurst W, Rosa F (2000) Coregulation of anterior and posterior mesendodermal development by a hairy-related transcriptional repressor. Genes Dev 14:1664–1677

Bamford RN, Roessler E, Burdine RD, Saplakoglu U, dela Cruz J, Splitt M, Goodship JA, Towbin J, Bowers P, Ferrero GB, Marino B, Schier AF, Shen MM, Muenke M, Casey B (2000) Loss-of-function mutations in the EGF-CFC gene CFC1 are associated with human left-right laterality defects. Nat Genet 26:365–369

Barbarino SML, Spada F, Cotelli F, Boncinelli E (1997) Inactivation of the zebrafish homologue of *Chx10* by antisense oligonucleotides causes eye malformations similar to the ocular retardation phenotype. Mech Dev 63:133–143

Barresi MJ, Stickney HL, Devoto SH (2000) The zebrafish *slow-muscle-omitted* gene product is required for hedgehog signal transduction and the development of slow muscle identity. Development 127:2189–2199

Barrantes ID, Elia AJ, Wunsch K, De Angelis MH, Mak TW, Rossant J, Conlon RA, Gossler A, de la Pompa JL (1999) Interaction between Notch signalling and Lunatic fringe during somite boundary formation in the mouse. Curr Biol 9:470–480

Barth KA, Wilson SW (1995) Expression of zebrafish *nk2.2* is influenced by sonic hedgehog/vertebrate hedgehog-1 and demarcates a zone of neuronal differentiation in the embryonic forebrain. Development 121:1755–1768

Barth KA, Kishimoto Y, Rohr KB, Seydler C, Schulte-Merker S, Wilson SW (1999) Bmp activity establishes a gradient of positional information throughout the entire neural plate. Development 126:4977–4987

Bashirullah A, Halsell S, Cooperstock R, Kloc M, Karaiskakis A, Fisher W, Fu W, Hamilton J, Etkin L, Lipshitz H (1999) Joint action of two RNA degradation pathways controls the timing of maternal transcript elimination at the midblastula transition in *Drosophila melanogaster*. EMBO J 18:2610–2620

Basinger S, Bok D, Hall M (1976) Rhodopsin in the rod outer segment plasma membrane. J Cell Biol 69:29–42

Batourina E, Gim S, Bello N, Shy M, Clagett-Dame M, Srinivas S, Costantini F, Mendelsohn C (2001) Vitamin A controls epithelial/mesenchymal interactions through Ret expression. Nat Genet 27:74–78

Bauer H, Meier A, Hild M, Stachel S, Economides A, Hazelett D, Harland RM, Hammerschmidt M (1998) Follistatin and Noggin are excluded from the zebrafish organizer. Dev Biol 204:488–507

Bauer H, Lele Z, Rauch GJ, Geisler R, Hammerschmidt M (2001) The type I serine/threonine kinase receptor Alk8/Lost-a-fin is required for Bmp2b/7 signal transduction during dorsoventral patterning of the zebrafish embryo. Development 128:849–858

Beattie CE, Hatta K, Halpern ME, Liu H, Eisen JS, Kimmel CB (1997) Temporal separation in the specification of primary and secondary motoneurons in zebrafish. Dev Biol 187:171–182

Beattie CB, Raible DW, Henion PD, Eisen JS (1999) Early pressure screens in zebrafish. In: Detrich HW, Westerfield M, Zon L (eds) Methods in cell biology. Academic Press, San Diego, pp 71–83

Beattie CE, Melancon E, Eisen JS (2000) Mutations in the *stumpy* gene define intermediate targets for zebrafish motor axons. Development 127:2653–2662

Becker DL, McGonnell I, Makarenkova HP, Patel K, Tickle C, Lorimer J, Green CR (1999) Roles for alpha 1 connexin in morphogenesis of chick embryos revealed using a novel antisense approach. Dev Genet 24:33–42

Becker TS, Burgess SM, Amsterdam AH, Allende ML, Hopkins N (1998) *not really finished* is crucial for development of the zebrafish outer retina and encodes a transcription factor highly homologous to human *Nuclear Respiratory Factor-1* and avian *Initiation Binding Repressor*. Development 125:4369–4378

Beddington RS, Robertson EJ (1999) Axis development and early asymmetry in mammals. Cell 96:195–209

Begemann G, Ingham PW (2000) Developmental regulation of Tbx5 in zebrafish embryogenesis. Mech Dev 90:299–304

Begemann G, Schilling TF, Rauch GJ, Geisler R, Ingham PW (2001) The zebrafish neckless mutation reveals a requirement for raldh2 in mesodermal signals that pattern the hindbrain. Development 128:3081–3094

Belecky-Adams T, Cook B, Adler R (1996) Correlations between terminal mitosis and differentiated fate of retinal precursor cells in vivo and in vitro: analysis with the "window-labeling" technique. Dev Biol 178:304–315

Bellefroid EJ, Bourguignon C, Hollemann T, Ma Q, Anderson DJ, Kintner C, Pieler T (1996) X-MyT1, a *Xenopus* C2HC-type zinc finger protein with a regulatory function in neuronal differentiation. Cell 87:1191–1202

Belting HG, Hauptmann G, Meyer D, Abdelilah-Seyfried S, Chitnis A, Eschbach C, Soll I, Thisse C, Thisse B, Artinger KB, Lunde K, Driever W (2001) *spiel ohne grenzen/pou2* is required during establishment of the zebrafish midbrain-hindbrain boundary organizer. Development 128:4165–4176

Bemis WGL (1992) Early development of the Actinopterygian head.1. External development and staging of the paddlefish Polyodon-Spathula. J Morphol 213:47–83

Bentley A, MacLennan B, Calvo J, Dearolf CR (2000) Targeted recovery of mutations in *Drosophila*. Genetics 156:1169–1173

Beppu H, Kawabata M, Hamamoto T, Chytil A, Minowa O, Noda T, Miyazono K (2000) BMP type II receptor is required for gastrulation and early development of mouse embryos. Dev Biol 221:249–258

Bernhardt RR (1999) Cellular and molecular bases of axonal pathfinding during embryogenesis of the fish central nervous system. J Neurobiol 38:137–160

Bernhardt RR, Chitnis AB, Lindamer L, Kuwada JY (1990) Identification of spinal neurons in the embryonic and larval zebrafish. J Comp Neurol 302:603–616

Bernhardt RR, Nguyen N, Kuwada J (1992a) Growth cone guidance by floor plate cells in the spinal cord of zebrafish embryos. Neuron 8:869–882

Bernhardt RR, Patel CK, Wilson SW, Kuwada JY (1992b) Axonal trajectories and distribution of GABAergic spinal neurons in wildtype and mutant zebrafish lacking floor plate cells. J Comp Neurol 326:263–272

Bernhardt RR, Goerlinger S, Roose M, Schachner M (1998) Anterior-posterior subdivision of the somite in embryonic zebrafish: implications for motor axon guidance. Dev Dynam 213:334–347

Betchaku T, Trinkaus JP (1978) Contact relations, surface activity, and cortical microfilaments of marginal cells of the enveloping layer and of the yolk syncytial and yolk cytoplasmic layers of *Fundulus* before and during epiboly. J Exp Zool 206:381–426

Bianco C, Kannan S, De Santis M, Seno M, Tang CK, Martinez-Lacaci I, Kim N, Wallace-Jones B, Lippman ME, Ebert AD, Wechselberger C, Salomon DS (1999) Cripto-1 indirectly stimulates the tyrosine phosphorylation of erb B-4 through a novel receptor. J Biol Chem 274:8624–8629

Biehs B, Francois V, Bier E (1996) The *Drosophila short gastrulation* gene prevents Dpp from autoactivating and suppressing neurogenesis in the neuroectoderm. Genes Dev 10:2922–2934

Biemar F, Argenton F, Schmidtke R, Epperlein S, Peers B, Driever W (2001) Pancreas development in the zebrafish: early dispersed appearance of endocrine hormone expressing cells and their convergence to form the definitive islet. Dev Biol 230:189–203

Bienz M (1998) TCF: transcriptional activator or repressor? Curr Opin Cell Biol 10:366–372

Bierkamp C, Campos-Ortega JA (1993) A zebrafish homologue of the *Drosophila* neurogenic gene *Notch* and its pattern of transcription during early embryogenesis. Mech Dev 43:87–100

Billin AN, Thirlwell H, Ayer DE (2000) Beta-catenin-histone deacetylase interactions regulate the transition of LEF1 from a transcriptional repressor to an activator. Mol Cell Biol 20:6882–6890

Bingham S, Nasevicius A, Ekker SC, Chandrasekhar A (2001) Sonic hedgehog and tiggy-winkle hedgehog cooperatively induce zebrafish branchiomotor neurons. Genesis 30:170–174

Bisazza A, Rogers LJ, Vallortigara G (1998) The origins of cerebral asymmetry: a review of evidence of behavioural and brain lateralization in fishes, reptiles and amphibians. Neurosci Biobehav Rev 22:411–426

Bisgrove BW, Raible DW, Walter V, Eisen JS, Grunwald DJ (1997) Expression of c-ret in the zebrafish embryo: potential roles in motoneuronal development. J Neurobiol 33:749–768

Bisgrove BW, Essner JJ, Yost HJ (1999) Regulation of midline development by antagonism of lefty and nodal signaling. Development 126:3253–3262

Bisgrove BW, Essner JJ, Yost HJ (2000) Multiple pathways in the midline regulate concordant brain, heart and gut left-right asymmetry. Development 127:3567–3579

Blader P, Strähle U (1998) Ethanol impairs migration of the prechordal plate in the zebrafish embryo. Dev Biol 201:185–201

Blader P, Fischer N, Gradwohl G, Guillemot F, Strähle U (1997a) The activity of *neurogenin1* is controlled by local cues in the zebrafish embryo. Development 124:4557–4569

Blader P, Rastegar S, Fischer N, Strähle U (1997b) Cleavage of the BMP-4 antagonist Chordin by zebrafish Tolloid. Science 278:1937–1940

Blagden CS, Currie PD, Ingham PW, Hughes SM (1997) Notochord induction of zebrafish slow muscle is mediated by Sonic hedgehog. Genes Dev 11:2163–2175

Bodick N, Levinthal C (1980) Growing optic nerve fibers follow neighbors during embryogenesis. Proc Natl Acad Sci USA 77:4374–4378

Bolker JA (1993) The mechanism of gastrulation in the white sturgeon. J Exp Zool 266:132–145

Bolker JA (1994) Comparison of gastrulation in frogs and fish. Am Zool 34:313–322

Bondurand N, Pingault V, Goerich DE, Lemort N, Sock E, Caignec CL, Wegner M, Goossens M (2000) Interaction among SOX10, PAX3 and MITF, three genes altered in Waardenburg syndrome. Hum Mol Genet 9:1907–1917

Bone Q (1978) Locomotor muscle. In: Hoar WS, Randall DJ (eds) Fish physiology, vol 7. Academic Press, New York, pp 361–424

Bone Q (1989) Evolutionary patterns of axial muscle systems in some invertebrates and fish. Am Zool 29:5–18

Boterenbrood EC, Nieuwkoop PD (1973) The formation of the mesoderm in Urodelan amphibians. V. Its regional induction by the endoderm. Wilhelm Roux's Arch 173:319–332

Bouwmeester T, Kim S, Sasai Y, Lu B, De Robertis EM (1996) Cerberus is a head-inducing secreted factor expressed in the anterior endoderm of Spemann's organizer. Nature 382:595–601

Bovalenta P, Dodd J (1990) Guidance of commissural growth cones at the floor plate in embryonic rat spinal cord. Development 109:435–447

Bozhkova V, Te Kronnie G, Timmermans LPM (1994) Mesoderm differentiation in explants of carp embryos. Roux's Arch Dev Biol 204:20–29

Braat A, Zandbergen T, van de Water S, Goos H, Zivkovic D (1999) Characterization of zebrafish primordial germ cells: morphology and early distribution of vasa RNA. Dev Dyn 216:153–167

Braat AK, van de Water S, Goos H, Bogerd J, Zivkovic D (2000) Vasa protein expression and localization in the zebrafish. Mech Dev 95:271–274

Braisted JE, Essman TF, Raymond PA (1994) Selective regeneration of photoreceptors in goldfish retina. Development 120:2409–2419

Branchek T, Bremiller R (1984) The development of photoreceptors in the zebrafish, *Brachydanio rerio*. I. Structure. J Comp Neurol 224:107–115

Brand M, Heisenberg CP, Warga RM, Pelegri F, Karlstrom RO, Beuchle D, Picker A, Jiang YJ, Furutani-Seiki M, van Eeden FJ, Granato M, Haffter P, Hammerschmidt M, Kane DA, Kelsh RN, Mullins MC, Odenthal J, Nüsslein-Volhard C (1996a) Mutations affecting development of the midline and general body shape during zebrafish embryogenesis. Development 123:129–142

Brand M, Heisenberg C-P, Jiang Y-J, Beuchle D, Lun K, Furutani-Seiki M, Granato M, Haffter P, Hammerschmidt M, Kane D, Kelsh R, Mullins M, Odenthal J, van Eeden FJM, Nüsslein-Volhard C (1996b) Mutations in zebrafish genes affecting the formation of the boundary between midbrain and hindbrain. Development 123:179–190

Branford WW, Essner JJ, Yost HJ (2000) Regulation of gut and heart left-right asymmetry by context-dependent interactions between *Xenopus* lefty and BMP4 signaling. Dev Biol 223:291–306

Brannon M, Gomperts M, Sumoy L, Moon RT, Kimelman D (1997) A beta-catenin/XTcf-3 complex binds to the siamois promoter to regulate dorsal axis specification in *Xenopus*. Genes Dev 11:2359–2370

Brannon M, Brown JD, Bates R, Kimelman D, Moon RT (1999) XCtBP is a XTcf-3 co-repressor with roles throughout *Xenopus* development. Development 126:3159–3170

Braus H (1899) Beitrage zur entwicklung der musculatur unddes peripheren nervensystems der selachier. Morphol J B 27:501–629

Breckenridge RA, Mohun TJ, Amaya E (2001) A role for BMP signalling in heart looping morphogenesis in *Xenopus*. Dev Biol 232:191–203

Brennan HC, Nijjar S, Jones EA (1998) The specification of the pronephric tubules and duct in *Xenopus laevis*. Mech Dev 75:127–137

Brennan HC, Nijjar S, Jones EA (1999) The specification and growth factor inducibility of the pronephric glomus in *Xenopus laevis*. Development 126:5847–5856

Brewster R, Lee J, Ruiz I, Altaba A (1998) Gli/Zic factors pattern the neural plate by defining domains of cell differentiation. Nature 393:579–583

Briscoe J, Ericson J (2001) Specification of neuronal fates in the ventral neural tube. Curr Opin Neurobiol 11:43–49

Broccoli V, Boncinelli E, Wurst W (1999) The caudal limit of otx2 expression positions the isthmic organizer. Nature 401:164–168

Brockerhoff SE, Hurley JB, Niemi GA, Dowling JE (1997) A new form of inherited red-blindness identified in zebrafish. J Neurosci 17:4236–4242

Brockerhoff SE, Dowling JE, Hurley JB (1998) Zebrafish retinal mutants. Vision Res 38:1335–1339

Bronner-Fraser M, Fraser S (1989) Developmental potential of avian trunk neural crest cells in situ. Neuron 3:755–766

Brose K, Tessier-Lavigne M (2000) Slit proteins: key regulators of axon guidance, axonal branching, and cell migration. Curr Opin Neurobiol 10:95–102

Brose K, Bland KS, Wang KH, Arnott D, Henzel W, Goodman CS, Tessier-Lavigne M, Kidd T (1999) Slit proteins bind Robo receptors and have an evolutionarily conserved role in repulsive axon guidance. Cell 96:795–806

Brown A, Yates PA, Burrola P, Ortuno D, Vaidya A, Jessell TM, Pfaff SL, O'Leary DD, Lemke G (2000a) Topographic mapping from the retina to the midbrain is controlled by relative but not absolute levels of EphA receptor signaling. Cell 102:77–88

Brown LA, Rodaway AR, Schilling TF, Jowett T, Ingham PW, Patient RK, Sharrocks AD (2000b) Insights into early vasculogenesis revealed by expression of the ETS-domain transcription factor Fli-1 in wild-type and mutant zebrafish embryos. Mech Dev 90:237–252

Brown NL, Patel S, Brzezinski J, Glaser T (2001) *Math5* is required for retinal ganglion cell and optic nerve formation. Development 128:2497–2508

Brueckner M (2001) Cilia propel the embryo in the right direction. Am J Med Genet 101:339–344

Brummel TJ, Twombly V, Marques G, Wrana JL, Newfeld SJ, Attisano L, Massague J, O'Connor MB, Gelbart WM (1994) Characterization and relationship of Dpp receptors encoded by the saxophone and thick veins genes in *Drosophila*. Cell 78:251–261

Buchberger A, Seidl K, Klein C, Eberhardt H, Arnold HH (1998) cMeso-1, a novel bHLH transcription factor, is involved in somite formation in chicken embryos. Dev Biol 199:201–215

Burgess S, Reim G, Chen W, Hopkins N, Brand M (2002) The zebrafish *spiel-ohne-grenzen* (*spg*) gene encodes the POU homeodomain protein Pou2 related to mammalian Oct4 and is essential for formation of the midbrain-hindbrain territory and early morphogenesis. Development 129:905–916

Burke AC, Nelson CE, Morgan BA, Tabin C (1995) *Hox* genes and the evolution of vertebrate axial morphology. Development 121:333–346

Burrill JD, Easter SS Jr (1994) The development of the retinofugal projections in the embryonic and larval zebrafish, (*Brachydanio rerio*). J Comp Neurol 346:583–600

Burrill JD, Easter SS Jr (1995) The first retinal axons and their microenvironment in zebrafish: cryptic pioneers and the pre-tract. J Neurosci 15:2935–2947

Burrow CR (2000) Retinoids and renal development. Exp Nephrol 8:219–225

Burstyn-Cohen T, Tzarfaty V, Frumkin A, Feinstein Y, Stoeckli E, Klar A (1999) F-Spondin is required for accurate pathfinding of commissural axons at the floor plate. Neuron 23:233–246

Cameron DA (2000) Cellular proliferation and neurogenesis in the injured retina of adult zebrafish. Vision Neurosci 17:789–797

Cameron DA, Carney LH (2000) Cell mosaic patterns in the native and regenerated inner retina of zebrafish: implications for retinal assembly. J Comp Neurol 416:356–367

Campos-Ortega JA (1995) Genetic mechanisms of early neurogenesis in Drosophila melanogaster. Mol Neurobiol 10:75–89

Capdevila J, Vogan KJ, Tabin CJ, Izpisua Belmonte JC (2000) Mechanisms of left-right determination in vertebrates. Cell 101:9–21

Carl M, Wittbrodt J (1999) Graded interference with fgf signalling reveals its dorsoventral asymmetry at the mid-hindbrain boundary. Development 12:5659–5667

Carmany-Rampey A, Schier AF (2001) Single-cell internalization during zebrafish gastrulation. Curr Biol 11:1261–1265

Carmeliet P, Ferreira V, Breier G, Pollefeyt S, Kieckens L, Gertsenstein M, Fahrig M, Vandenhoeck A, Harpal K, Eberhardt C, Declercq C, Pawling J, Moons L, Collen D, Risau W, Nagy A (1996) Abnormal blood vessel development and lethality in embryos lacking a single VEGF allele. Nature 380:435–439

Carroll TJ, Wallingford JB, Vize PD (1999) Dynamic patterns of gene expression in the developing pronephros of *Xenopus laevis* . Dev Genet 24:199–207

Casey B (2001) Genetics of human situs abnormalities. Am J Med Genet 101:356–358

Cavallo RA, Cox RT, Moline MM, Roose J, Polevoy GA, Clevers H, Peifer M, Bejsovec A (1998) *Drosophila* Tcf and Groucho interact to repress Wingless signalling activity. Nature 395:604–608

Cavodeassi F, Modolell J, Gomez-Skarmeta JL (2001) The Iroquois family of genes: from body building to neural patterning. Development 128:2847–2855

Chambers D, Mason I (2000) Expression of *sprouty2* during early development of the chick embryo is coincident with known sites of Fgf signalling. Mech Dev 91:361–364

Challa AK, Beattie CE, Seeger MA (2001) Identification and characterization of roundabout orthologs in zebrafish. Mech Dev 101:249–253

Chan J, Mably JD, Serluca FC, Chen JN, Goldstein NB, Thomas MC, Cleary JA, Brennan C, Fishman MC, Roberts TM (2001) Morphogenesis of prechordal plate and notochord requires intact eph/ephrin B signaling. Dev Biol 234:470–482

Chandrasekhar A, Moens C, Warren JTJ, Kimmel C, Kuwada J (1997) Development of branchiomotor neurons in zebrafish. Development 124:2633–2644

Chandrasekhar A, Warren JT Jr, Takahashi K, Schauerte HE, van Eeden FJM, Haffter P, Kuwada JY (1998) Role of sonic hedgehog in branchiomotor neuron induction in zebrafish. Mech Dev 76:101–115

Chandrasekhar A, Schauerte HE, Haffter P, Kuwada JY (1999) The zebrafish detour gene is essential for cranial but not spinal motor neuron induction. Development 126:2727–2737

Chang C, Hemmati-Brivanlou A (1998) Neural crest induction by Xwnt7B in *Xenopus*. Dev Biol 194:129–134

Chang C, Holtzman DA, Chau S, Chickering T, Woolf EA, Holmgren LM, Bodorova J, Gearing DP, Holmes WE, Brivanlou AH (2001) Twisted gastrulation can function as a BMP antagonist. Nature 410:483–487

Chang H, Huylebroeck D, Verschueren K, Guo Q, Matzuk MM, Zwijsen A (1999) Smad5 knockout mice die at mid-gestation due to multiple embryonic and extraembryonic defects. Development 126:1631–1642

Chen J, Ruley HE (1998) An enhancer element in the EphA2 (Eck) gene sufficient for rhombomere-specific expression is activated by HOXA1 and HOXB1 homeobox proteins. J Biol Chem 273:24670–24675

Chen J-N, Fishman MC (1996) Zebrafish *tinman* homolog demarcates the heart field and initiates myocardial differentiation. Development 122:3809–3816

Chen J-N, Fishman MC (2000) Genetics of heart development. Trends Genet 16:383–388

Chen S, Kimelman D (2000) The role of the yolk syncytial layer in germ layer patterning in zebrafish. Development 127:4681–4689

Chen Y, Schier AF (2001) The zebrafish Nodal signal Squint functions as a morphogen. Nature 411:607–610

Chen J-N, Haffter P, Odenthal J, Vogelsang E, Brand M, van Eeden FJM, Furutani-Seiki M, Granato M, Hammerschmidt M, Heisenberg CP, Jiang Y-J, Kane DA, Kelsh RN, Mullins MC, Nüsslein-Volhard C (1996) Mutations affecting the cardiovascular system and other internal organs in zebrafish. Development 123:293–302

Chen J-N, van Eeden FJM, Warren KS, Chin AS, Nüsslein-Volhard C, Haffter P, Fishman MC (1997) Left-right pattern of cardiac *BMP4* may drive asymmetry of the heart in zebrafish. Development 124:4373–4382

Chen W, Burgess S, Hopkins N (2001a) Analysis of the zebrafish smoothened mutant reveals conserved and divergent functions of hedgehog activity. Development 128:2385–2396

Chen YH, Lee WC, Liu CF, Tsai HJ (2001b) Molecular structure, dynamic expression, and promoter analysis of zebrafish (*Danio rerio*) *myf-5* gene. Genesis 29:22–35

Cheng SH, Chan PK, Wu RS (2001) The use of microangiography in detecting aberrant vasculature in zebrafish embryos exposed to cadmium. Aquat Toxicol 52:61–71

Chenn A, Zhang YA, Chang BT, McConnell SK (1998) Intrinsic polarity of mammalian neuroepithelial cells. Mol Cell Neurosci 11:183–193

Chiang C, Litingtung Y, Lee E, Young KE, Corden JL, Westphal H, Beachy PA (1996) Cyclopia and defective axial patterning in mice lacking sonic hedgehog gene function. Nature 383:407–413

Chin AJ, Tsang M, Weinberg ES (2000) Heart and gut chiralities are controlled independently from initial heart position in the developing zebrafish. Dev Biol 227:403–421

Chisholm A, Tessier-Lavigne M (1999) Conservation and divergence of axon guidance mechanisms. Curr Opin Neurobiol 9:603–615

Chitnis AB (1999) Control of neurogenesis – lessons from frogs, fish and flies. Curr Opin Neurobiol 9:18–25

Chitnis A, Kintner C (1996) Sensitivity of proneural genes to lateral inhibition affects the pattern of primary neurons in *Xenopus* embryos. Development 122:2295–2301

Chitnis AB, Kuwada JY (1990) Axonogenesis in the brain of zebrafish embryos. J Neurosci 10:1892–1905

Chitnis AB, Kuwada JY (1991) Elimination of a brain tract increases errors in pathfinding by follower growth cones in the zebrafish embryo. Neuron 7:277–285

Chitnis AB, Patel CK, Kim S, Kuwada JY (1992) A specific brain tract guides follower growth cones in two regions of the zebrafish brain. J Neurobiol 23:845–854

Cho KWY, Blumberg B, Steinbeisser H, De Robertis EM (1991) Molecular nature of Spemann's organizer: the role of the *Xenopus* homeobox gene goosecoid. Cell 67:1111–1120

Cho M, Hughes SM, Karsch-Mizrachi I, Travis M, Leinwand LA, Blau HM (1994) Fast myosin heavy chains expressed in secondary mammalian muscle fibres at the time of their inception. J Cell Sci 107:2361–2371

Christ B, Ordahl CP (1995) Early stages of chick somite development. Anat Embryol (Berl) 191:381–396

Christ B, Jacob M, Jacob HJ (1983) On the origin and development of the ventro-lateral trunk musculature in the avian embryo. An experimental and ultrastructural study. Anat Embryol (Berl) 166:87–101

Chuang JC, Raymond PA (2001) Zebrafish genes *rx1* and *rx2* help define the region of forebrain that gives rise to retina. Dev Biol 231:13–30

Chuang JC, Mathers PH, Raymond PA (1999) Expression of three *Rx* homeobox genes in embryonic and adult zebrafish. Mech Dev 84:195–198

Ciment G, Weston JA (1982) Early appearance in neural crest and crest-derived cells of an antigenic determinant present in avian neurons. Dev Biol 93:355–367

Cinnamon Y, Kahane N, Kalcheim C (1999) Characterisation of the early development of specific hypaxial muscles from the ventrolateral myotome. Development 126:4305–4315

Ciruna B, Rossant J (2001) FGF signaling regulates mesoderm cell fate specfication and morphogenetic movements at the primitive streak. Dev Cell 1:37–49

Clark TG, Conway SJ, Scott IC, Labosky PA, Winnier G, Bundy J, Hogan BLM, Greenspan DS (1999) The mammalian Tolloid-like 1 gene, Tll1, is necessary for normal septation and positioning of the heart. Development 126:2631–2642

Cleaver O, Krieg PA (1998) VEGF mediates angioblast migration during development of the dorsal aorta in *Xenopus*. Development 125:3905–3914

Cleaver O, Seufert DW, Krieg PA (2000) Endoderm patterning by the notochord: development of the hypochord in *Xenopus*. Development 127:869–879

Clements D, Friday RV, Woodland HR (1999) Mode of action of VegT in mesoderm and endoderm formation. Development 126:4903–4911

Clouthier DE, Hosoda K, Richardson JA, Williams SC, Yanagisawa H, Kuwaki T, Kumada M, Hammer RE, Yanagisawa M (1998) Cranial and cardiac neural crest defects in endothelin-A receptor-deficient mice. Development 125:813–824

Colamarino SA, Tessier-Lavigne M (1995) The role of the floor plate in axon guidance. Annu Rev Neurosci 18:497–529

Collignon J, Varlet I, Robertson EJ (1996) Relationship between asymmetric nodal expression and the direction of embryonic turning. Nature 381:155–158

Concha M, Adams R (1998) Oriented cell divisions and cellular morphogenesis in the zebrafish gastrula and neurula: a time-lapse analysis. Development 125:983–994

Concha M, Burdine RD, Russell C, Schier AF, Wilson SW (2000) A nodal signalling pathway regulates the laterality of neuroanatomical asymmetries in the zebrafish forebrain. Neuron 28:399–409

Concha M, Wilson SW (2001) Asymmetry in the epithalamus of vertebrates. J Anat 199:63–84

Conlon FL, Lyons KM, Takaesu N, Barth KS, Kispert A, Herrmann B, Robertson EJ (1994) A primary requirement for nodal in the formation and maintenance of the primitive streak in the mouse. Development 120:1919–1928

Connaughton VP, Nelson R (2000) Axonal stratification patterns and glutamate-gated conductance mechanisms in zebrafish retinal bipolar cells. J Physiol 524:135–146

Connors SA, Trout J, Ekker M, Mullins MC (1999) The role of tolloid/minifin in dorsoventral pattern formation of the zebrafish embryo. Development 126:3119–3130

Cooke J (1998) A gene that resuscitates a theory – somitogenesis and a molecular oscillator. Trends Genet 14:85–88

Cooke J, Zeeman EC (1976) A clock and wavefront model for control of the number of repeated structures during animal morphogenesis. J Theor Biol 58:455–476

Cooke JE, Xu Q, Wilson SW, Holder N (1997) Characterization of five novel zebrafish Eph-related receptor tyrosine kinases suggests roles in patterning neural plate. Dev Genes Evol 206:515–531

Cooke JE, Moens CB, Roth LW, Durbin L, Shiomi K, Brennan C, Kimmel CB, Wilson S, Holder N (2001) Eph signalling functions downstream of val to regulate cell sorting and boundary formation in the caudal hindbrain. Development 128:571–580

Cooper MS, D'Amico LA (1996) A cluster of noninvoluting endocytic cells at the margin of the zebrafish blastoderm marks the site of embryonic shield formation. Dev Biol 180:184–198

Cooper MS, Kimmel CB (1998) Morphogenetic cell behaviors and specification of cell fate during early teleost development. In: Soll DR (ed) Motion analysis of living cells. Wiley-Liss, New York, pp 177–220

Cooper MS, D'Amico LA, Henry CA (1999) Confocal microscopic analysis of morphogenetic movements. Methods Cell Biol 59:179–204

Cordes SP, Barsh GS (1994) The mouse segmentation gene kr encodes a novel basic domain-leucine zipper transcription factor. Cell 79:1025–1034

Cornell RA, Eisen JS (2000) Delta signaling mediates segregation of neural crest and spinal sensory neurons from zebrafish lateral neural plate. Development 127:2873–2882

Corning HK (1842) Über die ventralen urwirbelknopsen in der brustflosse der teleostier. Morphol J B 22:79–98

Cornish JA, Etkin LD (1993) The formation of the pronephric duct in *Xenopus* involves recruitment of posterior cells by migrating pronephric duct cells. Dev Biol 159:338–345

Coutelle O, Blagden CS, Hampson R, Halai C, Rigby PWJ, Hughes SM (2001) Hedgehog signalling is required for maintenance of *myf5* and *myoD* expression and timely terminal differentiation in zebrafish adaxial myogenesis. Dev Biol 236:136–150

Creton R, Speksnijder JE, Jaffe LF (1998) Patterns of free calcium in zebrafish embryos. J Cell Sci 111:1613–1622

Crosier PS, Bardsley A, Horsfield JA, Krassowska AK, Lavallie ER, Collins-Racie LA, Postlethwait JH, Yan YL, McCoy JM, Crosier KE (2001) In situ hybridization screen in zebrafish for the selection of genes encoding secreted proteins. Dev Dyn 222:637–644

Crossley P, Martinez S, Martin GR (1996) Midbrain development induced by fgf8 in the chick embryo. Nature 380:66–68

Crow MT, Stockdale FE (1986) Myosin expression and specialization among the earliest muscle fibres of the developing avian limb. Dev Biol 113:283–254

Culi J, Modolell J (1998) Proneural gene self-stimulation in neural precursors: an essential mechanism for sense organ development that is regulated by Notch signaling. Genes Dev 12: 2036–2047

Currie PD, Ingham PW (1996) Induction of a specific muscle cell type by a hedgehog-like protein in zebrafish. Nature 382:452–455

Currie PD, Ingham PW (1998) The generation and interpretation of positional information within the vertebrate myotome. Mech Dev 73:3–21

Czerny T, Halder G, Kloter U, Souabni A, Gehring WJ, Busslinger M (1999) *twin of eyeless*, a second *Pax-6* gene of *Drosophila*, acts upstream of *eyeless* in the control of eye development. Mol Cell 3:297–307

Dale JK, Vesque C, Lints TJ, Sampath TK, Furley A, Dodd J, Placzek M (1997) Cooperation of *bmp7* and *shh* in the induction of forebrain ventral midline cells by prechordal mesoderm. Cell 90:257–269

Dale L, Slack JMW (1987) Fate map for the 32-cell stage of *Xenopus laevis*. Development 99:527–551

Dale L, Wardle FC (1999) A gradient of Bmp activity specifies dorsal-ventral fates in early *Xenopus* embryos. Semin Dev Biol 10:319–326

Dale N, Roberts A, Ottersen OP, Storm-Mathisen J (1987a) The development of a population of spinal cord neurons and their axonal projections revealed by gaba immunocytochemistry in frog embryos. Proc R Soc Lond 232:205–215

Dale N, Roberts A, Ottersen OP, Storm-Mathisen J (1987b) The morphology and distribution of "kolmer-agduhr cells", a class of cerebrospinal-fluid-contacting neurons revealed in the frog embryo spinal cord by gaba immunocytochemistry. Proc Natl Acad Sci USA 232:193–203

D'Amico LA, Cooper MS (1997) Spatially distinct domains of cell behaviour in the zebrafish organizer region. Biochem Cell Biol 75:563–577

D'Amico LA, Cooper MS (2001) Morphogenetic domains in the yolk syncytial layer of axiating zebrafish embryos. Dev Dyn 222:611–624

Danielian PS, McMahon AP (1996) Engrailed-1 as a target of the wnt-1 signalling pathway in vertebrate midbrain development. Nature 383:332–334

Danos MC, Yost HJ (1996) Role of notochord in specification of cardiac left-right orientation in zebrafish and *Xenopus*. Dev Biol 177:96–103

Das P, Maduzia LL, Wang H, Finelli AL, Cho S-H, Smith MM, Padgett RW (1998) The *Drosophila* gene Medea demonstrates the requirement for different classes of Smads in dpp signaling. Development 125:1519–1528

David NB, Rosa FM (2001) Cell autonomous commitment to an endodermal fate and behaviour by activation of Nodal signalling. Development 128:3937–3947

Davies J (1950) The pronephros and the early development of the mesonephros in the chick. J Anat 84:95–103

Davis CA, Holmyard DP, Millen KJ, Joyner AL (1991) Examining pattern formation in mouse, chicken and frog embryos with an *En*-specific antiserum. Development 111:287–298

Davis GS, Phillips HM, Steinberg MS (1997) Germ-layer surface tensions and "tissue affinities" in *Rana pipiens* gastrulae: quantitative measurements. Dev Biol 192:630–644

Degenhardt K, Sassoon DA (2001) A role for Engrailed-2 in determination of skeletal muscle physiologic properties. Dev Biol 231:175–189

DeHaan RL (1965) Morphogenesis of the vertebrate heart. In: DeHaan H, Ursprung RL (eds) Organogenesis. Holt, Rinehart and Winston, New York, pp 377–419

Delarue M, Saez FJ, Johnson KE, Boucaut JC (1997) Fates of the blastomeres of the 32-cell stage Pleurodeles waltl embryo. Dev Dyn 210:236–248

Denetclaw WF, Ordahl CP (2000) The growth of the dermomyotome and formation of early myotome lineages in thoracolumbar somites of chicken embryos. Development 127:893–905

Denetclaw WF Jr, Christ B, Ordahl CP (1997) Location and growth of epaxial myotome precursor cells. Development 124:1601–10

DeRobertis EM, Sasai Y (1996) A common plan for dorsoventral patterning in Bilateria. Nature 380:27–40

De Robertis EM, Larrain J, Oelgeschlager M, Wessely O (2000) The establishment of Spemann's organizer and patterning of the vertebrate embryo. Nat Rev Genet 1:171–181

Deshpande G, Calhoun G, Yanowitz J, Schedl P (1999) Novel functions of nanos in downregulating mitosis and transcription during the development of the *Drosophila* germline. Cell 99:271–281

Detrich HW III, Kieran MW, Chan FY, Barone LM, Yee K, Rundstadler JA, Pratt S, Ransom D, Zon LI (1995) Intraembryonic hematopoietic cell migration during vertebrate development. Proc Natl Acad Sci USA 92:10713–10717

Devoto SH, Melancon E, Eisen JS, Westerfield M (1996) Identification of separate slow and fast muscle precursor cells in vivo, prior to somite formation. Development 122:3371–3380

Dick A, Meier A, Hammerschmidt M (1999) Smad1 and Smad5 have distinct roles during dorsoventral patterning of the zebrafish embryo. Dev Dyn 216:285–298

Dick A, Hild M, Bauer H, Imai Y, Maifeld H, Schier AF, Talbot WS, Bouwmeester T, Hammerschmidt M (2000) Essential role of Bmp7 (snailhouse) and its prodomain in dorsoventral patterning of the zebrafish embryo. Development 127:343–354

Dickinson ME, Krumlauf R, McMahon AP (1994) Evidence for a mitogenic effect of wnt-1 in the developing mammalian central nervous system. Development 120:1453–1471

Dickmeis T, Aanstad P, Clark M, Fischer N, Herwig R, Mourrain P, Blader P, Rosa F, Lehrach H, Strahle U (2001a) Identification of nodal signaling targets by array analysis of induced complex probes. Dev Dyn 222:571–580

Dickmeis T, Mourrain P, Saint-Etienne L, Fischer N, Aanstad P, Clark M, Strähle U, Rosa F (2001b) A crucial component of the endoderm formation pathway, Casanova, is encoded by a novel sox-related gene. Genes Dev 15:1487–1492

Dietrich S, Schubert FR, Healy C, Sharpe PT, Lumsden A (1998) Specification of the hypaxial musculature. Development 125:2235–2249

Dierick H, Bejsovec A (1999) Cellular mechanisms of wingless/Wnt signal transduction. Curr Topics Dev Biol 43:153–190

Dirksen ML, Jamrich M (1995) Differential expression of fork head genes during early *Xenopus* and zebrafish development. Dev Genet 17:107–116

Djiane A, Riou J, Umbhauer M, Boucaut J, Shi D (2000) Role of frizzled 7 in the regulation of convergent extension movements during gastrulation in *Xenopus laevis*. Development 127:3091–3100

Doerre G, Malicki J (2001) A mutation of early photoreceptor development, *mikre oko*, reveals cell-cell interactions involved in the survival and differentiation of zebrafish photoreceptors. J Neurosci 21:6745–6757

Doerre G, Malicki J (2002) Genetic analysis of photoreceptor cell development in the zebrafish retina. Mech Dev 110:125–138

Doe CQ, Technau GM (1993) Identification and cell lineage of individual neural precursors in the Drosophila CNS. Trends Neurosci 16:510–514

Dohrn A (1884) Die paarigen und unpaaren Flossen der Selachier. Mitt Zool Stat Neapel 5:161–195

Domingo C, Keller R (1995) Induction of notochord cell intercalation behavior and differentiation by progressive signals in the gastrula of *Xenopus laevis*. Development 121:3311–3321

Donovan A, Brownlie A, Zhou Y, Shepard J, Pratt SJ, Moynihan J, Paw BH, Drejer A, Barut B, Zapata A, Law TC, Brugnara C, Lux SE, Pinkus GS, Pinkus JL, Kingsley PD, Palis J, Fleming MD, Andrews NC, Zon LI (2000) Positional cloning of zebrafish ferroportin1 identifies a conserved vertebrate iron exporter. Nature 403:776–781

Dornseifer P, Takke C, Campos-Ortega JA (1997) Overexpression of a zebrafish homologue of the *Drosophila* neurogenic gene Delta perturbs differentiation of primary neurons and somite development. Mech Dev 63:159–171

Dorsky RI, Moon RT, Raible DW (1998) Control of neural crest cell fate by the Wnt signalling pathway. Nature 396:370–373.

Dorsky RI, Moon RT, Raible DW (2000a) Environmental signals and cell fate specification in premigratory neural crest. Bioessays 22:708–716

Dorsky RI, Raible, DW, Moon RT (2000b) Direct regulation of nacre, a zebrafish mitf homolog required for pigment cell formation, by the wnt pathway. Genes Dev 14:158–162

Dosch R, Gawantka V, Delius H, Blumenstock C, Niehrs C (1997) Bmp4 acts as a morphogen in dorsoventral patterning in *Xenopus*. Development 124:2325–2334

Dowling J (1987) The retina. Harvard University Press, Cambridge

Draper BW, Morcos PA, Kimmel CB (2001) Inhibition of zebrafish *fgf8* pre-mRNA splicing with morpholino oligos: a quantifiable method for gene knockdown. Genesis 30:154–156

Driever W (1995) Axis formation in zebrafish. Curr Opin Genet Dev 5:610–618

Driever W, Solnica-Krezel L, Schier AF, Neuhauss SCF, Malicki J, Stemple DL, Stainier DYR, Zwartkruis F, Abdelilah S, Rangini Z, Belak J, Boggs C (1996) A genetic screen for mutations affecting embryogenesis in zebrafish. Development 123:37–46

Drummond IA (2000) The zebrafish pronephros: a genetic system for studies of kidney development. Pediatr Nephrol 14:428–435

Drummond IA, Majumdar A, Hentschel H, Elger M, Solnica-Krezel L, Schier AF, Neuhauss SC, Stemple DL, Zwartkruis F, Rangini Z, Driever W, Fishman MC (1998) Early development of the zebrafish pronephros and analysis of mutations affecting pronephric function. Development 125:4655–4667

Du SJ, Devoto SH, Westerfield M, Moon RT (1997) Positive and negative regulation of muscle cell identity by members of the hedgehog and TGF-beta gene families. J Cell Biol 139:145–156

Dubrulle J, McGrew MJ, Pourquie O (2001) FGF signaling controls somite boundary position and regulates segmentation clock control of spatiotemporal *Hox* gene activation. Cell 106:219–232

Duchek P, Rørth P (2001) Guidance of cell migration by EGF receptor signaling during *Drosophila* oogenesis. Science 291:131–133

Dudley AT, Lyons KM, Robertson EJ (1995) A requirement for bone morphogenetic protein-7 during development of the mammalian kidney and eye. Genes Dev 9:2795–2807

Dufort D, Schwartz L, Harpal K, Rossant, R (1998) The transcription factor HNF3b is required in visceral endoderm for normal primitive streak morphogenesis. Development 125:3015–3025

Dunn KJ, Williams BO, Li Y, Pavan WJ (2000) Neural crest-directed gene transfer demonstrates Wnt1 role in melanocyte expansion and differentiation during mouse development. Proc Natl Acad Sci USA 97:10050–10055

Dupé V, Lumsden A (2001) Hindbrain patterning involves graded responses to retinoic acid signalling. Development 128:2199–2208

Dupé V, Ghyselinck NB, Wendling O, Chambon P, Mark M (1999) Key roles of retinoic acid receptors alpha and beta in the patterning of the caudal hindbrain, pharyngeal arches and otocyst in the mouse. Development 126:5051–5059

Dupin E, Baroffio A, Dulac C, Cameron-Curry P, Le Douarin N (1990) Schwann-cell differentiation in clonal cultures of the neural crest, as evidenced by the anti-Schwann cell myelin protein monoclonal antibody. Proc Natl Acad Sci USA 87:1119–1123

Dupin E, Glavieux C, Vaigot P, Le Douarin NM (2000) Endothelin 3 induces the reversion of melanocytes to glia through a neural crest-derived glial-melanocytic progenitor. Proc Natl Acad Sci USA 97:7882–7887

Durbin L, Brennan C, Shiomi K, Cooke J, Barrios A, Shanmugalingam S, Guthrie B, Lindberg R, Holder N (1998) Eph signaling is required for segmentation and differentiation of the somites. Genes Dev 12:3096–3109

Durbin L, Sordino P, Barrios A, Gering M, Thisse C, Thisse B, Brennan C, Green A, Wilson S, Holder N (2000) Anteroposterior patterning is required within segments for somite boundary formation in zebrafish. Development 127:1703–1713

Durston AJ, van der Wees J, Pijnappel WW, Godsave SF (1998) Retinoids and related signals in early development of the vertebrate central nervous system. Curr Topics Dev Biol 40:111–175

Dutton KA, Pauliny A, Lopes SS, Elworthy S, Carney TJ, Rauch J, Geisler R, Haffter P, Kelsh RN (2001) Zebrafish colourless encodes *sox10* and specifies non-ectomesenchymal neural crest fates. Development 128:4113–4125

Easter SS Jr, Nicola GN (1996) The development of vision in the zebrafish (*Danio rerio*). Dev Biol 180:646–663

Eaton RC, Farley RD, Kimmel CB, Schabtach E (1977) Functional development in the Mauthner cell system of embryos and larvae of the zebra fish. J Neurobiol 8:151–172

Eberhart CG, Maines JZ, Wasserman SA (1996) Meiotic cell cycle requirement for a fly homologue of human Deleted in Azoospermia. Nature 381:783–785

Edlund T, Jessell TM (1999) Progression from extrinsic to intrinsic signaling in cell fate specification: a view from the nervous system. Cell 96:211–224

Eisen JS (1991) Determination of primary motoneuron identity in developing zebrafish embryos. Science 252:569–572

Eisen JS (1992) The role of interactions in determining cell fate of two identified motoneurons in the embryonic zebrafish. Neuron 8:231–240

Eisen JS (1994) Development of motoneuronal phenotype. Annu Rev Neurosci 17:1–30

Eisen JS, Pike SH (1991) The *spt-1* mutation alters the segmental arrangement and axonal development of identified neurons in the spinal cord of the embryonic zebrafish. Neuron 6:767–776

Eisen JS, Weston JA (1993) Development of the neural crest in the zebrafish. Dev Biol 159:50–59

Eisen JS, Myers PZ, Westerfield M (1986) Pathway selection by growth cones of identified motoneurons in live zebrafish embryos. Nature 320:269–271

Eisen JS, Pike SH, Romancier B (1990) An identified motoneuron with variable fates in embryonic zebrafish. J Neurosci 10:34-43

Ekanayake S, Hall BK (1987) The development of acellularity of the vertebral bone of the Japanese medaka, *Oryzias latipes* (Teleostei; Cyprinidontidae). J Morphol 19:253–261

Ekanayake S, Hall BK (1988) Ultrastructure of the osteogenesis of acellular vertebral bone in the Japanese medaka, *Oryzias latipes* (Teleostei, Cyprinidontidae). Am J Anat 182:241–249

Ekker M, Wegner J, Akimenko MA, Westerfield M (1992) Coordinate embryonic expression of three zebrafish engrailed genes. Development 116:1001–1010

Ekker SC, Ungar AR, Greenstein P, von Kessler DP, Porter JA, Moon RT, Beachy PA (1995) Patterning activities of vertebrate hedgehog proteins in the developing eye and brain. Curr Biol 5:944–955

Elinson RP, Rowning B (1988) A transient array of parallel microtubules in frog eggs: potential tracks for a cytoplasmic rotation that specifies the dorso-ventral axis. Dev Biol 128:185–197

Ellies DL, Langille RM, Martin CC, Akimenko MA, Ekker M (1997) Specific craniofacial cartilage dysmorphogenesis coincides with a loss of *dlx* gene expression in retinoic acid-treated zebrafish embryos. Mech Dev 61:23–36

Elsdale T, Pearson M, Whitehead M (1976) Abnormalities in somite segmentation following heat shock to *Xenopus* embryos. J Embryol Exp Morphol 35:625–635

Elul T, Keller R (2000) Monopolar protrusive activity: a new morphogenic cell behavior in the neural plate dependent on vertical interactions with the mesoderm in *Xenopus*. Dev Biol 224:3–19

Elul T, Koehl MA, Keller R (1997) Cellular mechanism underlying neural convergent extension in *Xenopus laevis* embryos. Dev Biol 191:243–258

Ensini M, Tsuchida TN, Belting H-G, Jessell TM (1998) The control of rostrocaudal pattern in the developing spinal cord: specification of motor neuron subtype identity is initiated by signals from paraxial mesoderm. Development 125:969–982

Eriksson J, Lofberg J (2000) Development of the hypochord and dorsal aorta in the zebrafish embryo (*Danio rerio*). J Morphol 244:167–176

Erter CE, Solnica-Krezel L, Wright CV (1998) Zebrafish nodal-related 2 encodes an early mesendodermal inducer signaling from the extraembryonic yolk syncytial layer. Dev Biol 204:361–372

Erter CE, Wilm TP, Basler N, Wright CVE, Solnica-Krezel L (2001) Wnt8 is required in lateral mesendodermal precursors for neural posteriorization in vivo. Development 128:3571–3583

Essner JJ, Branford WW, Zhang J, Yost HJ (2000) Mesendoderm and left-right brain, heart and gut development are differentially regulated by *pitx2* isoforms. Development 127:1081–1093

Essner JJ, Wang X, Zhang J, Yost HJ (2001) The zebrafish *inv* gene is required for left-right brain and heart development. Dev Biol 235:181

Etheridge LA, Wu T, Liang JO, Ekker SC, Halpern ME (2001) Floor plate develops upon depletion of tiggy-winkle and sonic hedgehog. Genesis 30:164–169

Fadool JM, Brockerhoff SE, Hyatt GA, Dowling JE (1997) Mutations affecting eye morphology in the developing zebrafish (*Danio rerio*). Dev Genet 20:288–295

Fainsod A, Deißler K, Yelin R, Marom K, Epstein M, Pillemer G, Steinbeisser H, Blum M (1997) The dorsalizing and neural inducing gene follistatin is an antagonist of BMP-4. Mech Dev 63:39–50

Fan C-M, Tessier-Lavigne M (1994) Patterning of mammalian somites by surface ectoderm and notochord: evidence for sclerotome induction by a hedgehog homolog. Cell 79:1175–1186

Fashena D, Westerfield M (1999) Secondary motoneuron axons localize DM-GRASP on their fasciculated segments. J Comp Neurol 406:415–424

Faucourt M, Houliston E, Besnardeau L, Kimelman D, Lepage T (2001) The *pitx2* homeobox protein is required early for endoderm formation and Nodal signaling. Dev Biol 229:287–306

Fekany K, Yamanaka Y, Leung T, Sirotkin HI, Topczewski J, Gates MA, Hibi M, Renucci A, Stemple D, Radbill A, Schier AF, Driever W, Hirano T, Talbot WS, Solnica-Krezel L (1999) The zebrafish *bozozok* locus encodes Dharma, a homeodomain protein essential for induction of gastrula organizer and dorsoanterior embryonic structures. Development 126:1427–1438

Fekany-Lee K, Gonzalez E, Miller-Bertoglio V, Solnica-Krezel L (2000) The homeobox gene *bozozok* promotes anterior neuroectoderm formation in zebrafish through negative regulation of BMP2/4 and Wnt pathways. Development 127:2333–2345

Feldman B, Gates MA, Egan ES, Dougan ST, Rennebeck G, Sirotkin HI, Schier AF, Talbot WS (1998) Zebrafish organizer development and germ-layer formation require nodal-related signals. Nature 395:181–185

Feldman B, Dougan ST, Schier AF, Talbot WS (2000) Nodal-related signals establish mesendodermal fate and trunk neural identity in zebrafish. Curr Biol 10:531–534

Felsenfeld AL, Curry M, Kimmel CB (1991) The *fub-1* mutation blocks initial myofibril formation in zebrafish muscle pioneer cells. Dev Biol 148:23–30

Ferguson EL, Anderson KV (1992) Localized enhancement or repression of the activity of the TGF-β family member, decapentaplegic, is necessary for dorsal-ventral pattern formation in the *Drosophila* embryo. Development 114:583–597

Ferrara N, Carver-Moore K, Chen H, Dowd M, Lu L, O'Shea KS, Powell-Braxton L, Hillan KJ, Moore MW (1996) Heterozygous embryonic lethality induced by targeted inactivation of the VEGF gene. Nature 380:439–442

Field HH (1891) The development of the pronephros and segmental duct in Amphibia. Bull Mus Comp Zoöl Harvard 21:201–340

Fink RD, Cooper MS (1996) Apical membrane turnover is accelerated near cell-cell contacts in an embryonic epithelium. Dev Biol 174:180–189

Fisher A, Caudy M (1998) The function of hairy-related bHLH repressor proteins in cell fate decisions. BioEssays 20:298–306

Fisher S, Halpern ME (1999) Patterning the zebrafish axial skeleton requires early chordin function. Nat Genet 23:442–446

Fisher S, Amacher SL, Halpern ME (1997) Loss of cerebrum function ventralizes the zebrafish embryo. Development 124:1301–1311

Fishman MC, Chien KR (1997) Fashioning the vertebrate heart: earliest embryonic decisions. Development 124:2099–2117

Forsberg H, Crozet F, Brown NA (1998) Waves of mouse *Lunatic fringe* expression, in four-hour cycles at two-hour intervals, precede somite boundary formation. Curr Biol 8:1027–1030

Fouquet B, Weinstein BM, Serluca FC, Fishman MC (1997) Vessel patterning in the embryo of the zebrafish: guidance by notochord. Dev Biol 183:37–48

Fox H (1963) The amphibian pronephros. Quart Rev Biol 38:1–25

Franco D, Lamers WH, Moorman AF (1998) Patterns of expression in the developing myocardium: towards a morphologically integrated transcriptional model. Cardiovasc Res 38:25–53

François V, Bier E (1995) *Xenopus chordin* and *Drosophila short gastrulation* genes encode homologous proteins functioning in dorsal-ventral axis formation. Cell 80:19–20

Francois V, Solloway M, O'Neill JW, Emery J, Bier E (1994) Dorsal-ventral patterning of the Drosophila embryo depends on a putative negative growth factor encoded by the short gastrulation gene. Genes Dev 8:2602–2616

Franz T, Kothary R, Surani MAH, Halata Z, Grimm M (1993) The *Splotch* mutation interferes with muscle development in the limbs. Anat Embryol (Berl) 187:371–377

Fricke C, Lee J-S, Geiger-Rudolph S, Bonhoeffer F, Chien C-B (2001) *astray*, a zebrafish *roundabout* homolog required for retinal axon guidance. Science 292:507–510

Fritsche R, Schwerte T, Pelster B (2000) Nitric oxide and vascular reactivity in developing zebrafish, *Danio rerio*. Am J Physiol Regul Integr Comp Physiol 279:R2200–R2207

Fujisue M, Kobayakawa Y, Yamaha E (1993) Occurrence of dorsal axis-inducing activity around the vegetal pole of an uncleaved *Xenopus* egg and displacement to the equatorial region by cortical rotation. Development 118:163–170

Funayama N, Fagotto F, McCrea P, Gumbiner BM (1995) Embryonic axis induction by the armadillo repeat domain of beta-catenin: evidence for intracellular signaling. J Cell Biol 128:959–968

Funayama N, Sato Y, Matsumoto K, Ogura T, Takahashi Y (1999) Coelom formation: binary decision of the lateral plate mesoderm is controlled by the ectoderm. Development 126:4129–4138

Fürthauer M, Thisse C, Thisse B (1997) A role for FGF-8 in the dorsoventral patterning of the zebrafish gastrula. Development 124:4253–4264

Fürthauer M, Thisse B, Thisse C (1999) Three different *noggin* genes antagonize the activity of bone morphogenetic proteins in the zebrafish embryo. Dev Biol 214:181–196

Fürthauer M, Reifers F, Brand M, Thisse B, Thisse C (2001) sprouty4 acts in vivo as a feedback-induced antagonist of FGF signaling in zebrafish. Development 128:2175–2186

Furukawa T, Mukherjee S, Bao ZZ, Morrow EM, Cepko CL (2000) *rax*, *Hes1*, and *notch1* promote the formation of Müller glia by postnatal retinal progenitor cells. Neuron 26:383–394

Gaiano N, Nye JS, Fishell G (2000) Radial glial identity is promoted by Notch1 signaling in the murine forebrain. Neuron 26:395–404

Gamse JT, Shen YC, Thisse C, Thisse B, Raymond PA, Halpern ME, Liang JO (2002) *Otx5* regulates genes that show circadian expression in the zebrafish pineal complex. Nat Genet 30:117–121

Gavalas A, Krumlauf R (2000) Retinoid signalling and hindbrain patterning. Curr Opin Genet Dev 10:380–386

Gavalas A, Studer M, Lumsden A, Rijli FM, Krumlauf R, Chambon P (1998) *Hoxa1* and *Hoxb1* synergize in patterning the hindbrain, cranial nerves and second pharyngeal arch. Development 125:1123–1136

Gehring WJ, Ikeo K (1999) *Pax 6*: mastering eye morphogenesis and eye evolution. Trends Genet 15:371–377

Gendron-Maguire M, Mallo M, Zhang M, Gridley T (1993) *Hoxa-2* mutant mice exhibit homeotic transformation of skeletal elements derived from cranial neural crest. Cell 75:1317–1331

Geoffroy Saint-Hilaire E (1822) Considerations generales sur la vertebre. Mém du Mus Hist Nat 9:89–119

Gerhart J, Danilchik M, Doniach T, Roberts S, Rowning B, Stewart R (1989) Cortical rotation of the *Xenopus* egg: consequences for the anteroposterior pattern of embryonic dorsal development. Development 107:37–51

Gering M, Rodaway AR, Gottgens B, Patient RK, Green AR (1998) The SCL gene specifies haemangioblast development from early mesoderm. EMBO J 17:4029–4045

Germain S, Howell M, Esselmont GM, Hill CS (2000) Homeodomain and winged-helix transcription factors recruit activated Smads to distinct promoter elements via a common Smad interaction motif. Genes Dev 14:435–451

Gevers P, Dulos J, Schipper H, Timmermans L (1992a) Origin of primordial germ cells, as characterized by the presence of nuage, in embryos of the teleost fish *Barbus conchonius*. Eur J Morphol 30:195–204

Gevers P, Dulos J, van den Boogaart J, Timmermans LP (1992b) A study on cell lineage, especially the germ cell line, in embryos of the teleost fish, *Barbus conchonius*. Roux's Arch Dev Bio 201:275–283

Gilbert SF (1991) Developmental biology. Sinauer Associates, Sunderland

Gilland E, Miller AL, Karplus E, Baker R, Webb SE (1999) Imaging of multicellular large-scale rhythmic calcium waves during zebrafish gastrulation. Proc Natl Acad Sci USA 96:157–161

Gimlich RL (1986) Acquisition of developmental autonomy in the equatorial region of the Xenopus embryo. Dev Biol 115:340–352

Gimlich RL, Gerhart JC (1984) Early cellular interactions promote embryonic axis formation in *Xenopus laevis*. Dev Biol 104:117–130

Girdlestone J, Weston JA (1985) Identification of early neuronal subpopulations in avian neural crest cell cultures. Dev Biol 109:274–287

Glickman NS, Adams RJ, Jones MA, Walker C, Kimmel CB (2002) Shaping the zebrafish notochord. (submitted)

Glinka A, Wu W, Onichtchouk D, Blumenstock C, Niehrs C (1997) Head induction by simultaneous repression of Bmp and Wnt signalling in *Xenopus*. Nature 389:517–519

Glinka A, Wu W, Delius H, Monaghan AP, Blumenstock C, Niehrs C (1998) Dickkopf-1 is a member of a new family of secreted proteins and functions in head induction. Nature 391:357–362

Godel V, Iaina A, Nemet P, Lazar M (1980) Hereditary renal-retinal dysplasia. Doc Ophthalmol 49:347–359

Golding JP, Trainor P, Krumlauf R, Gassmann M (2000) Defects in pathfinding by cranial neural crest cells in mice lacking the neuregulin receptor ErbB4. Nat Cell Biol 2:103–109

Goldstein AM, Fishman MC (1998) Notochord regulates cardiac lineage in zebrafish embryos. Dev Biol 201:247–252

Goldstein AM, Ticho BS, Fishman MC (1998) Patterning the heart's left-right axis: from zebrafish to man. Dev Genet 22:278–287

Gomez-Skarmeta JL, de La Calle-Mustienes E, Modolell J (2001) The Wnt-activated *Xiro1* gene encodes a repressor that is essential for neural development and downregulates Bmp4. Development 128:551–560

Gonzalez EM, Fekany-Lee K, Carmany-Rampey A, Erter C, Topczewski J, Wright CV, Solnica-Krezel L (2000) Head and trunk in zebrafish arise via coinhibition of BMP signaling by *bozozok* and *chordino*. Genes Dev 14:3087–3092

Good PJ (1995) A conserved family of elav-like genes in vertebrates. Proc Natl Acad Sci USA 92:4557–4561

Goodman CS, Bastiani MJ, Doe CQ, du Lac S, Helfand SL, Kuwada JY, Thomas JB (1984) Neuronal recognition during development: cellular and molecular approaches. Science 225:1271–1279

Goodrich ES (1930) Studies on the structure and development of vertebrates. Macmillan, London

Goodrich ES (1958) Studies on the structure and development of vertebrates, vol 1. Dover Publ, New York/Constable and Company Ltd, London

Gossler A, Hrabe de Angelis M (1998) Somitogenesis. Curr Top Dev Biol 38:225–287

Goulding M, Lamar E (2000) Neuronal patterning: making stripes in the spinal cord. Curr Biol 10:R565–R568

Grammatopoulos GA, Bell E, Toole L, Lumsden A, Tucker AS (2000) Homeotic transformation of branchial arch identity after *Hoxa2* overexpression. Development 127:5355–5365

Granato M, van Eeden FJ, Schach U, Trowe T, Brand M, Furutani-Seiki M, Haffter P, Hammerschmidt M, Heisenberg CP, Jiang YJ, Kane DA, Kelsh RN, Mullins MC, Odenthal J, Nusslein-Volhard C (1996) Genes controlling and mediating locomotion behavior of the zebrafish embryo and larva. Development 123:399–413

Grandel H, Lun K, Rauch GJ, Piotrowski T, Houart C, Sordino P, Küchler A, Schulte-Merker S, Geisler R, Holder N et al (2002) Retinoic acid signaling in the zebrafish embryo is necessary to pattern the anteroposterior axis of the CNS and to induce a pectoral fin field. Development (submitted)

Gray M, Moens CB, Amacher SL, Eisen JS, Beattie CE (2001) Zebrafish *deadly seven* functions in neurogenesis. Dev Biol 237:306–323

Greenspoon S, Patel CK, Hashmi S, Bernhardt RR, Kuwada JY (1995) The notochord and floor plate guide growth cones in the zebrafish spinal cord. J Neurosci 15:5956–5965

Greenwald I, Rubin GM (1992) Making a difference: the role of cell-cell interactions in establishing separate identities for equivalent cells. Cell 68:271–281

Griffin KJ, Patient R, Holder N (1995) Analysis of FGF function in normal and no tail zebrafish embryos reveals separate mechanisms for formation of the trunk and the tail. Development 121:2983–2994

Griffin KJ, Amacher SL, Kimmel CB, Kimelman D (1998) Molecular identification of spadetail: regulation of zebrafish trunk and tail mesoderm formation by T-box genes. Development 125:3379–3388

Griffin KJP, Stoller J, Gibson M, Chen S, Yelon D, Stainier DYR, Kimelman D (2000) A conserved role for *H15*-related T-box transcription factors in zebrafish and *Drosophila* heart formation. Dev Biol 218:235–247

Grinblat Y, Gamse J, Patel M, Sive H (1998) Determination of the zebrafish forebrain: induction and patterning. Development 125:4403–4416

Gripp KW, Wotton D, Edwards MC, Roessler E, Ades L, Meinecke P, Richieri-Costa A, Zackai EH, Massague J, Muenke M, Elledge SJ (2000) Mutations in *tgif* cause holoprosencephaly and link nodal signalling to human neural axis determination. Nat Genet 25:205–208

Gritsman K, Zhang J, Cheng S, Heckscher E, Talbot WS, Schier AF (1999) The EGF-CFC protein one-eyed pinhead is essential for nodal signaling. Cell 97:121–132

Gritsman K, Talbot WS, Schier AF (2000) Nodal signaling patterns the organizer. Development 127:921–932

Gross MK, Moran-Rivard L, Velasquez T, Nakatsu MN, Jagla K, Goulding M (2000) *Lbx1* is required for muscle precursor migration along a lateral pathway into the limb. Development 127: 413–424

Gu Z, Reynolds EM, Song J, Lei H, Feijen A, Yu L, He W, MacLaughlin DT, van den Eijnden-van Raaij J, Donahoe PK, Li E (1999) The type I serine/threonine kinase receptor ActRIA (ALK2) is required for gastrulation of the mouse embryo. Development 126:2551–2561

Guo S, Brush J, Teraoka H, Goddard A, Wilson SW, Mullins MC, Rosenthal A (1999a) Development of noradrenergic neurons in the zebrafish hindbrain requires *bmp*, *fgf8* and the homeodomain protein *soulless/phox2a*. Neuron 24:555–566

Guo S, Wilson SW, Cooke S, Chitnis AB, Driever W, Rosenthal A (1999b) Mutations in the zebrafish unmask shared regulatory pathways controlling the development of catecholaminergic neurons. Dev Biol 208:473–487

Guo S, Yamaguchi Y, Schilbach S, Wada T, Lee J, Goddard A, French D, Handa H, Rosenthal A (2000) A regulator of transcriptional elongation controls vertebrate neuronal development. Nature 408:366–369

Hacker U, Lin X, Perrimon N (1997) The *Drosophila* sugarless gene modulates Wingless signaling and encodes an enzyme involved in polysaccharide biosynthesis. Development 124:3565–3573

Haddon C, Smithers L, Schneider-Maunoury S, Coche T, Henrique D, Lewis J (1998) Multiple *delta* genes and lateral inhibition in zebrafish primary neurogenesis. Development 125: 359–370

Haffter P, Granato M, Brand M, Mullins MC, Hammerschmidt M, Kane DA, Odenthal J, van Eeden FJM, Jiang Y-J, Heisenberg C-P, Kelsh RN, Furutani-Seiki M, Vogelsang E, Beuchle D, Schach U, Fabian C, Nüsslein-Volhard C (1996) The identification of genes with unique and essential functions in the development of the zebrafish, *Danio rerio*. Development 123:1–36

Hall BK (1980) Tissue interactions and the initiation of osteogenesis and chondrogenesis in the neural crest-derived mandibular skeleton of the embryonic mouse as seen in isolated murine tissues and in recombinations of murine and avian tissues. J Embryol Exp Morphol 58:251–264

Halloran M, Severance S, Yee CS, Gemza D, Kuwada JY (1998) Molecular cloning and expression of two novel zebrafish semaphorins. Mech Dev 76:165–168

Halloran M, Yee CS, Severance S, Gemza D, Raper JA, Kuwada JY (1999) Analysis of a zebrafish semaphorin reveals potential functions in vivo. Dev Dyn 214:13–25

Halloran MC, Sato-Maeda M, Warren JT, Su F, Lele Z, Krone PH, Kuwada JY, Shoji W (2000) Laser-induced gene expression in specific cells of transgenic zebrafish. Development 127:1953–1960

Halpern ME, Ho RK, Walker C, Kimmel CB (1993) The induction of muscle pioneers and floor plate is distinguished by the zebrafish mutation *no tail*. Cell 75:99–111

Halpern ME, Thisse C, Ho RK, Thisse B, Riggleman B, Trevarrow B, Weinberg ES, Postlethwait JH, Kimmel CB (1995) Cell-autonomous shift from axial to paraxial mesodermal development in zebrafish *floating head* mutants. Development 121:4257–4264

Hamaguchi S (1982) A light- and electron-microscopic study on the migration of primordial germ cells in the teleost, *Oryzias latipes*. Cell Tissue Res 227:139–151

Hamburger V (1988) The heritage of experimental embryology: Hans Spemann and the organizer. Oxford University Press, New York

Hammerschmidt M, Nüsslein-Volhard C (1993) The expression of a zebrafish gene homologous to *Drosophila snail* suggests a conserved function in invertebrate and vertebrate gastrulation. Development 119:1107–1118

Hammerschmidt M, Pelegri F, Mullins MC, Kane DA, Brand M, van Eeden FJM, Furutani-Seiki M, Granato M, Haffter P, Heisenberg C-P, Jiang Y-J, Kelsh RN, Odenthal J, Warga RM, Nüsslein-Volhard C (1996a) Mutations affecting morphogenesis during gastrulation and tail formation in the zebrafish, *Danio rerio*. Development 123:143–151

Hammerschmidt M, Pelegri F, Mullins MC, Kane DA, van Eeden FJ, Granato M, Brand M, Furutani-Seiki M, Haffter P, Heisenberg CP, Jiang YJ, Kelsh RN, Odenthal J, Warga RM, Nüsslein-Volhard C (1996b) *dino* and *mercedes*, two genes regulating dorsal development in the zebrafish embryo. Development 123:95–102

Hammerschmidt M, Serbedzija GN, McMahon AP (1996c) Genetic analysis of dorsoventral pattern formation in the zebrafish: requirement of a BMP-like ventralizing activity and its dorsal repressor. Genes Dev 10:2452–2461

Hammerschmidt M, Brook A, McMahon AP (1997) The world according to hedgehog. Trends Genet 13:14–21

Hanneman E, Trevarrow B, Metcalfe WK, Kimmel CB, Westerfield M (1988) Segmental pattern of development of the hindbrain and spinal cord of the zebrafish embryo. Development 103:49–58

Hansen U (1996) Electron microscopic study of possible sites of ultrafiltration in *Lumbricus terrestris* (Annelida, Oligochaeta). Tissue Cell 28:195–203

Harder W (1975) Anatomy of fishes (part I text and part II figures and plates). E Schweizerbart'sche Verlagsbuchhandlung (Nägele und Overmiller), Stuttgart

Hardistry MW (1979) Biology of the cyclostomes. Chapman and Hall, London

Harland R, Gerhart J (1997) Formation and function of Spemann's organizer. Annu Rev Cell Dev Biol 13:611–667

Harris JA, Guglielmotti V, Bentivoglio M (1996) Diencephalic asymmetries. Neurosci Biobehav Rev 20:637–643

Harris R, Sabatelli LM, Seeger MA (1996) Guidance cues at the Drosophila CNS midline: identification and characterization of two *Drosophila* netrin/UNC-6 homologs. Neuron 17:217–228

Harrison RG (1895) Die Entwicklung der unpaaren und paarigen Flossen der Teleostier. Arch Mikr Anat 46:500–578

Harvey RP (1990) The *Xenopus MyoD* gene: an unlocalised maternal mRNA predates lineage-restricted expression in the early embryo. Development 108:669–680

Hashimoto H, Itoh M, Yamanaka Y, Yamashita S, Shimizu T, Solnica-Krezel L, Hibi M, Hirano T (2000) Zebrafish Dkk1 functions in forebrain specification and axial mesendoderm formation. Dev Biol 217:138–152

Hatada Y, Stern CD (1994) A fate map of the epiblast of the early chick embryo. Development 120:2879–2889

Hatta K (1992) Role of the floor plate in axonal patterning in the zebrafish CNS. Neuron 9:629–642

Hatta K, Kimmel CB (1993) Midline structures and central nervous system coordinates in zebrafish. Perspect Dev Neurobiol 1:257–268

Hatta K, Korn H (1998) Tonic inhibition alternates in paired neurons that set direction of fish escape reaction. Proc Natl Acad Sci USA 96:12090–12095

Hatta K, Bremiller R, Westerfield M, Kimmel CB (1991a) Diversity of expression of engrailed-like antigens in zebrafish. Development 112:821–832

Hatta K, Kimmel CB, Ho RK, Walker C (1991b) The *cyclops* mutation blocks specification of the floor plate of the zebrafish central nervous system. Nature 350:339–341

Hatta K, Püschel AW, Kimmel CB (1994) Midline signaling in the primordium of the zebrafish anterior central nervous system. Proc Natl Acad Sci USA 91:2061–2065

Hay B, Jan L, Jan Y (1988) A protein component of *Drosophila* polar granules is encoded by vasa and has extensive sequence similarity to ATP-dependent helicases. Cell 55:577–587

He X, Saint-Jeannet JP, Woodgett JR, Varmus HE, Dawid IB (1995) Glycogen synthase kinase-3 and dorsoventral patterning in *Xenopus* embryos [published erratum appears in Nature (1995) 375:253]. Nature 374:617–622

Heasman J (1997) Patterning the *Xenopus* blastula. Development 124:4179–4191

Heasman J, Crawford A, Goldstone K, Garner-Hamrick P, Gumbiner B, McCrea P, Kintner C, Noro CY, Wylie C (1994) Overexpression of cadherins and underexpression of beta-catenin inhibit dorsal mesoderm induction in early *Xenopus* embryos. Cell 79:791–803

Heasman J, Kofron M, Wylie C (2000) Beta-catenin signaling activity dissected in the early *Xenopus* embryo: a novel antisense approach. Dev Biol 222:124–134

Hedgecock E, Culotti J, Hall D (1990) The *unc-5*, *unc-6*, and *unc-40* genes guide circumferential migrations of pioneer axons and mesodermal cells on the epidermis in *C. elegans*. Neuron 4:61–85

Hedgepeth CM, Conrad LJ, Zhang J, Huang HC, Lee VM, Klein PS (1997) Activation of the Wnt signaling pathway: a molecular mechanism for lithium action. Dev Biol 185:82–91

Heisenberg CP, Nüsslein-Volhard C (1997) The function of silberblick in the positioning of the eye anlage in the zebrafish embryo. Dev Biol 184:85–94

Heisenberg CP, Brand M, Jiang YJ, Warga R, Beuchle D, Eeden F, Furutani-Seiki M, Granato M, Haffter P, Hammerschmidt M, Kane D, Kelsh R, Mullins M, Odenthal J, Nüsslein-Volhard C (1996) Genes involved in forebrain development in the zebrafish, *Danio rerio*. Development 123:191–203

Heisenberg CP, Brennan C, Wilson SW (1999) Zebrafish aussicht mutant embryos exhibit widespread overexpression of *ace* (*fgf8*) and coincident defects in cns development. Development 126:2129–2140

Heisenberg CP, Tada M, Rauch GJ, Saude L, Concha ML, Geisler R, Stemple DL, Smith JC, Wilson SW (2000) Silberblick/Wnt11 mediates convergent extension movements during zebrafish gastrulation. Nature 405:76–81

Heisenberg CP, Houart C, Take-Uchi M, Rauch GJ, Young N, Coutinho P, Masai I, Caneparo L, Concha ML, Geisler R, Dale TC, Wilson SW, Stemple DL (2001) A mutation in the Gsk3-binding domain of zebrafish Masterblind/Axin1 leads to a fate transformation of telencephalon and eyes to diencephalon. Genes Dev 15:1427–1434

Helde KA, Wilson ET, Cretekos CJ, Grunwald DJ (1994) Contribution of early cells to the fate map of the zebrafish gastrula. Science 265:517–520

Heller N, Brandli AW (1997) *Xenopus Pax-2* displays multiple splice forms during embryogenesis and pronephric kidney development. Mech Dev 69:83–104

Heller N, Brandli AW (1999) *Xenopus Pax-2/5/8* orthologues: novel insights into *Pax* gene evolution and identification of Pax-8 as the earliest marker for otic and pronephric cell lineages. Dev Genet 24:208–219

Hemmati-Brivanlou A, Kelly OG, Melton DA (1994) Follistatin, an antagonist of activin, is expressed in the Spemann organizer and displays direct neuralizing activity. Cell 77:283–295

Henion P, Weston J (1997) Timing and pattern of cell fate restrictions in the neural crest lineage. Development 124:4351–4359

Henion PD, Raible DW, Beattie CE, Stoesser KL, Weston JA, Eisen JS (1996) Screen for mutations affecting development of zebrafish neural crest. Dev Genet 18:11–17

Henningfeld KA, Rastegar S, Adler G, Knöchel W (2000) Smad1 and Smad4 are components of the bone morphogenetic protein-4 (BMP-4)-induced transcription complex of the Xvent-2B promoter. J Biol Chem 275:21827–21835

Henry CA, Hall LA, Burr Hille M, Solnica-Krezel L, Cooper MS (2000) Somites in zebrafish doubly mutant for *knypek* and *trilobite* form without internal mesenchymal cells or compaction. Curr Biol 10:1063–1066

Henry GL, Melton DA (1998) Mixer, a homeobox gene required for endoderm development. Science 281:91–96

Hentschel H, Elger M (1996) Functional morphology of the developing pronephric kidney of zebrafish. J Am Soc Nephrol 7:1598

Herbarth B, Pingault V, Bondurand N, Kuhlbrodt K, Hermans-Borgmeyer I, Puliti A, Lemort N, Goossens M, Wegner M (1998) Mutation of the Sry-related *Sox10* gene in Dominant megacolon, a mouse model for human Hirschsprung disease. Proc Natl Acad Sci USA 95:5161–5165

Herbomel P, Thisse B, Thisse C (1999) Ontogeny and behaviour of early macrophages in the zebrafish embryo. Development 126:3735–3745

Hickman CP, Trump BF (1969) The kidney. In: Hoar WS, Randall DJ (eds) Fish physiology, vol 1. Academic Press, New York, pp 91–239

Higashijima S, Nose A, Eguchi G, Hotta Y Okamoto H (1997a) Mindin/F-spondin family: novel ECM proteins expressed in the zebrafish embryonic axis. Dev Biol 192:211–127

Higashijima S, Okamoto H, Ueno N, Hotta Y, Eguchi G (1997b) High-frequency generation of transgenic zebrafish which reliably express GFP in whole muscles or the whole body by using promoters of zebrafish origin. Dev Biol 192:289–299

Higashijima S, Hotta Y, Okamoto H (2000) Visualization of cranial motor neurons in live zebrafish expressing GFP under the control of the Islet-1 promoter/enhancer. J Neurosci 20:206–218

Hild M, Dick A, Rauch GJ, Meier A, Bouwmeester T, Haffter P, Hammerschmidt M (1999) The *smad5* mutation somitabun blocks Bmp2b signaling during early dorsoventral patterning of the zebrafish embryo. Development 126:2149–2159

Hinds J, Hinds P (1974) Early ganglion cell differentiation in the mouse retina: an electron microscopic analysis utilizing serial sections. Dev Biol 37:381–416

Hinds J, Hinds P (1979) Differentiation of photoreceptors and horizontal cells in the embryonic mouse retina: an electron microscopic, serial section analysis. J Comp Neurol 187:495–512

Hinds J, Hinds P (1983) Development of retinal amacrine cells in the mouse embryo: evidence for two modes of formation. J Comp Neurol 213:1–23

Hinds JW, Ruffett TL (1971) Cell proliferation in the neural tube: an electron microscopic and Golgi analysis in the mouse cerebral vesicle. Z Zellforsch Mikrosk Anat 115:226–264

Hirata T, Yamanaka Y, Ryu SL, Shimizu T, Yabe T, Hibi M, Hirano T (2000) Novel mix-family homeobox genes in zebrafish and their differential regulation. Biochem Biophys Res Commun 271:603–609

Ho CY, Houart C, Wilson SW, Stainier DYR (1999) A role for the extraembryonic yolk syncytial layer in patterning the zebrafish embryo suggested by properties of the hex gene. Curr Biol 9:1131–1134

Ho RK, Kane DA (1990) Cell-autonomous action of zebrafish spt-1 mutation in specific mesodermal precursors. Nature 348:728–730

Ho RK, Kimmel CB (1993) Commitment of cell fate in the early zebrafish embryo. Science 261:109–111

Hoadley L (1928) On the localization of the developmental potencies in the embryo of *Fundulus heteroclitus*. J Exp Zool 227:7–44

Hoch M, Pankratz MJ (1996) Control of gut signaling molecules in *Drosophila*. Mech Dev 58:3–14

Hodgkinson CA, Moore KJ, Nakayama A, Steingrimsson E, Copeland NG, Jenkins NA, Arnheiter H (1993) Mutations at the mouse microphthalmia locus are associated with defects in a gene encoding a novel basic-helix-loop-helix-zipper protein. Cell 74:395–404

Holder N, Hill J (1991) Retinoic acid modifies development of the midbrain-hindbrain border and affects cranial ganglion formation in zebrafish embryos. Development 113:1159–1170

Holder N, Klein R (1999) Eph receptors and ephrins: effectors of morphogenesis. Development 126:2033–2044

Holley SA, Nüsslein-Volhard C (2000) Somitogenesis in zebrafish. Curr Topics Dev Biol 47: 247–277

Holley SA, Jackson PD, Sasai Y, Lu B, De Robertis EM, Hoffmann FM, Ferguson EL (1995) A conserved system for dorsal-ventral patterning in insects and vertebrates involving sog and chordin. Nature 376:249–253

Holley SA, Geisler R, Nüsslein-Volhard C (2000) Control of *her1* expression during zebrafish somitogenesis by a Delta-dependent oscillator and an independent wave-front activity. Genes Dev 14:1678–1690

Holmes GP, Negus K, Burridge L, Raman S, Algar E, Yamada T, Little MH (1998) Distinct but overlapping expression patterns of two vertebrate slit homologs implies functional roles in CNS development and organogenesis. Mech Dev 79:57–72

Hopwood ND, Pluck A, Gurdon JB (1991) *Xenopus Myf-5* marks early muscle cells and can activate muscle genes ectopically in early embryos. Development 111:551–560

Horb ME, Slack JMW (2001) Endoderm specification and differentiation in *Xenopus* embryos. Dev Biol 236:330–343

Horne-Badovinac S, Lin D, Waldron S, Schwarz M, Mbamalu G, Pawson T, Jan YN, Stainier DYR, Abdelilah-Seyfried S (2001) Positional cloning of *heart and soul* reveals multiple roles of PKClambda in zebrafish organogenesis. Curr Biol 11:1492–502

Houart C, Westerfield M, Wilson SW (1998) A small population of anterior cells patterns the forebrain during zebrafish gastrulation. Nature 391:788–792

Houart C, Caneparo L, Heisenberg C-P, Take-Uchi M, Wilson SW (2002) Establishment of the telencephalon during gastrulation by local antagonism of Wnt signalling. Neuron (in press)

Houliston E, Elinson RP (1991) Evidence for the involvement of microtubules, ER, and kinesin in the cortical rotation of fertilized frog eggs. J Cell Biol 114:1017–1028

Houzelstein D, Auda-Boucher G, Cheraud Y, Rouaud T, Blanc I, Tajbakhsh S, Buckingham ME (1999) The homeobox gene *Msx1* is expressed in a subset of somites, and in muscle progenitor cells migrating into the forelimb. Development 126:2689–2701

Houston DW, King ML (2000a) Germ plasm and molecular determinants of germ cell fate. Curr Topics Dev Biol 50:155–181

Houston DW, King ML (2000b) A critical role for *Xdazl*, a germ plasm-localized RNA, in the differentiation of primordial germ cells in *Xenopus*. Development 127:447–456

Howard K (1998) Organogenesis: *Drosophila* goes gonadal. Curr Biol 8:R415–R417

Howell M, Hill CS (1997) XSmad2 directly activates the Activin-inducible, dorsal mesoderm gene XFKH1 in *Xenopus* embryos. EMBO J 16:7411–7421

Howland RB (1921) Experiments on the effect of the removal of the pronephros of *Ambystoma punctatum*. J Exp Zool 32:355–384

Hu M, Easter SS Jr (1999). Retinal neurogenesis: the formation of the initial central patch of postmitotic cells. Dev Biol 207:309–321

Hu N, Sedmera D, Yost HJ, Clark EB (2000) Structure and function of the developing zebrafish heart. Anat Rec 260:148–157

Huang H, Liu N, Lin S (2001) *Pdx-1* knockdown reduces insulin promoter activity in zebrafish. Genesis 30:134–136

Hudson C, Clements D, Friday RV, Stott D, Woodland HR (1997) *Xsox17a* and *-b* mediate endoderm formation in *Xenopus*. Cell 91:397–405

Hung TJ, Kemphues KJ (1999) PAR-6 is a conserved PDZ domain containing protein that colocalizes with PAR-3 in *Caenorhabditis elegans* embryos. Development 126:127–135

Hyatt BA, Yost HJ (1998) The left-right coordinator: the role of Vg1 in organizing left-right axis formation. Cell 93:37–46

Hyatt GA, Schmitt EA, Marsh-Armstrong NR, Dowling JE (1992) Retinoic acid-induced duplication of the zebrafish retina. Proc Natl Acad Sci USA 89:8293–8297

Hyatt GA, Schmitt EA, Marsh-Armstrong N, McCaffery P, Drager UC, Dowling JE (1996) Retinoic acid establishes ventral retinal characteristics. Development 122:195–204

Hyde CE, Old RW (2000) Regulation of the early expression of the *Xenopus* nodal-related 1 gene, *Xnr1*. Development 127:1221–1229

Hynes M, Poulsen K, Tessier LM, Rosenthal A (1995) Control of neuronal diversity by the floor plate: contact-mediated induction of midbrain dopaminergic neurons. Cell 80:95–101

Hyodo M, Aoki A, Ando C, Katsumata M, Nyui S, Motegi N, Morozumi T, Matsuhashi M (1996) Essential role of the yolk syncytial layer for the development of isolated blastoderms from medaka embryos. Dev Growth Differ 38:383–392

Ikeda S, Kishida S, Yamamoto H, Murai H, Koyama S, Kikuchi A (1998) Axin, a negative regulator of the Wnt signaling pathway, forms a complex with gsk-3beta and beta-catenin and promotes gsk3beta dependent phosphorylation of beta-catenin. EMBO J 17:1371–1384

Imai Y, Gates MA, Melby AE, Kimelman D, Schier AF, Talbot WS (2001) The homeobox genes *vox* and *vent* are redundant repressors of dorsal fates in zebrafish. Development 128:2407–2420

Imondi R, Wideman C, Kaprielian Z (2000) Complementary expression of transmembrane ephrins and their receptors in the mouse spinal cord: a possible role in constraining the orientation of longitudinally projecting axons. Development 127:1397–1410

Irving C, Mason I (2000) Signalling by Fgf8 from the isthmus patterns anterior hindbrain and establishes the anterior limit of hox gene expression. Development 127:177–186

Ishii N, Wadsworth WG, Stern BD, Culotti JG, Hedgecock EM (1992) UNC-6, a laminin-related protein, guides cell and pioneer axon migrations in *C. elegans*. Neuron 9:1–20

Ishitani T, Ninomiya-Tsuji J, Nagai S, Nishita M, Meneghini M, Barker N, Waterman M, Bowerman B, Clevers H, Shibuya H, Matsumoto K (1999) The TAK1-NLK-MAPK-related pathway antagonizes signalling between beta-catenin and transcription factor TCF. Nature 399:798–802

Isogai S, Horiguchi M, Weinstein BM (2001) The vascular anatomy of the developing zebrafish: an atlas of embryonic and early larval development. Dev Biol 230:278–301

Itoh K, Krupnik VE, Sokol SY (1998) Axis determination in *Xenopus* involves biochemical interactions of axin, glycogen synthase kinase 3 and beta-catenin. Curr Biol 8:591–594

Itoh M, Chitnis AB (2001) Expression of proneural and neurogenic genes in the zebrafish lateral line primordium correlates with selection of hair cell fate in neuromasts. Mech Dev 102: 263–266

Itoh M, Kudoh T, Dedekian M, Kim C-H, Chitnis A (2002) A role for zebrafish *iro1* and *iro7* in the establishment of the midbrain-hindbrain territory and the isthmic organizer. Development 129:2317–2327

Izraeli S, Lowe LA, Bertness VL, Good DJ, Dorward DW, Kirsch IR, Kuehn MR (1999) The SIL gene is required for mouse embryonic axial development and left-right specification. Nature 399:691–694

Izraeli S, Lowe LA, Bertness VL, Campaner S, Hahn H, Kirsch IR, Kuehn MR (2001) Genetic evidence that Sil is required for the sonic hedgehog response pathway. Genesis 31:72–77

Jacobson M (1991) Developmental neurobiology. Plenum, New York, pp 41–94

Jensen AM, Walker C, Westerfield M (2001) *mosaic eyes*: a zebrafish gene required in pigmented epithelium for apical localization of retinal cell division and lamination. Development 128: 95–105

Jessell TM (2000) Neuronal specification in the spinal cord: inductive signals and transcriptional codes. Nat Rev Genet 1:20–29

Jesuthasan S (1996) Contact inhibition/collapse and pathfinding of neural crest cells in the zebrafish trunk. Development 122:381–389

Jesuthasan S, Strähle U (1997) Dynamic microtubules and specification of the zebrafish embryonic axis. Curr Biol 7:31–42

Jiang YJ, Brand M, Heisenberg CP, Beuchle D, Furutani-Seiki M, Kelsh RN, Warga RM, Granato M, Haffter P, Hammerschmidt M, Kane DA, Mullins MC, Odenthal J, van Eeden FJ, Nüsslein-Volhard C (1996) Mutations affecting neurogenesis and brain morphology in the zebrafish, *Danio rerio*. Development 123:205–216

Jiang YJ, Aerne BL, Smithers L, Haddon C, Ish-Horowicz D, Lewis J (2000) Notch signalling and the synchronisation of the somite segmentation clock. Nature 408:475–479

Jin EJ, Erickson CA, Takada S, Burrus LW (2001) Wnt and BMP signaling govern lineage segregation of melanocytes in the avian embryo. Dev Biol 233:22–37

Johnson DW, Berg JN, Baldwin MA, Gallione CJ, Marondel I, Yoon SJ, Stenzel TT, Speer M, Pericak-Vance MA, Diamond A et al (1996) Mutations in the activin receptor-like kinase 1 gene in hereditary haemorrhagic telangiectasia type 2. Nat Genet 13:189–195

Johnson J, Rhee J, Parsons SM, Brown D, Olson EN, Rawls A (2001) The anterior/posterior polarity of somites is disrupted in paraxis-deficient mice. Dev Biol 229:176–187

Johns PR, Fernald RD (1981) Genesis of rods in teleost fish retina. Nature 293:141–142

Jones CM, Smith JC (1995) Revolving vertebrates. Curr Biol 5:574–576

Jones CM, Kuehn MR, Hogan BL, Smith JC, Wright CV (1995) Nodal-related signals induce axial mesoderm and dorsalize mesoderm during gastrulation. Development 121:3651–3662

Jones EA, Woodland HR (1987) The development of animal cap cells in *Xenopus*: a measure of the start of animal cap competent to form mesoderm. Development 101:557–564

Joseph EM, Melton DA (1997) *Xnr4*: a *Xenopus* nodal-related gene expressed in the Spemann organizer. Dev Biol 184:367–372

Juan H, Hamada H (2001) Roles of nodal-lefty regulatory loops in embryonic patterning of vertebrates. Genes Cells 6:923–930

Julian D, Ennis K, Korenbrot JI (1998) Birth and fate of proliferative cells in the inner nuclear layer of mature fish retina. J Comp Neurol 394:271–282

Juusola M, French AS, Uusitalo RO, Weckstrom M (1996) Information processing by graded-potential transmission through tonically active synapses. Trends Neurosci 19:292–297

Kageura H (1997) Activation of dorsal development by contact between the cortical dorsal determinant and the equatorial core cytoplasm in eggs of *Xenopus laevis*. Development 124:1543–1551

Kageura H (1990) Spatial distribution of the capacity to initiate a secondary embryo in the 32-cell embryo of *Xenopus laevis*. Dev Biol 142:432–438

Kahane N, Cinnamon Y, Kalcheim C (1998) The origin and fate of pioneer myotomal cells in the avian embryo. Mech Dev 74:59–73

Kalcheim C, Cinnamon Y, Kahane N (1999) Myotome formation: a multistage process. Cell Tissue Res 296:161–173

Kane DA, Kimmel CB (1993) The zebrafish midblastula transition. Development 119:447–456

Kane DA, Warga RM (1994) Domains of movement in the zebrafish gastrula. Semin Dev Biol 5:101–109

Kane DA, Warga RM, Kimmel CB (1992) Mitotic domains in the early embryo of the zebrafish. Nature 360:735–737

Kane DA, Hammerschmidt M, Mullins MC, Maischein H-M, Brand M, van Eeden FJM, Furutani-Seiki M, Granato M, Haffter P, Heisenberg C-P, Jiang Y-J, Kelsh RN, Odenthal J, Warga RM, Nüsslein-Volhard C (1996) The zebrafish epiboly mutants. Development 123:47–55

Kanekar S, Perron M, Dorsky R, Harris WA, Jan LY, Jan YN, Vetter ML (1997) Xath5 participates in a network of bHLH genes in the developing *Xenopus* retina. Neuron 19:981–994

Kanki JP, Ho RK (1997) The development of the posterior body in zebrafish. Development 124:881–893

Kanki JP, Chang S, Kuwada JY (1994) The molecular cloning and characterization of potential chick DM-GRASP homologs in zebrafish and mouse. J Neurobiol 25:831–845

Kao KR, Elinson RP (1988) The entire mesodermal mantle behaves as Spemann's organizer in dorsoanterior enhanced *Xenopus laevis* embryos. Dev Biol 127:64–77

Kapur RP (1999) Early death of neural crest cells is responsible for total enteric aganglionosis in Sox10(Dom)/Sox10(Dom) mouse embryos. Pediatr Dev Pathol 2:559–569

Kardong KV (1995) Vertebrates, comparative anatomy, function, evolution. Brown, Oxford

Karlstrom RO, Trowe T, Klostermann S, Baier H, Brand M, Crawford AD, Grunewald B, Haffter P, Hoffmann H, Meyer SU, Muller BK, Richter S, van Eeden FJ, Nüsslein-Volhard C, Bonhoeffer F (1996) Zebrafish mutations affecting retinotectal axon pathfinding. Development 123: 427–438

Karlstrom RO, Trowe T, Bonhoeffer F (1997) Genetic analysis of axon guidance and mapping in the zebrafish. Trends Neurosci 20:3–8

Karlstrom RO, Talbot WS, Schier AF (1999) Comparative synteny cloning of zebrafish you-too: mutations in the hedgehog target *gli2* affect ventral forebrain patterning. Genes Dev 15: 388–393

Kathiriya IS, Srivastava D (2000) Left-right asymmetry and cardiac looping: implications for cardiac development and congenital heart disease. Am J Med Genet 97:271–279

Kaufmann E, Knöchel W (1996) Five years on the wings of fork head. Mech Dev 57:3–20

Kawahara A, Wilm T, Solnica-Krezel L, Dawid IB (2000a) Antagonistic role of vegal and bozozok/dharma homeobox genes in organizer formation. Proc Natl Acad Sci USA 97:12121–12126

Kawahara A, Wilm T, Solnica-Krezel L, Dawid IB (2000b) Functional interaction of vega2 and goosecoid homeobox genes in zebrafish. Genesis 28:58–67

Kay JN, Finger-Baier KC, Roeser T, Staub W, Baier H (2001). Retinal ganglion cell genesis requires *lakritz*, a zebrafish *atonal* homolog. Neuron 30:725–736

Keller RE (1980) The cellular basis of epiboly: an SEM study of deep cell rearrangement during gastrulation in *Xenopus laevis*. J Embryol Exp Morph 60:201–234

Keller RE, Trinkaus JP (1987) Rearrangement of enveloping layer cells without disruption of the epithilial permeability barrier as a factor in *Fundulus* epiboly. Dev Biol 120:12–24

Keller RE, Davidson L, Edlund A, Elul T, Ezin M, Shook D, Skoglund P (2000) Mechanisms of convergence and extension by cell intercalation. Philos Trans R Soc Lond B Biol Sci 355:897–922

Kelly C, Chin AJ, Leatherman JL, Kozlowski DJ, Weinberg ES (2000) Maternally controlled (beta)-catenin-mediated signaling is required for organizer formation in the zebrafish. Development 127:3899–3911

Kelly GM, Moon RT (1995) Involvement of wnt1 and pax2 in the formation of the midbrain-hindbrain boundary in the zebrafish gastrula. Dev Genet 17:129–140

Kelly GM, Erezyilmaz DF, Moon RT (1995a) Induction of a secondary embryonic axis in zebrafish occurs following the overexpression of beta-catenin. Mech Dev 53:261–273

Kelly GM, Greenstein P, Erezyilmaz DF, Moon RT (1995b) Zebrafish *wnt8* and *wnt8b* share a common activity but are involved in distinct developmental pathways. Development 121: 1787–1799

Kelsh RN, Eisen JS (2000) The zebrafish *colourless* gene regulates development of non-ectomesenchymal neural crest derivatives. Development 127:515–525

Kelsh RN, Brand M, Jiang YJ, Heisenberg CP, Lin S, Haffter P, Odenthal J, Mullins MC, van Eeden FJ, Furutani-Seiki M, Granato M, Hammerschmidt M, Kane DA, Warga RM, Beuchle D, Vogelsang L, Nüsslein-Volhard C (1996) Zebrafish pigmentation mutations and the processes of neural crest development. Development 123:369–389

Kelsh RN, Dutton K, Medlin J, Eisen JS (2000a) Expression of zebrafish *fkd6* in neural crest-derived glia. Mech Dev 93:161–164

Kelsh RN, Schmid B, Eisen JS (2000b) Genetic analysis of melanophore development in zebrafish embryos. Dev Biol 225:277–293

Kennedy TE, Serafini T, de la Torre JR, Tessier-Lavigne M (1994) Netrins are diffusible chemotropic factors for commissural axons in the embryonic spinal cord. Cell 78:425–435

Keynes R, Stern C (1988) Mechanisms of vertebrate segmentation. Development 103:413–429

Kidd T, Brose K, Mitchell KJ, Fetter RD, Tessier-Lavigne M, Goodman CS, Tear G (1998) Roundabout controls axon crossing of the CNS midline and defines a novel subfamily of evolutionarily conserved guidance receptors. Cell 92:205–215

Kidd T, Bland KS, Goodman CS (1999) Slit is the midline repellent for the robo receptor in Drosophila. Cell 96:785–794

Kiecker C, Niehrs C (2001) The role of the prechordal mesendoderm in neural patterning. Curr Opin Neurobiol 11:27–33

Kikuchi Y, Segawa H, Tokumoto M, Tsubokawa T, Hotta Y, Uyemura K, Okamoto H (1997) Ocular and cerebellar defects in zebrafish induced by overexpression of the LIM domains of the *islet-3* LIM/homeodomain protein. Neuron 18:369–382

Kikuchi Y, Trinh LA, Reiter JF, Alexander J, Yelon D, Stainier DY (2000) The zebrafish *bonnie and clyde* gene encodes a Mix family homeodomain protein that regulates the generation of endodermal precursors. Genes Dev 14:1279–1289

Kikuchi Y, Agathon A, Alexander J, Thisse C, Waldron S, Yelon D, Thisse B, Stainier DY (2001) casanova encodes a novel Sox-related protein necessary and sufficient for early endoderm formation in zebrafish. Genes Dev 15:1493–1505

Kim CH, Ueshima E, Muraoka O, Tanaka H, Yeo SY, Huh TL, Miki N (1996) Zebrafish elav/HuC homologue as a very early neuronal marker. Neurosci Lett 216:109–112

Kim CH, Bae YK, Yamanaka Y, Yamashita S, Shimizu T, Fujii R, Park HC, Yeo SY, Huh TL, Hibi M, Hirano T (1997) Overexpression of neurogenin induces ectopic expression of HuC in zebrafish. Neurosci Lett 239:113–116

Kim CH, Oda T, Itoh M, Jiang D, Artinger KB, Chandrasekharappa SC, Driever W, Chitnis AB (2000) Repressor activity of Headless/Tcf3 is essential for vertebrate head formation. Nature 407:913–916

Kim SH, Park HC, Yeo SY, Hong SK, Choi JW, Kim CH, Weinstein BM, Huh TL (1998a) Characterization of two *frizzled8* homologues expressed in the embryonic shield and prechordal plate of zebrafish embryos. Mech Dev 78:193–201.

Kim SH, Yamamoto A, Bouwmeester T, Agius E, Robertis EM (1998b) The role of paraxial protocadherin in selective adhesion and cell movements of the mesoderm during *Xenopus* gastrulation. Development 125:4681–4690

Kim SH, Jen WC, De Robertis EM, Kintner C (2000) The protocadherin PAPC establishes segmental boundaries during somitogenesis in *Xenopus* embryos. Curr Biol 10:821–830

Kimelman D Griffin KJP (2000) Vertebrate mesendoderm induction and patterning. Curr Opin Gen Dev 10:350–356

Kimmel CB, Law RD (1985a) Cell lineage of zebrafish blastomeres. I. Cleavage pattern and cytoplasmic bridges between cells. Dev Biol 108:78–85

Kimmel CB, Law RD (1985b) Cell lineage of zebrafish blastomeres. II. Formation of the yolk syncytial layer. Dev Biol 108:86–93

Kimmel CB, Law RD (1985c) Cell lineage of zebrafish blastomeres: III. Clonal analyses of the blastula and gastrula stages. Dev Biol 108:94–101

Kimmel CB, Warga RM (1987a) Indeterminate cell lineage of the zebrafish embryo. Dev Biol 124:269–280

Kimmel CB, Warga RM (1987b) Cell lineages generating axial muscle in the zebrafish embryo. Nature 327:234–237

Kimmel CB, Westerfield M (1990) Primary neurons of the zebrafish. In: Edelman GM, Gall WE, Cowan WM (eds) Signals and sense: local and global order in perceptual maps. Wiley, New York, pp 561–588

Kimmel CB, Powell SL, Metcalfe WK (1982) Brain neurons which project to the spinal cord in young larvae of the zebrafish. J Comp Neurol 205:112–127

Kimmel CB, Metcalfe WK, Schabtach E (1985) T reticular interneurons: a class of serially repeating cells in the zebrafish hindbrain. J Comp Neurol 233:365–376

Kimmel CB, Sepich DS, Trevarrow B (1988) Development of segmentation in zebrafish. Development [Suppl] 104:197–207

Kimmel CB, Kane DA, Walker C, Warga RM, Rothman MB (1989) A mutation that changes cell movement and cell fate in the zebrafish embryo. Nature 337:358–362

Kimmel CB, Warga RM, Schilling TF (1990) Origin and organization of the zebrafish fate map. Development 108:581–594

Kimmel CB, Kane DA, Ho RK (1991) Lineage specification during early embryonic development of the zebrafish. In: Gerhard J (ed) Cell-cell interactions in early development. Wiley-Liss, New York, pp 203–225

Kimmel CB, Warga RM, Kane DA (1994) Cell cycles and clonal strings during formation of the zebrafish central nervous system. Development 120:265–276

Kimmel CB, Ballard WW, Kimmel SR, Ullmann B, Schilling TF (1995) Stages of embryonic development of the zebrafish. Dev Dyn 203:253–310

Kimmel CB, Miller CT, Kruze G, Ullmann B, BreMiller RA, Larison KD, Snyder HC (1998) The shaping of pharyngeal cartilages during early development of the zebrafish. Dev Biol 203: 245–263

Kimmel CB, Miller CT, Moens CB (2001) Specification and morphogenesis of the zebrafish larval head skeleton. Dev Biol 233:239–257

Kishida S, Yamamoto H, Hino S, Ikeda S, Kishida M, Kikuchi A (1999) Dix domains of dvl and axin are necessary for protein interactions and their ability to regulate beta-catenin stability. Mol Cell Biol 19:4414–4422

Kishimoto Y, Lee KH, Zon L, Hammerschmidt M, Schulte-Merker S (1997) The molecular nature of zebrafish swirl: BMP2 function is essential during early dorsoventral patterning. Development 124:4457–4466

Klar A, Baldassare M, Jessell TM (1992) F-spondin: a gene expressed at high levels in the floor plate encodes a secreted protein that promotes neural cell adhesion and neurite extension. Cell 69:95–110

Klein PS, Melton DA (1996) A molecular mechanism for the effect of lithium on development. Proc Natl Acad Sci USA 93:8455–8459

Klein R (2001) Excitatory Eph receptors and adhesive ephrin ligands. Curr Opin Cell Biol 13:196–203

Kljavin IJ (1987) Early development of photoreceptors in the ventral retina of the zebrafish embryo. J Comp Neurol 260:461–471

Knaut H, Pelegri F, Bohmann K, Schwarz H, Nüsslein-Volhard C (2000) Zebrafish *vasa* RNA but not its protein is a component of the germ plasm and segregates asymmetrically before germline specification. J Cell Biol 149:875–888

Knecht AK, Harland RM (1997) Mechanisms of dorsal-ventral patterning in noggin-induced neural tissue. Development 124:2477–2488

Kobayashi S, Kitamura T, Sasaki H, Okada M (1993) Two types of pole cells are present in the *Drosophila* embryo, one with and one without splicing activity for the third P-element intron. Development 117:885–893

Kobayashi T, Kajiura-Kobayashi H, Nagahama Y (2000) Differential expression of vasa homologue gene in the germ cells during oogenesis and spermatogenesis in a teleost fish, tilapia, *Oreochromis niloticus*. Mech Dev 99:139–142

Kodjabachian L, Dawid IB, Toyama R (1999) Gastrulation in zebrafish: what mutants teach us. Dev Biol 213:231–245

Kofron M, Demel T, Xanthos J, Lohr J, Sun B, Sive H, Osada S, Wright C, Wylie C, Heasman J (1999) Mesoderm induction in *Xenopus* is a zygotic event regulated by maternal VegT via TGFbeta growth factors. Development 126:5759–5770

Kojima D, Mano H, Fukada Y (2000) Vertebrate ancient-long opsin: a green-sensitive photoreceptive molecule present in zebrafish deep brain and retinal horizontal cells. J Neurosci 20:2845–2851

Kolodkin AL, Matthes DJ, Goodman CS (1993) The semaphorin genes encode a family of transmembrane and secreted growth cone guidance molecules. Cell 75:1389–1399

Kondoh H, Uchikawa M, Yoda H, Takeda H, Furutani-Seiki M, Karlstrom RO (2000) Zebrafish mutations in Gli-mediated hedgehog signaling lead to lens transdifferentiation from the adenohypophysis. Mech Dev 96:165–174

Koos DS, Ho RK (1998) The *nieuwkoid* gene characterizes and mediates a Nieuwkoop-center-like activity in the zebrafish. Curr Biol 8:1199–1206

Koos DS, Ho RK (1999) The nieuwkoid/dharma homeobox gene is essential for *bmp2b* repression in the zebrafish pregastrula. Dev Biol 215:190–207

Köprunner M, Thisse C, Thisse B Raz E (2001) A zebrafish nanos-related gene is essential for the development of primordial germ cells. Genes Dev 15:2877–2885

Korzh V, Edlund T, Thor S (1993) Zebrafish primary neurons initiate expression of the LIM homeodomain protein Isl-1at the end of gastrulation. Development 118:417–425

Korzh V, Sleptsova I, Liao J. He J, Gong Z (1998) Expression of zebrafish bHLH genes *ngn1* and *nrd* defines distinct stages of neural differentiation. Dev Dyn 213:92–104

Koshida S, Shinya M, Mizuno T, Kuroiwa A, Takeda H (1998) Initial anteroposterior pattern of the zebrafish central nervous system is determined by differential competence of the epiblast. Development 125:1957–1966

Köster RW, Fraser SE (2001a) Tracing transgene expression in living zebrafish embryos. Dev Biol 233:329–346

Köster R, Fraser SE (2001b) Direct imaging of in vivo neuronal migration in the developing cerebellum. Curr Biol 11:1858–1863

Kostomarova AA (1969) The differentiation capacity of isolated loach (Misgurnus fossilis L.) blastoderm. J Embryol Exp Morphol 22:407–430

Kozlowski DJ, Murakami T, Ho RK, Weinberg ES (1997) Regional cell movement and tissue patterning in the zebrafish embryo revealed by fate mapping with caged fluorescein. Biochem Cell Biol 75:551–562

Krauss SG, Concordet J-P, Ingham PW (1993) A functionally conserved homolog of the Drosophila segment polarity gene *hh* is expressed in tissues with polarizing activity in zebrafish embryos. Cell 75:1431–1444

Krauss SG, Johansen T, Korzh V, Fjose A (1991) Expression of the zebrafish paired box gene pax[zf-b] during early neurogenesis. Development 113:1193–1206

Krauss SG, Korzh V, Fjose A, Johansen T (1992) Expression of four wnt-related genes during embryogenesis. Development 116:249–259

Kramer SG, Kidd T, Simpson JH, Goodman CS (2001) Switching repulsion to attraction: changing responses to slit during transition in mesoderm migration. Science 292:737–740

Kudoh T, Dawid I (2001) Zebrafish *mab2112* is specifically expressed in the presumptive eye and tectum from early somitogenesis onwards. Mech Dev 109:95–98

Kudoh T, Tsang M, Hukriede NA, Chen X, Dedekian M, Clarke CJ, Kiang A, Schultz S, Epstein JA, Toyama R, Dawid IB (2001) A gene expression screen in zebrafish embryogenesis. Genome Res 11:1979–1987

Kuhl M, Sheldahl LC, Park M, Miller JR, Moon RT (2000) The Wnt/Ca2+ pathway: a new vertebrate Wnt signaling pathway takes shape. Trends Genet 16:279–283

Kuhl M, Geis K, Sheldahl LC, Pukrop T, Moon RT, Wedlich D (2001) Antagonistic regulation of convergent extension movements in Xenopus by Wnt/beta-catenin and Wnt/Ca(2+) signaling. Mech Dev 106:61–76

Kullander K, Croll SD, Zimmer M, Pan L, McClain J, Hughes V, Zabski S, DeChiara TM, Klein R, Yancopoulos GD, Gale NW (2001) Ephrin-B3 is the midline barrier that prevents corticospinal tract axons from recrossing, allowing for unilateral motor control. Genes Dev 15:877–888

Kupperman E, An S, Osborne N, Waldron S, Stainier DY (2000) A sphingosine-1-phosphate receptor regulates cell migration during vertebrate heart development. Nature 406:192–195

Kurihara Y, Kurihara H, Suzuki H, Kodama T, Maemura K, Nagai R, Oda H, Kuwaki T, Cao WH, Kamada N et al. (1994) Elevated blood pressure and craniofacial abnormalities in mice deficient in endothelin-1. Nature 368:703–710

Kuwada JY (1986) Cell recognition by neuronal growth cones in a simple vertebrate embryo. Science 233:740–746

Kuwada JY, Bernhardt RR, Chitnis AB (1990a) Pathfinding by identified growth cones in the spinal cord of zebrafish embryos. J Neurosci 10:1299–1308

Kuwada JY, Bernhardt RR, Nguyen N (1990b) Development of spinal neurons and tracts in the zebrafish embryo. J Comp Neurol 302:617–628

LaBonne C, Bronner-Fraser M (1998) Neural crest induction in *Xenopus*: evidence for a two-signal model. Development 125:2403–2414

Laessing U, Stuermer CA (1996) Spatiotemporal pattern of retinal ganglion cell differentiation revealed by the expression of neurolin in embryonic zebrafish. J Neurobiol 29:65–74

Laessing U, Giordano S, Stecher B, Lottspeich F, Stuermer CAO (1994) Molecular characterization of fish neurolin: a growth associated cell surface protein and member of the immunoglobin superfamily in the fish retinotectal system with similarities to chick protein DM-GRASP/SC-1/BEN. Differentiation 56:21–29

Lane MC, Keller R (1997) Microtubule disruption reveals that Spemann's organizer is subdivided into two domains by the vegetal alignment zone. Development 124:895–906

Lane MC, Smith WC (1999) The origins of primitive blood in *Xenopus*: implications for axial patterning. Development 126:423–34

Larabell CA, Rowning BA, Wells J, Wu M, Gerhart J (1996) Confocal microscopy of living *Xenopus* eggs and the mechanism of cortical rotation. Development 122:1281–1289

Larabell CA, Torres M, Rowning BA, Yost C, Miller JR, Wu M, Kimelman D, Moon RT (1997) Establishment of the dorso-ventral axis in *Xenopus* embryos is presaged by early asymmetries in beta-catenin that are modulated by the Wnt signaling pathway. J Cell Biol 136:1123–1136

Larison K, Bremiller R (1990) Early onset of phenotype and cell patterning in the embryonic zebrafish retina. Development 109:567–576

Larrain J, Bachiller D, Lu B, Agius E, Piccolo S, De Robertis EM (2000) BMP-binding modules in chordin: a model for signalling regulation in the extracellular space. Development 127:821–830

Larrivee B, Karsan A (2000) Signaling pathways induced by vascular endothelial growth factor. Int J Mol Med 5:447–456

Lasko P, Ashburner M (1988) The product of the *Drosophila* gene *vasa* is very similar to eukaryotic initiation factor-4A. Nature 335:611–617

Lauderdale JD, Davis NM, Kuwada JY (1997) Axon tracts correlate with netrin-1a expression in the zebrafish embryo. Mol Cell Neurosci 9:293–313

Lauderdale JD, Pasquali SK, Fazel R, van Eeden FJM, Schauerte HE, Hafter P, Kuwada JY (1998) Regulation of netrin-1a expression by hedgehog proteins. Mol Cell Neurosci 11:194–205

Laurent MN, Blitz IL, Hashimoto C, Rothbacher U, Cho KW (1997) The *Xenopus* homeobox gene twin mediates Wnt induction of goosecoid in establishment of Spemann's organizer. Development 124:4905–4916

LaVail MM, Rapaport DH, Rakic P (1991) Cytogenesis in the monkey retina. J Comp Neurol 309:86–114

Lawson KA (1999) Fate mapping the mouse embryo. Int J Dev Biol 43:773–775

Lawson ND, Scheer N, Pham VN, Kim CH, Chitnis AB, Campos-Ortega JA, Weinstein BM (2001) Notch signaling is required for arterial-venous differentiation during embryonic vascular development. Development 128:3675–3683

Le Douarin N, Dulac C, Dupin E, Cameron Curry P (1991) Glial cell lineages in the neural crest. Glia 4:175–184

Lee JE, Hollenberg SM, Snider L, Turner DL, Lipnick N, Weintraub H (1995) Conversion of Xenopus ectoderm into neurons by NeuroD, a basic helix-loop-helix protein. Science 268:836–844

Lee JS, Ray R, Chien CB (2001) Cloning and expression of three zebrafish roundabout homologs suggest roles in axon guidance and cell migration. Dev Dyn 221:216–230

Lee KH, Xu Q, Breitbart RE (1996) A new tinman-related gene, *nkx2.7*, anticipates the expression of *nkx2.5* and *nkx2.3* in zebrafish heart and pharyngeal endoderm. Dev Biol 180:722–731

Lee KJ, Jessell TM (1999) The specification of dorsal cell fates in the vertebrate central nervous system. Annu Rev Neurosci 22:261–294

Lee KJ, Mendelsohn M, Jessell TM (1998) Neuronal patterning by BMPs: a requirement for GDF7 in the generation of a discrete class of commissural interneurons in the mouse spinal cord. Genes Dev 12:3394–3407

Lee M, Goodall J, Verastegui C, Ballotti R, Goding CR (2000) Direct regulation of the microphthalmia promoter by Sox10 links Waardenburg-Shah syndrome (WS4)-associated hypopigmentation and deafness to WS2. J Biol Chem 275:37978–37983

Lee RK, Stainier DY, Weinstein BM, Fishman MC (1994) Cardiovascular development in the zebrafish. II. Endocardial progenitors are sequestered within the heart field. Development 120:3361–3366

Lehmann R, Nüsslein-Volhard C (1991) The maternal gene nanos has a central role in posterior pattern formation of the *Drosophila* embryo. Development 112:679–691

Lekven AC, Helde KA, Thorpe CJ, Rooke R, Moon RT (2000) Reverse genetics in zebrafish. Physiol Genom 2:37–48

Lekven AC, Thorpe CJ, Waxman JS, Moon RT (2001) Zebrafish wnt8 encodes two Wnt8 proteins on a bicistronic transcript and is required for mesoderm and neuroectoderm patterning. Dev Cell 1:130–114

Lele Z, Bakkers J, Hammerschmidt M (2001a) Morpholino phenocopies of the swirl, snailhouse, somitabun, minifin, silberblick, and pipetail mutations. Genesis 30:190–194

Lele Z, Nowak M, Hammerschmidt M (2001b) Zebrafish admp is required to restrict the size of the organizer and to promote posterior and ventral development. Dev Dyn 222:681–687

Lemaire P, Garrett N, Gurdon JB (1995) Expression cloning of Siamois, a *Xenopus* homeobox gene expressed in dorsal-vegetal cells of blastulae and able to induce a complete secondary axis. Cell 81:85–94

Lemaire P, Gurdon JB (1994) A role for cytoplasmic determinants in mesoderm patterning: cell-autonomous activation of the *goosecoid* and *Xwnt-8* genes along the dorsoventral axis of early *Xenopus* embryos. Development 120:1191–1199

Leppert CA, Diekmann H, Paul C, Laessing U, Marx M, Bastmeyer M, Stuermer CA (1999) Neurolin Ig domain 2 participates in retinal axon guidance and Ig domains 1 and 3 in fasciculation. J Cell Biol 144:339–349

Leve C, Gajewski M, Rohr KB, Tautz D (2001) Homologues of *c-hairy1* (*her9*) and lunatic fringe in zebrafish are expressed in the developing central nervous system, but not in the presomitic mesoderm. Dev Genes Evol 211:493–500

Levin M, Johnson RL, Stern CD, Kuehn M, Tabin C (1995) A molecular pathway determining left-right asymmetry in chick embryogenesis. Cell 82:803–814

Levin M, Mercola M (1998a) Gap junctions are involved in the early generation of left-right asymmetry. Dev Biol 203:90–105

Levin M, Mercola M (1998b) Evolutionary conservation of mechanisms upstream of asymmetric Nodal expression: reconciling chick and *Xenopus*. Dev Genet 23:185–193

Levin M, Mercola M (1999) Gap junction-mediated transfer of left-right patterning signals in the early chick blastoderm is upstream of Shh asymmetry in the node. Development 126:4703–4714

Lewis KE, Eisen JS (2001) Hedgehog signaling is required for primary motoneuron induction in zebrafish. Development 128:3485–3495

Lewis KE, Currie PD, Roy S, Schauerte H, Haffter P, Ingham PW (1999) Control of muscle cell type specification in the zebrafish embryo by hedgehog signalling. Dev Biol 216:469–480

Leyns L, Bouwmeester T, Kim SH, Piccolo S, DeRobertis E (1997) Frzb-1 is a secreted antagonist of Wnt signalling expressed in the Spemann organizer. Cell 88:747–756

Li L, Dowling JE (1997) A dominant form of inherited retinal degeneration caused by a non-photoreceptor cell-specific mutation. Proc Natl Acad Sci USA 94:11645–11650

Li W, Chen F, Nagarajan RP, Liu X, Chen Y (2001) Characterization of the DNA-binding property of Smad5. Biochem Biophys Res Commun 286:1163–1169

Li Z, Joseph NM, Easter SS Jr (2000a) The morphogenesis of the zebrafish eye, including a fate map of the optic vesicle. Dev Dyn 218:175–188

Li Z, Hu M, Ochocinska MJ, Joseph NM, Easter SS Jr (2000b) Modulation of cell proliferation during retinal morphogenesis of the zebrafish (*Danio rerio*). Dev Dyn 219:391–401

Liang D, Xu X, Chin AJ, Balasubramaniyan NV, Teo MA, Lam TJ, Weinberg ES, Ge R (1998) Cloning and characterization of vascular endothelial growth factor (VEGF) from zebrafish, *Danio rerio*. Biochim Biophys Acta 1397:14–20

Liang JO, Etheridge A, Hantsoo L, Rubinstein AL, Nowak SJ, Izpisua-Belmonte JC, Halpern M (2000) Asymmetric nodal signalling in the zebrafish diencephalon positions the pineal organ. Development 127:5101–5112

Liao EC, Paw BH, Oates AC, Pratt SJ, Postlethwait JH, Zon LI (1998) SCL/Tal-1 transcription factor acts downstream of cloche to specify hematopoietic and vascular progenitors in zebrafish. Genes Dev 12:621–626

Liao W, Bisgrove BW, Sawyer H, Hug B, Bell B, Peters K, Grunwald DJ, Stainier DY (1997) The zebrafish gene cloche acts upstream of a *flk-1* homologue to regulate endothelial cell differentiation. Development 124:381–389

Liao W, Ho CY, Yan YL, Postlethwait J, Stainier DY (2000) Hex and scl function in parallel to regulate early endothelial and blood differentiation in zebrafish. Development 127:4303–4313

Liem KF Jr, Tremml G, Roelink H, Jessell TM (1995) Dorsal differentiation of neural plate cells induced by BMP-mediated signals from epidermal ectoderm. Cell 82:969–979

Liem KF Jr, Tremml G, Jessell TM (1997) A role for the roof plate and its resident TGFbeta-related proteins in neuronal patterning in the dorsal spinal cord. Cell 91:127–138

Liem KF Jr, Jessell TM, Briscoe J (2000) Regulation of the neural patterning activity of Sonic hedgehog by secreted Bmp inhibitors expressed by notochord and somites. Development 127:4855–4866

Lin S, Long W, Chen J, Hopkins N (1992) Production of germ-line chimeras in zebrafish by cell transplants from genetically pigmented to albino embryos. Proc Natl Acad Sci USA 89:4519–4523

Lin X, Buff EM, Perrimon N, Michelson AM (1999) Heparan sulfate proteoglycans are essential for FGF receptor signaling during *Drosophila* embryonic development. Development 126: 3715–3723

Linask KK, Ludwig C, Han MD, Liu X, Radice GL, Knudsen KA (1998) N-cadherin/catenin-mediated morphoregulation of somite formation. Dev Biol 202:85–102

Lister JA, Robertson CP, Lepage T, Johnson SL, Raible DW (1999) *nacre* encodes a zebrafish microphthalmia-related protein that regulates neural-crest-derived pigment cell fate. Development 126:3757–3767

Lister JA, Close J, Raible DW (2001) Duplicate *mitf* genes in zebrafish: complementary expression and conservation of melanogenic potential. Dev Biol 237:333–344

Litingtung Y, Chiang C (2000) Specification of ventral neuron types is mediated by an antagonistic intreraction between Shh and Gli3. Nat Neurosci 3:979–985

Liu A, Losos K, Joyner AL (1999) Fgf8 can activate *gbx2* and transform regions of the rostral mouse brain into a hindbrain fate. Development 126:4827–4838

Liu A, Majumdar A, Schauerte HE, Haffter P, Drummond IA (2000) Zebrafish *wnt4b* expression in the floor plate is altered in *sonic hedgehog* and *gli-2* mutants. Mech Dev 91:409–413

Liu C, Liu W, Lu MF, Brown NA, Martin JF (2001) Regulation of left-right asymmetry by thresholds of *Pitx2c* activity. Development 128:2039–2048

Lohr JL, Danos MC, Yost HJ (1997) Left-right asymmetry of a nodal-related gene is regulated by dorsoanterior midline structures during *Xenopus* development. Development 124:1465–1472

Lohr JL, Danos MC, Groth TW, Yost HJ (1998) Maintenance of asymmetric nodal expression in Xenopus laevis. Dev Genet 23:194–202

Long WL (1983) The role of the yolk syncytial layer in determination of the plane of bilateral symmetry in the rainbow trout, *Salmmo gairdneri* Richardson. J Exp Zool 228:91–97

Long Q, Meng A, Wang H, Jessen JR, Farrell MJ, Lin S (1997) GATA-1 expression pattern can be recapitulated in living zebrafish using GFP reporter gene. Development 124:4105–4111

Loosli F, Winkler S, Wittbrodt J (1999) Six3 overexpression initiates the formation of ectopic retina. Genes Dev 13:649–654

Loosli F, Winkler S, Burgtof C, Wurmbach E, Ansorge W, Henrich T, Grabher C, Arendt D, Carl M, Krone A, Grzebisz E, Wittbrodt J (2001) *Rx3* links eye determination and growth. Development 128:4035–4044

Lorent K, Liu KS, Fetcho JR, Granato M (2001) The zebrafish *space cadet* gene controls axonal pathfinding of neurons that modulate fast turning movements. Development 128:2131–2142

Lowe LA, Supp DM, Sampath K, Yokoyama T, Wright CV, Potter SS, Overbeek P, Kuehn MR (1996) Conserved left-right asymmetry of nodal expression and alterations in murine situs inversus. Nature 381:158–161

Lowe LA, Yamada S, Kuehn MR (2001) Genetic dissection of nodal function in patterning the mouse embryo. Development 128:1831–1843

Lu Q, Sun EE, Klein RS, Flanagan JG (2001) Ephrin-B reverse signaling is mediated by a novel PDZ-RGS protein and selectively inhibits G protein-coupled chemoattraction. Cell 105:69–79

Lumsden A, Krumlauf R (1996) Patterning the vertebrate neuraxis. Science 274:1109–1115

Lun K, Brand M (1998) A series of *no isthmus* (*noi*) alleles of the zebrafish *pax2.1* gene reveals multiple signaling events in development of the midbrain-hindbrain boundary. Development 125:3049–3062

Luo G, Hofman C, Bronckers ALJJ, Sohocki M, Bradley A, Karsenty G (1995) BMP-7 is an inducer of nephrogenesis, and is also required for eye development and and skeletal patterning. Genes Dev 9:2808–2820

Luo R, An M, Arduini BL, Henion PD (2001) Specific pan-neural crest expression of zebrafish Crestin throughout embryonic development. Dev Dyn 220:169–74

Luo Y, Raible D, Raper JA (1993) Collapsin: a protein in brain that induces the collapse and paralysis of neuronal growth cones. Cell 75:217–227

Lustig KD, Kroll K, Sun E, Ramos R, Elmendorf H, Kirschner MW (1996) A *Xenopus* nodal-related gene that acts in synergy with noggin to induce complete secondary axis and notochord formation. Development 122:3275–3282

Lyons GE (1994) *In situ* analysis of the cardiac muscle gene program during embryogenesis. Trends Cardiovasc Med 3:184–190

Lyons MS, Bell B, Stainier D, Peters KG (1998) Isolation of the zebrafish homologues for the tie-1 and tie-2 endothelium-specific receptor tyrosine kinases. Dev Dyn 212:133–140

Ma C, Fan L, Ganassin R, Bols N, Collodi P (2001) Production of zebrafish germ-line chimeras from embryo cell cultures. Proc Natl Acad Sci USA 98:2461–2466

Ma Q, Kintner C, Anderson DJ (1996) Identification of neurogenin, a vertebrate neuronal determination gene. Cell 87:43–52

Macdonald R, Wilson SW (1997) Distribution of pax6 protein during eye development suggests discrete roles in proliferative and differentiated visual cells. Roux's Arch Dev Biol 206:363–369

Macdonald R, Xu Q, Barth KA, Mikkola I, Holder N, Fjose A, Krauss S, Wilson SW (1994) Regulatory gene expression boundaries demarcate sites of neuronal differentiation in the embryonic zebrafish forebrain. Neuron 13:1039–1053

Macdonald R, Barth KA, Xu Q, Holder N, Mikkola I, Wilson SW (1995) Midline signalling is required for Pax gene regulation and patterning of the eyes. Development 121:3267–3278

Macdonald R, Scholes J, Strähle U, Brennan C, Holder N, Brand M, Wilson SW (1997) The Pax protein Noi is required for commissural axon pathway formation in the rostral forebrain. Development 124:2397–2408

Maconochie M, Krishnamurthy R, Nonchev S, Meier P, Manzanares M, Mitchell PJ, Krumlauf R (1999) Regulation of Hoxa2 in cranial neural crest cells involves members of the AP-2 family. Development 126:1483–94

Manley NR, Capecchi MR (1998) Hox group 3 paralogs regulate the development and migration of the thymus, thyroid, and parathyroid glands. Dev Biol 195:1–15

Mao B, Wu W, Li Y, Hoppe D, Stannek P, Glinka A, Niehrs C (2001) LDL-receptor-related protein 6 is a receptor for Dickkopf proteins. Nature 411:321–325

Majumdar A, Drummond IA (1999) Podocyte differentiation in the absence of endothelial cells as revealed in the zebrafish avascular mutant, *cloche*. Dev Genet 24:220–229

Majumdar A, Drummond IA (2000) The zebrafish *floating head* mutant demonstrates podocytes play an important role in directing glomerular differentiation. Dev Biol 222:147–157

Majumdar A, Lun K, Brand M, Drummond IA (2000) Zebrafish *no isthmus* reveals a role for pax2.1 in tubule differentiation and patterning events in the pronephric primordia. Development 127:2089–2098

Malicki J (2000) Harnessing the power of forward genetics: analysis of neuronal diversity and patterning in the zebrafish retina. Trends Neurosci 23:531–541

Malicki J, Driever W (1999) *oko meduzy* mutations affect neuronal patterning in the zebrafish retina and reveal cell-cell interactions of the retinal neuroepithelial sheet. Development 126:1235–1246

Malicki J, Neuhauss SC, Schier AF, Solnica-Krezel L, Stemple DL, Stainier DY, Abdelilah S, Zwartkruis F, Rangini Z, Driever W (1996) Mutations affecting development of the zebrafish retina. Development 123:263–273

Mankoo BS, Collins NS, Ashby P, Grigorieva E, Pevny LH, Candia A, Wright CVE, Rigby PWJ, Pachnis V (1999) *Mox-2* is a component of the genetic hierarchy controlling limb muscle development. Nature 400:69–73

Mao J, Wang J, Liu B, Pan W, Farr GH, Flynn C, Yuan H, Takeda S, Kimelman D, Li L, Wu D (2001) Low-density-lipoprotein receptor-related protein-5 binds to axin and regulates the canonical Wnt signalling pathway. Mol Cell 7:801–809

Marchant L, Linker C, Ruiz P, Guerrero N, Mayor R (1998) The inductive properties of mesoderm suggest that the neural crest cells are specified by a BMP gradient. Dev Biol 198:319–329

Marcos-Gutierrez CV, Wilson SW, Holder N, Pachnis V (1997) The zebrafish homologue of the ret receptor and its pattern of expression during embryogenesis. Oncogene 14:879–889

Marcus RC, Delaney CL, Easter SS Jr (1999) Neurogenesis in the visual system of embryonic and adult zebrafish (*Danio rerio*). Vision Neurosci 16:417–424

Marin F, Puelles L (1994) Patterning of the embryonic avian midbrain after experimental inversions: a polarizing activity from the isthmus. Dev Biol 163:19–37

Marlow F, Zwartkruis F, Malicki J, Neuhauss SC, Abbas L, Weaver M, Driever W, Solnica-Krezel L (1998) Functional interactions of genes mediating convergent extension, *knypek* and *trilobite*, during the partitioning of the eye primordium in zebrafish. Dev Biol 203:382–399

Marlow F, Topczewski J, Sepich DS, Solnica-Krezel L (2002) Zebrafish Rho kinase 2 acts downstream of Wnt11 to mediate cell polarity and effective convergence and extension movements. Current Biology (in press)

Marsh-Armstrong N, McCaffery P, Gilbert W, Dowling JE, Drager UC (1994) Retinoic acid is necessary for development of the ventral retina in zebrafish. Proc Natl Acad Sci USA 91:7286–7290

Marshall EK, Smith HW (1930) The glomerular development of the vertebrate kidney in relation to habitat. Biol Bull 59:135–153

Martinez S, Alvarado-Mallart RM (1990) Expression of the homeobox chicken gene in chick/quail chimeras with inverted mes-metencephalic grafts. Dev Biol 139:432–436

Martinez S, Crossley P, Cobos I, Rubenstein JLR, Martin GR (1999) Fgf8 induces formation of an ectopic isthmic organizer and isthmocerebellar development via a repressive effect on *otx2* expression. Development 126:1189–1200

Martínez-Barberá JP, Toresso H, Da Rocha S, Krauss S (1997) Cloning and expression of three members of the zebrafish Bmp family: Bmp2a, Bmp2b and Bmp4. Gene 198:53–59

Masai I, Heisenberg C-P, Barth KA, Macdonald R, Adamek S, Wilson SW (1997) Floating head and masterblind regulate neurogenesis in the roof of the forebrain. Neuron 18:43–57

Masai I, Stemple DL, Okamoto H, Wilson SW (2000) Midline signals regulate retinal neurogenesis in zebrafish. Neuron 27:251–263

Massague J, Chen YG (2000) Controlling TGF-beta signaling. Genes Dev 14:627–644

Mathers PH, Grinberg A, Mahon KA, Jamrich M (1997) The *Rx* homeobox gene is essential for vertebrate eye development. Nature 387:603–607

Matsui Y, Zsebo K, Hogan B (1992) Derivation of pluripotential embryonic stem cells from murine primordial germ cells in culture. Cell 70:841–847

Matsuo I, Kuratani S, Kimura C, Takeda N, Aizawa S (1995) Mouse Otx2 functions in the formation and patterning of rostral head. Genes Dev 9:2646–2658

Matthieu J, Barth A, Rosa FM, Wilson SW, Peyreiras N (2002) Distinct and cooperative roles for Nodal & Medgehog signals during hypothalamic development. Development (in press)

Mauch TJ, Yang G, Wright M, Smith D, Schoenwolf GC (2000) Signals from trunk paraxial mesoderm induce pronephros formation in chick intermediate mesoderm. Dev Biol 220:62–75

McCallum CM, Comai L, Greene EA, Henikoff S (2000) Targeted screening for induced mutations. Nat Biotech 18:455–457

McClintock JM, Carlson R, Mann DM, Prince VE (2001) Consequences of *hox* gene duplication in the vertebrates: an investigation of the zebrafish hox paralogue group 1 genes. Development 128:2471–2484

McGrew MJ, Dale JK, Fraboulet S, Pourquie O (1998) The *lunatic fringe* gene is a target of the molecular clock linked to somite segmentation in avian embryos. Curr Biol 8:979–982

McLaughlin KA, Rones MS, Mercola M (2000) Notch regulates cell fate in the developing pronephros. Dev Biol 227:567–580

McMahon AP, Joyner AL, Bradley A, McMahon JA (1992) The midbrain-hindbrain phenotype of *wnt-1-/-* mice results from stepwise deletion of engrailed-expressing cells by 9.5 days post-coitum. Cell 69:581–595

McMahon JA, Takada S, Zimmerman LB, Fan C-M, Harland RM, McMahon AP (1998) Noggin-mediated antagonism of Bmp signaling is required for growth and patterning of the neural tube and somite. Genes Dev 12:1438–1452

Meier S (1980) Development of the chick embryo mesoblast: pronephros, lateral plate, and early vasculature. J Embryol Exp Morphol 55:291–306

Meinhardt H (1986) Models of segmentation. In: Bellairs R, Edde DA, Lash JW (eds) Somites in developing embryos. Plenum Press, New York, pp 179–191

Melançon E, Liu DW, Westerfield M, Eisen JS (1997) Pathfinding by identified zebrafish motoneurons in the absence of muscle pioneers. J Neurosci 17:7796–7804

Melby AE, Warga RM, Kimmel CB (1996) Specification of cell fates at the dorsal margin of the zebrafish gastrula. Development 122:2225–2237

Melby AE, Clements W, Kimelman D (1999) Regulation of dorsal gene expression in *Xenopus* by the ventralizing homeobox gene *Vox*. Dev Biol 211:293–305

Melby AE, Beach C, Mullins M, Kimelman D (2000) Patterning the early zebrafish by the opposing actions of *bozozok* and *vox/vent*. Dev Biol 224:275–285

Mellitzer G, Xu Q, Wilkinson DG (1999) Eph receptors and ephrins restrict cell intermingling and communication. Nature 400:77–81

Mendelsohn C, Batourina E, Fung S, Gilbert T, Dodd J (1999) Stromal cells mediate retinoid-dependent functions essential for renal development. Development 126:1139–1148

Menkes B, Sandor S (1977) Somitogenesis, regulation potencies, sequence determination and primordial interactions. In: Ede DA, Hinchcliffe DA, Balls M (eds) Vertebrate limb and somite morphogenesis. Br Soc Dev Biol Symp 3. Cambridge University Press, Cambridge, pp 405–419

Mennerich D, Schäfer K, Braun T (1998) Pax-3 is necessary but not sufficient for lbx1 expression in myogenic precursor cells of the limb. Mech Dev 73:147–158

Meno C, Shimono A, Saijoh Y, Yashiro K, Mochida K, Ohishi S, Noji S, Kondoh H, Hamada H (1998) lefty-1 is required for left-right determination as a regulator of lefty-2 and nodal. Cell 94: 287–297

Meno C, Gritsman K, Ohishi S, Ohfuji Y, Heckscher E, Mochida K, Shimono A, Kondoh H, Talbot WS, Robertson EJ, Schier AF, Hamada H (1999) Mouse Lefty2 and zebrafish antivin are feedback inhibitors of nodal signaling during vertebrate gastrulation. Mol Cell 4:287–298

Meno C, Takeuchi J, Sakuma R, Koshiba-Takeuchi K, Ohishi S, Saijoh Y, Miyazaki J, ten Dejike P, Ogura T, Hamada H (2001) Diffusion of Nodal signaling activity in the absence of the feedback inhibitor Lefty2. Dev Cell 1:127–138

Mercola M, Levin M (2001) Left-right asymmetry determination in vertebrates. Annu Rev Cell Dev Biol 17:779–805

Metcalfe WK, Mendelson B, Kimmel CB (1986) Segmental homologies among reticulospinal neurons in the hindbrain of the zebrafish larva. J Comp Neurol 251:147–159

Meyers EN, Lewandoski M, Martin GR (1998) An *fgf8* mutant allelic series generated by cre- and flp-mediated recombination. Nat Genet 18:136–141

Miklósi A, Andrew RJ (1999) Right eye use associated with decision to bite in zebrafish. Behav Brain Res 105:199–205

Miklósi A, Andrew RJ, Savage H (1998) Behavioural lateralisation of the tetrapod type in the zebrafish (*Brachydanio rerio*). Physiol Behav 63:127–135

Miller CT, Schilling TF, Lee K, Parker J, Kimmel CB (2000) *sucker* encodes a zebrafish Endothelin-1 required for ventral pharyngeal arch development. Development 127:3815–3828

Miller JR, Rowning BA, Larabell CA, Yang-Snyder JA, Bates RL, Moon RT (1999a) Establishment of the dorsal-ventral axis in *Xenopus* embryos coincides with the dorsal enrichment of *dishevelled* that is dependent on cortical rotation. J Cell Biol 146:427–437

Miller JR, Hocking AM, Brown JD, Moon RT (1999b) Mechanism and function of signal transduction by the Wnt/beta-catenin and Wnt/Ca^{2+} pathways. Oncogene 18:7860–7872

Miller-Bertoglio VE, Fisher S, Sanchez A, Mullins MC, Halpern ME (1997) Differential regulation of *chordin* expression domains in mutant zebrafish. Dev Biol 192:537–550

Miller-Bertoglio V, Carmany-Rampey A, Fürthauer M, Gonzalez EM, Thisse C, Thisse B, Halpern ME, Solnica-Krezel L (1999) Maternal and zygotic activity of the zebrafish ogon locus antagonizes BMP signaling. Dev Biol 214:72–86

Millet S, Campbell K, Epstein DJ, Losos K, Harris E, Joyner AL (1999) A role for *gbx2* in repression of *otx2* and positioning the mid/hindbrain organizer. Nature 401:161–164

Miner JH, Sanes JR (1994) Collagen IV alpha 3, alpha 4, and alpha 5 chains in rodent basal laminae: sequence, distribution, association with laminins, and developmental switches. J Cell Biol 127:879–891

Minowada G, Jarvis LA, Chi CL, Neubuser A, Sun X, Hacohen N, Krasnow MA, Martin GR (1999) Vertebrate *sprouty* genes are induced by Fgf signaling and can cause chondrodysplasia when overexpressed. Development 126:4465–4475

Mintzer KA, Lee MA, Runke G, Trout J, Whitman M, Mullins MC (2001) *Lost-a-fin* encodes a type I BMP receptor, Alk8, acting maternally and zygotically in dorsoventral pattern formation. Development 128:859–869

Mishina Y, Suzuki A, Ueno N, Behringer RR (1995) Bmpr encodes a type I bone morphogenetic protein receptor that is essential for gastrulation during mouse embryogenesis. Genes Dev 9:3027–3037

Mishina Y, Crombie R, Bradley A, Behringer RR (1999) Multiple roles of activin-like kinase-2 during mouse embryogenesis. Dev Biol 213:314–326

Mitchell KJ, Doyle JL, Serafini T, Kennedy TE, Tessier-Lavigne M, Goodman CS, Dickson BJ (1996) Genetic analysis of netrin genes in *Drosophila*: netrins guide CNS commissural axons and peripheral motor axons. Neuron 17:203–215

Mizuno T, Yamaha E, Wakahara M, Kuroiwa A, Takeda H (1996) Mesoderm induction in zebrafish. Nature 383:131–132

Mizuno T, Yamaha E, Yamazaki F (1997) Localized axis determinant in the early cleavage embryo of the goldfish, *Carassius auratus*. Dev Genes Evol 206: 389–396

Mizuno T, Shinya M, Takeda H (1999a) Cell and tissue transplantation in zebrafish embryos. Methods Mol Biol 127:15–28

Mizuno T, Yamaha E, Kuroiwa A, Takeda H (1999b) Removal of vegetal yolk causes dorsal deficencies and impairs dorsal-inducing ability of the yolk cell in zebrafish. Mech Dev 81:51–63

Mizuno T, Kawasaki M, Nakahira M, Kagamiyama H, Kikuchi Y, Okamoto H, Mori K, Yoshihara Y (2001) Molecular diversity in zebrafish NCAM family: three members with different VASE usage and distinct localization. Mol Cell Neurosci 18:119–130

Mlodzik M (1999) Planar polarity in the Drosophila eye: a multifaceted view of signaling specificity and cross-talk. EMBO J 18:6873–6879

Mochizuki T, Saijoh Y, Tsuchiya K, Shirayoshi Y, Takai S, Taya C, Yonekawa H, Yamada K, Nihei H, Nakatsuji N, Overbeek PA, Hamada H, Yokoyama T (1998) Cloning of inv, a gene that controls left/right asymmetry and kidney development. Nature 395:177–181

Moens CB, Fritz A (1999) Techniques in neural development. Methods Cell Biol 59:253–272

Moens CB, Yan Y-L, Appel B, Force AG, Kimmel CB (1996) *Valentino*: A zebrafish gene required for normal hindbrain segmentation. Development 122:3981–3990

Moens CB, Cordes S, Giorgianni M, Barsh G, Kimmel C (1998) Equivalence in the genetic control of hindbrain segmentation in fish and mouse. Development 125:381–391

Molenaar M, van de Wetering M, Oosterwegel M, Peterson-Maduro J, Godsave S, Korinek V, Roose J, Destree O, Clevers H (1996) XTcf-3 transcription factor mediates beta-catenin-induced axis formation in Xenopus embryos. Cell 86:391–399

Montross WT, Ji H, McCrea PD (2000) A beta-catenin/engrailed chimera selectively suppresses Wnt signaling. J Cell Sci 113:1759–1770

Moon RT, Kimelman D (1998) From cortical rotation to organizer gene expression: toward a molecular explanation of axis specification in *Xenopus*. Bioessays 20:536–545

Moore LA, Broihier HT, Van Doren M, Lunsford LB, Lehmann R (1998) Identification of genes controlling germ cell migration and embryonic gonad formation in *Drosophila*. Development 125:667–678

Moore MW, Klein RD, Farinas I, Sauer H, Armanini M, Phillips H, Reichardt LF, Ryan AM, Carver-Moore K, Rosenthal A (1996) Renal and neuronal abnormalities in mice lacking GDNF. Nature 382:76–79

Morgan R, Hooiveld MH, Pannese M, Dati G, Broders F, Delarue M, Thiery JP, Boncinelli E, Durston AJ (1999) Calponin modulates the exclusion of Otx-expressing cells from convergence extension movements. Nat Cell Biol 1:404–408

Morin-Kensicki EM, Eisen JS (1997) Sclerotome development and peripheral nervous system segmentation in embryoic zebrafish. Development 124:159–167

Morrison SJ, Perez SE, Qiao Z, Verdi JM, Hicks C, Weinmaster G, Anderson DJ (2000) Transient Notch activation initiates an irreversible switch from neurogenesis to gliogenesis by neural crest stem cells. Cell 101:499–510

Motoike T, Loughna S, Perens E, Roman BL, Liao W, Chau TC, Richardson CD, Kawate T, Kuno J, Weinstein BM et al. (2000) Universal GFP reporter for the study of vascular development. Genesis 28:75–81

Mullen RJ, LaVail MM (1976) Inherited retinal dystrophy: primary defect in pigment epithelium determined with experimental rat chimeras. Science 192:799–801

Müller F, Blader P, Rastegar S, Fischer N, Knöchel W, Strähle U (1999) Characterization of zebrafish smad1, smad2 and smad5: the amino-terminus of Smad1 and Smad5 is required for specific function in the embryo. Mech Dev 88:73–88

Müller F, Albert S, Blader P, Fischer N, Hallonet M, Strähle U (2000) Direct action of the nodal-related signal cyclops in induction of sonic hedgehog in the ventral midline of the CNS. Development 127:3889–3897

Müller J (1829) Ueber die Wolffschen Korper bei Embryonen der Frösche und Kröten. Meckel's Arch F Anat Physiol Jahrg 1829:65–70

Müller M, Von Weizsäcker E, Campos-Ortega JA (1996a) Expression domains of a zebrafish homologue of the *Drosophila* pair-rule gene *hairy* correspond to primordia of alternating somites. Development 122:2071–2078

Müller M, Von Weizsäcker E, Campos-Ortega JA (1996b) Transcription of a zebrafish gene of the *hairy-enhancer of split* family delineates the midbrain anlage in the neural plate. Dev Genes Evol 206:153–160

Mullins M (1998) Holy Tolloido: Tolloid cleaves SOG/Chordin to free DPP/BMPs. Trends Genet 14:127–129

Mullins MC (1999) Embryonic axis formation in the zebrafish. Methods Cell Biol 59:159–178

Mullins MC, Hammerschmidt M, Kane DA, Odenthal J, Brand M, van Eeden FJ, Furutani-Seiki M, Granato M, Haffter P, Heisenberg CP, Jiang YJ, Kelsh RN, Nüsslein-Volhard C (1996) Genes establishing dorsoventral pattern formation in the zebrafish embryo: the ventral specifying genes. Development 123:81–93

Myers D, Sepich DS, Solnica-Krezel L (2002) BMP activity gradient regulates convergent extension movements in zebrafish gastrula. Dev Biol 243:81–98

Myers PZ (1985) Spinal motoneurons of the larval zebrafish. J Comp Neurol 236:555–561

Myers PZ, Eisen JS, Westerfield M (1986) Development and axonal outgrowth of identified motoneurons in the zebrafish. J Neurosci 6:2278–2289

Nag AC, Nursall JR (1972) Histogenesis of white and red muscle fibres of trunk muscles of a fish Salmo gairdneri. Cytobios 6:227–246

Nakagawa O, Nakagawa M, Richardson JA, Olson EN, Srivastava D (1999) HRT1, HRT2, and HRT3: a new subclass of bHLH transcription factors marking specific cardiac, somitic, and pharyngeal arch segments. Dev Biol 216:72–84

Nakagawa O, McFadden DG, Nakagawa M, Yanagisawa H, Hu T, Srivastava D, Olson EN (2000) Members of the HRT family of basic helix-loop-helix proteins act as transcriptional repressors downstream of Notch signaling. Proc Natl Acad Sci USA 97:13655–13660

Nakao T (1965) Excretory organs of Amphioxus (*Branchiostoma*) *belcheri*. J Ultrastruct Res 12:1–12

Nakayama K, Ishida N, Shirane M, Inomata A, Inoue T, Shishido N, Horii I, Loh DY (1996) Mice lacking p27(Kip1) display increased body size, multiple organ hyperplasia, retinal dysplasia, and pituitary tumors. Cell 85:707–720

Nasevicius A, Ekker SC (2000) Effective targeted gene "knockdown" in zebrafish. Nat Genet 26:216–220

Nasevicius A, Hyatt T, Kim H, Guttman J, Walsh E, Sumanas S, Wang Y, Ekker SC (1998) Evidence for a frizzled-mediated Wnt pathway required for zebrafish dorsal mesoderm formation. Development 125:4283–4292

Nawrocki L (1985) Development of the neural retina in the zebrafish, *Brachydanio rerio*. PhD thesis. Oregon, Eugene

Neave B, Holder N, Patient R (1997) A graded response to BMP-4 spatially coordinates patterning of the mesoderm and ectoderm in the zebrafish. Mech Dev 62:183–195

Nelsen OE (1953) Comparative embryology of the vertebrates. PhD Thesis, University of Pennsylvania

Neuhauss SC, Solnica-Krezel L, Schier AF, Zwartkruis F, Stemple DL, Malicki J, Abdelilah S, Stainier DY, Driever W (1996) Mutations affecting craniofacial development in zebrafish. Development 123:357–367

Neuhauss SC, Biehlmaier O, Seeliger MW, Das T, Kohler K, Harris WA, Baier H (1999) Genetic disorders of vision revealed by a behavioral screen of 400 essential loci in zebrafish. J Neurosci 19:8603–8615

Neul JL, Ferguson EL (1998) Spatially restricted activation of the SAX receptor by SCW modulates DPP/TKV signaling in *Drosophila* dorsal-ventral patterning. Cell 95:483–494

Neumann CJ, Nüsslein-Volhard C (2000) Patterning of the zebrafish retina by a wave of *sonic hedgehog* activity. Science 289:2137–2139

Newfeld SJ, Chartoff EH, Graff JM, Melton DA, Gelbart WM (1996) Mothers against dpp encodes a conserved cytoplasmatic protein required in DPP/TGF-b responsive cells. Development 122:2099–2108

Newstead JD, Ford P (1960) Studies on the development of the kidney of the Pacific salmon, *Oncorhynchus forbuscha* (Walbaum). 1. The development of the pronephros. Can J Zool 36:15–21

Neyt C, Jagla K, Thisse C, Thisse B, Haines L, Currie PD (2000) Evolutionary origins of vertebrate appendicular muscle. Nature 408:82–86

Nguyen M, Park S, Marqués G, Arora K (1998a) Interpretation of a BMP activity gradient in Drosophila embryos depends on synergistic signaling by two type I receptors, SAX and TKV. Cell 95:495–506

Nguyen VH, Schmid B, Trout J, Connors SA, Ekker M, Mullins MC (1998b) Ventral and lateral regions of the zebrafish gastrula, including the neural crest progenitors, are established by a bmp2b/swirl pathway of genes. Dev Biol 199:93–110

Nguyen VH, Trout J, Connors SA, Andermann P, Weinberg E, Mullins MC (2000) Dorsal and intermediate neuronal cell types of the spinal cord are established by a BMP signaling pathway. Development 127:1209–1220

Niederreither K, Subbarayan V, Dollé P, Chambon P (1999) Embryonic retinoic acid synthesis is essential for early mouse post-implantation development. Nat Genet 21:444–448

Niederreither K, Vermot J, Schuhbaur B, Chambon P, Dollé P (2000) Retinoic acid synthesis and hindbrain patterning in the mouse embryo. Development 127:75–85

Niehrs C (1999) Head in the WNT. Tends Genet 15: 314–319

Niehrs C, Keller R, Cho KW, De Robertis EM (1993) The homeobox gene goosecoid controls cell migration in *Xenopus* embryos. Cell 72:491–503

Nieuwkoop PD (1969a) The formation of mesoderm in urodelean amphibians. II. The origin of the dorso-ventral polarity of the mesoderm. Wilhelm Roux' Arch Entw Mech Org 163:298–315

Nieuwkoop PD (1969b) The formation of the mesoderm in urodelean amphibians I. Induction by the endoderm. Wilhem Roux' Arch 162:341–373

Nieuwkoop PD, Faber J (1994) Normal table of *Xenopus laevis* (Daudin). Garland, New York

Nieuwkoop PD, Sutasurya LA (1979) Primordial germ cells in the chordates, Embryogenesis and phylogenesis. Cambridge University Press, Cambridge, pp 95–96

Nieuwkoop PD, Ubbels GA (1972) The formation of mesoderm in urodelean amphibians. IV. Quantitative evidence for the purely "ectodermal" origin of the entire mesoderm and of the pharyngeal endoderm. Roux's Arch 169:185–199

Nikaido M, Tada M, Saij T, Ueno N (1997) Conservation of BMP signaling in zebrafish mesoderm patterning. Mech Dev 61:75–88

Nikaido M, Tada M, Takeda H, Kuroiwa A, Ueno N (1999) In vivo analysis using variants of zebrafish BMPR-IA: range of action and involvement of BMP in ectoderm patterning. Development 126:181–190

Noden DM (1983) The role of the neural crest in patterning of avian cranial skeletal, connective, and muscle tissues. Dev Biol 96:144–165

Nonaka S, Tanaka Y, Okada Y, Takeda S, Harada A, Kanai Y, Kido M, Hirokawa N (1998) Randomization of left-right asymmetry due to loss of nodal cilia generating leftward flow of extraembryonic fluid in mice lacking KIF3B motor protein. Cell 95:829–837

Nonchev S, Vesque C, Maconochie M, Seitanidou T, Ariza-McNaughton L, Frain M, Marshall H, Sham MH, Krumlauf R, Charnay P (1996) Segmental expression of Hoxa-2 in the hindbrain is directly regulated by Krox-20. Development 122:543–554

Normura M, Li E (1998) Smad2 role in mesoderm formation, left-right patterning and craniofacial development. Nature 393:786–790

Nornes S, Clarkson M, Mikkola I, Pedersen M, Bardsley A, Martinez JP, Kraus S, Johansen T (1998) Zebrafish contains two *Pax6* genes involved in eye development. Mech Dev 77:185–196

Norris W, Neyt C, Ingham PW, Currie PD (2000) Slow muscle induction by Hedgehog signalling in vitro. J Cell Sci 113:2695–2703

Nusse R, Varmus HE (1992) Wnt genes. Cell 69:1073–1087

Nutt SL, Dingwell KS, Holt CE, Amaya E (2001) *Xenopus* Sprouty2 inhibits FGF-mediated gastrulation movements but does not affect mesoderm induction and patterning. Genes Dev 15:1152–1166

Oakley RA, Lasky CJ, Erickson CA, Tosney KW (1994) Glycoconjugates mark a transient barrier to neural crest migration in the chicken embryo. Development 120:103–114

Oates AC, Lackmann M, Power MA, Brennan C, Down LM, Do C, Evans B, Holder N, Boyd AW (1999) An early developmental role for eph-ephrin interaction during vertebrate gastrulation. Mech Dev 83:77–94

Ober EA, Schulte-Merker S (1999) Signals from the yolk cell induce mesoderm, neuroectoderm, the trunk organizer, and the notochord in zebrafish. Dev Biol 215:167–181

Odenthal J, Nüsslein-Volhard C (1998) fork head domain genes in zebrafish. Dev Genes Evol 208:245–258

Odenthal J, Haffter P, Vogelsang E, Brand M, van Eeden FJM, Furutani-Seiki M, Granato M, Hammershmidt M, Heisenberg C-P, Jiang Y-J, Kane DA, Kelsh RN, Mullins MC, Warga RM, Allende ML, Weinberg ES, Nüsslein-Volhard C (1996a) Mutations affecting the formation of the notochord in the zebrafish, *Danio rerio*. Development 123:103–115

Odenthal J, Rossnagel K, Haffter P, Kelsh R, Vogelsang E, Brand M, van Eeden F, Furutani-Seiki M, Granato M, Hammerschmidt M, Heisenberg C, Jiang Y, Kane D, Mullins MC, Nüsslein-Volhard C (1996b) Mutations affecting xanthophore pigmentation in the zebrafish, *Danio rerio*. Development 123:391–398

Odenthal J, van Eeden FJ, Haffter P, Ingham PW, Nüsslein-Volhard C (2000) Two distinct cell populations in the floor plate of the zebrafish are induced by different pathways. Dev Biol 219:350–363

Oelgeschläger M, Larrain J, Geissert D, De Robertis EM (2000) The evolutionarily conserved BMP-binding protein twisted gastrulation promotes BMP signalling. Nature 405:757–763

Oh SP, Seki T, Goss KA, Imamura T, Yi Y, Donahoe PK, Li L, Miyazono K, ten Dijke P, Kim S, Li E (2000) Activin receptor-like kinase 1 modulates transforming growth factor-beta 1 signaling in the regulation of angiogenesis. Proc Natl Acad Sci USA 97:2626–2631

Okada Y, Nonaka S, Tanaka Y, Saijoh Y, Hamada H, Hirokawa, N (1999) Abnormal nodal flow precedes situs inversus in *iv* and *inv* mice. Mol Cell 4:459–468

O'Leary DDM, Wilkinson DG (1999) Eph receptors and ephrins in neural development. Curr Opin Neurobiol 9:65–73

O'Leary DDM, Bicknese JA, De Carlos DA, Heffner CD, Koester SE, Kutka LJ, Terashima T (1990) Target selection by cortical axons: alternative mechanisms to establish axonal connections in the developing brain. Cold Spring Harbor Symp Quant Biol LV:453–468

Oliver G, Loosli F, Koster R, Wittbrodt J, Gruss P (1996) Ectopic lens induction in fish in response to the murine homeobox gene *Six3*. Mech Dev 60:233–239

Olsen L, Aasland R, Fjose A (1997) A *vasa*-like gene in zebrafish identifies putative primordial germ cells. Mech Dev 66:95–105

O'Malley DM, Kao YH, Fetcho JR (1996) Imaging the functional organization of zebrafish hindbrain segments during escape behaviors. Neuron 17:1145–1155

Onichtchouk D, Glinka A, Niehrs C (1998) Requirement for *Xvent-1* and *Xvent-2* gene function in dorsoventral patterning of *Xenopus* mesoderm. Development 125:1447–1456

Opdecamp K, Nakayama A, Nguyen MT, Hodgkinson CA, Pavan WJ, Arnheiter H (1997) Melanocyte development in vivo and in neural crest cell cultures: crucial dependence on the Mitf basic helix-loop-helix-zipper transcription factor. Development 124:2377–2386

Oppenheimer JM (1936) Transplantation experiments on developing teleosts (*Fundulus* and *Perca*). J Exp Zool 72:409–437

Ordahl CP, Le Douarin NM (1992) Two myogenic lineages within the developing somite. Development 114:339–353

Ott H, Diekmann H, Struemer CAO, Bastmeyer M (2001) Function of Neurolin (DM-GRASP/SC-1) in guidance of motor axons during zebrafish development. Dev Biol 235:86–97

Otteson DC, D'Costa AR, Hitchcock PF (2001) Putative stem cells and the lineage of rod photoreceptors in the mature retina of the goldfish. Dev Biol 232:62–76

Pagan-Westphal SM, Tabin CJ (1998) The transfer of left-right positional information during chick embryogenesis. Cell 93:25–35

Palmeirim I, Henrique D, Ish-Horowicz D, Pourquié O (1997) Avian *hairy* gene expression identifies a molecular clock linked to vertebrate segmentation and somitogenesis. Cell 91:639–648

Papalopulu N, Clarke JD, Bradley L, Wilkinson DG, Krumlauf R, Holder N (1991) Retinoic acid causes abnormal development and segmental patterning of the anterior hindbrain in *Xenopus* embryos. Development 113:1145–1158

Papan C, Campos-Ortega JA (1994) On the formation of the neural keel and neural tube in the zebrafish *Danio* (*Brachydanio*) *rerio*. Dev Biol 203:178–186

Papan C, Campos-Ortega JA (1997) A clonal analysis of spinal cord development in the zebrafish. Dev Genes Evol 207:71–81

Papan C, Campos-Ortega JA (1999) Region-specific cell clones in the developing spinal cord of the zebrafish. Dev Genes Evol 209:135–144

Parichy DM, Rawls JF, Pratt SJ, Whitfield TT, Johnson SL (1999) Zebrafish *sparse* corresponds to an orthologue of c-kit and is required for the morphogenesis of a subpopulation of melanocytes, but is not essential for hematopoiesis or primordial germ cell development. Development 126:3425–3436

Parichy DM, Mellgren EM, Rawls JF, Lopes SS, Kelsh RN, Johnson SL (2000a) Mutational analysis of endothelin receptor b1 (rose) during neural crest and pigment pattern development in the zebrafish, *Danio rerio*. Dev Biol 227:294–306

Parichy DM, Ransom DG, Paw B, Zon LI, Johnson SL (2000b) An orthologue of the kit-related gene fms is required for development of neural crest-derived xanthophores and a subpopulation of adult melanocytes in the zebrafish, *Danio rerio*. Development 127:3031–3044

Park HC, Mehta A, Richardson JS, Appel B (2002) *olg2* is required for zebrafish primary motor neuron and oligodendrocyte development. (submitted)

Parker L, Stainier DY (1999) Cell-autonomous and non-autonomous requirements for the zebrafish gene *cloche* in hematopoiesis. Development 126:2643–2651

Parker LH, Zon LI, Stainier DY (1999) Vascular and blood gene expression. Methods Cell Biol 59:313–336

Pasini A, Henrique D, Wilkinson DG (2001) The zebrafish *hairy/enhancer-of-split*-related gene *her6* is segmentally expressed during the early development of hindbrain and somites. Mech Dev 100:317–321

Pasqualetti M, Ori M, Nardi I, Rijli FM (2000) Ectopic *Hoxa2* induction after neural crest migration results in homeosis of jaw elements in *Xenopus*. Development 127:5367–5378

Passini MA, Levine EM, Canger AK, Raymond PA, Schechter N (1997) *Vsx-1* and *Vsx-2*: differential expression of two paired-like homeobox genes during zebrafish and goldfish retinogenesis. J Comp Neurol 388:494–505

Pasteels J (1936) Etudes sur la gastrulation dés vertebres meroblasteques. I. Teleosteens. Arch Biol 47:205–308

Patten BM (1951) Early embryology of the chick. McGraw-Hill, New York

Patten I, Placzek M (2000) The role of sonic hedgehog in neural tube patterning. Cell Mol Life Sci 57:1695–1708

Patton EE, Zon LI (2001) The art and design of genetic screens: zebrafish. Nat Rev Genet 2:956–966

Payne TL, Postlethwait JH, Yelick PC (2001) Functional characterization and genetic mapping of *alk8*. Mech Dev 100:275–289

Peifer M, Polakis P (2000) Wnt signaling in oncogenesis and embryogenesis – a look outside the nucleus. Science 287:1606–1609

Pelegri F, Maischein HM (1998) Function of zebrafish beta-catenin and TCF-3 in dorsoventral patterning Mech Dev 77:63–74

Pelegri F, Schulte-Merker S (1999) A gynogenesis-based screen for maternal-effect genes in zebrafish, *Danio rerio*. Methods Cell Biol 60:1–20

Pelegri F, Knaut H, Maischein H, Schulte-Merker S, Nüsslein-Volhard C (1999) A mutation in the zebrafish maternal-effect gene *nebel* affects furrow formation and *vasa* RNA localization. Curr Biol 9:1431–1440

Pelster B, Burggren WW (1996) Disruption of hemoglobin oxygen transport does not impact oxygen-dependent physiological processes in developing embryos of zebra fish (*Danio rerio*). Circ Res 79:358–362

Perz-Edwards A, Hardison NL, Linney E (2001) Retinoic acid-mediated gene expression in transgenic reporter zebrafish. Dev Biol 229:89–101

Peters JM, McKay RM, McKay JP, Graff JM (1999) Casein kinase I transduces Wnt signals. Nature 401:345–350

Peterson RT, Link BA, Dowling JE, Schreiber SL (2000) Small molecule developmental screens reveal the logic and timing of vertebrate development. Proc Natl Acad Sci USA 97:12965–12969

Peterson RT, Mably JD, Chen J-N, Fishman MC (2001) Convergence of distinct pathways to heart patterning revealed by the small molecular concentramide and the mutation *heart-and-soul*. Curr Biol 11:1481–1491

Peyrieras N, Strähle U, Rosa F (1998) Conversion of zebrafish blastomeres to an endodermal fate by TGF-beta-related signaling. Curr Biol 8:783–786

Pfaff SL, Mendelsohn M, Stewart CL, Edlund T, Jessell TM (1996) Requirement for lim homeobox gene isl1 in motor neuron generation reveals a motor neuron-dependent step in interneuron differentiation. Cell 84:309–320

Pfeffer PL, Gerster T, Lun K, Brand M, Busslinger M (1998) Characterization of three novel members of the zebrafish Pax2/5/8 family: dependency of Pax5 and Pax8 expression on the Pax2.1 (noi) function. Development 125:3063–3074

Pham VN, Roman BL, Weinstein BM (2001) Isolation and expression analysis of three zebrafish angiopoietin genes. Dev Dyn 221:470–474

Piccolo S, Agius E, Lu B, Goodman S, Dale L, DeRobertis EM (1997) Cleavage of Chordin by Xolloid metalloprotease suggests a role for proteolytic processing in the regulation of Spemann organizer activity. Cell 91:407–416

Piccolo SY, Sasai Y, Lu B, De Robertis EM (1996) A possible molecular mechanism for Spemann organizer function: inhibition of ventral signals by direct binding of Chordin to BMP-4. Cell 85:589–598

Piccolo SY, Agius E, Leyns L, Bhattacharyya S, Grunz H, Bouwmeester T, De Robertis EM (1999) The head inducer Cerberus is a multifunctional antagonist of Nodal, BMP and Wnt signals. Nature 397:707–710

Pichel JG, Shen L, Sheng HZ, Granholm AC, Drago J, Grinberg A, Lee EJ, Huang SP, Saarma M, Hoffer BJ, Sariola H, Westphal H (1996) Defects in enteric innervation and kidney development in mice lacking GDNF. Nature 382:73–76

Pick L (1998) Segmentation: painting stripes from flies to vertebrates. Dev Genet 23:1–10

Picker A, Brennan C, Reifers F, Clarke JDW, Holder N, Brand M (1999) Requirement for the zebrafish mid-hindbrain boundary in midbrain polarisation, mapping and confinement of the retinotectal projection. Development 126:2967–2978

Pierce SB, Kimelman D (1995) Regulation of Spemann organizer formation by the intracellular kinase Xgsk-3. Development 121:755–765

Pike SH, Melancon EF, Eisen JS (1992) Pathfinding of zebrafish secondary motoneurons in the absence of normal pioneer axons. Development 114:825–831

Piotrowski T, Nüsslein-Volhard C (2000) The endoderm plays an important role in patterning the segmented pharyngeal region in zebrafish (*Danio rerio*). Dev Biol 225:339–356

Piotrowski T, Schilling T, Brand M, Jiang Y, Heisenberg C, Beuchle D, Grandel H, van Eeden F, Furutani-Seiki M, Granato M, Haffter P, Hammerschmidt M, Kane D, Kelsh R, Mullins M, Odenthal J, Warga R, Nüsslein-Volhard C (1996) Jaw and branchial arch mutants in zebrafish. II. Anterior arches and cartilage differentiation. Development 123:345-356

Placzek M, Skaer H (1999) Airway patterning: a paradigm for restricted signalling. Curr Biol 9:R506–R510

Placzek M, Tessier-Lavigne M, Yamada T, Jessell T, Dodd J (1990) Mesodermal control of neural cell identity: floor plate induction by the notochord. Science 250:985–988

Pogoda HM, Solnica-Krezel L, Driever W, Meyer D (2000) The zebrafish forkhead transcription factor FoxH1/Fast1 is a modulator of nodal signaling required for organizer formation. Curr Biol 10:1041–1049

Pöpperl H, Rikhof H, Chang H, Haffter P, Kimmel CB, Moens CB (2000) Lazarus is a novel pbx gene that globally mediates hox gene function in zebrafish. Mol Cell 6:255–267

Porcher C, Liao EC, Fujiwara Y, Zon LI, Orkin SH (1999) Specification of hematopoietic and vascular development by the bHLH transcription factor SCL without direct DNA binding. Development 126:4603–4615

Posner M, Kantorow M, Horwitz J (1999) Cloning, sequencing and differential expression of alphaB-crystallin in the zebrafish, *Danio rerio*. Biochim Biophys Acta 1447:271-277

Postlethwait J, Yan Y, Gates M, Horne S, Amores A, Brownlie A, Donovan A, Egan E, Force A, Gong Z, Goutel C, Fritz A, Kelsh R, Knapik E, Liao E, Paw B, Ransom D, Singer A, Thomson M, Abduljabbar T, Yelick P, Beier D, Joly J, Larhammar D, Talbot W et al (1998) Vertebrate genome evolution and the zebrafish gene map. Nat Genet 18:345–349

Potterf SB, Furumura M, Dunn KJ, Arnheiter H, Pavan WJ (2000) Transcription factor hierarchy in Waardenburg syndrome: regulation of MITF expression by SOX10 and PAX3. Hum Genet 107:1–6

Pourquie O (1999) Notch around the clock. Curr Opin Genet Dev 9:559–565

Pourquie O (2000) Segmentation of the paraxial mesoderm and vertebrate somitogenesis. Curr Top Dev Biol 47:81–105

Primmett DRN, Stern CD, Keynes RJ (1988) Heat shock causes repeated segmental anomalies in the chick embryo. Development 104:331–339

Primmett DRN, Norris WE, Carlson GJ, Keynes RJ, Stern CD (1989) Periodic segmental anomalies induced by heat shock in the chick embryo are associated with the cell cycle. Development 105:119–130

Prince VE, Joly L, Ekker M, Ho RK (1998a) Zebrafish *hox* genes: genomic organization and modified colinear expression patterns in the trunk. Development 125:407–420

Prince VE, Moens C, Kimmel C, Ho R (1998b) Zebrafish hox genes: expression in the hindbrain region of wild-type and mutants of the segmentation gene, valentino. Development 125:393–406

Prince VE, Holley SA, Bally-Cuif L, Prabhakaran B, Oates AC, Ho RK, Vogt TF (2001) Zebrafish lunatic fringe demarcates segmental boundaries. Mech Dev 105:175–180

Pujic Z, Malicki J (2001) Mutation of the zebrafish *glass onion* locus causes early cell-nonautonomous loss of neuroepithelial integrity followed by severe neuronal patterning defects. Dev Biol 234:454–469

Püschel AW, Westerfield M, Dressler GR (1992) Comparative analysis of Pax-2 protein distributions during neurulation in mice and zebrafish. Mech Dev 38:197–208

Raible DW, Eisen JS (1994) Restriction of neural crest cell fate in the trunk of the embryonic zebrafish. Development 120:495–503

Raible DW, Eisen JS (1996) Regulative interactions in zebrafish neural crest. Development 122:501–507

Raible DW, Wood A, Hodsdon W, Henion P, Weston JA, Eisen JS (1992) Segregation and early dispersal of neural crest cells in the embryonic zebrafish. Dev Dyn 195:29–42

Raible F, Brand M (2001) Tight transcriptional control of the *ets* domain factors erm and *pea3* by fgf signaling during early zebrafish development. Mech Dev 107:105–117

Rajagopalan S, Nicolas E, Vivancos V, Berger J, Dickson BJ (2000) Crossing the midline: roles and regulation of Robo receptors. Neuron 28:637–640

Ramsdell AF, Yost HJ (1998) Molecular mechanisms of vertebrate left-right development. Trends Genet 14:459–465

Ransom DG, Haffter P, Odenthal J, Brownlie A, Vogelsang E, Kelsh RN, Brand M, van Eeden FJ, Furutani-Seiki M, Granato M, Hammerschmidt M, Heisenberg CP, Jiang YJ, Kane DA, Mullins MC, Nüsslein-Volhard C (1996) Characterization of zebrafish mutants with defects in embryonic hematopoiesis. Development 123:311–319

Raper JA (2000) Semaphorins and their receptors in vertebrates and invertebrates. Curr Opin Neurobiol 10:88–94

Raper JA, Kapfhammer J (1990) The enrichment of a neuronal growth cone collapsing activity from embryonic chick brain. Neuron 4:21–29

Ratcliffe MJ, Itoh K, Sokol SY (2000) A positive role for the PP2A catalytic subunit in Wnt signal transduction. J Biol Chem 275:35680–35683

Rawls A, Wilson-Rawls J, Olson EN (2000) Genetic regulation of somite formation. Curr Top Dev Biol 47:131–154

Raymond PA, Rivlin PK (1987) Germinal cells in the goldfish retina that produce rod photoreceptors. Dev Biol 122:120–138

Raymond P, Barthel L, Rounsifer M, Sullivan S, Knight J (1993) Expression of rod and cone visual pigments in goldfish and zebrafish: a rhodopsin-like gene is expressed in cones. Neuron 10:1161–1174

Raymond P, Barthel L, Curran G (1995) Developmental patterning of rod and cone photoreceptors in embryonic zebrafish. J Comp Neurol 359:537–550

Raz E (2000) The function and regulation of *vasa*-like genes in germ-cell development, Genome Biol 1:1017.1–1017.6

Rebagliati MR, Toyama R, Fricke C, Haffter P, Dawid IB (1998a) Zebrafish nodal-related genes are implicated in axial patterning and establishing left-right asymmetry. Dev Biol 199:261–272

Rebagliati MR, Toyama R, Haffter P, Dawid IB (1998b) *cyclops* encodes a nodal-related factor involved in midline signaling. Proc Natl Acad Sci USA 95:9932–9937

Reifers F, Böhli H, Walsh EC, Crossley PH, Stainier DY, Brand M (1998) Fgf8 is mutated in zebrafish *acerebellar* (*ace*) mutants and is required for maintenance of midbrain-hindbrain boundary development and somitogenesis. Development 125:2381–2395

Reifers F, Adams J, Mason I, Schulte-Merker S, Brand M (2000a) Overlapping and distinct functions provided by *fgf17*, a new zebrafish member of the *fgf8/17/18* subgroup of fgfs. Mech Dev 99:39–49

Reifers F, Walsh EC, Leger S, Stainier DYR, Brand M (2000b) Induction and differentiation of the zebrafish heart requires fibroblast growth factor 8 (*fgf8/acerebellar*). Development 127: 225–235

Reim G, Brand M (2002) *spiel-ohne-grenzen/pou2* mediates regional competence to respond to Fgf8 during zebrafish early neural development. Development 129:917–933

Reiter JF, Alexander J, Rodaway A, Yelon D, Patient R, Holder N, Stainier DYR (1999) Gata5 is required for the development of the heart and endoderm in zebrafish. Genes Dev 13: 2983–2995

Reiter JF, Kikuchi Y, Stainier DYR (2001a) Multiple roles for Gata5 in zebrafish endoderm formation. Development 128:125–135

Reiter JF, Verkade H, Stainier DY (2001b) Bmp2b and Oep promote early myocardial differentiation through their tegulation of *gata5*. Dev Biol 234:330–338

Rhinn M, Brand M (2001) The midbrain-hindbrain boundary organizer. Curr Opin Neurobiol 11:34–42

Ribatti D, Vacca A, Roncali L, Dammacco F (2000) Hematopoiesis and angiogenesis: a link between two apparently independent processes. J Hematother Stem Cell Res 9:13–19

Rijli FM, Mark M, Lakkaraju S, Dierich A, Dolle P, Chambon P (1993) A homeotic transformation is generated in the rostral branchial region of the head by disruption of *Hoxa-2*, which acts as a selector gene. Cell 75:1333–1349

Roberts A, Dale N, Ottersen OP, Storm-Mathisen J (1988) Development and characterization of commissural interneurones in the spinal cord of *Xenopus laevis* embryos revealed by antibodies to glycine. Development 103:447–461

Robinson V, Smith A, Flenniken AM, Wilkinson DG (1997) Roles of Eph receptors and ephrins in neural crest pathfinding. Cell Tissue Res 290:265–274

Rodaway A, Takeda H, Koshida S, Broadbent J, Price B, Smith JC, Patient R, Holder N (1999) Induction of the mesendoderm in the zebrafish germ ring by yolk cell-derived TGF-beta family signals and discrimination of mesoderm and endoderm by FGF. Development 126:3067–3078

Rodieck RW (1973) The vertebrate retina. Principles of structure and function. Freeman, San Francisco

Roehl H, Nüsslein-Volhard C (2001) Zebrafish *pea3* and *erm* are general targets of FGF8 signaling. Curr Biol 11:503–507

Roelink H, Porter JA, Chiang C, Tanabe Y, Chang DT, Beachy PA, Jessell TM (1995) Floor plate and motor neuron induction by different concentrations of the amino-terminal cleavage product of sonic hedgehog autoproteolysis. Cell 81:445–455

Roesseler E, Muenke M (2001) Midline and laterality defects: left and right meet in the middle. Bioessays 23:888–900

Rohr KB, Concha ML (2000) Expression of nk2.1a during early development of the thyroid gland in zebrafish. Mech Dev 95:267–270

Rohr KB, Barth KA, Varga Z, Wilson SW (2001) The nodal pathway acts upstream of hedgehog signalling to specify ventral telencephalic identity. Neuron 29:341–351

Romanoff AL (1960) The urogenital system. In: The avian embryo. Macmillan, New York

Rongo C, Lehmann R (1996) Regulated synthesis, transport and assembly of the *Drosophila* germ plasm. Trends Genet 12:102–109

Roos M, Schachner M, Bernhardt RR (1999) Zebrafish semaphorin Z1b inhibits growing motor axons in vivo. Mech Dev 87:103–117

Roose J, Molenaar M, Peterson J, Hurenkamp J, Brantjes H, Moerer P, van de Wetering M, Destree O, Clevers H (1998) The *Xenopus* Wnt effector XTcf-3 interacts with Groucho-related transcriptional repressors. Nature 395:608–612

Ross JJ, Shimmi O, Vilmos P, Petryk A, Kim H, Gaudenz K, Hermanson S, Ekker SC, O'Connor MB, Marsh L (2001) Twisted gastrulation is a conserved extracellular BMP antagonist. Nature 410:479–483

Rothberg JM, Hartley DA, Walther Z, Artavanis-Tsakonas S (1988) slit: an EGF-homologous locus of D. melanogaster involved in the development of the embryonic central nervous system. Cell 55:1047–1059

Rottbauer W, Baker K, Wo ZG, Mohideen MPK, Cantiello HF, Fishman MC (2001) Growth and function of the embryonic heart depend upon the cardiac-specific L-type calcium channel $\alpha 1$ subunit. Dev Cell 1:265–275

Rowning BA, Wells J, Wu M, Gerhart JC, Moon RT, Larabell CA (1997) Microtubule-mediated transport of organelles and localization of beta-catenin to the future dorsal side of *Xenopus* eggs. Proc Natl Acad Sci USA 94:1224–1229

Roy MN, Prince VE, Ho RK (1999) Heat shock produces periodic somitic disturbances in the zebrafish embryo. Mech Dev 85:27–34

Roy S, Wolff C, Ingham PW (2001) The u-boot mutation identifies a Hedgehog-regulated myogenic switch for fiber-type diversification in the zebrafish embryo. Genes Dev 15:1563–1576

Rubenstein JLR, Shimamura K, Martinez S, Puelles L (1998) Regionalization of the prosencephalic neural plate. Annu Rev Neurosci 21:445–477

Ruberte E, Marty T, Nellen D, Affolter M, Basler K (1995) An absolute requirement for both the type II and type I receptors, punt and thick veins, for dpp signaling in vivo. Cell 80:889–897

Rupp RA, Weintraub H (1991) Ubiquitous MyoD transcription at the midblastula transition precedes induction-dependent MyoD expression in presumptive mesoderm of *X. laevis*. Cell 65:927–937

Ruppert EE (1994) Evolutionary origin of the vertebrate nephron. Am Zool 34:542–553

Ruppert EE, Smith PR (1988) The functional organization of filtration nephridia. Biol Rev 63:231–258

Ruvinsky I, Silver LM, Ho RK (1998) Characterization of the zebrafish *tbx16* gene and evolution of the vertebrate T-box family. Dev Genes Evol 208:94–99

Ruvinsky I, Oates AC, Silver LM, Ho RK (2000) The evolution of paired appendages in vertebrates: T-box genes in the zebrafish. Dev Genes Evol 210:82–91

Ryu SL, Fujii R, Yamanaka Y, Shimizu T, Yabe T, Hirata T, Hibi M, Hirano T (2001) Regulation of dharma/bozozok by the Wnt pathway. Dev Biol 231:397–409

Saffman E, Lasko P (1999) Germline development in vertebrates and invertebrates. Cell Mol Life Sci 55:1141–1163

Saga Y, Hata N, Kobayashi S, Magnuson T, Seldin MF, Taketo MM (1996) MesP1: a novel basic helix loop helix protein expressed in the nascent mesodermal cells during mouse gastrulation. Development 122:2769–2778

Saga Y, Hata N, Koseika H, Taketo MM (1997) *MesP2*: a novel mouse gene expressed in the pre-segmented mesoderm and essential for segmentation initiation. Genes Dev 11:1827–1839

Sahly I, Andermann P, Petit C (1999) The zebrafish *eya1* gene and its expression pattern during embryogenesis. Dev Genes Evol 209:399–410

Saint-Amant L, Drapeau P (1998) Time course of the development of motor behaviors in the zebrafish embryo. J Neurobiol 37:622–632

Saint-Amant L, Drapeau P (2001) Synchronization of an embryonic network of identified spinal interneurons solely by electrical coupling. Neuron 31:1035–1046

Saint-Jeannet JP, He X, Varmus HE, Dawid IB (1997) Regulation of dorsal fate in the neuraxis by Wnt-1 and Wnt-3a. Proc Natl Acad Sci USA 94:13713–13718

Sakai M (1996) The vegetal determinants required for the Spemann organizer move equatorially during the first cell cycle. Development 122:2207–2214

Sakaguchi T, Kuroiwa A, Takeda H (2001) A novel *sox* gene, *226D7*, acts downstream of Nodal signaling to specify endoderm precursors in zebrafish. Mech Dev 107:25–38

Sakanaka C, Leong P, Xu L, Harrison SD, Williams LT (1999) Casein kinase iepsilon in the wnt pathway: regulation of beta-catenin function. Proc Natl Acad Sci USA 96:12548–12552

Salomon DS, Bianco C, De Santis M (1999) Cripto: a novel epidermal growth factor (EGF)-related peptide in mammary gland development and neoplasia. Bioessays 21:61–70

Sampath K, Cheng AM, Frisch A, Wright CVE (1997) Functional differences among *Xenopus* nodal-related genes in left-right axis determination. Development 124:3293–3302

Sampath K, Rubinstein AL, Cheng AM, Liang JO, Fekany K, Solnica-Krezel L, Korzh V, Halpern ME, Wright CV (1998) Induction of the zebrafish ventral brain and floorplate requires cyclops/nodal signalling. Nature 395:185–189

Sanchez MP, Silos-Santiago I, Frisen J, He B, Lira SA, Barbacid M (1996) Renal agenesis and the absence of enteric neurons in mice lacking GDNF. Nature 382:70–73

Sasai Y (2001) Regulation of neural determination by evolutionarily conserved signals: anti-BMP factors and what next? Curr Opin Neurobiol 11:22–26

Sasai Y, De Robertis EM (1997) Ectodermal patterning in vertebrate embryos. Dev Biol 182:5–20

Sasai Y, Lu B, Steinbeisser H, Geissert D, Gont LK, De Robertis EM (1994) *Xenopus* chordin: a novel dorsalizing factor activated by organizer-specific homeobox genes. Cell 79:779–790

Sasai Y, Lu B, Steinbeisser H, De Robertis EM (1995) Regulation of neural induction by the chd and Bmp-4 antagonistic patterning signals in *Xenopus*. Nature 376:333–336

Sasai Y, Lu B, Piccolo S, De Robertis EM (1996) Endoderm induction by the organizer-secreted factors chordin and noggin in *Xenopus* animal caps. EMBO 15:4547–4555

Satin J, Fujii S, DeHaan RL (1988) Development of cardiac beat rate in early chick embryos is regulated by regional cues. Dev Biol 129:103–113

Saúde L, Woolley K, Martin P, Driever W, Stemple DL (2000) Axis-inducing activites and cell fates of the zebrafish organizer. Development 127:3407–3417

Sawada A, Fritz A, Jiang YJ, Yamamoto A, Yamasu K, Kuroiwa A, Saga Y, Takeda H (2000) Zebrafish Mesp family genes, *mesp-a* and *mesp-b*, are segmentally expressed in the presomitic mesoderm, and Mesp-b confers the anterior identity to the developing somites. Development 127:1691–1702

Sawada A, Shinya M, Jiang Y-J, Kawakami A, Kuroiwa A, Takeda H (2001) Fgf/MAPK signalling is a crucial positional cue in somite boundary formation. Development 128:4873–4880

Saxén L (1987) Organogenesis of the kidney. Cambridge University Press, Cambridge

Schäfer K, Braun T (1999) Early specification of limb muscle precursor cells by the homeobox gene *Lbx1h*. Nat Genet 23:213–216

Schauerte HE, van Eeden FJ, Fricke C, Odenthal J, Strähle U, Haffter P (1998) Sonic hedgehog is not required for the induction of medial floor plate cells in the zebrafish. Development 125:2983–2993

Scheer N, Campos-Ortega JA (1999) Use of the GAL4-UAS technique for targeted gene expression in the zebrafish. Mech Dev 80:153–158

Scheer N, Groth A, Hans S, Campos-Ortega J (2001) An instructive function for *Notch* in promoting gliogenesis in the zebrafish retina. Development 128:1099–1107

Schier AF (2001) Axis formation and patterning in zebrafish. Curr Opin Gen Dev 11:393–404

Schier AF, Shen MM (2000) Nodal signalling in vertebrate development. Nature 403:385–389

Schier AF, Talbot WS (2001) Nodal signaling and the zebrafish organizer. Int J Dev Biol 45:289–297

Schier AF, Neuhauss SC, Harvey M, Malicki J, Solnica-Krezel L, Stainier DY, Zwartkruis F, Abdelilah S, Stemple DL, Rangini Z, Yang H, Driever W (1996) Mutations affecting the development of the embryonic zebrafish brain. Development 123:165–178

Schier AF, Neuhauss SC, Helde KA, Talbot WS, Driever W (1997) The *one-eyed pinhead* gene functions in mesoderm and endoderm formation in zebrafish and interacts with no tail. Development 124:327–342

Schilling TF, Concordet JP, Ingham PW (1999) Regulation of left-right asymmetries in the zebrafish by Shh and BMP4. Dev Biol 210:277–287

Schilling TF, Kimmel CB (1994) Segment and cell type lineage restrictions during pharyngeal arch development in the zebrafish embryo. Development 120:483–494

Schilling TF, Knight RD (2001) Origins of anteroposterior patterning and hox gene regulation during chordate evolution. Philos Trans R Soc Lond B 356:1599–1613

Schilling TF, Piotrowski T, Grandel H, Brand M, Heisenberg CP, Jiang YJ, Beuchle D, Hammerschmidt M, Kane DA, Mullins MC, van Eeden FJ, Kelsh RN, Furutani-Seiki M, Granato M, Haffter P, Odenthal J, Warga RM, Trowe T, Nüsslein-Volhard C (1996a) Jaw and branchial arch mutants in zebrafish. I. Branchial arches. Development 123:329–344

Schilling TF, Walker C, Kimmel CB (1996b) The chinless mutation and neural crest cell interactions in zebrafish jaw development. Development 122:1417–1426

Schilling TF, Prince V, Ingham PW (2001) Plasticity in zebrafish hox expression in the hindbrain and cranial neural crest. Dev Biol 231:201–216

Schmid B, Fürthauer M, Connors SA, Trout J, Thisse B, Thisse C, Mullins MC (2000) Equivalent genetic roles for *bmp7/snailhouse* and *bmp2b/swirl* in dorsoventral pattern formation. Development 127:957–967

Schmidt J, Francois V, Bier E, Kimelman D (1995) Drosophila short gastrulation induces an ectopic axis in *Xenopus*: evidence for conserved mechanisms of dorsal-ventral patterning. Development 121:4319–4328

Schmitt EA, Dowling JE (1994) Early eye morphogenesis in the zebrafish, *Brachydanio rerio*. J Comp Neurol 344:532–542

Schmitt EA, Dowling JE (1996) Comparison of topographical patterns of ganglion and photoreceptor cell differentiation in the retina of the zebrafish, *Danio rerio*. J Comp Neurol 371:222–234

Schmitt EA, Dowling JE (1999) Early retinal development in the zebrafish, *Danio rerio*: light and electron microscopic analyses. J Comp Neurol 404:515–536

Schmitz B, Campos-Ortega JA (1994) Dorso-ventral polarity of the zebrafish embryo is distinguishable prior to the onset of gastrulation. Roux's Arch Dev Biol 203:374–380

Schmitz B, Papan C, Campos-Ortega JA (1993) Neurulation in the anterior trunk region of the zebrafish *Brachydanio rerio*. Roux's Arch Dev Biol 202:250–259

Schneider A, Mijalski T, Schlange T, Dai W, Overbeek P, Arnold HH, Brand T (1999) The homeobox gene NKX3.2 is a target of left-right signalling and is expressed on opposite sides in chick and mouse embryos. Curr Biol 9:911–914

Schneider S, Steinbeisser H, Warga RM, Hausen P (1996) beta-Catenin translocation into nuclei demarcates the dorsalizing centers in frog and fish embryos. Mech Dev 57:191–198

Scholpp S, Brand M (2001) Morpholino-induced knockdown of zebrafish engrailed genes *eng2* and *eng3* reveals redundant and unique functions in midbrain-hindbrain boundary development. Genesis 30:129–133

Schüpbach T, Wieschaus E (1986) Maternal-effect mutations altering the anterior-posterior pattern of the *Drosophila* embryo. Roux's Arch Dev Biol 195:302–317

Schuchardt A, D'Agati V, Larsson-Blomberg L, Costantini F, Pachnis V (1994) Defects in the kidney and enteric nervous system of mice lacking the tyrosine kinase receptor Ret. Nature 367:380–383

Schuchardt A, D'Agati V, Larsson-Blomberg L, Costantini F, Pachnis V (1995) RET-deficient mice: an animal model for Hirschsprung's disease and renal agenesis. J Int Med 238:327–332

Schulte-Merker S, Ho RK, Herrmann BG, Nüsslein-Volhard C (1992) The protein product of the zebrafish homologue of the mouse T gene is expressed in nuclei of the germ ring and the notochord of the early embryo. Development 116:1021–1032

Schulte-Merker S, van Eeden FJ, Halpern ME, Kimmel CB, Nüsslein-Volhard C (1994) *no tail* (*ntl*) is the zebrafish homologue of the mouse *T* (*Brachyury*) gene. Development 120:1009–1015

Schulte-Merker S, Lee KJ, McMahon AP, Hammerschmidt M (1997) The zebrafish organizer requires chordino. Nature 387:862–863

Schwarz-Ramond T, Asbrand C, Bakkers J, Kuehl M, Schaeffer H-J, Huelsken J, Behrens J, Hammerschmidt M, Birchmeier W (2002) The ankyrin repeat protein Diversin acts in the Wnt signaling pathway and controls axis formation in zebrafish embryos. (submitted)

Schwerte T, Pelster B (2000) Digital motion analysis as a tool for analysing the shape and performance of the circulatory system in transparent animals. J Exp Biol 203 (Pt 11):1659–1669

Scott IC, Blitz IL, Pappano WN, Maas SA, Cho KW, Greenspan DS (2001) Homologues of Twisted gastrulation are extracellular cofactors in antagonism of BMP signalling. Nature 410:475–478

Segawa H, Miyashita T, Hirate Y, Higashijima S, Chino N, Uyemura K, Kikuchi Y, Okamoto H (2001) Functional repression of Islet-2 by disruption of complex with Ldb impairs peripheral axonal outgrowth in embryonic zebrafish. Neuron 30:423–436

Seidensticker MJ, Behrens J (2000) Biochemical interactions in the wnt pathway. Biochim Biophys Acta 1495:168–182

Semenov MV, Tamai K, Brott BK, Kuhl M, Sokol S, He X (2001) Head inducer Dickkopf-1 is a ligand for the Wnt coreceptor LRP6. Curr Biol 11:951–961

Seo H-C, Drivenes Ø, Ellingsen S, Fjose A (1998) Expression of two zebrafish homologues of the murine *Six3* gene demarcates the initial eye primordia. Mech Dev 73:45–57

Seo H-C, Curtiss J, Mlodzik M, Fjose A (1999) *Six* class homeobox genes in *Drosophila* belong to three distinct families and are involved in head development. Mech Dev 83:127–139

Sepich DS, Myers DC, Short R, Topczewski J, Marlow F, Solnica-Krezel L (2000) Role of the zebrafish trilobite locus in gastrulation movements of convergence and extension. Genesis 27:159–173

Serafini T, Kennedy TE, Galko, MJ, Mirzayan C, Jessell TM, Tessier-LavigneM (1994) The netrins define a family of axon outgrowth-promoting proteins homologous to *C. elegans* UNC-6 protein. Cell 78:409–424

Serafini T, Colamarino SA, Leonardo ED, Wang H, Beddington R, Skarnes WC, Tessier-Lavigne M (1996) Netrin-1 is required for commissural axon guidance in the developing vertebrate nervous system. Cell 87:1001–1114

Serbedzija GN, Chen J-N, Fishman MC (1998) Regulation in the heart field of zebrafish. Development 125:1095–1101

Serluca FC, Fishman MC (2001) Pre-pattern in the pronephric kidney field of zebrafish. Development 128:2233–2241

Serluca F, Drummond IA, Fishman MC (2002) The role of vascular sheer stress in glomerular angiogenesis. (submitted)

Seufert DW, Brennan HC, DeGuire J, Jones EA, Vize PD (1999) Developmental basis of pronephric defects in *Xenopus* body plan phenotypes. Dev Biol 215:233–242

Seydoux G, Dunn M (1997) Transcriptionally repressed germ cells lack a subpopulation of phosphorylated RNA polymerase II in early embryos of *Caenorhabditis elegans* and *Drosophila melanogaster*. Development 124:2191–2201

Seydoux G, Fire A (1994) Soma-germline asymmetry in the distributions of embryonic RNAs in *Caenorhabditis elegans*. Development 120:2823–2834

Shah NM, Anderson DJ (1997) Integration of multiple instructive cues by neural crest stem cells reveals cell-intrinsic biases in relative growth factor responsiveness. Proc Natl Acad Sci USA 94:11369–11374

Shah NM, Groves AK, Anderson DJ (1996) Alternative neural crest cell fates are instructively promoted by TGFbeta superfamily members. Cell 85:331–343

Shalaby F, Rossant J, Yamaguchi TP, Gertsenstein M, Wu XF, Breitman ML, Schuh AC (1995) Failure of blood-island formation and vasculogenesis in Flk-1-deficient mice. Nature 376:62–66

Shanmugalingam S, Houart C, Picker A, Reifers F, Macdonald R, Barth A, Griffin K, Brand M, Wilson SW (2000) Ace/fgf8 is required for forebrain commissure formation and patterning of the telencephalon. Development 127:2549–2561

Shawlot W, Behringer RR (1995) Requirement for *Lim1* in head-organizer function. Nature 374:425–430

Sheffield J, Moscona A (1970) Electron microscopic analysis of aggregation of embryonic cells: the structure and differentiation of aggregates of neural retina cells. Dev Biol 23:36–61

Shen MM, Schier AF (2000) The EGF-CFC gene family in vertebrate development. Trends Genet 16:303–309

Shepherd IT, Beattie CE, Raible DW (2001) Functional analysis of zebrafish GDNF. Dev Biol 231:420–435

Shih J, Fraser SE (1995) Distribution of tissue progenitors within the shield region of the zebrafish gastrula. Development 121:2755–2765

Shih J, Fraser SE (1996) Characterizing the zebrafish organizer: microsurgical analysis at the early-shield stage. Development 122:1313–1322

Shih J, Keller R (1992) Cell motility driving mediolateral intercalation in explants of *Xenopus laevis*. Development 116:901–914

Shih J, Keller R (1994) Gastrulation in *Xenopus laevis*: involution – a current view. Semin Dev Biol 5:85–90

Shimamura K, Rubenstein JL (1997) Inductive interactions direct early regionalization of the mouse forebrain. Development 124:2709–2718

Shimizu T, Yamanaka Y, Ryu S, Hashimoto H, Yabe T, Hirata T, Bae Y, Hibi M, Hirano T (2000) Cooperative roles of Bozozok/Dharma and nodal-related proteins in the formation of the dorsal organizer in zebrafish. Mech Dev 91:293–303

Shin CS, Lecanda F, Sheikh S, Weitzmann L, Cheng SL, Civitelli R (2000) Relative abundance of different cadherins defines differentiation of mesenchymal precursors into osteogenic, myogenic, or adipogenic pathways. J Cell Biochem 78:566–577

Shinomiya A, Tanaka M, Kobayashi T, Nagahama Y, Hamaguchi S (2000) The *vasa*-like gene, *olvas*, identifies the migration path of primordial germ cells during embryonic body formation stage in the medaka, *Oryzias latipes*. Dev Growth Differ 42:317–326

Shinya M, Eschbach C, Clark M, Lehrach H, Furutani-Seiki M (2000) Zebrafish Dkk1, induced by the pre-MBT Wnt signaling, is secreted from the prechordal plate and patterns the anterior neural plate. Mech Dev 98:3–17

Shiratori H, Sakuma R, Watanabe M, Hashiguchi H, Mochida K, Sakai Y, Nishino J, Saijoh Y, Whitman M, Hamada H (2001) Two-step regulation of left-right asymmetric expression of Pitx2: initiation by nodal signaling and maintenance by Nkx2. Mol Cell 7:137–149

Shoji W, Yee CS, Kuwada JY (1998) Zebrafish Semaphorin Z1a collapses specific growth cones and alters their pathway in vivo. Development 125:1275–1283

Simeone A (1998) Otx1 and otx2 in the development and evolution of the mammalian brain. EMBO J 17:6790–6798

Simpson JH, Kidd T, Bland KS, Goodman CS (2000) Short-range and long-range guidance by slit and its Robo receptors. Robo and Robo2 play distinct roles in midline guidance. Neuron 28:753–766

Sirotkin HI, Dougan ST, Schier AF, Talbot WS (2000a) *bozozok* and *squint* act in parallel to specify dorsal mesoderm and anterior neuroectoderm in zebrafish. Development 127:2583–2592

Sirotkin HI, Gates MA, Kelly PD, Schier AF, Talbot WS (2000b) Fast1 is required for the development of dorsal axial structures in zebrafish. Curr Biol 10:1051–1054

Skeath JB, Carroll SB (1992) Regulation of proneural gene expression and cell fate during neuroblast segregation in the *Drosophila* embryo. Development 114:939–946

Slack JM, Tannahill D (1992) Mechanism of anteroposterior axis specification in vertebrates. Lessons from the amphibians. Development 114:285–302

Sleptsova-Friedrich I, Li Y, Emelyanov A, Ekker M, Korzh V, Ge R (2001) Fgfr3 and regionalization of anterior neural tube in zebrafish. Mech Dev 102:213–217

Smith A, Robinson V, Patel K, Wilkinson DG (1997) The EphA4 and EphB1 receptor tyrosine kinases and ephrin-B2 ligand regulate targeted migration of branchial neural crest cells. Curr Biol 7:561–570

Smith J (1999) T-box genes: what they do and how they do it. Trends Genet 15:154–158

Smith JC, Slack JM (1983) Dorsalization and neural induction: properties of the organizer in *Xenopus laevis*. J Embryol Exp Morphol 78:299–317

Smith HW (1953) From fish to philosopher. Little Brown, Boston

Smith WC, Harland RM (1992) Expression cloning of noggin, a new dorsalizing factor localized in the Spemann organizer in *Xenopus* embryos. Cell 70:829–840

Smith WC, McKendry R, Ribisi S Jr, Harland RM (1995) A nodal-related gene defines a physical and functional domain within the Spemann organizer. Cell 82:37–46

Smithers L, Haddon C, Jiang Y, Lewis J (2000) Sequence and embryonic expression of *deltaC* in the zebrafish. Mech Dev 90:119–123

Sokol SY (1996) Analysis of Dishevelled signalling pathways during *Xenopus* development. Curr Biol 6:1456–1467

Sokol SY (1999) Wnt signaling and dorso-ventral axis specification in vertebrates. Curr Opin Genet Dev 9:405–410

Sokol SY, Christian JL, Moon RT, Melton D (1991) Injected Wnt RNA induces a complete body axis in *Xenopus* embryos. Cell 67:741–752

Sokol SY, Klingensmith J, Perrimon N, Itoh K (1995) Dorsalizing and neuralizing properties of Xdsh, a maternally expressed *Xenopus* homolog of dishevelled. Development 121:1637–1647

Solnica-Krezel L, Driever W (1994) Microtubule arrays of the zebrafish yolk cell: organization and function during epiboly. Development 120:2443–2455

Solnica-Krezel L, Driever W (2001) The role of the homeodomain protein Bozozok in zebrafish axis formation. Int J Dev Biol 45:299–310

Solnica-Krezel L, Stemple DL, Driever W (1995) Transparent things: cell fates and cell movements during early embryogenesis of zebrafish. BioEssays 17:931–939

Solnica-Krezel L, Stemple DL, Mountcastle-Shah E, Rangini Z, Neuhauss SC, Malicki J, Schier AF, Stainier DY, Zwartkruis F, Abdelilah S, Driever W (1996) Mutations affecting cell fates and cellular rearrangements during gastrulation in zebrafish. Development 123:67–80

Sommer L, Ma Q, Anderson DJ (1996) Neurogenins, a novel family of atonal related bhlh transcription factors, are putative mammalian neuronal determination genes that reveal progenitor cell heterogeneity in the developing cns and pns. Mol Cell Neurosci 8:221–241

Sorvano VA, Rainoldi C, Bisazza A, Vallortigara G (1999) Roots of brain specializations: preferential left-eye use during mirror-image inspection in six species of teleost fish. Behav Brain Res 106:175–180

Southard-Smith EM, Kos L, Pavan WJ (1998) *Sox10* mutation disrupts neural crest development in DOM Hirschsprung mouse model. Nat Genet 18:60–64

Sparrow DB, Jen WC, Kotecha S, Towers N, Kintner C, Mohun TJ (1998) *Thylacine 1* is expressed segmentally within the paraxial mesoderm of the *Xenopus* embryo and interacts with the Notch pathway. Development 125:2041–2051

Spemann H (1938) Embryonic development and induction. Yale University Press, New Haven

Spemann H, Mangold H (1924) Über Induktion von Embryonalanlagen durch Implantation artfremder Organisatoren. Wilhelm Roux' Arch EntwMech Org 100:599–638

Sperry RW (1963) Chemoaffinity in the orderly growth of nerve fiber patterns and connections. Proc Natl Acad Sci USA 50:703–710

Stachel SE, Grunwald DJ, Myers PZ (1993) Lithium perturbation and *goosecoid* expression identify a dorsal specification pathway in the pregastrula zebrafish. Development 117:1261–1274

Stainier DYR, Fishman MC (1992) Patterning the zebrafish heart tube: acquisition of anteroposterior polarity. Dev Biol 153:91–101

Stainier DYR, Fishman MC (1994) The zebrafish as a model system to study cardiovascular development. Trends Cardiovasc Med 4:207–212

Stainier DYR, Lee RK, Fishman MC (1993) Cardiovascular development in the zebrafish. I. Myocardial fate map and heart tube formation. Development 119:31–40

Stainier DYR, Weinstein BM, Detrich HW, Zon LI, Fishman MC (1995) *Cloche*, an early acting zebrafish gene, is required by both the endothelial and hematopoietic lineages. Development 121:3141–3150

Stainier DYR, Fouquet B, Chen JN, Warren KS, Weinstein BM, Meiler S, Mohideen MAPK, Neuhauss SC, Solnica-Krezel L, Schier AF, Zwartkruis F, Stemple DL, Malicki J, Driever W, Fishman MC (1996) Mutations affecting the formation and function of the cardiovascular system in the zebrafish embryo. Development 123:285–292

Stein E, Tessier-Lavigne M (2001) Hierarchical organization of guidance receptors: silencing of netrin attraction by slit through a Robo/DCC receptor complex. Science 291:1928–1938

Stemple DL, Anderson DJ (1992) Isolation of a stem-cell for neurons and glia from the mammalian neural crest. Cell 71:973–985

Stemple DL, Solnica-Krezel L, Zwartkruis F, Neuhauss SC, Schier AF, Malicki J, Stainier DY, Abdelilah S, Rangini Z, Mountcastle-Shah E, Driever W (1996) Mutations affecting development of the notochord in zebrafish. Development 123:117–128

Stenkamp DL, Frey RA, Prabhudesai SN, Raymond PA (2000) Function for *hedgehog* genes in zebrafish retinal development. Dev Biol 220:238–252

Stennard F (1998) VegT gets specific. Curr Biol 8:R928–R930

Stern C, Keynes R (1987) Interactions between somitic cells: the formation and maintenance of segment boundaries in the chick embryo. Development 99:261–272

Stern CD, Fraser SE, Keynes RJ, Primmett DR (1988) A cell lineage analysis of segmentation in the chick embryo. Development [Suppl] 104:231–244

Stickney HL, Barresi MJ, Devoto SH (2000) Somite development in zebrafish. Dev Dyn 219:287–303

Stiemke MM, Landers RA, al-Ubaidi MR, Rayborn ME, Hollyfield JG (1994) Photoreceptor outer segment development in *Xenopus laevis*: influence of the pigment epithelium. Dev Biol 162:169–180

Strähle U, Jesuthasan S (1993) Ultraviolet irradiation impairs epiboly in zebrafish embryos: evidence for a microtubule-dependent mechanism of epiboly. Development 119:909–919

Strähle U, Blader P, Henrique D, Ingham PW (1993) *Axial*, a zebrafish gene expressed along the developing body axis, shows altered expression in *cyclops* mutant embryos. Genes Dev 7:1436–1446

Strähle U, Fischer N, Blader P (1997a) Expression and regulation of netrin homologue in the zebrafish embryo. Mech Dev 62:147–160

Strähle U, Jesuthasan S, Blader P, Garcia-Villalba P, Hatta K, Ingham P (1997b) *One-eyed pinhead* is required for development of the ventral midline of the zebrafish (*Danio rerio*) neural tube. Genes Funct 1:131–148

Streisinger G, Singer F, Walker C, Knauber D, Dower N (1986) Segregation analyses and gene-centromere distances in zebrafish. Genetics 112:311–319

Streisinger G, Coale F, Taggart C, Walker C, Grunwald DJ (1989) Clonal origins of cells in the pigmented retina of the zebrafish eye. Dev Biol 131:60–69

Streit A, Lee KJ, Woo I, Roberts C, Jessell TM, Stern CD (1998) Chordin regulates primitive streak development and the stability of induced neural cells, but is not sufficient for neural induction in the chick embryo. Development 125:507–519

Stuermer CA (1988) Retinotopic organization of the developing retinotectal projection in the zebrafish embryo. J Neurosci 8:4513–4530

Sullivan SA, Barthel LK, Largent BL, Raymond PA (1997) A goldfish *Notch-3* homologue is expressed in neurogenic regions of embryonic, adult, and regenerating brain and retina. Dev Genet 20:208–223

Subramaniam K, Seydoux G (1999) *nos-1* and *nos-2*, two genes related to *Drosophila* nanos, regulate primordial germ cell development and survival in *Caenorhabditis elegans*. Development 126:4861–4871

Sumanas S, Strege P, Heasman J, Ekker SC (2000) The putative Wnt receptor *Xenopus* Frizzled-7 functions upstream of beta-catenin in vertebrate dorsoventral mesoderm patterning. Development 127:1981–1990

Sumanas S, Kim HJ, Hermanson S, Ekker SC (2001) Zebrafish *frizzled-2* morphant displays defects in body axis elongation. Genesis 30:114–118

Summerton J (1999) Morpholino antisense oligomers: the case for an RNase H-independent structural type. Biochim Biophys Acta 1489:141–158

Sumoy L, Keasey JB, Dittman TD, Kimelman D (1997) A role for notochord in axial vascular development revealed by analysis of phenotype and the expression of VEGR-2 in zebrafish flh and ntl mutant embryos. Mech Dev 63:15–27

Sumoy L, Kiefer J, Kimelman D (1999) Conservation of intracellular Wnt signaling components in dorsal-ventral axis formation in zebrafish. Dev Genes Evol 209:48–58

Sun X, Meyers EN, Lewandoski M, Martin GR (1999) Targeted disruption of *Fgf8* causes failure of cell migration in the gastrulating mouse embryo. Genes Dev 13:1834–1846

Supp DM, Witte DP, Potter SS, Brueckner M (1997) Mutation of an axonemal dynein affects left-right asymmetry in inversus viscerum mice. Nature 389:963–966

Sutherland RJ (1982) The dorsal diencephalic conduction system: a review of the anatomy and functions of the habenular complex. Neurosci Biobehav Rev 6:1–13

Suzuki N, Labosky PA, Furuta Y, Hargett L, Dunn R, Fogo AB, Takahara K, Peters DM, Greenspan DS, Hogan BL (1996) Failure of ventral body wall closure in mouse embryos lacking a procollagen C-proteinase encoded by Bmp1, a mammalian gene related to *Drosophila* tolloid. Development 122:2587–3595

Svoboda KR, Linares AE, Ribera AB (2001) Activity regulates neuronal cell death. Development 128:3511–3520

Tabuse Y, Izumi Y, Piano F, Kemphues KJ, Miwa J, Ohno S (1998) Atypical protein kinase C cooperates with PAR-3 to establish embryonic polarity in *Caenorhabditis elegans*. Development 125:3607–3614

Tada M, Smith JC (2000) *Xwnt11* is a target of *Xenopus brachyury*: regulation of gastrulation movements via Dishevelled, but not through the canonical Wnt pathway. Development 127:2227–2238

Taira M, Otani H, Jamrich M, Dawid IB (1994) Expression of the LIM class homeobox gene Xlim-1 in pronephros and CNS cell lineages of *Xenopus* embryos is affected by retinoic acid and exogastrulation. Development 120:1525–1536

Takahashi S, Yokota C, Takano K, Tanegashima K, Onuma Y, Goto J, Asashima M (2000a) Two novel nodal-related genes initiate early inductive events in *Xenopus* Nieuwkoop center. Development 127:5319–5329

Takahashi Y, Koizumi K, Takagi A, Kitajima S, Inoue T, Koseiki H, Saga Y (2000b) Mesp2 initiates somite segmentation throuh the Notch signalling pathway. Nat Genet 25:390–396

Takeda K, Yasumoto K, Takada R, Takada S, Watanabe K, Udono T, Saito H, Takahashi K, Shibahara S (2000) Induction of melanocyte-specific microphthalmia-associated transcription factor by Wnt-3a. J Biol Chem 275:14013–14016

Takeda S, Yonekawa Y, Tanaka Y, Okada Y, Nonaka S, Hirokawa N (1999) Left-right asymmetry and kinesin superfamily protein KIF3A: new insights in determination of laterality and mesoderm induction by kif3A-/- mice analysis. J Cell Biol 145:825–836

Takemaru KI, Moon RT (2000) The transcriptional coactivator CBP interacts with beta-catenin to activate gene expression. J Cell Biol 149:249–254

Takke C, Campos-Ortega J (1999) *her1*, a zebrafish pair-rule like gene, acts downstream of notch signalling to control somite development. Development 126:3005–3014

Takke C, Dornseifer P, von Weizsacker E, Campos-Ortega JA (1999) *her4*, a zebrafish homologue of the *Drosophila* neurogenic gene *E(spl)*, is a target of Notch signalling. Development 126:1811–1821

Talbot WS, Hopkins N (2000) Zebrafish mutations and functional analysis of the vertebrate genome. Genes Dev 14:755–762

Talbot WS, Trevarrow B, Halpern ME, Melby AE, Farr G, Postlethwait JH, Jowett T, Kimmel CB, Kimelman D (1995) A homeobox gene essential for zebrafish notochord development. Nature 378:150–157

Tallafus A, Wilm TP, Crozatier M, Pfeffer P, Wassef M, Bally-Cuif L (2001) The zebrafish button-head-like factor *bts1* is an early regulator of *pax2.1* expression during mid-hindbrain development. Development 128:4021–4034

Tam PP, Beddington RSP (1986) The metameric organization of the presomitic mesoderm and somite specification in the mouse embryo. In: Bellairs R, Ede DA, Lash JW (eds) Somites in developing embryos. Plenum Press, New York, pp 17–36

Tanabe Y, Jessell TM (1996) Diversity and pattern in the developing spinal cord. Science 274:1115–1123

Tanaka SS, Toyooka Y, Akasu R, Katoh-Fukui Y, Nakahara Y, Suzuki R, Yokoyama M, Noce T (2000) The mouse homolog of *Drosophila vasa* is required for the development of male germ cells. Genes Dev 14:841–853

Tautz D, Sommer RJ (1995) Evolution of segmentation genes in insects. Trends Genet 11:23–27

Taylor JS, van de Peer Y, Braasch I, Meyer A (2001) Comparative genomics provides evidence for an ancient genome duplication event in fish. Philos Trans R Soc Lond B Biol Sci 356:1661–1679

Technau G, Campos-Ortega J (1986) Lineage analysis of transplanted individual cells in embryos of *Drosophila melanogaster.* III. Commitment and proliferation capabilities of pole cells and midgut progenitors. Roux's Arch Dev Biol 195:489–498

Terman JR, Kolodkin AL (1999) Attracted or repelled? Look within. Neuron 23:193–200

Tessier-Lavigne M, Goodman CS (1996) The molecular biology of axon guidance. Science 274:1123–1133

Thisse C, Thisse B (1999) Antivin, a novel and divergent member of the TGFbeta superfamily, negatively regulates mesoderm induction. Development 126:229–240

Thisse C, Thisse B, Schilling TF, Postlethwait JH (1993) Structure of the zebrafish *snail* gene and its expression in wild-type, *spadetail* and *no tail* mutant embryos. Development 119: 1203–1215

Thisse B, Wright CVE, Thisse C (2000) Activin- and Nodal-related factors control antero-posterior patterning of the zebrafish embryo. Nature 403:425–428

Thomas PQ, Brown A, Beddington RS (1998) *Hex*: a homeobox gene revealing peri-implantation asymmetry in the mouse embryo and an early transient marker of endothelial cell precursors. Development 125:85–94

Thompson MA, Ransom DG, Pratt SJ, MacLennan H, Kieran MW, Detrich HW III, Vail B, Huber TL, Paw B, Brownlie AJ, Oates AC, Fritz A, Gates MA, Amores A, Bahary N, Talbot WS, Her H, Beier DR, Postlethwait JH, Zon LI (1998) The *cloche* and *spadetail* genes differentially affect hematopoiesis and vasculogenesis. Dev Biol 197:248–269

Timmermans L, Taverne N (1989) Segregation of primordial germ cells: their numbers and fate during early development of *Barbus conchonius* (*Cyprinidae, Teleosti*) as indicated by 3H-thymidine incorporation. J Morphol 202:225–237

Tokumoto M, Gong Z, Tsubokawa T, Hew CL, Uyemura K, Hotta Y, Okamoto H (1995) Molecular heterogeneity among primary motoneurons and within myotomes revealed by the differential mrna expression of novel *islet-1* homologs in embryonic zebrafish. Dev Biol 171:578–589

Tomita K, Ishibashi M, Nakahara K, Ang SL, Nakanishi S, Guillemot F, Kageyama R (1996) Mammalian *hairy and Enhancer of split homolog 1* regulates differentiation of retinal neurons and is essential for eye morphogenesis. Neuron 16:723–734

Tongiorgi E (1999) Tenascin-C expression in the trunk of wild-type, *cyclops* and *floating head* zebrafish embryos. Brain Res Bull 48:79–88

Topczewska JM, Topczewski J, Solnica-Krezel L, Hogan BL (2001a) Sequence and expression of zebrafish *foxc1a* and *foxc1b*, encoding conserved forkhead/winged helix transcription factors. Mech Dev 100:343–347

Topczewska JM, Topczewski J, Kume T, Shostak A, Solnica-Krezel L, Hogan BLM (2001b) The winged helix transcription factor, Foxc1a, is essential for somitogenesis in zebrafish. Genes Dev 15:2483–2493

Topczewski J, Sepich DS, Myers DC, Amores A, Walker C, Lele Z, Hammerschmidt M, Postlethwait J, Solnica-Krezel L (2001) Positional cloning of the zebrafish *knypek* gene reveals a requirement for glypican in polarized cell morphology during convergent extension. Dev Cell 1:251–264

Torrey RW (1965) Morphogenesis of the vertebrate kidney. In: DeHaan RL, Ursprung H (eds) Organogenesis. Holt, Rinehart and Winston, New York

Toyama R, Dawid IB (1997) *lim6*, a novel LIM homeobox gene in the zebrafish: comparison of its expression pattern with *lim1*. Dev Dyn 209:406–417

Toyama R, O'Connell ML, Wright CV, Kuehn MR, Dawid I (1995) Nodal induces ectopic *goosecoid* and *lim1* expression and axis duplication in zebrafish. Development 121:383–391

Trainor P, Krumlauf R (2000) Plasticity in mouse neural crest cells reveals a new patterning role for cranial mesoderm. Nat Cell Biol 2:96–102

Trainor PA, Tam PP (1995) Cranial paraxial mesoderm and neural crest cells of the mouse embryo: co-distribution in the craniofacial mesenchyme but distinct segregation in branchial arches. Development 121:2569–2582

Tremblay KD, Hoodless PA, Bikoff EK, Robertson EJ (2000) Formation of the definitive endoderm in mouse is a Smad2-dependent process. Development 127:3079–3090

Trembley KD, Dunn NR, Robertson EJ (2001) Mouse embryos lacking Smad1 signals display defects in extra-embryonic tissues and germ cell formation. Development 128:3609–3621

Trevarrow B, Marks DL, Kimmel CB (1990) Organization of hindbrain segments in the zebrafish embryo. Neuron 4:669–679

Trinkaus JP (1951) A study of the mechanism of epiboly in the egg of *Fundulus heteroclitus*. J Exp Zool 118:269–320

Trinkaus JP (1984a) Mechanism of *Fundulus* epiboly – a current view. Am Zool 24:673–688

Trinkaus JP (1984b) Cells into organs, the forces that shape the embryo. Prentice-Hall, Englewood Cliffs, New Jersey

Trinkaus JP (1992) The midblastula transition, the YSL transition and the onset of gastrulation in *Fundulus*. Dev Suppl 75–80

Trinkaus JP (1993) The yolk syncytial layer of *Fundulus*: its origin and history and its significance for early embryogenesis. J Exp Zool 265: 258–284

Trinkaus JP (1996) Ingression during early gastrulation of *Fundulus*. Dev Biol 177:356–70

Trinkaus JP, Trinkaus M, Fink RD (1992) On the convergent cell movements of gastrulation in *Fundulus* [published erratum appears in J Exp Zool 262:353–355]. J Exp Zool 261:40–61

Trowe T, Klostermann S, Baier H, Granato M, Crawford AD, Grunewald B, Hoffmann H, Karlstrom RO, Meyer SU, Muller B, Richter S, Nüsslein-Volhard C, Bonhoeffer F (1996) Mutations disrupting the ordering and topographic mapping of axons in the retinotectal projection of the zebrafish, *Danio rerio*. Development 123:439–450

Tsang TE, Shawlot W, Kinder SJ, Kobayashi A, Kwan KM, Schughart K, Kania A, Jessell TM, Behringer RR, Tam PP (2000) *Lim1* activity is required for intermediate mesoderm differentiation in the mouse embryo. Dev Biol 223:77–90

Tsunekawa N, Naito M, Sakai Y, Nishida T, Noce T (2000) Isolation of chicken *vasa* homolog gene and tracing the origin of primordial germ cells. Development 127:2741–2750

Tung TC, Chan CY, Tung YFY (1945) Experiments on the developmental potencies of blastoderms and fragments of teleostean eggs separated latitudinally. Proc Zool Soc Lond 115:175–189

Tytler P (1988) Morphology of the pronephros of the juvenile brown trout, *Salmo trutta*. J Morphol 195:189–204

Tytler P, Ireland J, Fitches E (1996) A study of the structure and function of the pronephros in the larvae of the turbot (*Scophthalmus maximus*) and the herring (*Clupea harengus*). Mar Fresh Behav Physiol 28:3–18

Tzarfati-Majar V, Burstyn-Cohen T, Klar A (1999) F-spondin is a contact-repellent molecule for embryonic motor neurons. Proc Natl Acad Sci USA 98:4722–4727

Underwood EM, Caulton JH, Allis CD, Mahowald AP (1980) Developmental fate of pole cells in *Drosophila melanogaster*. Dev Biol 77:303–314

Ungar A, Kelly GM, Moon RT (1995) Wnt4 affects morphogenesis when misexpressed in the zebrafish embryo. Mech Dev 52:1–12

Vallortigara G, Rogers LJ, Bisazza A (1999) Possible evolutionary origins of cognitive brain lateralization. Brain Res Rev 30:164–175

van de Water S, van de Wetering M, Joore J, Esseling J, Bink R, Clevers H, Zivkovic D (2001) Ectopic Wnt signal determines the eyeless phenotype of zebrafish *masterblind* mutant. Development 128:3877–3888

van Doren M, Brohier HT, Moore LA, Lehmann R (1998a) HMG-CoA reductase guides migrating primordial germ cells. Nature 396:466–469

van Doren M, Williamson AL, Lehmann R (1998b) Regulation of zygotic gene expression in *Drosophila* primordial germ cells. Curr Biol 8:243–246

van Eeden FJ, Granato M, Schach U, Brand M, Furutani-Seiki M, Haffter P, Hammerschmidt M, Heisenberg CP, Jiang YJ, Kane DA, Kelsh RN, Mullins MC, Odenthal J, Nüsslein-Volhard C (1996) Mutations affecting somite formation and patterning in the zebrafish, *Danio rerio*. Development 123:153–164

van Eeden FJ, Holley SA, Haffter P, Nüsslein-Volhard C (1998) Zebrafish segmentation and pair-rule patterning. Dev Genet 23:65–76

Van Vactor D, Flanagan JG (1999) The middle and the end: slit brings guidance and branching together in axon pathway selection. Neuron 22:649–652

Varga ZM, Wegner J, Westerfield M (1999) Anterior movement of ventral diencephalic precursors separates the primordial eye field in the neural plate and requires *cyclops*. Development 126:5533–5546

Varga ZM, Amores A, Lewis KE, Yan YL, Postlethwait JH, Eisen JS, Westerfield M (2001) Zebrafish *smoothened* functions in ventral neural tube specification and axon tract formation. Development 128:3497–3509

Verastegui C, Bille K, Ortonne JP, Ballotti R (2000) Regulation of the microphthalmia-associated transcription factor gene by the Waardenburg syndrome type 4 gene, *SOX10*. J Biol Chem 275:30757–30760

Vetter ML, Moore KB (2001) Becoming glial in the neural retina. Development 221:146–153

Vihtelic TS, Doro CJ, Hyde DR (1999) Cloning and characterization of six zebrafish photoreceptor *opsin* cDNAs and immunolocalization of their corresponding proteins. Vision Neurosci 16:571–585

Vize PD, Jones EA, Pfister R (1995) Development of the *Xenopus* pronephric system. Dev Biol 171:531–540

Vize PD, Seufert DW, Carroll TJ, Wallingford JB (1997) Model systems for the study of kidney development: use of the pronephros in the analysis of organ induction and patterning. Dev Biol 188:189–204

Vogel AM, Weinstein BM (2000) Studying vascular development in the zebrafish. Trends Cardiovasc Med 10:352–360

Wagner D, Mullins MC (2002) Modulation of BMP activity in dorsal-ventral pattern formation by the Chordin and Ogon antagonists. Dev Biol 245:109–123

Walker C, Streisinger G (1983) Induction of mutations by γ-rays in pregonial germ cells of zebrafish embryos. Genetics 103:125–136

Wallingford JB, Harland RM (2001) *Xenopus* Dishevelled signaling regulates both neural and mesodermal convergent extension: parallel forces elongating the body axis. Development 128:2581–2592

Wallingford JB, Rowning BA, Vogeli KM, Rothbacher U, Fraser SE, Harland RM (2000) Dishevelled controls cell polarity during *Xenopus* gastrulation. Nature 405:81–85

Wallingford JB, Ewald AJ, Harland RM, Fraser SE (2001a) Calcium signaling during convergent extension in *Xenopus*. Curr Biol 11:652–661

Wallingford JB, Vogeli KM, Harland RM (2001b) Regulation of convergent extension in *Xenopus* by Wnt5a and Frizzled-8 is independent of the canonical Wnt pathway. Int J Dev Biol 45:225–227

Walsh EC, Stainier DYR (2001) UDP-glucose dehydrogenase required for cardiac valve formation in zebrafish. Science 293:1670–1673

Walsh FS, Doherty P (1997) Neural cell adhesion molecules of the immunoglobulin superfamily: role in axon growth and guidance. Annu Rev Cell Dev Biol 13:425–456

Wang C, Lehmann R (1991) Nanos is the localized posterior determinant in *Drosophila*. Cell 66:637–647

Wang SW, Kim BS, Ding K, Wang H, Sun D, Johnson RL, Klein WH, Gan L (2001) Requirement for math5 in the development of retinal ganglion cells. Genes Dev 15:24–29

Warga RM, Kimmel CB (1990) Cell movements during epiboly and gastrulation in zebrafish. Development 108:569–580

Warga RM, Nüsslein-Volhard C (1998) spadetail-dependent cell compaction of the dorsal zebrafish blastula. Dev Biol 203:116–121

Warga RM, Nüsslein-Volhard C (1999) Origin and development of the zebrafish endoderm. Development 126:827–838

Warren JT, Chandrasekhar A, Kanki JP, Rangarajan R, Furley AH, Kuwada JY (1999) Molecular cloning and developmental expression of a zebrafish homolog of the axonal glyocoprotein TAG-1. Mech Dev 80:197–201

Warren KS, Fishman MC (1998) "Physiological genomics": mutant screens in zebrafish. Am J Physiol 275:H1–H7

Warren KS, Baker K, Fishman MC (2001) The *slow mo* mutation reduces pacemaker current and heart rate in adult zebrafish. AJP Heart Circ Physiol 1:H1711–1719

Waskiewicz AJ, Rikhof H, Hernandez RE, Moens CB (2001) Zebrafish *meis* functions to stabilize pbx proteins and regulate hindbrain patterning. Development 128:4139–4151

Wassarman KM, Lewandoski M, Campbell K, Joyner AL, Rubenstein JL, Martinez S, Martin GR (1997) Specification of the anterior hindbrain and establishment of a normal mid/hindbrain organizer is dependent on *gbx2* gene function. Development 124:2923–2934

Watanabe T, Raff M (1990) Rod photoreceptor development in vitro: intrinsic properties of proliferating neuroepithelial cells change as development proceeds in the rat retina. Neuron 2:461–467

Waterman RE (1969) Development of the lateral musculature in the teleost, *Brachydanio rerio*: a fine structure study. Am J Anat 125:457–493

Wawersik S, Maas RL (2000) Vertebrate eye development as modeled in *Drosophila*. Hum Mol Gen 9:917–925

Wehrle-Haller B, Weston JA (1995) Soluble and cell-bound forms of steel factor activity play distinct roles in melanocyte precursor dispersal and survival on the lateral neural crest migration pathway. Development 121:731–742

Wei X, Malicki J (2002) nagie oko, encoding a MAGUK-family protein, is essential for cellular patterning of the retina. Nat Genet (in press)

Weidinger G, Wolke U, Koprunner M, Klinger M, Raz E (1999) Identification of tissues and patterning events required for distinct steps in early migration of zebrafish primordial germ cells. Development 126:5295–307

Weidinger G, Wolke U, Köprunner M, Thisse C, Thisse B, Raz E (2002) Regulation of zebrafish primordial germ cell migration by attraction towards an intermediate target. Development 129:25–36

Weigel D, Jürgens G, Küttner F, Seifert E, Jäckle H (1989) The homeotic gene fork head encodes a nuclear protein and is expressed in the terminal regions of the *Drosophila* embryo. Cell 57:645–658

Weinstein BM, Stemple DL, Driever W, Fishman MC (1995) *gridlock*, a localized heritable vascular patterning defect in the zebrafish. Nat Med 1:1143–1147

Weinstein BM, Schier AF, Abdelilah S, Malicki J, Solnica-Krezel L, Stemple DL, Stainier DY, Zwartkruis F, Driever W, Fishman MC (1996) Hematopoietic mutations in the zebrafish. Development 123:303–309

Weinberg ES, Allende ML, Kelly CS, Abdelhamid A, Murakami T, Andermann P, Doerre OG, Grunwald DJ, Riggleman B (1996) Developmental regulation of zebrafish MyoD in wild-type, *no tail* and *spadetail* embryos. Development 122:271–280

Wessely O, Agius E, Oelgeschlager M, Pera EM, De Robertis EM (2001) Neural induction in the absence of mesoderm: beta-catenin-dependent expression of secreted BMP antagonists at the blastula stage in *Xenopus*. Dev Biol 234:16–73

Westerfield M (1995) The zebrafish book. University of Oregon Press, Eugene

Westerfield M, McMurray J, Eisen JS (1986) Identified motoneurons and their innervation of axial muscles in the zebrafish. J Neurosci 6:2267–2277

Westin J, Lardelli M (1997) Three novel *Notch* genes in zebrafish: implications for vertebrate *Notch* gene evolution and function. Dev Genes Evol 207:51–63

Weston JA (1982) Neural crest cell development. In: Burger MM, Weber R (eds) Embryonic development: IX Congress of the International Society of Developmental Biologists, Aug 28–Sept 1, 1981, Basel, Switzerland. Liss, New York, pp 359–379

Weston JA (1983) Regulation of neural crest cell migration and differentiation. In: Yamada KM (ed) Cell interactions and development: molecular mechanisms. Wiley, New York, pp 153–184

Weston JA (1991) Sequential segregation and fate of developmentally restricted intermediate cell populations in the neural crest lineage. Curr Top Dev Biol 25:133–153

Wharton KA, Ray RP, Gelbert WM (1993) An activity gradient of decapentaplegic is necessary for the specification of dorsal pattern elements in the *Drosophila* embryo. Development 117:807–822

Wharton RP, Struhl G (1991) RNA regulatory elements mediate control of *Drosophila* body pattern by the posterior morphogen nanos. Cell 67:955–967

Whitlock KE, Westerfield M (2000) The olfactory placodes of the zebrafish form by convergence of cellular fields at the edge of the neural plate. Development 127:3645–3653

Whitman M (1998) Smads and early developmental signaling by the TGFbeta superfamily. Genes Dev 12:2445–2462

Whitman M (2001) Nodal signaling in early vertebrate embryos: themes and variations. Dev Cell 1:605–617

Wilkinson DG (2001) Multiple roles of EPH receptors and ephrins in neural development. Nat Rev Neurosci 2:155–164

Williams BA, Ordahl CP (1994) Pax-3 expression in segmental mesoderm marks early stages in myogenic cell specification. Development 120:785–796

Williams JA, Barrios A, Gatchalian C, Rubin L, Wilson SW, Holder N (2000) Programmed cell death in zebrafish rohon beard neurons is influenced by TrkC1/NT-3 signaling. Dev Biol 226:220–230

Williamson A, Lehmann R (1996) Germ cell development in *Drosophila*. Annu Rev Cell Dev Biol 12:365–91

Willot V, Mathieu J, Lu Y, Schmid B, Sidi S, Yan Y-L, Postlethwait JH, Mullins M, Rosa F, Peyriéras N (2001) Cooperative action of ADMP and BMP mediated pathways in regulating cell fates in the zebrafish gastrula. Dev Biol 241:59–78

Wilson ET, Cretekos CJ, Helde KA (1995) Cell mixing during early epiboly in the zebrafish embryo. Dev Genet 17:6–15

Wilson HV (1889) The embryology of the sea bass. Bull US Fish Comm 9:209–278

Wilson JG, Roth CB, Warkany J (1953) An analysis of the syndrome of malformation induced by maternal vitamin A deficiency. Effects of restoration of vtiamin A at various timees during gestation. Am J Anat 92:189–217

Wilson SW, Easter SJ (1991) A pioneering growth cone in the embryonic zebrafish brain. Proc Natl Acad Sci USA 88:2293–2296

Wilson SW, Rubenstein JL (2000) Induction and dorsoventral patterning of the telencephalon. Neuron 28:641–651

Wilson SW, Ross LS, Parrett T, Easter SJ (1990) The development of a simple scaffold of axon tracts in the brain of the embryonic zebrafish, *Brachydanio rerio*. Development 108:121–145

Winklbauer R, Schurfeld M (1999) Vegetal rotation, a new gastrulation movement involved in the internalization of the mesoderm and endoderm in *Xenopus*. Development 126:3703–3713

Winkler C, Moon RT (2001) Zebrafish mdk2, a novel secreted midkine, participates in posterior neurogenesis. Dev Biol 229:102–118

Winnier G, Blessing M, Labosky PA, Hogan BLM (1995) Bone morphogenetic protein-4 is required for mesoderm formation and patterning in the mouse. Genes Dev 9:2105–2116

Winter CG, Wang B, Ballew A, Royou A, Karess R, Axelrod JD, Luo L (2001) Drosophila Rho-associated kinase (Drok) links Frizzled-mediated planar cell polarity signaling to the actin cytoskeleton. Cell 105:81–91

Winterbottom R (1974) A descriptive synomy of the striated muscle of the Teleostei. Proc Acad Nat Sci USA 125:225–317

Wodarz A, Nusse R (1998) Mechanisms of wnt signalling in development. Annu Rev Cell Dev Biol 14:59–88

Wodarz A, Ramrath A, Grimm A, Knust E (2000) *Drosophila* atypical protein kinase C associates with Bazooka and controls polarity of epithelia and neuroblasts. J Cell Biol 150:1361–1374

Wolff T, Martin KA, Rubin GM, Zipursky SL (1997) The development of the *Drosophila* visual system. In: Cowan WM, Jessell TM, Zipursky SL (eds) Molecular and cellular approaches to neural development. Oxford University Press, New York, pp 474–508

Wolke U, Weidinger G, Köprunner M, Raz E (2002) Multiple levels of post-transcriptional control lead to germ line specific gene expression in the zebrafish. Curr Biol 12:289–294

Woo K, Fraser SE (1995) Order and coherence in the fate map of the zebrafish nervous system. Development 121:2595–2609

Woo K, Fraser SE (1997) Specification of the zebrafish nervous system by nonaxial signals. Science 277: 254–257

Woo K, Fraser SE (1998) Specification of the hindbrain fate in the zebrafish. Dev Biol 197:283–296

Wood A, Thorogood P (1994) Patterns of cell behaviour underlying somitogenesis and notochord formation in intact vertebrate embryos. Dev Dyn 201:151–167

Wood A, Timmermans LPM (1988) Teleost epiboly: a reassessment of deep cell movement in the germ ring. Development 102:575–585

Wright CVE (2001) Mechanisms of left-right asymmetry: what's right and what's left? Dev Cell 1:179–186

Wu W, Wong K, Chen J, Jiang Z, Dupuis S, Wu JY, Rao Y (1999) Directional guidance of neuronal migration in the olfactory system by the protein Slit. Nature 400:331–336

Wullimann MF, Rupp B, Reichert H et al (1996) Neuroanatomy of the zebrafish brain. A topological atlas. Birkhauser, Basel, 160 pp

Wurst W, Bally-Cuif L (2001) Neural plate patterning upstream and downstream of the isthmic organizer. Nat Rev Genet 2:99–108

Wylie CC (1999) Germ cells. Cell 96:165–74

Wylie CC (2000) Germ cells. Curr Opin Genet Dev 10:410–413

Wylie CC, Snape A, Heasman J, Smith JC (1987) Vegetal pole cells and commitment to form endoderm in *Xenopus laevis*. Dev Biol 119:496–502

Wylie CC, Kofron M, Payne C, Anderson R, Hosobuchi M, Joseph E, Heasman J (1996) Maternal beta-catenin establishes a "dorsal signal" in early *Xenopus* embryos. Development 122: 2987–96

Xanthos JB, Kofron M, Wylie CC, Heasman J (2001) Maternal VegT is the initiator of a molecular network specifying endoderm in *Xenopus laevis*. Development 128:167–180

Xu C, Liguori G, Graziella Persico M, Adamson ED (1999) Abrogation of the Cripto gene in mouse leads to failure of postgastrulation morphogenesis and lack of differentiation of cardiomyocytes. Development 126:483–494

Xu J, Liu Z, Ornitz DM (2000) Temporal and spatial gradients of Fgf8 and Fgf17 regulate proliferation and differentiation of midline cerebellar structures. Development 127:1833–1843

Xu Q, Holder N, Patient R, Wilson SW (1994) Spatially regulated expression of three receptor tyrosine kinase genes during gastrulation in the zebrafish. Development 120:287–299

Xu Q, Mellitzer G, Robinson V, Wilkinson DG (1999) In vivo cell sorting in complementary segmental domains mediated by eph receptors and ephrins. Nature 399:267–271

Xu QL, Alldus G, Holder N, Wilkinson DG (1995) Expression of truncated sek-1 receptor tyrosine kinase disrupts the segmental restriction of gene expression in the *Xenopus* and zebrafish hindbrain. Development 121:4005–4016

Xu X, Meiler SE, Zhong TP, Mohideen M, Crossley DA, Burggren WW, Fishman MC (2002) Cardiomyopathy in zebrafish due to mutation in an alternatively spliced exon of *titin*. Nat Genet 30:205–209

Yaginuma H, Homma S, Kunzi R, Oppenheim RW (1991) Pathfinding by growth cones of commissural interneurons in the chick embryo spinal cord: a light and electron microscopic study. J Comp Neurol 304:78–102

Yamada T, Placzek M, Tanaka H, Dodd J, Jessell TM (1991) Control of cell pattern in the developing nervous system: polarizing activity of the floor plate and notochord. Cell 64:635–647

Yamada T, Pfaff SL, Edlund T, Jessell TM (1993) Control of cell pattern in the neural tube: motor neuron induction by diffusible factors from notochord and floor plate. Cell 73:673–686

Yamaguchi TP (2001) Heads or tails: Wnts and anterior-posterior patterning. Curr Biol 11:R713–R724

Yamaha E, Mizuno T, Hasebe Y, Takeda H, Yamazaki F (1998) Dorsal specification in blastoderm at the blastula stage in the goldfish, *Carassius auratus*. Dev Growth Differ 40:267–267

Yamamoto A, Amacher SL, Kim SH, Geissert D, Kimmel CB, De Robertis EM (1998a) Zebrafish paraxial protocadherin is a downstream target of spadetail involved in morphogenesis of gastrula mesoderm. Development 125:3389–3397

Yamamoto H, Kishida S, Uochi T, Ikeda S, Koyama S, Asashima M, Kikuchi A (1998b) Axil, a member of the Axin family, interacts with both glycogen synthase kinase 3beta and beta-catenin and inhibits axis formation of *Xenopus* embryos. Mol Cell Biol 18:2867–2875

Yamomoto Y, Jeffery WR (2000) Central role for the lens in cavefish eye degeneration. Science 289:631–633

Yamanaka Y, Mizuno T, Sasai Y, Kishi M, Takeda H, Kim CH, Hibi M, Hirano T (1998) A novel homeobox gene, *dharma*, can induce the organizer in a non-cell-autonomous manner. Genes Dev 12:2345–2353

Yan YL, Jowett T, Postlethwait JH (1998) Ectopic expression of *hoxb2* after retinoic acid treatment or mRNA injection: disruption of hindbrain and craniofacial morphogenesis in zebrafish embryos. Dev Dyn 213:370–385

Yan YT, Gritsman K, Ding J, Burdine RD, Corrales JD, Price SM, Talbot WS, Schier AF, Shen MM (1999) Conserved requirement for EGF-CFC genes in vertebrate left-right axis formation. Genes Dev 13:2527–2537

Yanagisawa H, Yanagisawa M, Kapur RP, Richardson JA, Williams SC, Clouthier DE, de Wit D, Emoto N, Hammer RE (1998) Dual genetic pathways of endothelin-mediated intercellular signaling revealed by targeted disruption of *endothelin converting enzyme-1* gene. Development 125:825–836

Yancopoulos GD, Davis S, Gale NW, Rudge JS, Wiegand SJ, Holash J (2000) Vascular-specific growth factors and blood vessel formation. Nature 407:242–248

Yang A, Schweitzer R, Sun D, Kaghad M, Walker N, Bronson RT, Tabin C, Sharpe A, Caput D, Crum C, McKeon F (1999) p63 is essential for regenerative proliferation in limb, craniofacial and epithelial development. Nature 398:714–718

Yang X, Castilla LH, Xu X, Li C, Gotay J, Weinstein M, Liu PP, Deng C-X (1999) Angiogenesis defects and mesenchymal apoptosis in mice lacking SMAD5. Development 126:1571–1580

Yee CS, Chandrasekhar A, Halloran MC, Shoji W, Warren JT, Kuwada JY (1999) Molecular cloning, expression, and activity of zebrafish semaphorin Z1a. Brain Res Bull 48:581–593

Yelick PC, Abduljabbar TS, Stashenko P (1998) zALK-8, a novel type I serine/threonine kinase receptor, is expressed throughout early zebrafish development. Dev Dyn 211:352–361

Yelon D (2001) Cardiac patterning and morphogenesis in zebrafish. Dev Dyn 222:552–563

Yelon D, Horne SA, Stainier DY (1999) Restricted expression of cardiac myosin genes reveals regulated aspects of heart tube assembly in zebrafish. Dev Biol 214:23–37

Yelon D, Ticho B, Halpern ME, Ruvinsky I, Ho RK, Silver LM, Stainier DYR (2000) The bHLH transcription factor Hand2 plays parallel roles in zebrafish heart and pectoral fin development. Development 127:2573–2582

Yeo C-Y, Whitman M (2001) Nodal signals to Smads through Cripto-dependent and Cripto-independent mechanisms. Cell 7:949–957

Yeo SY, Little MH, Yamada T, Miyashita T, Halloran MC, Kuwada JY, Huh TL, Okamoto H (2001) Overexpression of a slit homologue impairs convergent extension of the mesoderm and causes cyclopia in embryonic zebrafish. Dev Biol 230:1–17

Yokoyama N, Romero MI, Cowan CA, Galvan P, Helmbacher F, Charnay P, Parada LF, Henkemeyer M (2001) Forward signaling mediated by ephrin-B3 prevents contralateral corticospinal axons from recrossing the spinal cord midline. Neuron 29:85–97

Yost C, Farr GH III, Pierce SB, Ferkey DM, Chen MM, Kimelman D (1998) GBP, an inhibitor of GSK-3, is implicated in *Xenopus* development and oncogenesis. Cell 93:1031–1041

Yoon C, Kawakami K, Hopkins N (1997) Zebrafish *vasa* homologue RNA is localized to the cleavage planes of 2- and 4-cell-stage embryos and is expressed in the primordial germ cells. Development 124:3157–3165

Yoshizaki G, Sakatani S, Tominaga H, Takeuchi T (2000) Cloning and characterization of a vasa-like gene in rainbow trout and its expression in the germ cell lineage. Mol Reprod Dev 55:364–371

Yost HJ, Harris EB, Essner JJ (2001) Pitx2c-gfp trangenic zebrafish identify regions of asymmetric gene expression in the central nervous system. Dev Biol 235:186

Young RW (1985) Cell differentiation in the retina of the mouse. Anat Rec 212:199–205

Yue Y, Su J, Cerretti DP, Fox GM, Jing S, Zhou R (1999) Selective inhibition of spinal cord neurite outgrowth and cell survival by the Eph family ligand ephrin-A5. J Neurosci 19:10026–10035

Yuge M, Kobayakawa Y, Fujisue M, Yamana K (1990) A cytoplasmic determinant for dorsal axis formation in an early embryo of *Xenopus* laevis. Development 110:1051–1056

Zallen JA, Yi BA, Bargmann CI (1998) The conserved immunoglobulin superfamily member SAX-3/Robo directs multiple aspects of axon guidance in *C. elegans*. Cell 92:217–227

Zalik SE, Lewandowski E, Kam Z, Geiger B (1999) Cell adhesion and the actin cytoskeleton of the enveloping layer in the zebrafish embryo during epiboly. Biochem Cell Biol 77:527–542

Zardoya R, Abouheif E, Meyer A (1996) Evolution and orthology of *hedgehog* genes. TIG 12:496–497

Zeller J, Granato M (1999) The zebrafish *diwanka* gene controls an early step of motor growth cone migration. Development 126:3461–3472

Zalokar M (1976) Autoradiographic study of protein and RNA formation during early development of *Drosophila* eggs. Dev Biol 49:425–437

Zeltser LM, Larsen CW, Lumsden A (2001) A new developmental compartment in the forebrain regulated by lunatic fringe. Nat Neurosci 4:683–684

Zeng L, Fagotto F, Zhang T, Hsu W, Vasicek TJ, Perry WL III, Lee JJ, Tilghman SM, Gumbiner BM, Costantini F (1997) The mouse *Fused* locus encodes Axin, an inhibitor of the Wnt signaling pathway that regulates embryonic axis formation. Cell 90:181–192

Zhang H, Bradley A (1996) Mice deficient for BMP2 are nonviable and have defects in amnion/chorion and cardiac development. Development 122:2977–2986

Zhang J, Granato M (2000) The zebrafish *unplugged* gene controls motor axon pathway selection. Development 127:2099–2111

Zhang J, Houston DW, King ML, Payne C, Wylie C, Heasman J (1998a) The role of maternal VegT in establishing the primary germ layers in *Xenopus* embryos. Cell 94:515–524

Zhang J, Talbot WS, Schier AF (1998b) Positional cloning identifies zebrafish one-eyed pinhead as a permissive EGF-related ligand required during gastrulation. Cell 92:241–251

Zhong TP, Rosenberg M, Mohideen MA, Weinstein B, Fishman MC (2000) *gridlock*, an HLH gene required for assembly of the aorta in zebrafish. Science 287:1820–1824

Zhou X, Sasaki H, Lowe L, Hogan BL, Kuehn MR (1993) Nodal is a novel TGF-beta-like gene expressed in the mouse node during gastrulation. Nature 361:543–547

Zimmerman LB, Harland RM (1996) Bmp-4 function is blocked by high affinity binding to the Spemann organizer signal Noggin. Cell 85:599–606

Subject Index